DATE DUE

GAYLORD

PRINTED IN U.S.A.

D1087183

MACROMOLECULES

Some Other IUPAC Titles of Interest from Pergamon Press

Books

ANGUS: International Thermodynamic Tables of the Fluid State, 7-Propylene

BATTINO: Oxygen and Ozone

BROWN & DAVIES: Organ-Directed Toxicity—Chemical Indices and Mechanisms

CIARDELLI & GIUSTI: Structural Order in Polymers

EGAN & WEST: Collaborative Interlaboratory Studies in Chemical Analysis

FREIDLINA & SKOROVA: Organic Sulfur Chemistry

FUWA: Recent Advances in Analytical Spectroscopy

GOETHALS: Polymeric Amines and Ammonium Salts

GOODWIN & BRITTON: Carotenoids

HÖGFELDT: Stability Constants of Metal-Ion Complexes, Part A: Inorganic Ligands

KORNHAUSER, RAO & WADDINGTON: Chemical Education in the Seventies

LAIDLER: Frontiers of Chemistry

LAURENT: Coordination Chemistry—21

PERRIN: Stability Constants of Metal-Ion Complexes, Part B: Organic Ligands

RIGAUDY & KLESNEY: Nomenclature of Organic Chemistry

ST-PIERRE & BROWN: Future Sources of Organic Raw Materials

STEC: Phosphorus Chemistry Directed Towards Biology

TROST & HUTCHINSON: Organic Synthesis—Today and Tomorrow

YOUNG: Hydrogen and Deuterium

YOUNG: Oxides of Nitrogen

Journals

CHEMISTRY INTERNATIONAL, the news magazine for chemists in all fields of specialization in all countries of the world.

PURE AND APPLIED CHEMISTRY, the international research journal publishing proceedings of IUPAC conferences, nomenclature rules and technical reports.

INTERNATIONAL UNION OF PURE AND APPLIED CHEMISTRY

(Macromolecular Division)

in conjunction with

Centre National de la Recherche Scientifique
Université Louis Pasteur de Strasbourg

MACROMOLECULES

Main Lectures Presented at the
27th International Symposium on Macromolecules
Strasbourg, France, 6-9 July 1981

Edited by

H. BENOIT and P. REMPP

Centre de Recherches sur les Macromolecules
Strasbourg, France

PERGAMON PRESS

OXFORD · NEW YORK · TORONTO · SYDNEY · PARIS · FRANKFURT

U.K.	Pergamon Press Ltd., Headington Hill Hall, Oxford OX3 0BW, England
U.S.A.	Pergamon Press Inc., Maxwell House, Fairview Park, Elmsford, New York 10523, U.S.A.
CANADA	Pergamon Press Canada Ltd., Suite 104, 150 Consumers Rd., Willowdale, Ontario M2J 1P9, Canada
AUSTRALIA	Pergamon Press (Aust.) Pty. Ltd., P.O. Box 544, Potts Point, N.S.W. 2011, Australia
FRANCE	Pergamon Press SARL, 24 rue des Ecoles, 75240 Paris, Cedex 05, France
FEDERAL REPUBLIC OF GERMANY	Pergamon Press GmbH, 6242 Kronberg-Taunus, Hammerweg 6, Federal Republic of Germany

First edition 1982

Library of Congress Cataloging in Publication Data

International Symposium on Macromolecules (27th: 1981: Strasbourg, France)
Macromolecules: main lectures presented at the 27th International Symposium on Macromolecules, Strasbourg, France, 6-9 July 1981.
(IUPAC symposium series)
1. Macromolecules—Congresses. I. Benoit, Henri, 1921— II. Rempp, P. III. International Union of Pure and Applied Chemistry. IV. Université Louis Pasteur de Strasbourg. V. Title. VI. Series.
QD380.I58   1981   547.7   81-23481
AACR2

British Library Cataloguing in Publication Data

International Symposium on Macromolecules
(27th: 1981: Strasbourg)
Macromolecules.—(IUPAC symposium series)
1. Polymers and polymerization—Congresses
I. Benoit, H. II. Rempp, P. III. International Union of Pure and Applied Chemistry. *Macromolecular Division* IV. Series
547.7   QD380
ISBN 0-08-026226-0

In order to make this volume available as economically and as rapidly as possible the authors' typescripts have been reproduced in their original forms. This method unfortunately has its typographical limitations but it is hoped that they in no way distract the reader.

Printed in Great Britain by A. Wheaton & Co. Ltd., Exeter

CONTENTS

Scientific Committee   vii

Preface   ix

On Common Tendencies of Non-Equilibrium Polycondensation   1
    V. V. KORSHAK

Twenty-Five Years of Living Polymers   15
    M. SZWARC

New Developments in Ring-Opening Polymerization   27
    T. SAEGUSA

Emulsion Polymerization: I. Mechanisms of Particle Formation. II. Chemistry
at the Interface   39
    R. M. FITCH

Some Problems of the Theory of Polymeranalogous and Intramolecular Reactions of
Macromolecules   65
    N. A. PLATÉ, O. V. NOAH and L. B. STROGANOV

Chemical Modifications of Chlorinated Polymers   85
    E. MARECHAL

Catalytic Asymmetric Synthesis Using Polymer Supported Optically Active Catalysts   99
    J. K. STILLE

Recent Developments in the Use of Polymeric Reagents   113
    A. PATCHORNIK

The Effect of Macromolecular Structure on the Behaviour of Polymer Catalysts   125
    F. CIARDELLI, M. AGLIETTO, C. CARLINI, S. D'ANTONE, G. RUGGERI and R. SOLARO

Intermolecular Excimer Interactions in Polymers   139
    R. QIAN

Relaxation Methods for Studying Macromolecular Motion in the Bulk   155
    H. SILLESCU

Some New Aspects of Crystallization Modes in Polymers   171
    A. KELLER

Neutron Scattering Studies on the Crystallization of Polymers   191
    E. W. FISCHER

Molecular Mechanisms in Polymer Fracture   211
    H. H. KAUSCH

Polyelectrolytes - Counterion Interactions   225
    G. WEILL

Structures and Properties of Conducting Polymers   233
    J. C. W. CHIEN

A Comprehensive Theory of the High Ionic Conductivity of Macromolecular Networks   251
    H. CHERADAME

The Structure and Properties of Ultra High Modulus Films and Fibres   265
    I. M. WARD

Structure-Property Relationships in Composite Matrix Resins   275
    F. N. KELLEY, B. J. SWETLIN and D. TRAINOR

The Adhesion of Polymers to High Energy Solids   289
    J. SCHULTZ and A. CARRÉ

Biodegradation of Polymers for Biomedical Use   305
    J. KOPEČEK

Synthetic Polymer Biomaterials in Medicine - A Review   321
    A. S. HOFFMAN

SCIENTIFIC COMMITTEE

Honorary Chairmen

G. Champetier (Deceased), C. Sadron

Chairman

H. Benoit

Vice-Chairmen

P. Sigwalt, C. Wippler

Members

A. Banderet, E. Bouchez, M. Carrega, A. Chapiro, J.-B. Donnet,
E. Guillet, A. Guyot, A. Kepes, A.-J. Kovacs, J. Minoux,
J. Neel, C. Pinazzi, C. Quivoron, P. Rempp,
M. Rinaudo, B. Sillion, A. de Vries

INTERNATIONAL UNION OF PURE AND APPLIED CHEMISTRY

IUPAC Secretariat: Bank Court Chambers, 2-3 Pound Way,
Cowley Centre, Oxford OX4 3YF, UK

PREFACE

This book is a collection of the Main Lectures which were delivered during the 27th IUPAC Symposium on Macromolecules that was held in Strasbourg , July 6 to 9, 1981.

The purpose of these lectures was to present and illustrate each of the themes selected by the scientific committee. These fields were chosen for their importance in Macromolecular Science and with the feeling that they were representative of the present trends in Polymer Research.

Qualified lecturers were selected for their own contributions to the corresponding domains. They were asked beforehand to report not only on their own work, but also on the present state of knowledge and on the possible developements and potential applications in the field in which they are involved. The authors of Main Lectures originate from countries in which Polymer Science is most actively investigated, and a world-wide distribution of invited speakers was achieved.

These considerations will explain the broad field covered by this book. It contains topics ranging from Pure Polymer Chemistry to Polymer Physics and Polymer Technology. This illustrates the diversity of the interest of scientists involved in Polymer Science and shows once more the pluridisciplinarity of our field.

In a certain way, this is not the kind of book in which a specialist of a given topic will gain new information or results in his domain of interest. The purpose of this book is to allow anybody interested to get precise and up-to-date information on the present state of research in the domains selected. It should be specially valuable for scientists who try to keep contact with progress in Polymer Science beyond their own field of activity. It should also provide for new ideas, encourage collaboration, and possibly lead to new applications.

Besides the Main Lectures some 320 short communications have been presented at the IUPAC Symposium in Strasbourg. These short communications will not be published as such, but two volumes of preprints have been made available for the participants. Additional copies can be obtained from the Centre de Recherches sur les Macromolécules by anybody wanting more details on the Meeting.

Twenty-nine years ago, a similar IUPAC Symposium was held in Strasbourg, gathering fewer than 200 scientists, but most of the pioneers of Polymer Science. In 1981 there were about 1000 people attending, on selected topics. One can rejoice in the huge developements of our scientific discipline, for it illustrate the growing interest in Polymer Science and Technology and the increasing importance of Scientific Research in the present world.

We want to express our deep appreciation to the authors of the Main Lectures for having prepared and made available a written version of their presentation to the IUPAC Symposium. We know that this takes time and requires much care, and we feel indebted to them.

M - A*

We would also like to express our warmest thanks to all the researchers, technicians and students of the Centre de Recherches sur les Macromolécules, Strasbourg, and of the Ecole d'Application des Hauts Polymères, Strasbourg for having participated actively and willingly in the organisation of the Meeting. Without their time-consuming help it would have been impossible to set up such a Symposium. Last, but not least, we would like to acknowledge the financial support of Franch Public Research Bodies (CNRS, DGRST, DRET) and of numerous French Industries devoted to Polymer production and processing.

H. BENOIT   P. REMPP

Centre de Recherches sur les Macromolécules
6,rue Boussingault - 67083 STRASBOURG(France)

Septembre 1981

ON COMMON TENDENCIES OF NON-EQUILIBRIUM POLYCONDENSATION

V.V. Korshak

Institute of Organo-Element Compounds, Acad. Sci. USSR,
117813 Moscow, USSR

Abstract - The chemistry of polycondensation processes has
gained a rather rapid development at the present time. All
known polycondensation reactions can be combined into two
groups: 1) equilibrium and 2) non-equilibrium polycondensation. They differ in the equilibrium constant of the process
under investigation. Equilibrium polycondensation includes
reactions with $K_p < 10^3$, reactions with $K_p > 10^3$ belong to
non-equilibrium process. The difference between these two
polycondensation types shows up in the properties of the resulting polymers, the mechanism of elementary reactions, the
nature and mechanism of catalyst action, the kinetics, the
molecular weight distribution, the mechanism of copolycondensation, the structure of copolymers and the like. Various
side reactions also occur. As a result, anomalous units are
formed in the macromolecule which gives rise to "raznozvenny" (different-unit) polymers. Their structure can be des-
cribed by the following formula: $\left[(M)_{\overline{m}} (A)_{\overline{n}} \right]_x$ with M as
normal unit and A as anomalous one. Anomalous units and,
as a consequence, polymer raznozvennost (different-unit
structure) have been shown to present in different classes
of polymers prepared by polycondensation methods. Catalysts
strongly influence the non-equilibrium polycondensation process. The use of tertiary amines makes it possible to control the structure of the resulting polymers and to produce
conformational-specific polymers.

In 1833 Gay-Lussak and Pelouze obtained the first synthetic polyester by
polycondensation of the hydroxycarboxylic acid (1). In the following years
we observed an ever-increasing number of investigations concerning the polycondensation field (2-4).

Thus at the present time the polycondensation as a synthetic process has given a large number of polymeric structures to science and engineering and
played an important role in the development of basic concepts of polymer
science (2-4). The rapid progress of investigations of polymer synthesis
by polycondensation has enriched the polymer science with new reactions for
polymer preparation, made it possible to understand the mechanism of polycondensation processes and enlarged a store of science with many new polymers.

Some Specific Features of Polycondensation

Polycondensation processes essentially differ from polymerization processes
(2 & 3). First of all, attention should be given to great universality of
polycondensation processes and, consequently, to great variety of structures
prepared with their help. The structures obtained by polymerization are less
diversified than those obtained by polycondensation. Among the polymers obtained by polycondensation is a great number of polymers having high thermostability, conductive polymers, physiologically active polymers and the like.
Although the polycondensation is studied nowadays by a more narrow circle of
workers than the polymerization, the effectiveness of their synthetic investigations is rather high (4). Thus, when considering various polymeric compounds reported in scientific papers one can find that nearly 50% of new polymers described in the literature are polymers produced by means of poly-

condensation reactions (4). And finally, I would say, the most important difference consists in that these two processes strongly differ from each other in the number of chemical reactions used in both cases. In polymerization we make use only of two chemical reactions: addition to double and triple bonds between two atoms or addition to cycles. As to polycondensation, some dozens of chemical reactions are used in it already at the present time, and the number of reactions being drawn into these transformations increases from year to year. This is also true of a very important field such as biopolymers which are also mainly formed by means of polycondensation methods. It should be emphasized that polycondensation is of great importance as a method of natural polymer synthesis, since many significant biopolymers such as proteins, nucleic acids, natural caoutchouc, cellulose, starch, glycogen, chitin, pentosanes and many other polymers, as well as ferments, enzymes, and hormones are formed in living organisms by means of various polycondensation processes, that is, this process is widely represented in nature. And such still quite a new, but very promising domain of polymer science as inorganic polymers is almost completely the area of polycondensation application, because most heterochain inorganic polymers are prepared just by polycondensation (5 & 6).

After such brief characteristic of specific features of polycondensation we dwell on the modern state of this field, giving particular consideration to distinguishing features of non-equilibrium polycondensation, to causes of the rise of raznozvennost in non-equilibrium polycondensation, as well as to the process of macromolecule formation and the effect of reaction conditions, catalysts and solvents, that is, to questions which nowadays determine the progress in the polycondensation that finally determines the perspectives of developments in this field.

Types of Polycondensation

Investigation of common tendencies of polycondensation processes has led us to the conclusion that all known polycondensation reactions can be combined into two large groups called equilibrium and non-equilibrium polycondensation (2,3,7,8). The difference between them is largely determined by the value of equilibrium constant (K_p) of the process under investigation. Equilibrium polycondensation includes reactions with $K_p < 10^3$, reactions with $K_p > 10^3$ belong to non-equilibrium polycondensation (7 & 8).

The difference between these two types of polycondensation shows up also in the properties of the resulting polymers, the mechanism of the proceeding reactions, the nature and mechanism of catalysts, the kinetics, molecular weight distribution, the mechanism of copolycondensation, the structure of copolymers, and other specific features of the process that will be shown below.

The non-equilibrium process has been studied not so comprehensively, and this put the task of a detailed investigation of laws governing its course. The imortance of this trend is caused by that just non-equilibrium polycondensation makes it possible to synthesize polymers at a higher rate and to prepare rather high molecular weight polymers with new unit structures and interesting complex of physical and chemical properties (7,9,10).

Functionality of Monomers

Carothers (11) formulated the rule of monomer functionality stating that linear polymers are formed when the monomers are bifunctional. If they are three-functional and higher, a gel formation occurs and a steric, insoluble and infusible polymer is formed. This rule needs radical changes, however, because there are many cases when this relationship is not confirmed, and three- and tetrafunctional monomers proved to be capable of forming polymers with linear macromolecules. This is possible in case of significant difference in reactivity of functional groups, when the more reactive groups react earlier forming a linear polymer. This can be exemplified by the reaction of glycerine with phthalic anhydride.

Reaction conditions are also of importance. For instance, the non-equilibrium polycondensation of tetramine with terephthaloyl chloride gives a three-dimensional insoluble polymer. If the reaction proceeds in pyridine hydrochloride or in the salt of another amine, a linear polymer is formed, although the monomer is tetrafunctional. But in this medium it acts as a difunctional compound (13):

$$
\text{H}_2\text{N} \underset{\text{H}_2\text{N}}{\overset{\text{NH}_2}{\bigcirc}} \text{NH}_2 \quad \xrightarrow{\text{ClCORCOCl}} \quad \text{H}_2\text{N} \underset{-\text{CORCO-NH}}{\overset{\text{NH-CORCO-}}{\bigcirc}} \text{NH-CORCO-}
$$

$$
\text{H}_2\text{N} \underset{\text{H}_2\text{N}}{\overset{\text{NH}_2}{\bigcirc}} \text{NH}_2 \quad \xrightarrow{\text{HCl} \cdot \text{C}_5\text{H}_5\text{N}} \quad \text{HCl} \cdot \text{H}_2\text{N} \underset{\text{H}_2\text{N}}{\overset{\text{NH}_2 \cdot \text{HCl}}{\bigcirc}} \text{NH}_2 \quad \xrightarrow{\text{ClCORCOCl}}
$$

$$
\longrightarrow \quad \text{HCl} \cdot \text{H}_2\text{N} \underset{-\text{CORCO-NH}}{\overset{\text{NH}_2 \cdot \text{HCl}}{\bigcirc}} \text{NH-CORCO-}
$$

$$(\underline{1})$$

There is a distinct difference between potential or structural and real or
reactive functionality (13). Steric factors can be of rather essential importance. This can be exemplified by non-equilibrium polycondensation reactions, for instance, by the formation of various polyheteroarylenes, when
tetrafunctional compounds such as tetraamines, dihydroxydiamines, dithiadia-
mines, tetracarboxylic acids, diaminodicarboxylic acids and other similar
compounds of the general formula:

$$
\underset{Y}{\overset{X}{\bigcirc}} \underset{Y}{\overset{X}{}} \quad , \quad \underset{X}{\overset{X}{\bigcirc}} -Z- \underset{X}{\overset{X}{\bigcirc}} \quad , \quad \underset{X}{\overset{X}{\bigcirc}} \underset{Y}{\overset{X}{}} \quad , \quad X - \underset{X}{\overset{X}{\bigcirc}} - X
$$

$$(\underline{2})$$

are used as monomers. Important in this case is the mutual position of functional groups. If the position enables the formation of five- and six-mem-
bered cycles, linear macromolecules of heterocyclochain polymers are formed
due to a steric effect of functional groups. This fact demonstrates the possibility of preparing linear polymers from three- and tetrafunctional monomers that earlier was considered to be impossible. Therefore, it should be
concluded that important for the formation of linear polymers is not only
the common functionality of monomers, but also the corresponding favourable
mutual position of functional groups in the structural unit or monomers as
well as the effect of reaction medium and reaction conditions in the course
of polycondensation. This makes one to pose differently the problem of the
importance of monomer functionality in a non-equilibrium polycondensation
(4).

The Rule of Non-Equivalence of Functional Groups

As known, the rule of non-equivalence of functional groups holds for the
reactions of dicarboxylic acids with glycols or diamines and generally for
all cases of polycondensation when the initial substances are bifunctional
(2,13). This principle provides a means of controlling the molecular weight
of polymers prepared by equilibrium polycondensation when different quantities of monofunctional substances or an excess of one of the monomers are
added (13).

Figure 1 shows the changes of found and calculated molecular weight as a
function of the excess of one of the monomers. Investigations of this relation for non-equilibrium polycondensation with initial substances having
higher functionality showed that in most cases here again the excess of one
of the initial substances has similar effect in the reactions of tetrafunctional monomers decreasing the molecular weight proportionally to the
excess of one of the monomers. Such a dependence was observed in many cases
of non-equilibrium polycondensation.

The reaction of various tetrafunctional nucleophiles with tetracarboxylic
acids or their derivatives follows the rule of non-equivalency of functional groups, as shown in the general form in Fig. 2.

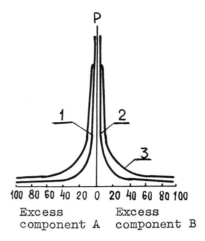

Fig. 1. Dependence of the molecular weight of polyamides on
the excess of one of the monomers. 1,2) calculated, 3) found
molecular weight. P - coefficient of polymerization.

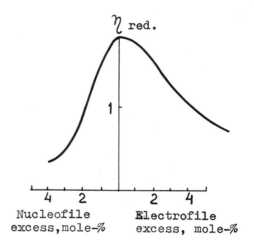

Fig. 2. Influence of excess monomers on the solution viscosity of polymers prepared from various tetrafunctional monomers
(dihydrazides of dicarboxylic acids, bis-o-aminophenols, di-
anilinodiamines, diamides of bis-anthranyl acids) with poly-
carboxylic acids.

Figure 3 shows the dependence of the viscosity of polyimide solution on the
excess of initial substances in the reaction of diamine with pyromellitic
dianhydride (16).

The same occurs in the reaction of carboranedicarboxylic acid dichloride
with tetraamines, as shown in Fig. 4 (16).

Thus, the above examples evidence rather clearly a distinct dependence of
the molecular weight of polymer to be synthesized on the excess of one of
the monomers also in the case of tetrafunctional monomers and, consequently,
"the rule of non-equivalency of functional groups" is also valid for many-
functional monomers in a non-equilibrium polycondensation (3 & 13).

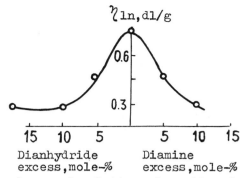

Fig. 3. Dependence of solution viscosity of polypyromellite-imide of anilinephthaleine in dimethylformamide on the ratio of initial substances.

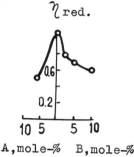

Fig. 4. Influence of monomer ratio on the reduced viscosity of polyamido-m-carborane in dimethylformamide.

Raznozvennost of Polymers Formed by Non-Equilibrium Polycondensation

Raznozvenny polymers (different-unit polymers) are formed in a non-equilibrium polycondensation as a result of the formation of anomalous units in the complex of polycondensation reactions shown in Fig. 5 (17-19).

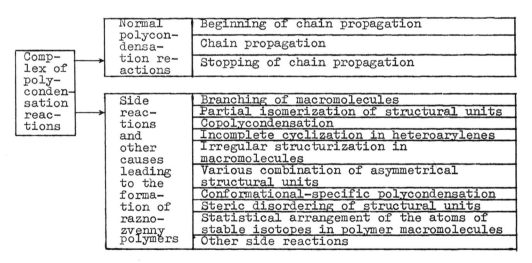

Fig. 5. Complex of polycondensation reactions.

The resulting anomalous units differ in their structure from the main macromolecule units, as can be seen in Fig. 6.

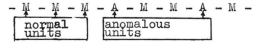

Fig. 6. The structure of a raznozvenny polymer.

As a result of anomalous unit formation, so-called "raznozvenny" polymers
arise (4,5,9,17-20). The structure of raznozvenny polymers is proposed to
express by a general formula

$$\left[\,(M)_{\boxed{m}}\,(A)_{\boxed{n}}\,\right]_x$$

(3)

where M is the normal unit and A is the anomalous unit, and the coefficients $\boxed{m}$ and $\boxed{n}$ squared denote statistical arrangement of units M and A.
Therefore, the real formula of such polymer as polyimide will be as follows:

normal unit   anomalous unit

Raznozvenny polyimide

(4)

In this case the anomalous units are such units in which no cycle closure
occurs. In other cases various side reactions of different type proceed. The
presence of anomalous units substantially effects the total chemical and
physical properties inherent in the given polymer. Therefore, the investigation of conditions leading to the formation of anomalous units and, consequently, to the rise of raznozvenny polymers is one of the most important
tasks of synthetic polymer chemistry, since it would otherwise be impossible
to understand and to explain extensively the dependence of polymer properties on their structure (4,5,9,17,18,20 & 21). The presence of anomalous units
was evidenced for various classes of polymers prepared by non-equilibrium
polycondensation methods (5).

Since the presence of anomalous units in raznozvenny polymers essentially
effects the chemical and physical properties, this fact should be taken into
account when considering the relationship between the structure and properties of polymers (17,18,21-23).

Kinetics and Mechanism of Polycondensation Reactions
Kinetics of equilibrium polycondensation was studied long ago with many substances (2). Kinetics of non-equilibrium polycondensation has been investigated not so comprehensively.

Kinetic investigations of cardo polyimide formation proved to be possible
because of their solubility resulting from the use as initial substances of
diamines with bulk substituents such as aniline phthaleine (24 & 25). The
same   result is reached by introducting in the polyimide unit the polar
groups by means of monomers such as 4,4'-dihydroxy-3,3-diaminodiphenyl (26).

Kinetics of the reaction of terephthaloyl chloride with phenolphthaleine in
the presence of triethylamine is shown in Fig. 7 (27).

It is interesting that the temperature dependence of viscosity of the resulting polyarylate is described by two maxima as shown in Fig. 8 (28). This
can be explained by a complex temperature dependence of the rate constant
(28).

Kuchanov (29) calculated the effect of dosing out the monomers on the molecular weight distribution of polymers obtained by non-equilibrium polycondensation. He drew a conclusion that the number-average coefficient of polycondensation was inversely proportional to the square root of the monomer addition rate.

The one-step synthesis of polyimide from pyromellitic anhydride and aniline
phthaleine in nitrobenzene proceeds as shown in Fig. 9 (15).

An interesting specific feature of this process, as seen from the Fig. 10,
is the fact that the activation energy increases with increasing conversion
degree of cyclization (30). This is evidently caused by increasing rigidity
of the macromolecule chain and by difficulties in steric order of heterocyc-

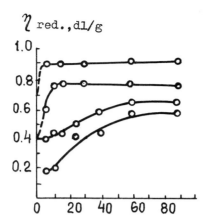

Fig. 7. Change in viscosity of polyarylate Φ-2 solution (polyphenolphthaleineterephthalate) in the course of its synthesis by acceptor-catalytic polycondensation at various temperatures.

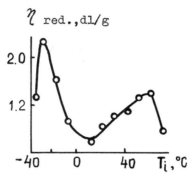

Fig. 8. Temperature dependence of the viscosity of polyarylate solution in the non-equilibrium polycondensation of terephthaloyl chloride with 1,1-bis(4-hydroxyphenyl)-1-ethanol.

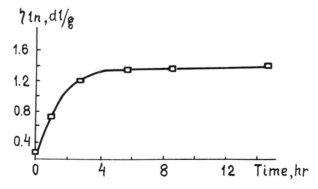

Fig. 9. Dependence of logarithmic viscosity of pyromellite-imide of aniline phthaleine solution in nitrobenzene at 210°C on the reaction time.

lization.

It is found that the order of mixing the initial substances greatly effects the structure of the resulting polymer in the case of asymmetric monomers (31).

It should be noted that the mechanism of different polycondensation reacti-

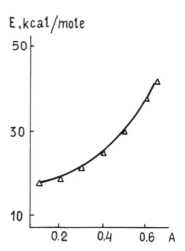

Fig. 10. Change in activation energy (U) of heterocyclization
of polyhydrazides: (P - conversion degree)

$$\left[\begin{array}{c}\text{...}\end{array}\right]_x$$

(5)

ons is clarified in detail only for some cases. The mechanism of non-equilibrium polycondensation is investigated mainly with several polycondensation reactions in solution (3).

Investigation of high-temperature non-equilibrium polycondensation of dicarboxylic acid chlorides with bisphenols in solution has led to conclusion
that the reaction proceeds according to an ionic mechanism (32):

$$\text{Cl-CO-}\bigcirc\text{-COCl} \rightleftharpoons \text{ClCO-}\bigcirc\text{-CO}^+ + \text{Cl}^-$$
$$\text{ClCO-}\bigcirc\text{-CO}^+ + \text{HOR} \longrightarrow \text{ClCO-}\bigcirc\text{-COOR} + \text{H}^+$$
$$\text{ROCO-}\bigcirc\text{-COCl} \rightleftharpoons \text{ROCO-}\bigcirc\text{-CO}^+ + \text{Cl}^-$$
$$\text{ROCO-}\bigcirc\text{-CO}^+ + \text{HOR} \rightleftharpoons \text{ROCO-}\bigcirc\text{-COOR} + \text{H}^+$$
$$\text{H}^+ + \text{Cl}^- \longrightarrow \text{HCl}$$

(6)

Investigation of non-equilibrium polycondensation in solution in the case
of acceptor-catalytic polyesterification in the presence of tertiary amines
has led to conclusion that the reaction can proceed both   according to the
mechanism of nucleofilic catalysis:

$$\text{R-CO-Cl} + \text{R}_3\text{N} \rightleftharpoons \text{R-CO-NR}_3^+...\text{Cl}^- \xrightarrow{\text{Ar-OH}} \text{R-CO-OAr} + \text{R}_3\text{N·HCl,}$$

and to that of total basic catalysis:

$$\text{Ar-OH} + \text{R}_3'\text{N} \rightleftharpoons \text{Ar-OH...NR}_3' \rightleftharpoons$$
$$\rightleftharpoons \text{Ar-O}^-...\text{HNR}_3'^+ \longrightarrow \text{Ar-OCOR} + \text{R}_3\text{N·HCl}$$

(7)

The reaction course in this case depends on the basicity of tertiary amine
and, consequently, on its ability to form a hydrogen bond with phenol (33 &
34).

A new method of controlling the structure of the resulting macromolecule should be mentioned. The polycondensation of aromatic tetramines with m-carboranedicarboxylic acid chloride gives a three-dimensional insoluble polymer. And the addition of pyridine hydrochloride allows the preparation of a linear polymer (12).

Among the last tendencies the use of thermodynamic methods for studying polycondensation processes should be mentioned. Thus, the thermodynamics of polyheteroarylene synthesis was studied comprehensively. As a result, temperature limits of the reaction stages were found, thermodynamic functions of polymers and equilibrium constants of the processes were determined, which showed that the reactions are non-equilibrium polycondensation processes (35).

Catalysis in Polycondensation Reactions

Catalysis in non-equilibrium polycondensation can proceed as cationic, anionic, ion-coordinative, and free-radical processes depending on the nature of initial substances (4). The schemes of intermediate complexes are given below:

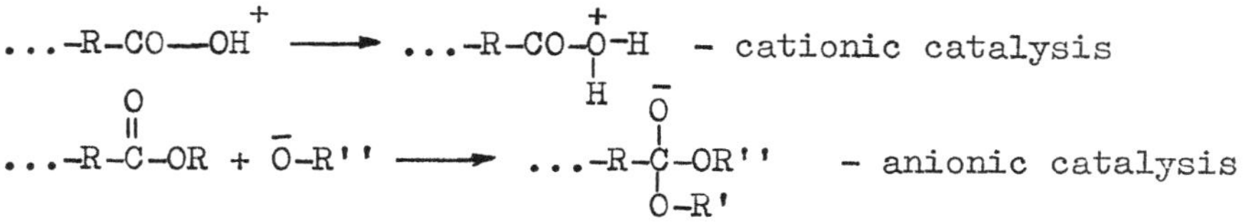

$$...-R-CO-\overset{+}{O}H \longrightarrow ...-R-CO-\overset{+}{O}-H \quad - \text{ cationic catalysis}$$

$$...-R-\overset{O}{\overset{\|}{C}}-OR + \overset{-}{O}-R'' \longrightarrow ...-R-\overset{O-R'}{\underset{O-R'}{C}}-OR'' \quad - \text{ anionic catalysis}$$

$$(8)$$

Ion-coordinative catalysis occurs, for instance, in the presence of tertiary amines to form apparently the following intermediate complex:

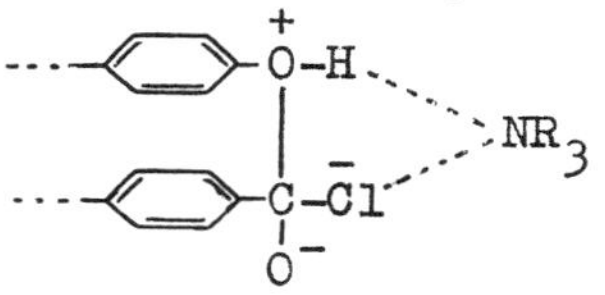

$$(9)$$

At the present time various compounds of the following elements has been investigated as polycondensation catalysts: H,Li,Na,K,Cu,Ag,Be,Mg,Ca,Zn,Sr,Cd, Ba,Hg,B,Al,Ce,C,Si,Ti,Ge,Zr,Sn,Pb,Th,N,P,As,Sb,Ta,Bi,O,S,Cr,Mo,U,Cl,Mn,Fe, Co,Ni,Pd,Pt.

The free-radical catalysis under action of copper/amine complexes occurs, for instance, in the oxidative process of acetylene polycondensation giving carbine, the third linear form of carbon (36):

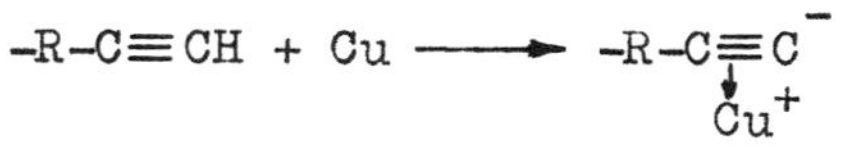

$$-R-C\equiv CH + Cu \longrightarrow -R-C\equiv C^- \quad \underset{Cu^+}{}$$

$$(10)$$

Stereoregular Polycondensation

The non-equilibrium polycondensation of dicarboxylic acid chlorides with bisphenols in the presence of catalysts such as tertiary amines yielded polyarylates, the properties of which were different depending on the reaction conditions and the catalyst nature.

The mechanism of acceptor-catalytic polyesterification proceeding in this case can be described by the following scheme (37):

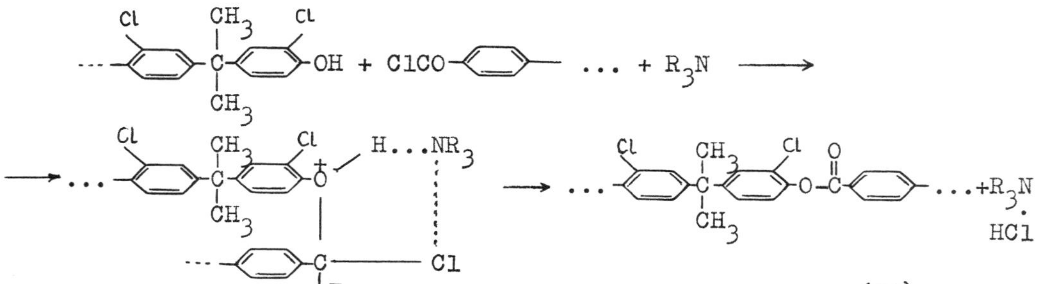

$$(11)$$

The reaction can give a polyester containing two unit conformations differing in mutual arrangement of the halogen atom (37). Thus, the polycondensation of bis-(4-hydroxy-3-chlorophenyl)-2,2-propane with terephthaloyl chloride leads to the formation of polyarylates with different properties, melting points, and solubilities depending on the reaction conditions (38). Investigation of the resulting polyarylates showed that they are conformation-regular polymers with various conformation resulting from hindered rotation (37 & 38). Their structure is as follows:

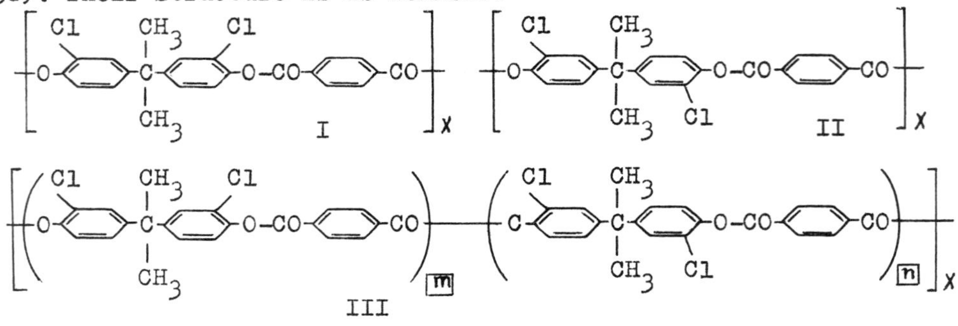

(12)

Formulas I and II express two extreme structures: I - cis and II - trans. They can be considered as analogs of isotactic and syndiotactic polyolefine structures. Formula III corresponds to an atactic polymer containing both conformations distributed statistically or blockwise in the polyarylate macromolecule (39).

Conformation-regular polymers could be prepared from other diane derivatives containing two methyl or nitro groups (40). All such polymers seem to have blocks of regularly arranged units.

Influence of the Solvent Nature on Polycondensation

The solvent nature was found to effect essentially the molecular weight and the yield of polyamides forming by the non-equilibrium polycondensation of acid dichlorides with aromatic diamines (41).

Sokolov found that the addition of water makes it possible to change rather markedly the conditions of synthesis of polyamides from dicarboxylic difluoroanhydrides and diamines (42).

Herlinger and coworkers (43) showed that the non-equilibrium polycondensation in amide solvents proceeded through the stage of a reactive complex of acid chloride with amine:

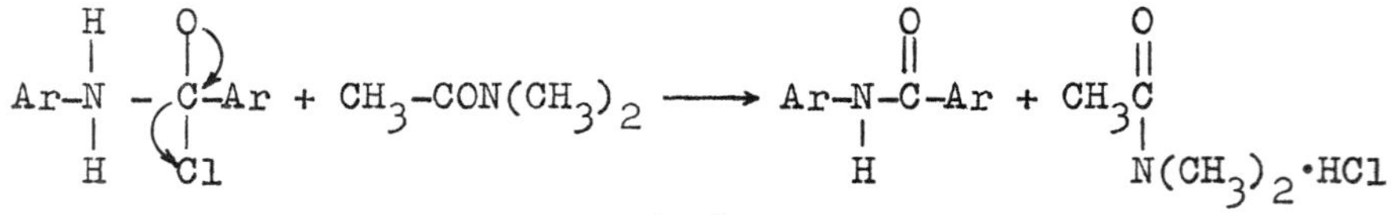

(13)

Then an amide bond and dimethylacetamide hydrochloride were formed.

Polycondensation of dicarboxylic acid dichlorides with diamines in dimethylformamide was also showed to give reactive adducts of acid chlorides with dimethylformamide (44). They react with amines to form N,N-dimethyl-N-phenylformamidine:

$$RCO-O-CH=\overset{+}{N}(CH_3)_2Br^-$$

(14)

The unreacted amine can be acid acceptor which effects the process course changing the ratio of initial substances. The course of polycondensation depends, therefore, on the equilibrium.

$$RNH_2 \cdot HCl \rightleftharpoons DMA \cdot HCl \quad (43).$$

(15)

Besides, a reamidation reaction can occur:

$$HCO-N(CH_3)_2 + ClCOAr \longrightarrow (CH_3)_2N-CO-Ar + HCOCl \longrightarrow CH_3COCl + ArCON(CH_3)_2$$
$$\text{(unstable)} \quad \text{stable}$$
$$\downarrow$$
$$CO + HCl$$

(16)

This reaction investigated by using radioactive dimethylacetamide $-C^{14}$ leads to chain termination according to the scheme:

$$H_2N-polymer-COCl + CH_3\overset{14}{C}ON(CH_3)_2 \rightleftharpoons H_2N-polymer-CO-N(CH_3)_2 + CH_3\overset{14}{C}OCl$$

$$H_2N-polymer-\overset{14}{C}ON(CH_3)_2 + CH_3\overset{14}{C}OCl \rightleftharpoons CH_3-CO-NH-polymer-CO-N(CH_3)_2 + HCl$$

(17)

The use of radioactive atoms (C^{14}) showed that the rate of polymeric chain propagation and termination is in the ratio 280:1. Kinetic investigations showed that 70-80° is the upper temperature limit above which no high polymer can be prepared because of high reamidation rate and, consequently, chain termination (43).

Copolycondensation

Investigations of copolycondensation revealed the influence of monomer reactivity and the order of the reaction course on the structure of the resulting polymer. Turska and coworkers derived a kinetic equation describing the course of copolycondensation of three bifunctional monomers A,B,C. These monomers are so selected that monomer A reacts with monomers B and C, while the latter do not react with each other (45). Turska and coworkers studied the influence of the reactivity of functional groups of monomers on the composition and the chain structure of the macromolecule in a non-equilibrium polycondensation (46). They determined the isokinetic temperature (T_i) for each monomer pair at which the rate constants for both reactions are equal.

$$T_i = \frac{E_{AB} - E_{AC}}{R(B_{AB}-B_{AC})} \quad , \tag{1}$$

where E_{AB} and E_{AC} are activation energies of the corresponding polycondensation reactions, B_{AB} and B_{AC} are the preexponent factors of the Arrhenius equation, and R is the universal gas constant.

If at temperatures lower than T_i $K_{AB} > K_{AC}$, then at a temperature above T_i the reactivity of monomers B and C is converted, and the latter becomes more active than monomer B. Consequently, the distribution of units B and C along the macromolecule chain differs with varying temperature. At the same time, at a reaction temperature T_i a product of invariant composition with a statistical distribution of units B and C is formed independent of the conversion degree. This fact makes it possible to control the copolycondensation process by means of suitable selection of the reaction temperature (46). Copolycondensation can be carried out at will to give a statistical or a block copolymer. If, as usual, both copolymers are mixed with the intermonomer, a statistical copolymer is always formed. If the intermonomer is gradually introduced into the comonomer mixture, a block copolymer is obtained (47). In this case, the slower the intermonomer addition occurs, the more pronounced is block formation process in the resulting copolymer (48).

Rather interesting is the specific feature of polycondensation which provides a means for preparing polyblock copolymers unlike the polymerization, which gives block copolymers only with di- and threeblock structures.

The nature of solvent for polyarylate synthesis effects the structure of
the resulting copolymer. Distribution of units in the copolymer obtained
by the reaction of isophthaloyl chloride with 4,4'-diaminodiphenyloxide-
-4,4'-diaminodiphenylsulfone was studied under different conditions of polycondensation (49).

Sulfolane was found to reveal a pronounced tendency to block formation. But
in the presence of propylene oxide, a weaker acid acceptor, it was possible
to prepare copolymers with a considerable content of alternating structures
(49).

Rather important is also the order of reactant mixing. Thus, if one diol is
reacted with triethylamine, then dicarboxylic acid chloride is added, and
only afterwards the second diol is introduced, a block copolymer is formed
(31 & 50). But if triethylamine is added to the solution of acid chloride
and diol and only then the second diol is introduced, a statistical copolymer is obtained (31).

Perspective of Polycondensation Development

Undoubtedly, that further developments of polycondensation synthesis enrich
the polymeric science with many new polymers possessing a complex of valuable properties. Among those we can primarily expect thermostability, in-
combustibility, electroconductivity, semi-conductive and photoelectrical
properties, high mechanical strength and others. Rather promising is the
synthesis of copolymers, among which various block copolymers should be
emphasized, interesting as initial materials for preparing separating membranes. Of great interest is the synthesis of various polymers duplicating
natural and biopolymers, especially for medical application.

Characteristic of polycondensation development in the last years is the
wide use of a number of elements for synthetic purposes (4,20 & 51). Many
elements of the Periodic system can be found in various polymers obtained
by polycondensation methods (51): H,Be,B,C,N,O,F,Mg,Al,Si,P,S,Cl,Ca,Ti,V,
Cr,Mn,Fe,Co,Ni,Cu,Zn,Ga,Ge,As,Se,Br,Sr,Zr,Nb,Mo,Ru,Rh,Pd,Ag,Cd,Sn,Sb,Te,I,
Ba,Ce,Hf,Ta,W,Os,Ir,Pt,Hg,Tl,Pb,Bi,Th,U.

As can be seen,there is great variation in structures of organo-element polymers prepared by polycondensation, since at the present time already 55
elements are used in these polymers. Obviously, we have every reason to believe that further development of investigations in the field of organo-
element polymers enriches the science and practice with a great number of
new interesting polymers.

Directly in the field of polycondensation processes we have every reason to
expect in the nearest future new progress in controlling the polycondensation methods of synthesis. Among the problems of this kind the question
of controlling the molecular weight to prepare high polymers should be
mentioned, as well as the search for new reactions and new methods of polycondensation processes. Important task is such a control of synthesis processes, which will allow the regulation of structure and raznozvennost of
the resulting polymers and hence the preparation of polymers with a desired
optimum combination of properties. One of the most important ways of solving these problems is a widespread development of investigations in the
field of catalysis in polycondensation to find new approaches to the control of processes of polymer synthesis by polycondensation and the expansion of assortment of such reactions.

In conclusion it should be noted that, as may be seen from the above-mentio-
ned examples, polycondensation chemistry is in good progress now. New reactions of polymer synthesis are being developped, and the known synthetic
processes are being investigated and elaborated. Inquisitive mind of investigators gets to know common tendencies of polycondensation processes and
finds the ways for controlling the synthesis of polymers. As a result, the
science becomes enriched with numerous new polymers, and a great number of
known polymers prepared by improved methods acquires a combination of more
valuable properties. The development of synthetic aspect is characterized
by more and more profound investigations of increasingly complex chemical
reactions and by preparation of new structurally complex polymeric materials which is indicative of the advanced chemical skill in polymer synthe-

sis. The ways giving rise to the formation of anomalous units leading to polymer raznozvennost are being found, which is very important in determining the relationship between the polymer structure and its properties. And after all, very characteristic of polycondensation synthesis is the possibility to synthesize not only constructive polymers, but also, to an ever greater extent, compounds with specific properties such as thermostable, semi-conductive, electroconductive, photoactive, bioactive polymers, biopolymers, catalysts, ionites etc.

REFERENCES

1. J. Gay-Lussac, _Ann._ _7_, 40 (1833).
2. V.V. Korshak and S.V. Vinogradova, _Equilibrium Polycondensation_, Nauka, Moscow (1968).
3. V.V. Korshak and S.V. Vinogradova, _Non-Equilibrium Polycondensation_, Nauka, Moscow (1971).
4. V.V. Korshak, _Vysokomol. soed._ A21, 3 (1979).
5. V.V. Korshak, _Vysokomol. soed._ A15, 198 (1973).
6. V.V. Korshak and K.K. Mozgova _Uspekhi khimii_ 28, 783 (1959).
7. V.V. Korshak, _J. prakt. Chem._ 313, 422 (1971).
8. V.V. Korshak, _Pure and Appl. Chem._ 12, 101 (1966).
9. V.V. Korshak, _Chemical Structure and Temperature Properties of Polymers_, p.376, Nauka, Moscow (1970).
10. V.V. Korshak, _Thermostable Polymers_, Nauka, Moscow (1969).
11. W.H. Carothers, _Chem. Rev._ 8, 353 (1931).
12. V.V. Korshak, N.I. Bekasova, V.V. Vagin and A.A. Izyneev, _Vysokomol. soed._ 15, 6 (1973); _Doklady AN SSSR_ 210, 110 (1973).
13. V.V. Korshak, _Chemistry of High Polymers_, p. 299, Izdat. AN SSSR, Moscow-Leningrad (1950).
14. V.V. Korshak and A.L. Rusanov, _Polymery-Tworzywa Wielkoczasteczkowe_ 401 (1970).
15. S.V. Vinogradova, Ya.S. Vygodskii and V.V. Korshak, _Vysokomol. soed._ A12, 1987 (1970).
16. V.V. Korshak, N.I. Bekasova and L.G. Komarova, _Vysokomol. soed._ A12, 1966 (1970).
17. V.V. Korshak, _Vysokomol. soed._ A19, 1179 (1977).
18. V.V. Korshak, _Raznozvennost of Polymers_. Nauka. Moskow (1977).
19. V.V. Korshak, _Doklady AN SSSR_ 247, 138 (1979).
20. V.V. Korshak, _Vysokomol. soed._ A18, 1443 (1976).
21. V.V. Korshak, _Uspekhi khimii_ 42, 695 (1973).
22. S.T. Alekseeva, S.V. Vinogradova, V.D. Vorob'yov, Ya.S. Vygodskii, V.V.Korshak, I.Ya. Slonim, T.I. Spirina, Ya.G. Urman and L.I. Chudina, _Vysokomol. soed._ A21, 2207 (1979).
23. V.V. Korshak and S.V. Vinogradova, D.R. Tur and N.N. Kazarava, _Acta Polymerica_ 31, 669 (1980).
24. V.V. Korshak, S.V. Vinogradova and Ya.S. Vygodskii, _J. Macromol.Sci.-Rev. Macromol. Sci._ C11, 45 (1974).
25. S.V. Vinogradova, Z.V. Gerashchenko and Ya.S. Vygodskii, _Doklady AN SSSR_ 203, 821 (1972).
26. V.V. Korshak, G.M. Zeitlin, O.S. Zhuravkov, F.B. Sherman and V.A. Klimova, _Vysokomol. soed._ 16, 99 (1974).
27. S.V. Vinogradova, V.A. Vasnyov and V.V. Korshak, _Vysokomol. soed._ 9, 522 (1967).
28. V.V. Korshak, E.A. Turska, M. Siniarska-Kapustinska, V.A. Vasnjov, S.V. Vinogradova, S.A. Pavlova, A.V. Vasil'yev and L.V. Dubrovina, _Vysokomol. soed._ 16, 147 (1974).
29. S.I. Kuchanov, _Vysokomol. soed._ 16, 136 (1974).
30. I.P. Bragina, V.V. Korshak, G.L. Berestneva, S.V. Vinogradova, D.R. Tur, V.A. Khomutov and V.V. Krylova, _Vysokomol. soed._ A18, 2318 (1976).
31. V.V. Korshak, S.V. Vinogradova, V.A. Vasnyow, G.D. Markova and T.V. Lecae, _J. Polymer Sci._ 13, 2741 (1975).
32. R.S. Velichkova, V.V. Korshak, S.V. Vinogradova, I.V. Ivanov, A.I. Ponomarenko and I.S. Enikolopov, _Izvest. AN SSSR_, ser. khim. 1969, 858.
33. S.V. Vinogradova, V.V. Korshak, L.I. Komarova, V.A. Vasnyov and T.I.Mitaishvili, _Vysokomol. soed._ A14, 2591 (1972).
34. V.V. Korshak, S.V. Vinogradova and V.A. Vasnyov, _Doklady AN SSSR_ 191, 614 (1970).
35. N.V. Karyakin, I.B. Rabinovich, V.V. Korshak, A.L. Rusanov, D.S. Tugushi, A.N. Mochalov and V.N. Sapozhnikov, _Vysokomol. soed._ A16, 691 (1974).

36. V.V. Korshak, Yu.P. Kudryavtsev and A.M. Sladkov, <u>Vestnik AN SSSR</u> No.1, 70 (1978).
37. V.V. Korshak, S.V. Vinogradova, V.A. Vasnyov, E.B. Musaeva, A.P. Gorshkov, G.K. Syomin and L.N. Gvozdeva, <u>Doklady AN SSSR</u> <u>226</u>, 350 (1976).
38. V.V. Korshak, S.V. Vinogradova, V.A. Vasnyov, M.G. Keshelava, M.M. Dzhananashvili and L.N. Gvozdeva, <u>Vysokomol. soed.</u> <u>A19</u>, 2625 (1977).
39. V.V. Korshak, S.V. Vinogradova, V.A. Vasnyov and A.V. Vasil'yev, <u>Polym. Letters</u> <u>10</u>, 429 (1972).
40. V.V. Korshak, S.V. Vinogradova and V.A. Vasnyov, <u>Faserforsch. u. Textiltechn.</u> <u>28</u>, 491 (1977).
41. V.V. Korshak, P.K. Vorob'yov, N.I. Bekasova, E.A. Chizhova and L.G. Komarova, <u>Vysokomol. soed.</u> <u>A18</u>, 632 (1976).
42. V.I. Lokunova, L.B. Sokolova and V.M. Savinova, <u>Vysokomol. soed.</u> <u>A18</u>, 450 (1976).
43. H. Herlinger, H.-P. Hörner, F. Druschke, H. Knöll and F. Haiber, <u>Angew. makromol. Chem.</u> <u>29-30</u>, 227 (1973).
44. S.N. Kharkov, V.P. Kabanov and L.P. Grechishnikova, <u>Vysokomol. soed.</u> <u>A15</u>, 2045 (1974).
45. E. Turska, S. Boryniec and L. Pietrzak, <u>J. Appl. Polym. Sci.</u> <u>18</u>, 667 (1974).
46. E. Turska, S. Boryniec and A. Dems, <u>J. Appl. Polym. Sci.</u> <u>18</u>, 671 (1974).
47. V.V. Korshak, V.A. Vasnyov, S.V. Vinogradova, P.O. Okulevich and Yu.I. Perfilov, <u>Doklady AN SSSR</u> <u>204</u>, 1129 (1972).
48. V.V. Korshak, S.V. Vinogradova, V.A. Vasnyov, Yu.I. Perfilov and P.O. Okulevich, <u>J. Polymer Sci., Polym. Chem. Ed.</u> <u>11</u>, 2209 (1973).
49. P.A. Curnuck, M.E. Michael, <u>Brit. Polymer J.</u> <u>5</u>, 21 (1973).
50. S.V. Vinogradova, V.V. Korshak, P.O. Okulevich, Yu.I. Perfilov and V.A. Vasnyov, <u>Vysokomol. soed.</u> 15, 470 (1973).
51. In book <u>Progress in the Field of Synthesis of Organo-Element Polymers</u>, Nauka, Moscow (1980).

TWENTY-FIVE YEARS OF LIVING POLYMERS

Michael Szwarc

Polymer Research Center of State University of New York, Syracuse, N.Y.,
13210

(1) Early developments in anionic polymerization.

     The modern concept of linear polymers was developed in 1920 by Staudinger, who
introduced the idea of chain addition reaction yielding long molecules composed of
monomeric units linked by covalent bonds. He was also the first to understand the anionic
character of formaldehyde polymerization initiated by bases such as sodium methoxide [1].
Indeed, studies of this reaction led him to the notion of linear macromolecules. Polymerization of ethylene oxide initiated by alkali metals and reported as early as 1878 [2]
could also be interpreted in these terms.
     Subsequent developments in the field of polymerization were centered on radical poly-
addition. Its mechanism was firmly established in the 1930's and attracted much attention.
The interest in anionic polymerization was marginal and the activities in this field were
centered at that time around Ziegler in Germany and Lebedev in Russia. Both groups were
interested in polymerization of styrene and dienes initiated by sodium metal and their
work led to industrial production of synthetic rubber marketed by I. G. Farbenindustrie as
Buna.
     Systematic studies of Ziegler allowed him to formulate the initiation step as the
addition of two sodium atoms to a monomer with formation of two covalent Na-carbon bonds
[3], e.g.,

$$CH_2:CH.CH:CH_2 + 2Na \rightarrow Na - CH_2.CH:CH.CH_2 - Na \qquad .$$

     The concept of carbanions and ion-pairs was at its infancy at that time. Although the
isolation of such adducts eluded him, he argued for their existence by demonstrating the
formation of butene-2 in a reaction performed in the presence of an excess of methylaniline
[3a], a compound that does not react directly with sodium metal. Since two moles of sodium
amide were obtained for each mole of the butene formed by this process, Ziegler described
its course by the scheme:

$$Na.CH_2.CH:CH.CH_2.Na + 2PhNH.CH_3 \rightarrow CH_3.CH:CH.CH_3 + 2NaN(CH_3)Ph \qquad .$$

Our modern description follows the line:

$$CH_2:CH.CH:CH_2 + Na \rightarrow (CH_2:CH.CH:CH_2)^{\overline{\cdot}},Na^+ \xrightarrow{PhNHCH_3} CH_3.CH:CH.CH_2.$$

$$CH_3.CH:CH.CH_2. + Na \rightarrow CH_3.CH:CH.CH_2^-,Na^+ \xrightarrow{PhNHCH_3} CH_3.CH:CH.CH_3$$

with simultaneous formation of 2 molecules of the sodium anilide.
     The propagation of the ensuing polymerization was then described as an insertion of a
monomer into the C-Na bond, yielding another carbon-sodium bond -- a description which
differs insignificantly from our present formulation. Nor surprisingly, it was superfluous
to postulate a termination step in this mechanism; the polymer still possessed the active
C-Na bond. Indeed, polymerization was resumed when fresh monomer was added to the reactor
[4].
     An alternative interpretation of this polymerization was advocated by Schlenk and
Bergmann [5]. In his early work reported in 1914 [6], Schlenk described the addition of
alkali metals to aromatic hydrocarbons leading to intensely colored solutions. The concepts
of free radicals or radical-ions were unknown at that time, hence Schlenk referred to the
adduct as a complex. He also showed that a similar reaction of 1,1-diphenyl ethylene
resulted in its colored dimer which yielded 2,2,5,5,-tetraphenyl adipic acid on carboxyla-
tion. When the concept of free radicals was established, Schlenk and Bergmann argued that
the initially formed adduct is a free radical, e.g.,

$$CH_2:CH.CH:CH_2 + Na \rightarrow NaCH_2.CH:CH.CH_2. \qquad ,$$

and it initiates a radical type monomer addition. Similar reasoning accounted for the formation of the dimeric di-adduct of 1,1-diphenyl ethylene arising from radical dimerization. Interestingly, the latter interpretation comes close to our present description of this reaction -- the recombination of radical-anions.

The success of radical theories of polymerization added credibility to the Schlenk and Bergmann mechanism and even caused some regression. For example, Schulz in 1938 [7] and Bolland as late as 1941 [8] were still upholding the idea of radical nature of diene polymerization induced by alkali metals. Nevertheless, slow progress was made. Abkin and Medvedev [9] demonstrated the long-life nature of the growing centers, implying that they are not radicals.

In the following years, several processes were attributed to anionic polymerization. Blomquist [10] reported anionic polymerization of nitro-olefines initiated by hydroxyl ions. Beaman [11] recognized the anionic character of methacrylonitrile polymerization initiated in ether solution by Grignard reagents or by triphenylmethyl sodium. Robertson and Marion [12] reinvestigated the sodium initiated polymerization of butadiene in toluene and isolated oligomers having benzyl moiety as their end-groups. The characteristic red color developed in the course of that reaction implied the formation of sodium benzyl, presumably through transfer reaction. Studies of Higginson and Wooding [13] of homogeneous polymerization initiated by alkali metals or their amides in liquid ammonia left no doubt about their anionic character.

However, the final impetus for vigorous studies of anionic polymerization came in 1956 when three papers were published: The description of homogeneous electron-transfer initiated polymerization of styrene and isoprene, yielding living polymers [14] and the discovery of 1,4-cis polymerization of isoprene initiated by metallic lithium in hydrocarbon solvents [15]. Since then the interest in this field has grown tremendously and its development has progressed in a truly exponential fashion. In fact, during the following 25 years, more than 2,000 papers were published on this subject.

(2) Living polymers.

The extensive studies of radical polymerization carried out in the period 1935-1950 firmly established the basic mechanism of the poly-addition [16]. Termination steps have been essential in accounting for the numerous observations and their existence was unquestionable [17].

The imperative requirement of termination in radical polymerization arises from the nature of the interaction taking place between two radicals as they encounter each other. Coupling or disproportionation are the results, and in either case the interacting radicals are annihilated. However, the encounter between two ionically growing macromolecules does not annihilate them. Neither coupling nor disproportionation is feasible as two cations or two anions encounter each other. However, exclusion of the bimolecular termination involving two growing polymers does not exclude other kinds of termination or transfer, and the success of the conventional polymerization scheme created the impression that a terminationless polymerization is highly improbable, if not impossible.

Polymerization schemes free of termination had been considered in earlier days. For example, the kinetic scheme of terminationless polymerization was developed by Dostal and Mark [18] in 1935. Similarly, terminationless sodium-initiated polymerization of butadiene was visualized by Ziegler, and in fact, the need of a termination step was not appreciated at that time. Later, several examples of terminationless polymerization were considered by Flory [19], who also discussed the ramifications of such schemes.

However, it was not until 1956 that Szwarc and his associates [14] conclusively demonstrated the terminationless character of anionic polymerization of vinyl monomers in the absnece of impurities. They proposed the term "living polymers" for those macro-molecules which spontaneously resume their growth whenever fresh monomer is supplied to the system. It should be stressed that living polymers, although not named in this way, were described earlier by Ziegler [4]. However, while the heterogeneous nature of these systems obscured some aspects of his work, and the ramifications were not emphasized strongly enough, Szwarc' studies were performed in homogeneous solution and all the important ramifications were clearly outlined. The characteristic features of living polymers are clearly revealed by their experiments. Living polymers resume their growth whenever monomer is added to the system. They do not die but remain active and wait for the next prey. If the monomer added is different from the one previously used, a block polymer results. This, indeed, is the most versatile technique for synthesizing block polymers [19].

Living polymers do not become infinitely long. Any system producing living polymers contains a finite amount of monomer. The system also contains some specified concentration of growing centers, or living polymers, and consequently all of the available monomer becomes partitioned among them. Hence, the number average degree of polymerization, $\overline{DP}_n$, is simply given by the ratio, (total no. of moles of added monomer)/(total no. of moles of living polymers), if endowed with one active group, or twice as high if both ends are active.

The lack of natural death, i.e., of spontaneous termination, does not imply immortality either. The reactive end-groups of living polymers may be annihilated by suitable reagents, a process known as "killing" of living polymers. It is desirable, indeed, to distinguish between a "killing" reaction and a spontaneous termination. The latter is governed by the law of probabilities and its course is set by the conditions existing _during_ polymerization. On the other hand, the moment of the "killing" reaction is determined by the free choice of the experimenter, usually after completion of the polymerization. Moreover, he is free also to choose at will the reagent which converts active end-groups into dead ones. For example, in the anionic polymerization of styrene propagated by carbanions, addition of water, or of any other proton-donating substance, converts the active $\sim\!\!\sim\!\!\sim\!CH(Ph)^-$ end-groups into

$>\!\!-C\!-\!H$, whereas addition of carbon dioxide converts $>\!\!-C^-$ into $>\!\!-C\cdot COO^-$ and

eventually into $>\!\!-C\cdot COOH$. Similarly a terminal hydroxyl group may be formed upon

addition of ethylene oxide. It is important to realize that all these end-groups cannot propagate a non-polar vinyl polymerization, and in this sense a dead polymer is produced in each case. We should stress that this technique allows us to introduce valuable functional end-groups into a macromolecule, as was shown by Rempp and his associates [20], giving, therefore, a novel and interesting product.

The functional polymers have many practical applications, especially the bi-functional ones. This point was stressed in our first papers dealing with this subject. By a proper choice of reagent, e.g., using

$$BrCH_2 \cdot C_6H_4 \cdot CH:CH_2$$

as a terminating reagent, a polymer or oligomer is converted into a species behaving like a monomer. Such macromers, a term proposed by Milkovich, could be used in co-polymerization allowing the preparation of polymers possessing well defined branches, not necessarily composed of the same monomeric units as the main chain.

The other two important applications of living polymers are associated with lack of termination. As pointed out by Flory, living polymers, if properly prepared, yield mono-dispersed polymers, i.e., macromolecules having Poisson molecular distribution. Such polymers allow for more precise studies of the effects of molecular weight distribution on the rheological properties of polymer solutions and solid polymers.

The second advantage arising from the lack of termination is associated with the feasibility of preparing block polymers. Whenever living polymer of monomer A is capable of initiating polymerization of monomer B, a block polymer,

$$A\sim\!\!\sim\!\!\sim\!\!AA.B\sim\!\!\sim\!\!\sim\!BB \quad ,$$

free of homo-polymer, could be prepared. Moreover, the length of each block is easily controlled by the amounts of the monomers fed into the system.

Provided that living end-group of B can initiate polymerization of monomer A, a tri-block could be prepared:

$$A\sim\!\!\sim\!\!\sim\!AAB\sim\!\!\sim\!\!\sim\!BB.A\sim\!\!\sim\!\!\sim\!AA \quad .$$

In fact, this technique permits any desired distribution of two, or more, monomers along a chain of a predetermined size. The spectacular discovery by Ralph Milkovich [21] of the triblock polymers yielding thermo-plastic rubber, and the unexpected difference in the properties of triblock and diblock polymers added further impetus to those investigations. Indeed, a number of monographs and quite a few symposia were devoted to the synthesis and properties of block-polymers prepared through the living polymer technique. The industrial production of those macro-molecules led to many new products of considerable value and unique in their applications.

Preparation of star-shaped, comb-shaped, etc., polymers exemplifies another variant of block-polymer synthesis. These materials have also been found to have interesting and important industrial applications. A review dealing with the synthesis and properties of star-shaped polymers was published recently by Bywater [22].

The first examples of living polymers were furnished by anionic polymerization. However, living cationic polymerization is well known today, as well as living co-ordination polymerization. Furthermore, interesting techniques that allow conversion of anionic living polymers into cationic ones, or _vice versa_, were recently developed. It is impossible to discuss all these subjects in this paper, and I will therefore limit myself to just a few examples.

(3) _Review of some living polymer systems_.

The living character of ethylene oxide polymerization in tetrahydrofuran was demon-strated by Deffieux and Boileau [23] when the alkali counter-ion (K^+) was chelated by the (2,2,2)-cryptate. Kinetic and conductance data revealed the participation of free alkoxide ions and ion-pairs in the propagation and a lack of further aggregation, provided that the concentration of living polymers is lower than 6.10^{-4} M. Similar results were obtained in a recent study [24] where chelated Cs^+ ions were the counter-ions.

Further improvement of oxiranes polymerization was reported by Inoue [25]. Following his earlier studies of tetraphenyl-porphyrine-Al complexes, which revealed their usefulness as catalysts for propylene oxide polymerization and its co-polymerization with carbon dioxide [26], he modified the original complex by substituting $AlEt_2Cl$ for $AlEt_3$. The active species has the structure:

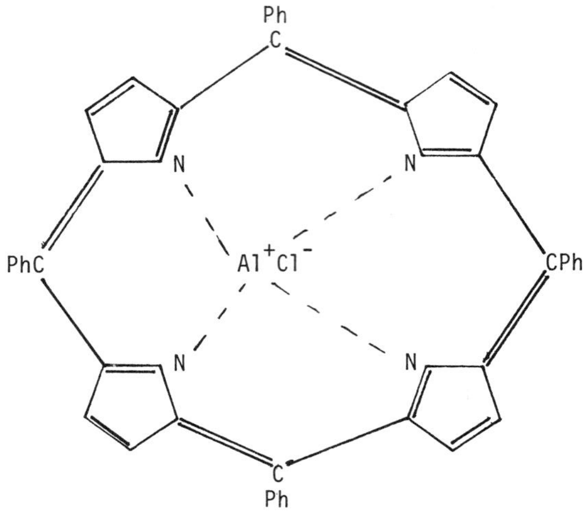

= TPhPORAlCl

and it rapidly initiates polymerization of ethylene-, propylene-, or 1,2-butene-oxides. The reaction involves the insertion into the Al-Cl bond, i.e.,

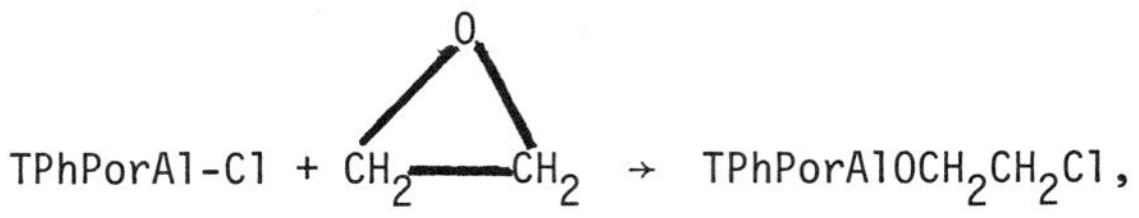

followed by the analogously proceeding propagation. Indeed, $ClCH_2CH_2(CH_3)OH$ was isolated when an equimolar mixture of the catalyst and propylene oxide were eventually hydrolyzed [27].

The resulting polymer showed a very narrow molecular weight distribution, and further evidence for a truly living character of this reaction was provided by the preparation of di- and tri-block polymers of the oxiranes, with exclusion of the homo-polymers.

The nature of the catalysts resulting from the controlled hydrolysis of organometallic compounds was extensively studied by Teyssié. Following the previous observations of Tsuruta [28] and Vandenberg [29], who produced useful catalysts by hydrolysis of zinc and aluminum alkyls, he investigated two-step condensation between metal acetates and alkoxides. This procedure led to a well-defined catalyst of the following structure [30]:

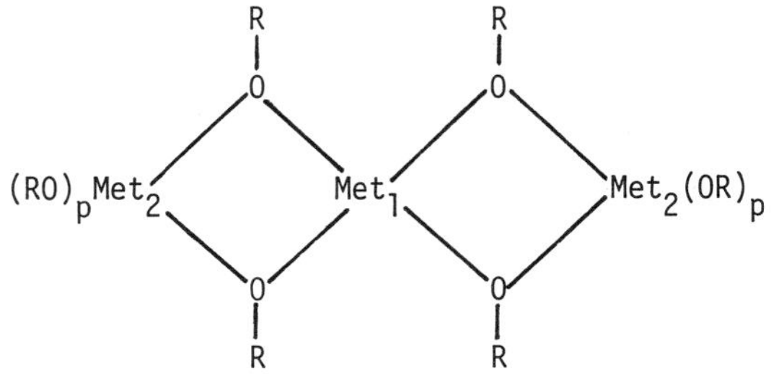

,

where Met_1 stands for Zn^{II}, Cr^{II}, Mo^{II}, Fe^{II}, or Mn^{II}, while Met_2 represents Al^{III} or Ti^{IV} and the subscript p is 2 for Al^{III} and 3 for Ti^{IV}. These mixed bimetallic alkoxides are aggregated, e.g., in benzene or cyclohexane they form octamers, although they are monomeric in butanol [31]. The aggregation probably explains their remarkable solubility in hydrocarbons arising from a compact oxide structure surrounded by a lipophilic layer of the alkoxide groups.

The mixed bimetallic alkoxides rank among the best catalysts of ring-opening polymerization of oxiranes, thiiranes, and lactones. The one formed from $Al(OBu)_3$ and zinc acetate seems to be superior. Kinetic and structural data suggest a coordinave anionic mechanism of initiation and propagation. The monomer is inserted into an A-OR bond simultaneously with its opening, like in the proposed flip-flop mechanism of Vandenberg [32]. The reactivity of the catalyst and its selectivity in co-polymerization could be greatly altered by varying solvents and the nature of the R groups.

Polymerization of ε-caprolactone initiated by the above catalyst shows all the characteristic features of living polymer system [33]. The degree of polymerization increases with conversion and the reaction is resumed on addition of fresh monomer. If carried out in butanol, which dissociates the aggregates, the product has a narrow molecular weight distribution with the M_w/M_n ratio approaching 1.05 [34]. Under these conditions all four terminal butoxy groups participate in the reaction and hence the number average degree of polymerization is given by the ratio of moles of the polymerized monomer to 4 Zn. On hydrolysis of the polymer-catalyst complex, one obtains the hydroxyl terminated polymers possessing a -CO.OBu group on their other end -- a result expected on the basis of the proposed mechanism.

Finally, the living character of that polymerization permits the preparation of block polymers. Consecutive addition of two different lactones, e.g., caprolactone and β-propiolactone, readily led to the expected two-block polymers. The virtually quantitative conversion and the exclusion of homo-polymers were attained under complete dissociation of the aggregates. Otherwise, as the second monomer is added, some inactive, sterically hindered OR groups may become available for the initiation. This would yield some homo-polymers. Alternatively, addition of caprolactone followed by propylene oxide resulted in preparation of the respective two-block polymers.

Anionic polymerization of thiiranes, e.g., 2-methyl-2-ethyl thiirane [34], initiated by carbazyl or fluorenyl salts in tetrahydrofuran, also yields living polymers.

Some coordination polymerizations also exhibit the character of a living polymerization process. For example, Natta [35] describes a heterogeneous catalyst yielding block polymers when the initially present monomer was replaced by another one. In that system the lifetime of a growing polymer exceeds one-half hour, i.e., the termination caused by the hydride transfer to the catalytic center was very slow. Similar results were reported by Bier [36] and by Kontos et al. [37].

The most spectacular coordination system free of termination and transfer was described recently by Doi et al. [38]. The homogeneous vanadium acetyl acetonate complex with $AlEt_2Cl$ rapidly initiated polymerization of propylene at -60°C. The molecular weight of the product increased with conversion and the ratio M_w/M_n was only slightly greater than 1.0. These facts confirm the rapidity of initiation and the lack of termination or chain transfer. Interestingly, the above catalyst was described previously by Natta [39], who utilized it for synthesis of syndiotactic poly-propylene.

Finally, two more examples of living polymer systems should be mentioned. Polymerization of diazomethane initiated by boron trifluoride gives a stable compound [40]

$$\sim\sim CH_2.BF_2$$

which reacts with diazomethane and perpetuates further growth of the polymer, i.e.,

$$\sim\sim CH_2BF_2 + CH_2N_2 \rightarrow \sim\sim CH_2CH_2BF_2 + N_2 \quad .$$

Polymerization of chloral initiated by organo-Sn compounds yields [41]

$$\sim\sim \overset{\displaystyle |}{\underset{\displaystyle CCl_3}{CH}} - O - Sn(CH_3)_3 \quad .$$

The latter reacts with chloral by insertion into the Sn-O bonds,

$$\sim\sim \overset{|}{\underset{CCl_3}{CH}} - O - Sn(CH_3)_3 + \overset{|}{\underset{CCl_3}{CHO}} \rightarrow \sim\sim \overset{|}{\underset{CCl_3}{CH}}.O - \overset{|}{\underset{CCl_3}{CH}} - O - Sn(CH_3)_3 \quad ,$$

hence the growth continues as chloral is supplied to the reactor.

(4) Living and dormant polymers.

In some systems most of the polymers potentially capable of growing are inert and do not contribute to propagation. The polymerization involves a small fraction of active polymers remaining in dynamic equilibrium with the inert ones. We refer to the latter as the dormant polymers in equilibrium with the living ones. A few examples illustrate this behavior:

(a) In anionic polymerization most of the polymers are often present in the form of tight ion-pairs, virtually not contributing to the propagation that is carried out by free ions. The latter are then the living polymers in equilibrium with the dormant ion-pairs.

(b) Lithium salts of living polystyrene, polybutadiene, etc., are present in hydro-carbon solvents as inactive dormant aggregates, e.g., dimers. The propagation arises from the presence of a minute fraction of active unaggregated polymers, the living ones. An even more complex situation is encountered in the presence of lithium alkoxides. According to Roovers and Bywater [42], the following equilibrium is established:

$$(\text{\textasciitilde styrene}^-,\text{Li}^+)_2 \underset{\text{LiOR}}{\xrightleftharpoons} (\text{\textasciitilde styrene}^-,\text{Li}^+,\text{LiOR})_2 \quad - \quad \text{dormant}$$

$$2(\text{\textasciitilde styrene}^-,\text{Li}^+) \qquad\qquad (\text{\textasciitilde styrene}^-,\text{Li}^+,\text{OR}) \quad - \quad \text{living}$$

more reactive   less reactive

(c) In cationic polymerization of tetrahydrofuran with $CF_3SO_3^-$ counter-ions, an equilibrium is established between the inert macro-esters,

$$\text{\textasciitilde}(CH_2)_4 O.SO_2CF_3 \quad ,$$

and the reactive macro-ions,

$$\text{\textasciitilde}(CH_2)_4 \overset{+}{O}\!\!\pentagon \quad , \qquad CF_3SO_3^- \quad .$$

Similarly, in cationic polymerization of oxazoline initiated by methyl-iodide, the inert polymer terminated by the $-CH_2I$ group is in equilibrium with its reactive ionized form.

The presence of dormant polymers does not affect the character of propagation but slows it down. Whenever the lifetime of the dormant polymer is short compared with the time between two consecutive monomer additions, this phenomenon does not affect molecular weight distribution, otherwise it leads to its broadening.

The existence of a variety of states of ionically growing polymeric end-groups calls for clarification of the term "concentration of living polymers". What is meant is the total concentration of all the inter-convertible species, whether living or dormant, that ultimately participate in the growth. Other polymers, endowed with end-groups, inert with respect to propagation, and that are <u>not convertible</u> into active ones, are referred to as dead polymers.

Ideally living polymers remain active forever. In practice this is not the case. Some slow and irreversible reactions ultimately convert them into dead polymers or lead to their degradation into unpolymerizable molecules. However, whenever such conversions are imperceptible during a time needed for completion of a desired task, e.g., a preparation of a block or star-shaped polymer or the performance of a kinetic study, etc., we may refer to such polymers as truly living.

(4) <u>Some mechanistic studies in living polymer systems</u>.

Mechanisms of polymerization were particularly fascinating problems to me. Hence, I propose to devote the rest of this review to a discussion of some mechanistic aspects of ionic polymerizations. I discussed in many papers and review articles the role of free ions and of a variety of ion-pairs in propagation of ionic polymerization. Therefore, this subject is omitted in the present review. Instead, I would like to discuss some problems concerned with triple ions and with aggregates formed by living polymers with lithium counter-ions in hydrocarbon solvents.

Triple ions form rather unusual species. They contribute little, if at all, to the propagation but their formation may have a buffering effect. The intramolecular triple ions' formation in the cesium polystyrene system illustrates this phenomenon [43]. The mechanism is outlined in Scheme 1. The formation of triple ions increases the concentration of Cs^+ ions and buffers, therefore, the dissociation of ion-pairs into free styryl ions which are the main contributors to the propagation.

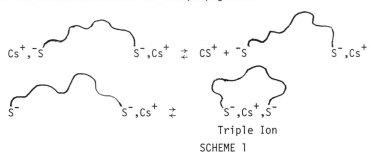

Triple Ion

SCHEME 1

S^- denotes a terminal styryl carbanion.

Generally, two distinct processes could be responsible for the formation of triple ions, namely,

$$2Cat^+,An^- \rightleftarrows Cat^+ + An^-,Cat^+An^- \qquad (a)$$

or

$$2Cat^+,An^- \rightleftarrows An^- + Cat^+,An^-,Cat^+ \qquad . \qquad (b)$$

Whenever equilibrium (a), which is concentration-independent, is more important than the conventional dissociation of ion-pairs, then its effect, in conjunction with the ion-pairs dissociation, decreases the anticipated concentration of anions. The reverse is true when equilibrium (b) prevails on (a).

A somewhat similar problem was encountered in the anionic propagation involving the barium salt of living poly-styrene possessing only one growth-propagating end per macromolecule [44]. The pseudo-first order rate constant of propagation was found to be independent of the salt concentration. This result was accounted for by a mechanism outlined in Scheme 2 with the assumption that the free poly-styrene anions are the only species contributing to the growth.

$$2Ba^{2+}, (S^- \sim\sim)_2 \rightleftarrows Ba^{2+}, (S^- \sim\sim) + Ba^{2+}, (S^- \sim\sim)_3$$

$$Ba^{2+}, (S^- \sim\sim)_2 \rightleftarrows Ba^{2+}, (S^- \sim\sim) + S^- \sim\sim \qquad .$$

SCHEME 2

Denoting by K_{triple} and K_{diss} the respective equilibrium constants and by k_p the bimolecular rate constant of propagation by the free poly-styrene anions, one finds the observed pseudo-first order rate constant to be given by

$$k_p K_{diss}/K_{triple}^{1/2} \qquad ,$$

i.e., it is independent of the salt concentration because within the investigated concentration range the relation

$$K_{triple} \gg K_{diss}/[Ba^{2+},(S^- \sim\sim)_2]$$

was found to be valid.

The understanding of the mechanism of propagation of polymers involving lithium cations and proceeding in hydrocarbons stems from the pioneering studies of Worsfold and Bywater [45]. Not only was the living character of these polymerizations unambiguously established, but the kinetics of propagation of lithium poly-styrene in benzene revealed the involvement in this process of dormant, inactive dimeric poly-styrenes which are in equilibrium with a minute fraction of growing monomeric lithium poly-styrenes. This conclusion was deduced from the observed square-root dependence of the propagation rate on the concentration of the polymers, i.e.,

$$\text{the rate of propagation} = k_p (K_{diss}/2)^{1/2}[Li\text{-}polyS]^{1/2}[S] \qquad ,$$

where k_p is the propagation constant of the monomeric polymers and K_{diss} is the dissociation constant of the dimers. Since this relation applies even at polymer concentrations as low as 10^{-5}M, Dr. Bywater pointed out in his talks that K_{diss} has to be of the order of 10^{-6}M or less.

The above mechanism served as a prototype accounting for other similar polymerizations. However, the kinetics of propagation of lithium polydienes revealed a dependence on the polymer concentration lower than 1/2, apparently 1/4 or less order. This prompted the assumption that the lithium polydienyls form higher aggregates than dimers.

A useful quasi-quantitative technique allowing determination of the degree of association of such polymers was reported by Professor Morton [46]. The viscosity of concentrated solutions of high-molecular weight polymers is proportional to a power of their weight average molecular weight, M_W, viz.,

$$\eta \approx M_W^\alpha \qquad ,$$

where α, although not precisely known, is within the range of 3.3 - 3.5 (see thorough review by Porter and Johnson [47]). On "killing" the aggregated polymers, say by adding a drop of methanol, the aggregates are converted into monomeric dead polymers, and the viscosity of the solution drastically decreases. Thus, the ratio of viscosities before and after methanol addition is given by N^α, where N denotes the average degree of association of the living polymers.

In spite of some uncertainty in the precise value of α, this method gives N with reasonable accuracy. For example, if the ratio of viscosities is 10, then N = 2.01, for $\alpha = 3.3$, and N = 1.93 for $\alpha = 3.5$.

Using this approach, Professor Morton and his associates confirmed the dimeric nature of lithium polystryl in benzene, in agreement with the deduction of Worsfold and Bywater, but they claimed a dimeric nature also for lithium polydienyls in hydrocarbon media. The latter claim conflicted with the kinetic evidence (about 1/4 order) which, after initial denials [48], eventually was confirmed by the Akron group [49].

Controversy arose because the viscosity studies of other workers [50] led to different conclusions. The extension of these investigations, involving light-scatter studies, again led to conflicting results when performed in one laboratory or the other. Moreover, Dr. Fetters and Professor Morton were unable to account for the 1/4 order of propagation of the presumably dimeric polydienyls. Strangely enough, they stated [51] that "plausible mechanisms" accounting for the dimeric nature of the aggregates and 1/4 order of propagation were presented. However, they did not propose mechanisms accounting for these observations but quoted instead 12 irrelevant references in support of their claim.

Let me suggest a way by which the kinetic and viscosity results could be reconciled, although I have reservations about the validity of the mechanisms I am proposing. Let us assume that the polymers with lithium cations are dimeric in hydrocarbon solvents. Their dissociation into <u>free</u> lithium cations is unlikely in hydrocarbon media, although I would not exclude the feasibility of forming the macro-anions. Hence, I reject the dissociation

$$(\text{Li}^+, \text{poly-dienyl}^-)_2 \;\rightleftarrows\; 2\text{Li}^+ + 2\text{poly-dienyl}^- \quad .$$

However, an alternative mode of dissociation of the dimers is possible, namely,

$$(\text{Li}^+, \text{poly-dienyl}^-)_2 \;\rightleftarrows\; (2\text{Li}^+, \text{poly-dienyl}^-) + (\text{poly-dienyl}^-), \qquad K_2 \quad .$$

This equilibrium in conjunction with the equilibria

$$(\text{Li}^+, \text{poly-dienyl}^-)_2 \;\rightleftarrows\; 2(\text{Li}^+, \text{poly-dienyl}^-), \qquad\qquad K_1 \quad ,$$

and

$$(\text{Li}^+, \text{poly-dienyl}^-) + (\text{poly-dienyl}^-) \;\rightleftarrows\; \text{Li}^+, (\text{poly-dienyl}^-)_2, \qquad K_3 \quad ,$$

leads to the following rate expression, provided that the free (poly-dienyl$^-$) ions are the <u>only</u> species contributing to the propagation. Thus,

$$\text{Rate of propagation} = k_p\,[(\text{poly-dienyl}^-)][M], \qquad \text{i.e.,}$$

$$\text{Rate of propagation}/[M] = k_p K_2^{1/2}[(\text{Li}^+,\text{P}^-)_2]^{1/2}/\{1 + K_1^{1/2}K_3[(\text{Li}^+,\text{P}^-)_2]^{1/2}\}^{1/2} \quad ,$$

where Li$^+$,P$^-$ is short for Li$^+$, poly-dienyl. For $K_1^{1/2}\cdot K_3[(\text{Li}^+, \text{poly-dienyl})_2]^{1/2} \ll 1$, this expression yields propagation rate proportional to the square-root of the living polymers concentration; however, for

$$K_1^{1/2}\cdot K_3[(\text{Li}^+, \text{poly-dienyl}^-)_2]^{1/2} \gg 1,$$

the rate becomes proportional to 1/4 power of living polymers concentration. Since K_3 is large while K_1 is very small, both extremes could be possible.

The heat of dissociation, ΔH, of the presumably dimeric lithium poly-isoprenyl is claimed by Morton and Fetters to be 37.4±0.8 kcal/mole [52], a value impossibly high. A conservative estimate leads to $\Delta H < 15$ kcal/mole. The errors associated with that work were discussed elsewhere [53]. It is significant that the value of 37.4±0.8 was derived through viscometric studies in which the exponent α relating the viscosity to weight average molecular weight, i.e.,

$$\eta = M^\alpha \quad ,$$

had to be 3.40. The two values of α reported in that work [52] are 3.37 and 3.42; and on substituting them for $\alpha = 3.40$, one derives $\Delta H \sim 25$ kcal/mole and $\Delta H \sim 50$ kcal/mole, respectively [53]. Interestingly, Dr. Fetters drew my attention to his later paper [52], reporting a redetermined value of α, namely 3.39 and 3.41. Unfortunately, the data given in that paper, in a table reproduced below, show that the actual α values are 3.36 and 3.48. This discrepancy as well as other questionable claims discussed in [53] are perturbing.

Table II, from Morton, Fetters, <u>et al.</u>; Macromolec. 3, 330 (1970) [54]

$M_v,10^{-4}$ g/mole	Volume Fraction of the Polymer	Flow Time; Min.	α	The True α
9.8	40	58		$\ln(147/58)/\ln(12.8/9.8) = 3.48$
12.8	40	147	3.41	
7.7	50	51.5		$\ln(284/51.5)/\ln(12.8/7.7) = 3.36$
12.8	50	284.0	3.39	

Let me pass now to some problems of co-polymerization of lithium salts of living polymers. With Dr. Zdenek Laita we studied the rate of cross-over reaction converting lithium poly-styryl into 1,1-diphenyl ethylene⁻, D⁻ end-groups in benzene [55]. The stoichiometry of the reaction is

$$\sim\sim S^-,Li^+ + D \rightarrow \sim\sim SD^-,Li^+ \quad ,$$

S^- denoting the styryl unit. Since lithium poly-styryl in benzene is dimeric and the reaction is presumably carried out by the minute fraction of monomeric poly-styryl, we expected 1/2 order of the conversion, provided the excess of D is large. However, the conversion obeyed a first order law, i.e.,

$$d \ln(\sim\sim S^-,Li^+)/dt = k_i[D]; \quad [D] = [D]_0 = const.$$

Moreover, the pseudo-first order rate constant, k_i, was found to be inversely proportional to the square-root of the initial concentration of $\sim\sim S^-,Li^+$. The peculiar features were accounted for by the following mechanism.

The polymers are present in the form of homo- and mixed dimers, $(\sim\sim S^-,Li^+)_2$, $(\sim\sim SD^-,Li^+)_2$ and $(\sim\sim S^-,Li^+,\sim\sim SD^-,Li^+)$. These are in equilibrium with minute concentrations of the monomer polymers, viz.

$$(\sim\sim S^-,Li^+) = u \text{ and } (\sim\sim SD^-,Li^+) = v, \text{ with } v/u = f.$$

Hence,

$$(\sim\sim S^-,Li^+)_2 = (1/2)K_1u^2, \quad (\sim\sim SD^-,Li^+)_2 = (1/2)K_2v^2 \quad ,$$

and

$$(\sim\sim S^-,Li^+,\sim\sim SD^-Li^+) = K_{12}uv.$$

The rate of conversion is given by k_aDu, $D = [CH_2=CPh_2] = const.$ Denoting the initial concentration of lithium poly-styryl by C_0 and by g the fraction of lithium poly-styryl present in any form at time t, one finds

$$C_0 = u^2(K_1+2K_{12}f+K_2f^2), \quad g = (K_1+K_{12}f)/(K_1+2K_{12}f+K_2f^2)$$

and hence

$$d\ln g/dt = (k_aD/C_0^{1/2})(K_1+2K_{12}f+K_2f^2)^{1/2}/(K_1+K_{12}f) \quad .$$

For $K_{12} = (K_1K_2)^{1/2}$, this expression is reduced to

$$d\ln.g/dt = k_aD/K_1^{1/2}C_0^{1/2} \quad ,$$

being in agreement with the experimental findings of pseudo-first order conversion and the pseudo-first order rate constant being inversely proportional to the square-root of the initial concentration of lithium poly-styryl, i.e.,

$$C_0^{-1/2} \quad .$$

For $k_{12} \neq (K_1K_2)^{1/2}$ the plot of ln.g versus time becomes curved, but its initial slope would still be inversely proportional to [lithium poly-styryl]$_0^{1/2}$.

The interesting feature of this kind of reaction is their memory. The observed rate at a chosen concentration of poly-styryl depends on its initial concentration. This is evident from inspection of Fig. 1; the effect arises from the influence of products of the reaction on its rate. This is a general phenomenon. In the system discussed here, the formation of the mixed dimer decreases the concentration of the reactive monomeric lithium poly-styryl and thus retards the reaction.

Mixed dimerization also affects the kinetics of anionic co-polymerization of styrene and para-methyl styrene, a system studied by O'Driscoll [56]. The reaction was performed in benzene with lithium counter-ions and the plots of ln.[styrene] or ln[para-methyl styrene] vs. time were both linear. Their slopes, denoted by λ_1 and λ_2, were, however, differnt from each other. This is seen in Fig. 2. These results were accounted for by Yamagishi and myself [57] in the terms of the previous treatment. Denoting again by u and v the concentration of the monomeric living polymers terminated by styryl or para-methyl styryl unit, respectively, and by C_0 the concentration of all the polymers, we find

$$K_1u^2 + 2K_{12}uv + K_2v^2 = C_0 \quad ,$$

$$k_{11}u + k_{21}v = \lambda_1 \quad \text{and} \quad k_{12}u + k_{22}v = \lambda_2 \quad ,$$

with λ_1 and λ_2 being constant and the other symbols having their usual meaning. Since u and v are variables but are not related to each other through the equation,

$$k_{21}v[styrene] = k_{12}u[para-methyl styrene] \quad ,$$

the first three equations have to be identical. This leads to the conditions

$$K_{12} = (K_1 K_2)^{1/2}$$

and

$$k_{11}/k_{12} = k_{21}/k_{22} \quad \text{i.e.,} \quad r_1 r_2 = 1.$$

Moreover

$$\lambda_1 = \gamma_1 C_0^{1/2} \quad \text{and} \quad \lambda_2 = \gamma_2 C_0^{1/2}$$

with γ_1 and γ_2 as two _different_ constants, independent of the concentrations or composition of the monomers and the concentration of the initiator, i.e., independent of C_0. Let it be stressed again that all these conclusions are mathematical consequences of the empirical findings revealed in Fig. 2, and the assumptions of the existence of inert homo- and mixed dimers in equilibrium with the active growing monomeric polymers.

In conclusion, I thank the National Science Foundation for the years of continuous support of our work.

REFERENCES

1. H. Staudinger; Chem. Ber. 53, 1073 (1920).
2. A. Wurtz; C.r. held. Seanc. Acad. Sci. Paris; 86, 1176 (1878).
3. K. Ziegler et al.; Ann. 473, 36 (1929); 479, 150 (1930), 511, 64 (1934).
4. K. Ziegler, Aneg. Chem. 49, 499 (1936).
5. W. Schlenk and E. Bergmann; Ann. 464, 1 (1928); 479, 42, 58, 78 (1930).
6. W. Schlenk et al.; Chem. Ber. 47, 473 (1914).
7. G. V. Schulz; Erg. Exact. Naturw. 17, 405 (1938).
8. J. L. Bolland; Proc. Roy. Soc. (London), A178, 24 (1941).
9. A. Abkin and S. Medvedev; Trans. Faraday Soc. 32, 286 (1936).
10. A. T. Blomquist, W. J. Tapp and J. R. Johnston; J. Amer. Chem. Soc. 67, 1519 (1945).
11. R. G. Beaman; J. Amer. Chem. Soc. 70, 3115 (1948).
12. R. E. Robertson and L. Marion; Can. J. Research 26B, 657 (1948).
13. W. C. E. Higginson and N. S. Wooding; J. Chem. Soc. p. 760 (1952).
14. (a) M. Szwarc, M. Levy and R. Milkovich; J. Amer. Chem. Soc. 78, 2656 (1956).
    (b) M. Szwarc; Nature 178, 1168 (1956).
15. F. W. Stavley et al.; Ind. Eng. Chem. 48, 778 (1956).
16. P. J. Flory; "Principles of Polymer Chemistry", Cornell Univ. Press (1953).
17. C. H. Bamford et al.; "The Kinetics of Vinyl Polymerisation by Radical Mechanism", Butterworth, London (1958).
18. H. Dostal and H. Mark; Z. Phys. Chem. B29, 299 (1935).
19. P. J. Flory; J. Amer. Chem. Soc. 65, 372 (1943).
19. (a) M. Szwarc; Polymer Eng. and Sci. 13, 1 (1973).
20. (a) P. Rempp et al.; H. M. Loncheux; Bull. Soc. Chim. (France), p. 1497 (1958).
    (b) C. A. Uraneck et al.; U. S. Pat. 3,135,716 (1964).
21. R. Milkovitch; Brit. Pat. 1,000,090 (1965); 1,035,873 (1966).
22. S. Bywater; Adv. Polymer Chem. 30, 89 (1979).
23. A. Deffieux and S. Boileau; Polymer 18, 1047 (1977).
24. A. Deffieux, E. Graff and S. Boileau; Polymer 21, 549 (1981).
25. T. Aide and S. Inoue; Makromolek. Chem.
26. S. Inoue et al; Makromolek. Chem. 179, 1377 (1978); 130, 210 (1969).
27. T. Aide and S. Inoue; Makromolek. Chem., in press.
28. R. Sakata and T. Tsuruta; Makromolek. Chem. 40, 64 (1960).
29. E. J. Vanderberg; J. Polymer Sci. 47, 486 (1960).
30. M. Osgan and Ph. Teyssié; Polymer Lett B5, 789 (1967).
31. Ph. Teyssié et al.; Inorg. Chim. Acta 19, 203 (1976).
32. E. J. Vandenberg; J. Polymer Sci. A7, 525 (1969).
33. Ph. Teyssie et al.; Macromolec. 6, 651 (1973).
34. S. Boileau et al.; Eur. Polymer J. 14, 581 (1978).
35. G. Natta and I. Pasquon; Adv. Cat. 11, 1 (1959).
36. G. Bier; Ange. Chem. 73, 186 (1961).
37. E. G. Kontos et al.; J. Polymer Sci. 61, 69 (1962).
38. Y. Doi, S. Neki and T. Keii; Macromolec. 12, 814 (1979).
39. G. Natta, I. Pasquon and A. Zambelli; J. Amer. Chem. Soc. 84, 1488 (1962).
40. C. E. H. Bawn, A. Ledwith and P. Mathies; J. Polymer Sci. 24, 93 (1959).
41. A. J. Bloodworth and A. G. Davies; Proc. Chem. Soc. (London), p. 315 (1963).
42. J. E. L. Roovers and S. Bywater; Trans. Faraday Soc. 62, 1876 (1966).
43. D. N. Bhattacharyya, J. Smid and M. Szwarc; J. Amer. Chem. Soc. 86, 5024 (1964).
44. B. DeGroof, M. Van Beylen and M. Szwarc; Macromolec. 8, 396 (1975).
45. D. J. Worsfold and S. Bywater; Can. J. Chem. 38, 1891 (1960).
46. M. Morton, E. E. Bostick, and R. Livigni; Rubber Plastic Age 42, 397 (1961).

47.   R. S. Porter and R. Johnson; Chem. Rev. 66, 1 (1966).
48.   M. Morton and L. J. Fetters; J. Polymer Sci. A2, 3111 (1964).
49.   M. Morton and L. J. Fetters; J. Rubber Chem. Techn. 48, 359 (1975).
50.   D. J. Worsfold and S. Bywater; Macromolec. 5, 393 (1972).
51.   L. J. Fetters and M. Morton; Macromolec. 7, 552 (1974).
52.   See ref. [48].
53.   M. Szwarc; J. Polymer Sci. B18, 493 (1980).
54.   M. Morton and L. J. Fetters, et al.; Macromolec. 3, 330 (1970).
55.   Z. Laite and M. Szwarc; Macromolec. 2, 412 (1969).
56.   K. F. O'Driscoll and R. Patsiga; J. Polymer Sci. A3, 1037 (1965).
57.   A. Yamagishi and M. Szwarc; Macromolec. 11, 1091 (1978).

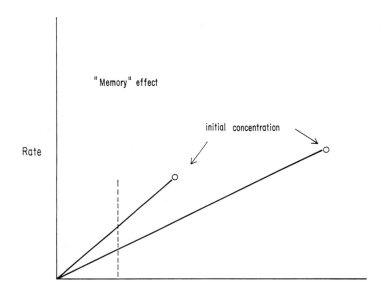

Fig. 1.   Concentration of living polymers

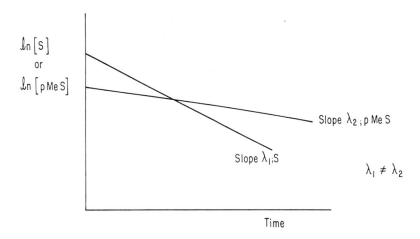

Fig. 2. Copolymerisation of styrene and
p—methyl styrene initiated by RLi in benzene

New Developments in Ring-Opening Polymerization

Takeo Saegusa

Department of Synthetic Chemistry, Kyoto University, Kyoto
606, Japan

Abstract-Several topics among the new developments in the
field of ring-opening polymerization are described. The
topics have been selected according to the three fundamentals
of polymer synthesis, i.e., (1) the exploration of new polymerization catalysts, (2) the molecular design of cyclic
monomers, and (3) the exploration of new polymerization
reactions. The topics are
1. Catalysts for photo-initiated cationic polymerization
2. Free radical ring-opening polymerization
3. Preparation of polyphosphine——Polymerization of
    deoxophostone followed by reduction
4. No catalyst copolymerization via zwitterion inter-
    mediates

INTRODUCTION

Ring-opening polymerization is expressed by the following general equation
(Eq. (1)), which involves a large variety of monomers, reaction catalysts
(initiators), and reaction mechanisms.

$$Cn \bigcirc X \longrightarrow \left(Cn - X \right)_p \qquad (1)$$

The reactivity characteristics of monomers change markedly according to the
nature of functional group -X-, the ring size, and the substituents attached
to the ring. In ring-opening polymerization, several groups of reaction
mechanisms have been established, i.e., cationic (electrophilic), anionic
(nucleophilic), coordinatae anionic, and free radical polymerizations.
The chemistry of ring-opening polymerization is making now a steady progress.
In this article, new developments in ring-opening polymerization are described from the viewpoint of polymer synthesis. The topics have been selected.
according to the three fundamentals of polymer synthesis, i.e., (1) the
exploration of new polymerization catalysts, (2) the molecular design of
cyclic monomers, and (3) the exporation of new polymerization reactions.

1. Catalysts for Photo-initiated Cationic Polymerization

As a topic for the exploration of polymerization catalysts, a new group of
catalysts are described here, which are activated by UV irradiation to cause
cationic polymerization of various heterocyclic monomers as well as some
vinyl monomers.
It has long been known that UV-irradiation causes free radical polymerization.
Monomer or some other reagents are activated to generate a free radical which
is responsible for free radical polymerization. Recently Crivello etal[1]
descovered a group of photo-initiators for cationic polymerization. They are
diaryliodonium salts (**1**) [2], triarylsalfonium salts (**2**) [3-5], triarylseleno-
nium salts (**3**) [6], dialkyl(penacyl)sulfonium salts (**4**) [7], and dialkyl(4-
hydroxyphenyl)sulfonium salts (**5**) [8].

$$\begin{matrix} Ar \\ Ar' \end{matrix} \!\!> I^+ \; X^- \qquad\qquad Ar - \overset{\overset{Ar'}{|}}{\underset{\underset{Ar''}{|}}{S}}{}^{\!+} \; X^- \qquad\qquad Ar - \overset{\overset{Ar'}{|}}{\underset{\underset{Ar''}{|}}{Se}}{}^{\!+} \; X^-$$

$$\mathbf{1} \qquad\qquad\qquad\qquad \mathbf{2} \qquad\qquad\qquad\qquad \mathbf{3}$$

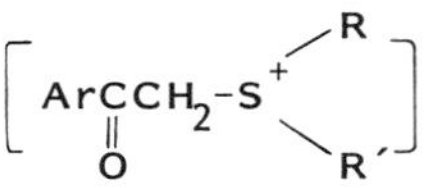

$$\left[\text{ArCCH}_2\text{-S}^+\underset{\text{R}'}{\overset{\text{R}}{<}} \right] \text{X}^-$$

(structure 4, with ArCCH₂ bearing =O)

$$\text{HO}\underset{\text{R}_3\ \text{R}_4}{\overset{\text{R}_1\ \text{R}_2}{\bigcirc}}\overset{+}{\text{S}}\underset{\text{R}_6}{\overset{\text{R}_5}{<}}\ \text{X}^-$$

4 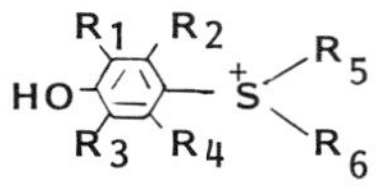   **5**

$$\text{X}^- : \text{BF}_4^-,\ \text{PF}_6^-,\ \text{AsF}_6^-,\ \text{SbF}_6^-$$

In the absence of UV light, these salts are stable, which do not exhibit catalyst activity. Upon UV irradiation, they generate a strong acid HX which initiates cationic polymerization.

These five salts are divided into two groups according to the mechanism of the generation of HX. The salts **1-3** are decomposed by UV irradiation through the homolytic cleavage of the carbon to hateroatom bonds (Ar-I, Ar-S, and Ar-Se bonds) to give cation radicals which abstract hydrogen from monomer and /or solvent (YH) to generate hydrogen onium salts ((ARI$^+$H)X$^-$, (Ar$_2$S$^+$H)X$^-$, and (Ar$_2$Se$^+$H)X$^-$). The hydrogen onium salts decompose to give a strong acid HX. The reaction scheme of the decomposition of **2** is shown below in order to illustrate the course of the decomposition of these three salts.

$$\text{Ar}_3\text{S}^+\text{X}^- \xrightarrow{h\nu} \text{Ar}_2\text{S}^+\cdot + \text{Ar}\cdot + \text{X}^- \tag{2}$$

$$\text{Ar}_2\text{S}^+\cdot + \text{YH} \longrightarrow \text{Ar}_2\overset{+}{\text{S}}\text{H} + \text{Y}\cdot \tag{3}$$

$$(\text{Ar}_2\overset{+}{\text{S}}\text{H})\text{X}^- \longrightarrow \text{HX} + \text{Ar}_2\text{S} \tag{4}$$

The nature of the counter *anion* X$^-$ is very important. As has been indicated above, X$^-$ should be a stable anion having low nucleophilic reactivity, otherwise the cationic propagating species are extinguished (terminated) by covalent bonding with X$^-$. Various monomers including cyclic ethers (epoxide, oxetane, tetrahydrofuran, and trioxane), cyclic sulfides (propylene sulfide and thietane), lactones, and spiro bicyclic orthoesters as well as vinyl monomers (styrene, α-methylstyrene and vinyl ether) were polymerized by these salts.

As is seen in the above scheme (Eqs. (2) to (4)), free radical species (Ar$_2$S$^+\cdot$, Ar$\cdot$, Y$\cdot$) are also generated, which are able to initiate free radical polymerization. The amphi-functional cahracter of **2** was demonstrated by the following series of experiments. When a mixture of 1,4-cyclohexene oxide and methyl methacrylate including Ph$_3$S$^+\cdot$SbF$_6^-$ was irradiated, the homopolymers of the two monomers were produced independently. Thus, cationic (1,4-cyclohexene oxide) and free radical (methyl methacrylate) polymerizations took place separately. In the presence of a radical inhibitor of 2,6-di-t-butylphenol, only 1,4-cyclohexene oxide was polymerized. In the presence of triethylamine which inhibits cationic polymerization, only methyl methacrylate was polymerized. Glycidyl acrylate and methacrylate, which are monomers having two different functional groups capable of separate cationic and free radical polymerizations were converted by the irradiation of **2** into cross-linked insoluble polymers.

Photo-irradiation of **4** and **5** generates a strong acid HX (Eqs. (5) and (6)).

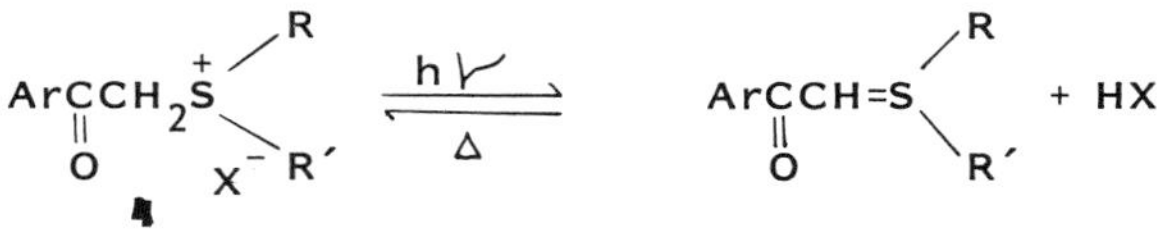

$$\text{ArCCH}_2\overset{+}{\text{S}}\underset{\text{X}^-\ \text{R}'}{\overset{\text{R}}{<}} \underset{\Delta}{\overset{h\nu}{\rightleftharpoons}} \text{ArCCH}=\text{S}\underset{\text{R}'}{\overset{\text{R}}{<}} + \text{HX} \tag{5}$$

4

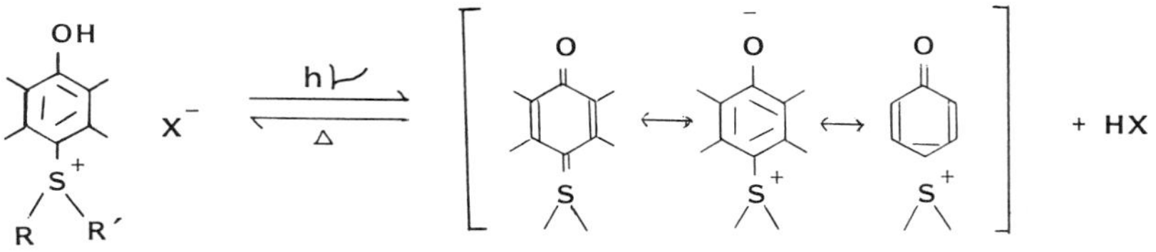

$$\text{(6)}$$

The above reactions are reversible, and cationic polymerization occurs only when the irradiation was performed in the presence of monomer.
The above findings of Crivello et al have broadened the scope of monomers in the technology of photo-curable coatings which used to be carried out only with vinyl monomers utilizing free radical polymerization.

2. Free Radical Ring-Opening Polymerization

Usually ring-opening polymerization proceeds through ionic mechanisms, i.e., cationic, anionic, and coordinate anionic mechanisms. Ring-opening polymer-ization via free radical mechanism has recently been developed, in which the so-called "molecular design of monomers" is essential.
Free radical polymerizations of vinylcyclopropane and its substituted deriva-tives **6** (Eq. (7))[9-11], the o-xylylene dimer **7** (Eq. (8))[12], and tetrafluoro-ethylene sulfide **8** (Eq. (9))[13] are seen occasionally in the literatures.

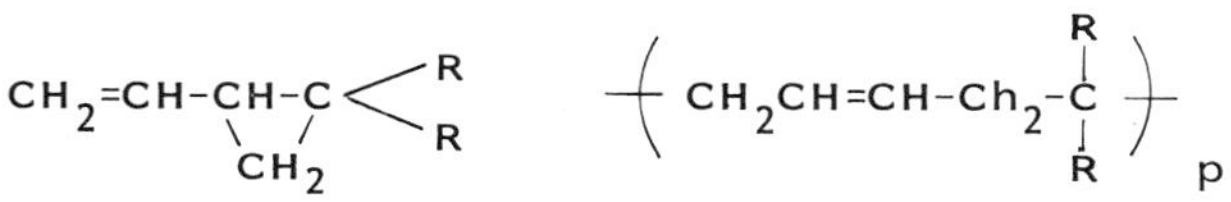

$$\text{(7)}$$

6 (R=H,[9] CO_2Et,[10] CN[11])

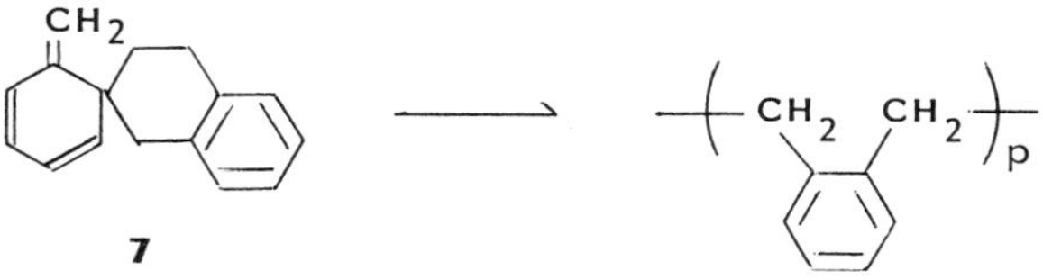

$$\text{(8)}$$

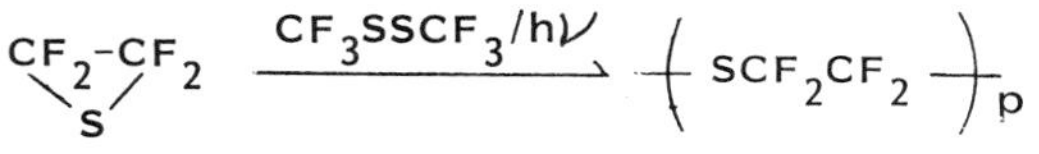

$$\text{(9)}$$

Recently Bailey and Endo have made a series of systematic studies on free radical ring-opening polymerizations.[14-15] They have proposed the following three patterns (Eqs. (10)-(12)), and explored several polymerizations mainly on the basis of Pattern I (Eq. (10)).

Pattern I

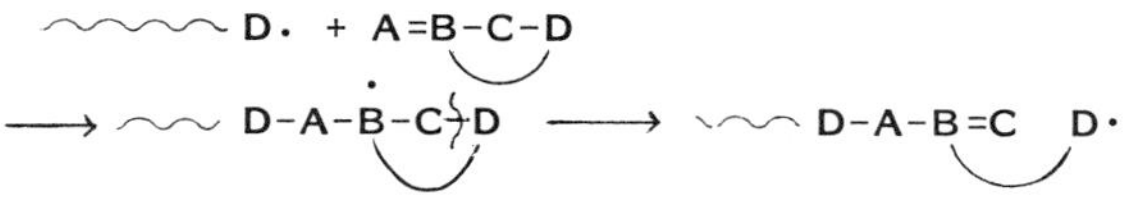

$$\text{(10)}$$

Pattern II

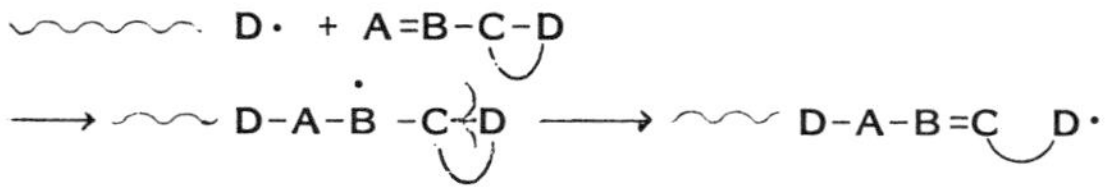

$$\text{(11)}$$

Pattern III

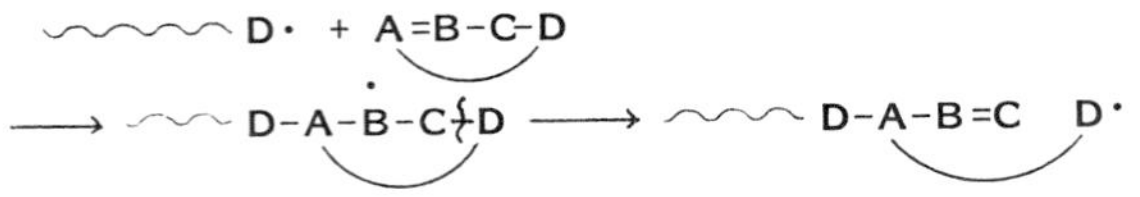

$$\begin{aligned} \sim\!\!\sim\!\!\sim D\cdot \;+\; A=B-C-D \\ \longrightarrow \;\sim\!\!\sim D-A-B-\overset{\cdot}{C}\!\!\leftarrow\!\!D \longrightarrow \;\sim\!\!\sim\!\!\sim D-A-B=C \quad D\cdot \end{aligned} \tag{12}$$

In all patterns, A and B are carbon atoms. The propagating radical first adds to the carbon-carbon double bond, and then the ring of the monomer is opened by the cleavage of the C-D bond which is situated at the ß- position from the newly produced radical at atom B. Thus, the ring-opening is being assisted by the C=C bond. 2-Methylene-1,3-dioxolane **9** is polymerized by peroxide initiator into a polyester **10** which corresponds to the hypothetical ring-opened polymer of a non-polymerizable compound, γ-butyrolactone.

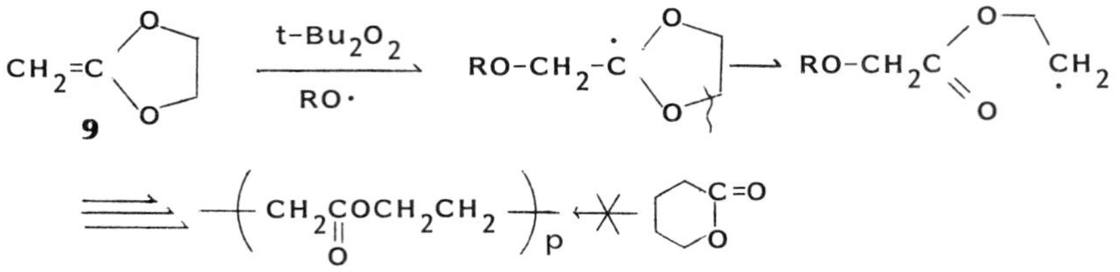

2-Methylenetetrahydrofuran **11** is polymerized in a similar way to produce a polymeric ketone **12**, which may be regarded as a 2:1 sequence-ordered co-polymer between ethylene and carbon monoxide.

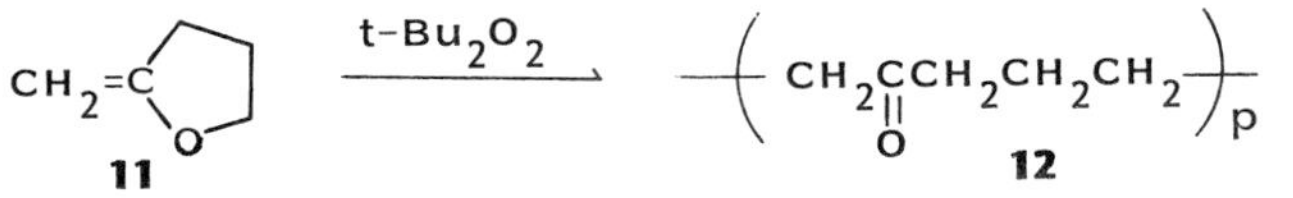

A bis-methylene spiro-bicyclic compound **13** is polymerized into a ether-carbonate polymer **14** via the following scheme of free radical rearrangements.[16]

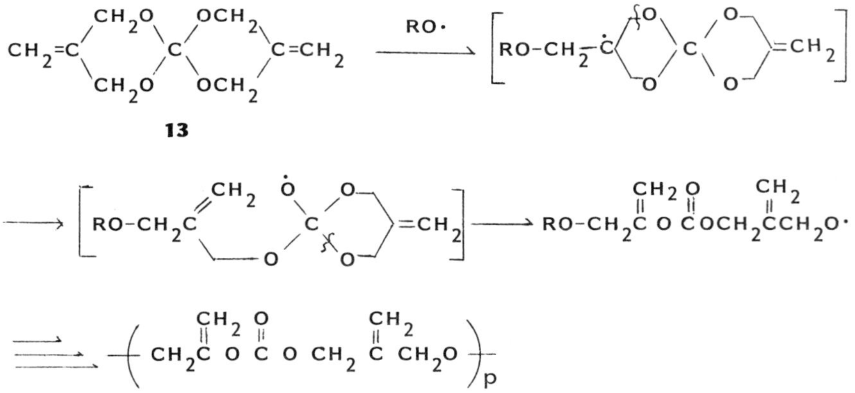

The monomer **13** is polymerized also by cationic mechanism to produce the same polymer **14**. The **13** → **14** polymerization is accompanied by volume expansion. In a similar way, another methylene spiro-bicyclic monomer **15** is polymerized by peroxide initiator. [17]

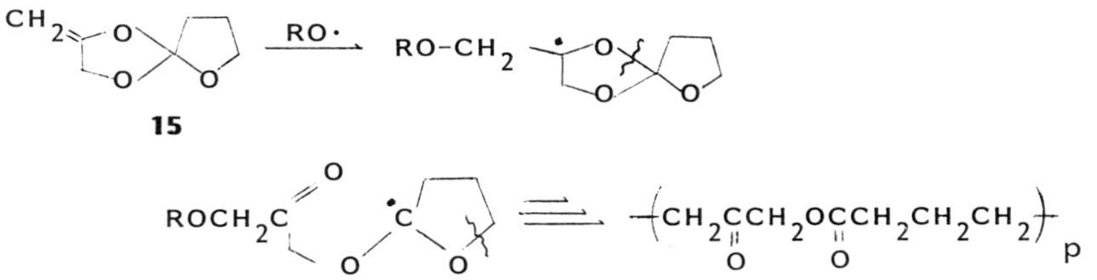

These ring-opening polymerizations can be combined with the radical polymerization of conventional vinyl compounds. In other words, the above cyclic
monomers can be copolymerized with vinyl compounds to produce polymers
containing carbonyl, ester or carbonate linkages in the backbone chains.
Thus, the variety of the polymers produced by radical polymerization has been
enlarged. From the view point of polymerization chemistry, radical copolymerization between heterocyclic and vinyl monomers is significant because the
copolymerization between these two types of monomers through ionic mechanism
has been known almost impossible.

3. Preparation of Polyphosphine

Many interesting and important characters are expected with polymeric
phosphine whose backbone consists of phosphine repeating units. The synthesis
of polyphosphine through ring-opening polymerization was recently accomplished
by us, which was largely due to the molecular design of monomer. Deoxopho-
stone **16** was adopted as the monomer, which was found to be polymerized by
cationic initiators. The product polymer **17**, polyphosphine oxide, was
quantitatively reduced to polyphosphine **18** by the chlorination with oxalyl
chloride followed by the reduction with i-Bu$_2$AlH.

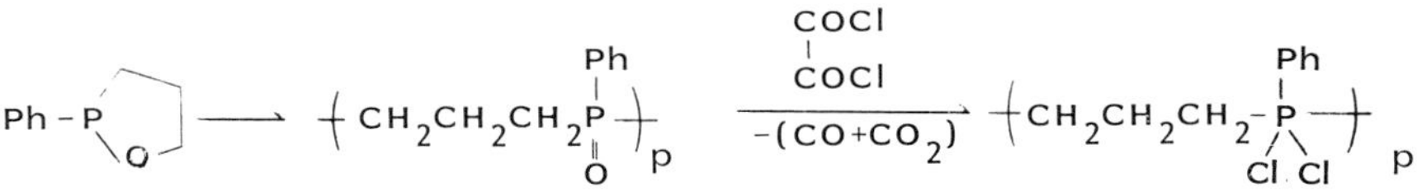

The polymerization of **16** using methyl triflate as initiator is explained by
the following scheme of reactions.

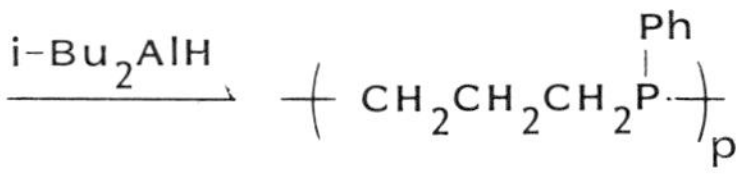

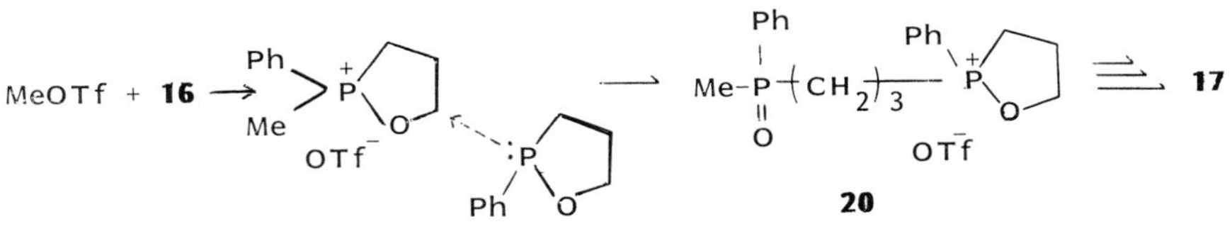

The propagating species (**20** and higher homologues) are of ionic character,
i.e., cyclic phosphonium triflate. This structure of the ionic propagating
end has been established by ^{31}P and ^{19}F NMR spectroscopy.[20]
Another mechanism of polymerization of **16** has been realized, in which benzyl
chlorid was used as initiator. This second type of polymerization proceeds
via species having a covalent electrophilic end-group of alkyl chloride [20]

(**22** and **24**).

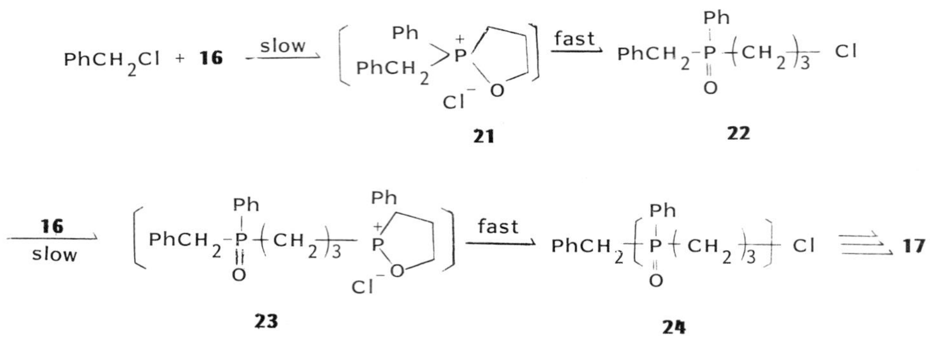

The species having cyclic phosphonium groups **21** and **23** are transiently formed, which are quickly opened by the nucleophilic attack of the chloride counter anion.
The difference of mechanism between the above two polymerizations with the initiators of MeOTf and benzyl chloride is ascribed to the difference of the nucleophilic reactivity between triflate anion (stable, less reactive) and chloride anion (less stable, more reactive). The two types of polymerization of **16** have been differentiated from each other by a kinetic method, too.[20]
A similar phenomena has recently been observed also in the polymerizations of cyclic imino ethers[21-28] and cyclic ethers.[29-36] For example, two types of polymerization of 2-oxazoline **25** by using methyl tosylate (via ionic propagating species, **26** and **27**) and methyl iodide (via covalent propagating species **28** and **29**), respectively, as initiator are shown below.

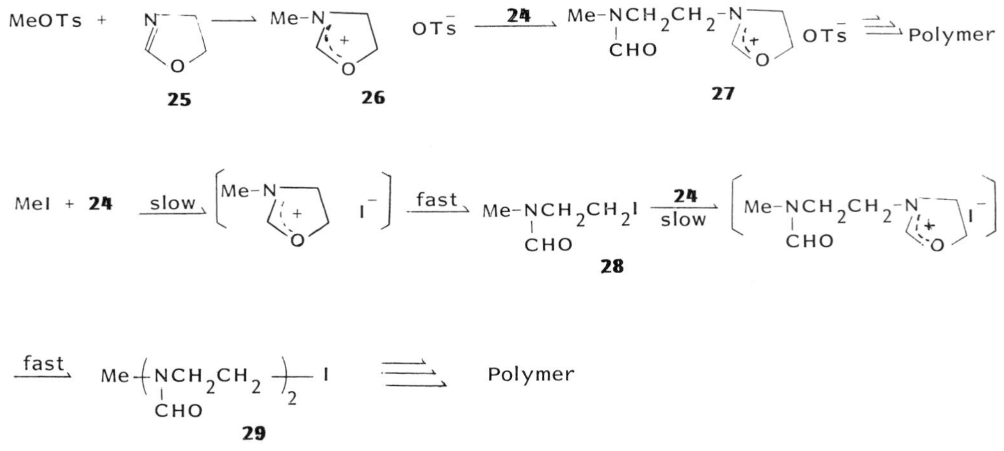

The complete conversion of polyphosphine oxide **17** into polyphosphine **19** as mentioned above has been achieved after several trials using different reagents.[20] The conversion and the extent of reaction were conveniently monitored by ^{31}P NMR spectroscopy.

4. No Catalyst Alternating Copolymerizatio via Zwitterion Intermediates

In order to induce polymerization reaction, catalyst (or initiator) or high energy radiation is required in almost all instances. In these several years, we have developed a novel type of copolymerization which proceeds spontaneously without any added catalyst.[37] For example,[38] an equimolar mixture of 2-oxazoline **25** and ß-propiolactone **30** in an aprotic polar solvent kept at room temperature gives rise to the production of an 1:1 alternating copolymer **31**.

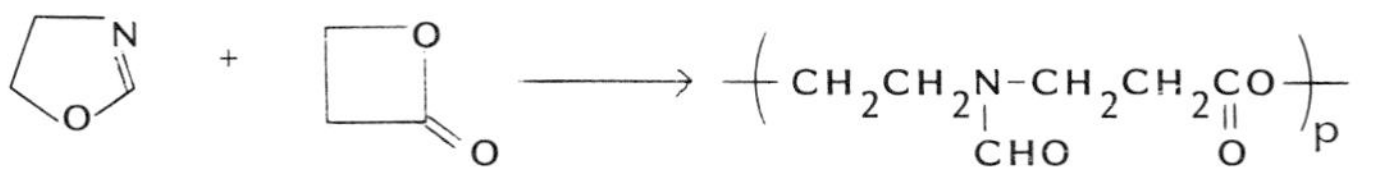

$$25 + 30 \longrightarrow +\!\!\left(CH_2CH_2\underset{\underset{CHO}{|}}{N}\!-\!CH_2CH_2\underset{\underset{O}{\|}}{C}O\right)_{\!p}$$

25 **30**   **31**

The key intermediate in the above copolymerization is a zwitterion **32** which is generated by the ring-opening of **30** by the nucleophilic attack of **25**.

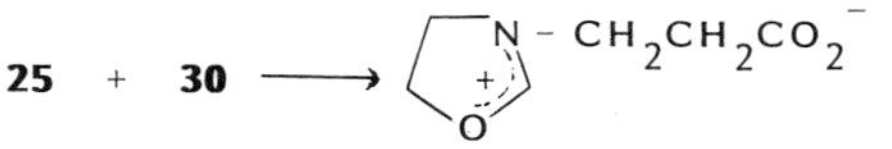

$$25 + 30 \longrightarrow \overset{+}{N}-CH_2CH_2CO_2^-$$

32

The opening of the oxazolinium ring of one moleculs of **32** by the nucleophilic attack of carboxylate group of another molecule of **32** constitutes the growth of copolymer molecule.

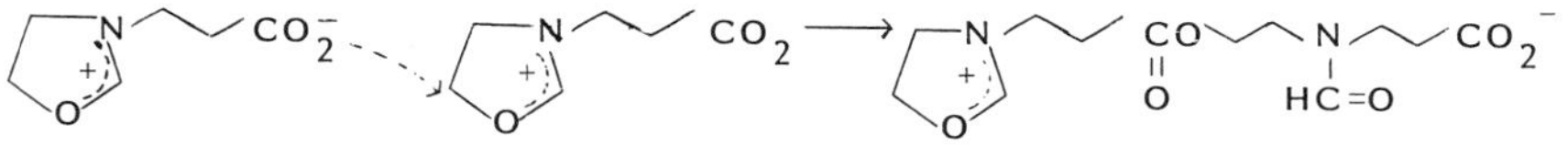

In general,

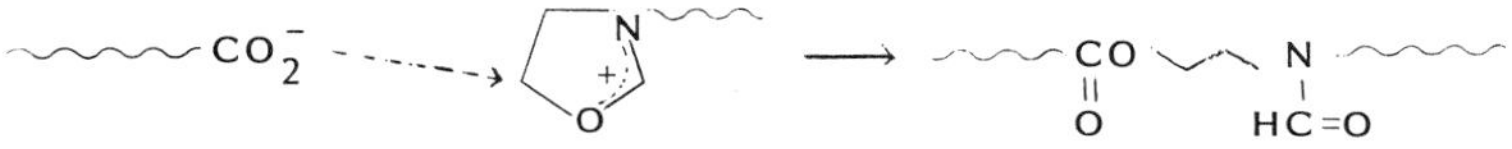

In the generation of zwitterion **32** in the above copolymerization, **25** behaves as a nucleophile and **30** as an electrophile.

$$M_N + M_E \longrightarrow {}^+M_N\!-\!M_E{}^-$$

33

The above copolymerization has been extended to a general concept of co-polymerization in which a polymerizable compound of nucleophilic reactivity is combined with another polymerizable compound having electrophilic reactivity. The general scheme of the copolymerization, in which M_N and M_E designate respectively nucleopholic and electrophilic monomers.

$$M_N + M_E \longrightarrow {}^+M_N\!-\!M_E{}^-$$

33

$$33 + 33 \longrightarrow {}^+M_N\!-\!M_E M_N\!-\!M_E{}^-$$

34

$$34 + 33 \longrightarrow {}^+M_N\!\!\left(M_E M_N\right)_{\!2}\!\!-\!M_E{}^-$$

35

For the growth of zwitterions **34** and **35**, two types of reactions have been assumed, i.e.,

$${}^+M_N\!\!\left(M_E M_N\right)_{\!n}\!\!M_E{}^- + 33 \longrightarrow {}^+M_N\!\!\left(M_E M_N\right)_{\!n+1}\!\!M_E{}^-$$

$${}^+M_N\!\!\left(M_E M_N\right)_{\!m}\!\!M_E{}^- + {}^+M_N\!\!\left(M_E M_N\right)_{\!n}\!\!M_E{}^-$$

$$\longrightarrow {}^+M_N\!\!\left(M_E M_N\right)_{\!m+n+1}\!\!M_E{}^-$$

On the basis of the above concept, we have opened up a new field of ring-opening polymerization. In the development of this new copolymerization, the most

important elements of research are the exploration of the reactions and the molecular design of M_N and M_E monomers. Table 1 shows typical M_N and M_E monomers.

Table 1. Typical M_N and M_E monomers.

M_N monomers

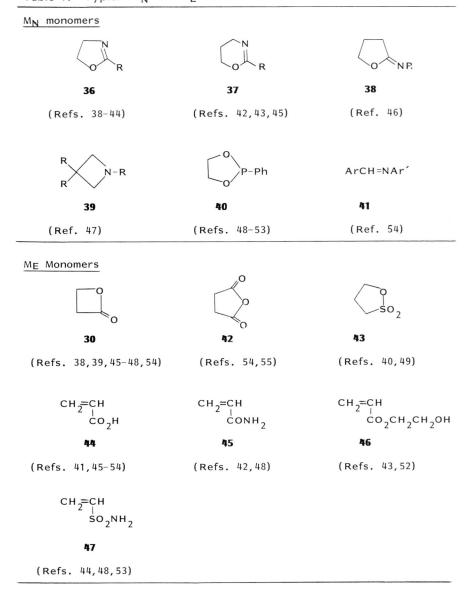

36   **37**   **38**

(Refs. 38-44)   (Refs. 42,43,45)   (Ref. 46)

39   **40**   **41**

(Ref. 47)   (Refs. 48-53)   (Ref. 54)

M_E Monomers

30   **42**   **43**

(Refs. 38,39,45-48,54)   (Refs. 54,55)   (Refs. 40,49)

$$CH_2=CH \\ \quad CO_2H$$ $$CH_2=CH \\ \quad CONH_2$$ $$CH_2=CH \\ \quad CO_2CH_2CH_2OH$$

44   **45**   **46**

(Refs. 41,45-54)   (Refs. 42,48)   (Refs. 43,52)

$$CH_2=CH \\ \quad SO_2NH_2$$

47

(Refs. 44,48,53)

The number of possible combinations of copolymerizations using six M_N and seven M_E monomers in Table 1 is 6 X 7 = 42. Among these combinations, co-polymerization of over twenty combinations were examined, which were shown to take place without any added catalyst.[37-55] In most cases, alternating copolymers were obtained. Sometimes, copolymers having biased compositions were produced by the occurrence of homo-propagation of M_N or M_E concurrently with the alternating propagation which are being shown in the general scheme. A few highlights selected from the combinations of M_N and M_E monomers in Table 1 are described here. Acrylic acid **44** has been copolymerized with M_N monomers. For example, the 1:1 alternating copolymerization of acrylic acid **44** (M_E) with 2-oxazoline **25** (M_N), which occurs spontaneously, is schematized as follows.[41]

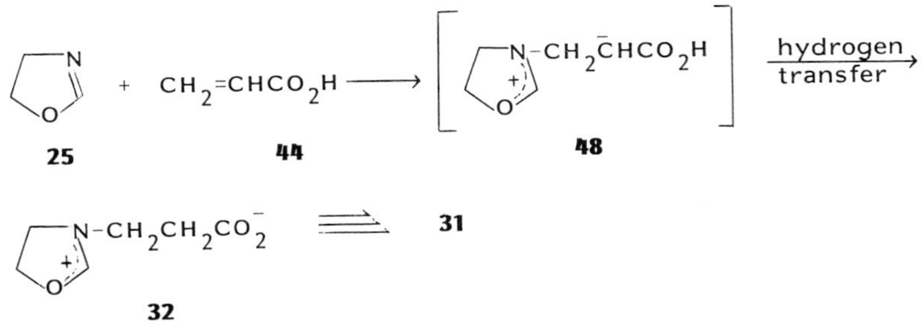

In the first formed zwitterion **48**, hydrogen transfer gives rise to the formation of zwitterion **32** which is the same as that produced from 2-oxazoline **25** and ß-propiolactone **30**. Consequently, the produced copolymer from **25** and **44** is the same as that from **25** and **30**.
The addition of a nucleophilic monomer onto an electron-deficient olefinic bond of M_E monomer followed by the hydrogen transfer reaction has been observed also with acrylamide **45**, hydroxyalkyl acrylate **46**, and ethlenesulfonamide **47**. Zwitterious having the formulae **48 — 50** are the key intermediates of the copolymerizations of these derivatives of acrylic acid.

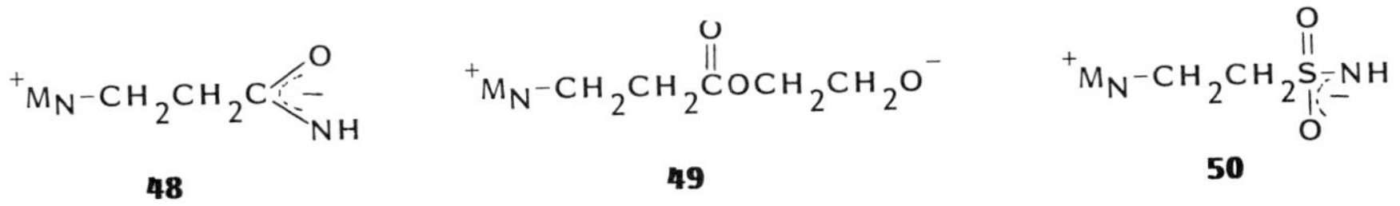

Copolymerizations of a cyclic phosphonite **40** with several M_E monomers have offered a group of new synthetic methods of phosphorus-containing polymers. For example, the two copolymerizations of **40** with ß-propiolactone **30** and with acrylic acid **44** produced at temperatures above 120°C the identical copolymer of an 1:1 alternating structure **51**.

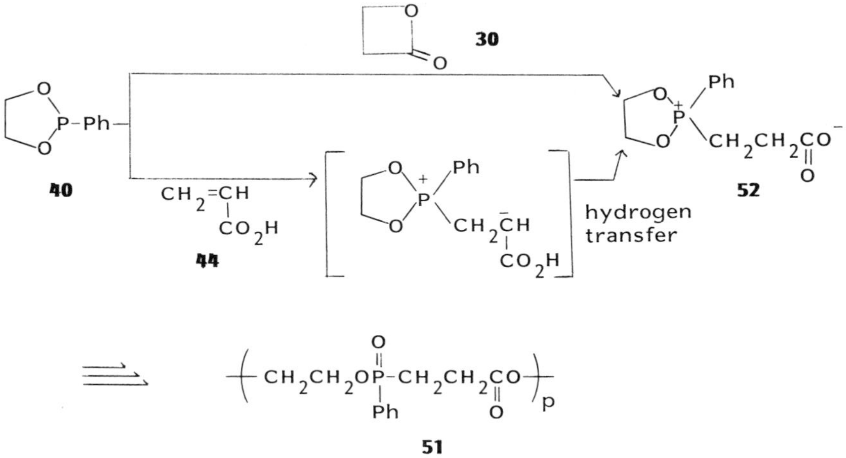

These two copolymerizations proceed through the common key intermediate of zwitterion **52**, and consequently they give the identical copolymer **51**. In the above scheme, the opening of cyclic phosphonium group in one zwitterion by the nucleophilic attack of carboxylate anion in another zwitterion constitutes the propagation.

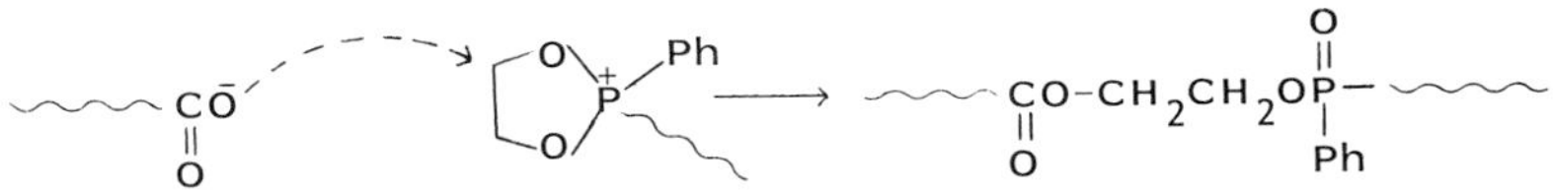

The intermediacy of **52** has been supported by th production cf a spiroacyloxy-phosphorane **53** in lower temperature (below 50°C) reactions of the **40/30** and **40/44** combinations. The production of **53** in the **40/44** combination is almost quantitative. In addition, **53** is polymerized to give **51** at higher tempralures (above 120°C) through a zwitterion intermediate **52**. Thus, **53** may be regarded as a stabilized form of **52**.

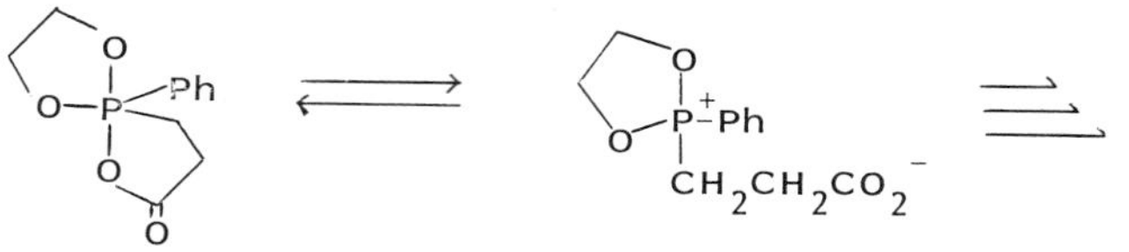

There have been found several interesting copolymerizations involving monomers which are not being included in Table 1. One of interesting and important copolymerizations has been found in the combination of a cyclic phosphite **54** and an α-keto acid **55**, in which an 1:1 alternating copolymer consisting of a phosphate ester unit **56** was produced at higher reaction temperature (above 120°C)[56,57]

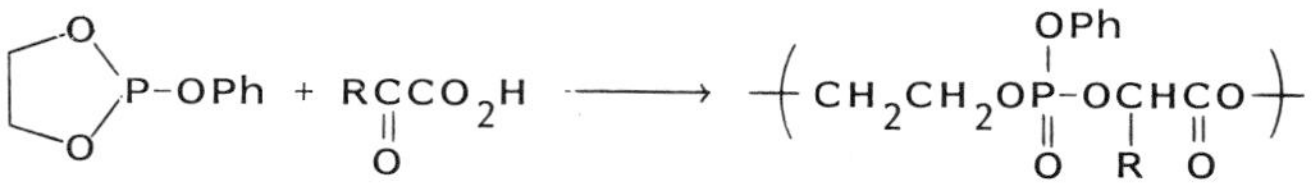

In the above copolymerization, a phosphite monomer of **54** has been oxidized to a phosphate unit in **56**, whereas an α-keto acid monomer **55** has been reduced to an ester unit of a hydroxy acid in **56**. Thus, the term of "Redox Copolymerization" has been proposed to designate this type of copolymerization.
The key intermediate in the above redox copolymerization is a zwitterion **57**, which is generated by the following scheme of reactions.

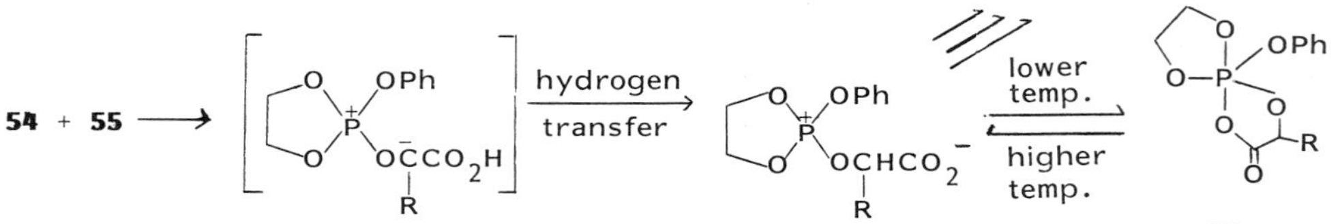

As in a zwitterion **52**, **57** is ring-closed to a spiroacylpentaoxyphosphorane **58** at lower temperatures (e.g. room temperature). In addition, **58** itself is polymerized at higher temperatures (e.g. above 120°C) to produce **56**.
p-Benzoquinone **59** is a reactive and highly polarizable oxidant which has successfully been copolymerized with vatious P(III) compounds.[58-60] For example, **59**, has been copolymerized with salicyl phenylphosphonite **60** at room temperature to produce an 1:1 alternating copolymer **61**. A zwitterion **62** has been assumed as the key intermediate in which the opening of phosphonium ring occurs readily.

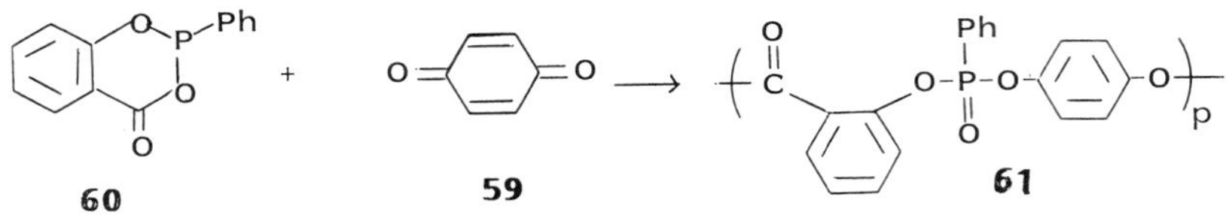

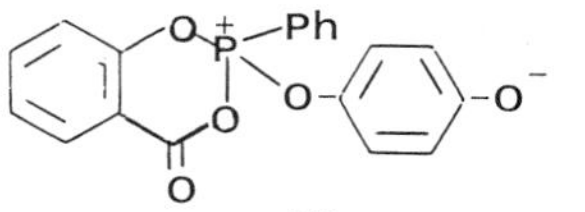

REFERENCES

1) A review article, J. V. Crivello, *Chem. Tech.* 7, 625 (1980)
2) J. V. Crivello and J. H. W. Lam, *macromolecules*, 10, 1307 (1977); Fourth International Symposium on Cationic Polymerization, Akron, U.S.A., June, 1976; *J. Polym. Sci., Polymer Symposium 56*, 383 (1976)
3) J. V. Crivello and J. H. W. Lam, *J. Polym. Sci., Polym. Chem. Ed*, 17, 977 (1979)
4) J. V. Crivello and J. H. W. Lam, *J. Polym. Sci., Polym. Lett. Ed.*, 17, 759 (1979)
5) J. V. Crivello and J. H. W. Lam, *J. Polym. Sci., Polym Chem. Ed.*, 18, 2677 (1980)
6) J. V. Crivello and J. H. W. Lam, *J. Polym. Sci., Polym. Chem. Ed.*, 17, 1047 (1979)
7) J. V. Crivello and J. H. W. Lam, *J. Polym. Sci., Polym. Chem. Ed.*, 17, 2877 (1979)
8) J. V. Crivello and J. H. W. Lam, *J. Polym. Sci., Polym. Chem. Ed.*, 18, 1021 (1980)
9) T. Takahashi, *J. Polym. Sci., A-1*, 6, 403 (1968)
10) I. Chow and K. D. Ahm, *J. Polym. Sci., Polym. Lett. Ed.*, 15, 751 (1977)
11) I. Chow and Ju-Y. Lee, *J. Polym. Sci., Polym. Lett. Ed.*, 18, 639 (1980)
12) L. A. Errede, *J. Polym. Sci.*, 49, 253 (1961)
13) W. R. Brasen et al, *J. Org. Chem.*, 30, 4188 (1965)
14) W. J. Bailey, International Symposium on Ring-Opening Polymerization, Karlovy Vary (Czechoslovakia), Sept., 1980
15) T. Endo and W. J. Bailey, *Kobunshi (in Japanese)*, 30, 331 (1981)
16) T. Endo and W. J. Bailey, *J. Polym. Sci., Polym. Lett. Ed.*, 13, 193 (1975)
17) T. Endo and W. J. Bailey, *J. Polym. Sci., Polym. Lett. Ed.*, 18, 25 (1980)
18) S. Kobayashi, M. Suzuki, and T. Saegusa, *Polymer Bull.*, 4, 315 (1981)
19) S. Kobayashi, M. Suzuki, and T. Saegusa, Presented at 30th annual Meeting of Soc. Polymer Sci., Japan, May 1981, Kyoto, Japan; *Polymer Preprints Japan (in Japanese)*, 30, 124 (1981)
20) S. Kobayashi, M. Suzuki, and T. Saegusa, Presented at the Symposium of Polymer Science, Japan, October, 1981, Tokyo
21) T. Saegusa, H. Ikeda, and H. Fujii, *Macromolecules*, 5, 359 (1972)
22) T. Saegusa, H. Ikeda, and H. Fujii, *Macromolecules*, 6, 315 (1973)
23) T. Saegusa and H. Ikeda, *Macromolecules*, 6, 808 (1973)
24) T. Saegusa, S. Kobayashi, and Y. Nagura, *Macromolecules*, 7, 265 (1974)
25) T. Saegusa, S. Kobayashi, and Y. Nagura, *Macromolecules*, 7, 272 (1974)
26) T. Saegusa, S. Kobayashi, and Y. Nagura, *Macromolecules*, 7, 713 (1974)
27) T. Saegusa, *Pure and Appl. Chem. 39*, 81 (1975)
28) T. Saegusa, S. Kobayashi, and A. Yamada, *Makromol. Chem.*, 177, 2271 (1975)
29) S. Kobayashi, H. Danda, and T. Saegusa, *Bull. Chem. Soc. Japan*, 46, 3214 (1973)
30) S. Smith and A. J. Hubin, *J. Macromol, Sci.*, A7, 1399 (1973)
31) K. Matyjaszewski and S. Penczek, *J. Polymer Sci., Polym. Chem. Ed.*, 12, 1905 (1974)
32) S. Kobayashi, H. Danda, and T. Saegusa, *Macromoleceles 7*, 415 (1974)
33) T. K. Wu and G. Pruckmayr, *Macromolecules*, 8, 77 (1975)
34) G. Pruckmayr, and T. K. Wu, *Macromolecules*, 8, 954 (1975)
35) T. Saegusa and S. Kobayashi, *J. Polymer Sci., Polymer Symposium 56*, 241 (1976)
36) S. Penczek, P. Kubisa, and K. Matyjaszewski, *Advances in Polymer Sci.*, 37, 77 (1980)
37) Review articles of "No Catalyst Copolymerizations "
    (a) T. Saegusa, *Chem. Tech.*, 5, 295 (1975)
    (b) T. Saegusa, S. Kobayashi, and Y. Kimura, *Pure and Appl. Chem.*, 48 3071 (1976)
    (c) T. Saegusa, *Angew. Chem.*, 89, 867 (1977); *Angew. Chem., Internat. Ed. (English) 16*, 826 (1977)
    (d) T. Saegusa and S. Kobayashi, *J. Polymer Sci., Polymer Symposium*, 62, 79 (1978)
    (e) T. Saegusa and S. Kobayashi, *Pure and Appl. Chem.*, 50, 281 (1978)
    (f) T. Saegusa, *Macromol. Chem., Suppl. 3*, 157 (1979)
38) T. Saegusa, H. Ikeda, and H. Jujii, *Macromolecules*, 5, 354 (1972)
39) T. Saegusa, S. Kobayashi, Y. Kimura, *Macromolecules*, 7, 1 (1974)
40) T. Saegusa, H. Ikeda, S. Hirayanagi, Y. Kimura, and S. Kobayashi, *Macromolecules*, 8, 259 (1975)
41) T. Saegusa, S. Kobayashi, and Y. Kimura, *Macromolecules*, 7, 139 (1974)
42) T. Saegusa, S. Kobayashi, and Y. Kimura, *Macromolecules*, 8, 374 (1975)
43) T. Saegusa, Y. Kimura and S. Kobayashi, *Macromolecules*, 10, 239 (1977)
44) T. Saegusa, S. Kobayashi, and J. Furukawa, *Macromolecules*, 9, 728 (1976)
45) T. Saegusa, Y. Kimura, and S. Kobayashi, *Macromolecules*, 10, 236 (1977)
46) T. Saegusa, Y. Kimura, K. Sano, and S. Kobayashi, *Macromolecules*, 7, 546 (1974)

47) T. Saegusa, Y. Kimura, S. Sawada, and S. Kobayashi, *Macromolecules, 7*, 956 (1974)

48) t. Saegusa, Y. Kimura, N. Ishikawa, and S. Kobayashi, *Macromolecules, 9*, 724 (1976)

49) T. Saegusa, S. Kobayashi, and J. Furukawa, *Macromolecules, 10*, 73 (1977)

50) T. Saegusa, T. Yokoyama, Y. Kimura, and S. Kobayashi, *Polymer Bull., 1*, 91 (1978)

51) T. Saegusa, T. Kobayashi, and S. Kobayashi, *Polymer Bull., 1*, 259 (1979)

52) T. Saegusa, M. Niwano, and S. Kobayashi, *Polymer Bull., 2*, 249 (1980)

53) T. Saegusa, S. Kobayashi, and J. Furukawa, *Macromolecules, 11*, 1027 (1978)

54) T. Saegusa, S. Kobayashi, and J. Furukawa, *Macromolecules, 8*, 703 (1975)

55) S. Kobayashi, M. Isobe, M. Niwano, and T. Saegusa, *Macromolecules, in press.*

56) T. Saegusa, T. Yokoyama, Y. Kimura, and S. Kobayashi, *Macromolecules, 10*, 791 (1979)

57) T. Saegusa, T. Yokoyama, and S. Kobayashi, *Polymer Bull., 1*, 55 (1978)

58) T. Saegusa, T. Kobayashi, T-Yuen Chow, and S. Kobayashi, *Macromolecules, 12*, 533 (1979)

59) S. Kobayashi, T. Kobayashi, and T. Saegusa, *Polymer Preprints, Japan, 28*, 762 (1979)

60) S. Kobayashi, T. Kobayashi, and T. Saegusa, *Macromolecules, May/June (1981)*

EMULSION POLYMERIZATION: I. MECHANISMS OF PARTICLE FORMATION. II.   CHEMISTRY AT THE INTERFACE

Robert M. Fitch

Department of Chemistry and Institute of Materials Science
The University of Connecticut, Storrs, CT 06268, USA

Abstract - Two aspects of recent developments in the fields of emulsion polymerization and the derived polymer colloids are dealt with. The first involves theoretical attempts to develop a comprehensive scheme for the mechanisms of particle formation as well as experimental attempts at obtaining the necessary physical parameters. The theory starts with concepts of homogeneous nucleation, kinetics of free radical capture by particles and coagulation of polymer particles during the early stages of reaction. The last process is shown to be very important at low surfactant concentrations and with rates which are very dependent upon these concentrations. This regime is ordinarily reserved for monomers of appreciable solubility in the continuous medium. For insoluble monomers, solubilization in surfactant micelles is often employed accompanied by heterogeneous nucleation of polymer particles, although copolymerization with water-soluble monomers may also be used. Experimental methods involving the time-dependence of the intensity of scattered light are being employed to follow the kinetics of these processes and to obtain the desired parameters.

The second aspect involves the synthesis of model polymer colloids bearing chemically functional groups at the particle surfaces and the study of chemical reactions occuring at the interface. New methods have been devised for the synthesis of monodisperse polystyrene colloids containing only RSO_3^- groups fixed on the surface of the particles, and with both RSO_3^- and ROH groups in controlled amounts and at various ratios over a large range of particle sizes. In the acid form these colloids have been found to behave as effective heterogeneous catalysts for the hydrolysis of various acetate esters and of sucrose. The surface strong acid concentration coupled with a hydrophobic interaction between catalyst and substrate are primarily responsible for differences in rates and for rates which are in some cases considerably greater than those of other acids, such as ion-exchange resin beads, soluble polystyrene sulfonic acid, and sulfuric acid. The autocatalyzed hydrolysis of acrylic polymer colloids by surface strong acid groups has been shown to occur according to zero-order overall kinetics, suggesting possible applications in controlled release systems.

I.   MECHANISMS OF PARTICLE FORMATION

INTRODUCTION

One of the most important areas in which fundamental progress has recently been made in the field of emulsion polymerization concerns the mechanisms of particle formation. In 1948 Smith and Ewart presented a quantitative theory of particle formation based on Harkins' earlier concept of nucleation in monomer-swollen surfactant micelles (1). This process may be considered one of heterogeneous nucleation since the system contained pre-formed nuclei in the form of the micelles. Two years earlier Baxendale, Evans and coworkers published a qualitative theory of homogeneous nucleation in which surfactant

micelles played no role, but in which surfactant was important for stabilizing
the polymer particles against coagulation as they were formed (2). Priest
formulated a more detailed qualitative theory of homogeneous nucleation in
1952 (3). Thus contending schools of thought were established, but that of
Smith and Ewart generally dominated the scene, primarily because it was
elegant, quantitative, associated with their important theory for the kinetics
of emulsion polymerization, and supported by considerable experimental
evidence. Nevertheless a large body of work was developing which did not
agree with the Smith-Ewart predictions. Generally it appeared that
differences arose primarily on the basis of monomer solubility in the continuous medium (nearly always water): homogeneous nucleation could be observed
with the more soluble monomers, micellar nucleation was required for
insoluble ones.

Recent developments have shown how the two theories may be unified into a
single, overall scheme. In order to see how this may come about, it is best
to start with the theory of homogeneous nucleation and end with micellar
systems.

QUANTITATIVE PREDICTION OF THE PARTICLE NUMBER; HOMOGENEOUS NUCLEATION

The first attempt to formulate a homogeneous nucleation theory to predict the
absolute number-concentration of particles, N, was made by Fitch and Tsai in
1970 (4). This was supported by a large number of experiments on the polymerization of methyl methacrylate, MMA. It was further developed the following year (5), and was based primarily on the scheme of Priest (3) with an idea
from Gardon (6). The latter suggested that the rate of capture of oligomeric
radicals in solution by pre-existing particles, Rc, should be proportional to
the collision cross-section, or the square of the radius of the particles, r^2.
This has been called the "collision theory" of radical capture. In 1975 Fitch
and Shih measured capture rates in MMA seeded polymerizations and came to the
conclusion that Rc was proportional to the first power of the radius, as
would be predicted by Fick's theory of diffusion (7). In his book,
K. J. Barrett also pointed out that diffusion must govern the motions of these
species in condensed media (8).

The theory says that the number of particles increases at a rate which is equal
to the rate of generation of free radicals, R_i, reduced by the rate at which
the radicals are captured by particles, R_c. Coagulation will further reduce
the number of particles at a rate R_f. If the mutual termination of radicals
in solution is neglected, then

$$dN/dt = R_i - R_c - R_f.$$ (1)

This assumes that only individual radicals nucleate ("self-nucleation")
primary particles. If aggregation of radicals is involved, then the first
term on the right of Equation 1 should be modified to bR_i, where b is the
reciprocal of the aggregation number in a primary particle. This term may
also represent the fraction of radicals terminated in the aqueous phase prior
to nucleation. The absolute number of particles is obtained by integration:

$$N = \int_0^t (R_i - R_c - R_f)dt,$$ (2)

The individual terms must be evaluated in order to obtain numerical values for
N. In the simplest case the rate of coagulation, in the presence of adequate
stabilizer, will be equal to zero. There is some question as to whether this
can be accomplished in practice because of the extreme instability of primary
particles which are 1 to 2nm in size (9, 10). The rate of initiation may be
obtained from literature values or from independent measurements. The
initiator efficiency, f, must also be obtained. For the rate of capture, an
approximate solution to Fick's first law may be employed.
        This gives the capture rate proportional to the product of the
particle number and the radius:

$$R_c \simeq 4\pi D_{op}C_s Nr ,$$ (3)

where D_{op} is the average mutual diffusion coefficient of oligoradical and
polymer particle, C_s is the steady state concentration of radicals in solution,
and r is the radius of the particle. Experimentally one can obtain absolute
values of the capture rates by conducting a series of seeded polymerizations

in which N and r are varied. If conditions are set so that no coagulation takes place (R_f = 0), then Equations 1 and 3 combined give

$$dN/dt = R_i - 4\pi\, D_{op}C_sNr. \tag{4}$$

At low seed concentrations, $R_i > R_c$; dN/dt >0, and new particles will be formed. As the number of seed particles is increased at a given size r, a condition will be reached at which no new particles are formed, i.e. dN/dt = 0. At this point $R_i = R_c$. If R_i is known, R_c is immediately obtained. In seeded polymerizations with MMA, Fitch and Shih (7), and with vinyl chloride, Gatta and coworkers (11) found the rate of capture proportional to N_r, in support of Equation 3. A problem with Equation 3 arises from the fact that the concentration of free radicals in solution, C_s, cannot be determined, so that absolute values of R_c cannot be predicted. A way around this difficulty was given by Ugelstad and will be discussed later.

In unseeded polymerizations, the number of particles increases and they grow in size by polymerization and imbition of monomer, so that both N and r are functions of time. The value of N is computed by iterative numerical integration of Equation 2, whilst the value of r must be calculated from the rate of polymerization: Since the particles are spheres,

$$r = (3v/4\pi)^{1/3} \tag{5}$$

where v is the volume of a particle. The volume of polymer in a particle is given by

$$v_p = V_p/N = (N\rho)^{-1} \int_o^t Rpdt, \tag{6}$$

where V_p is the total volume of polymer in a dm^3 of latex, ρ is the polymer density and Rp is the overall rate of polymerization. The particles are swollen with monomer at a rate which is fast compared to the rate of consumption of monomer by polymerization, especially when the particles are very small (12). Thus the problem is treated in terms of swelling equilibrium (13). The volume of the monomer-swollen particle is thus

$$V = V_p/(1 - \phi_m) \quad , \tag{7}$$

where ϕ_m is the equilibrium volume fraction of monomer, from which the particle radius is obtained upon substitution into Equation 9.

ABSOLUTE CAPTURE RATES

Prior to nucleation the free radicals generated in the continuous phase propagate by reaction with dissolved monomer. Propagation continues stepwise until the radicals have reached the critical chain length for nucleation, jcr, at which point phase separation occurs and primary particles are formed. The rate of primary particle formation may therefore be expressed in terms of the polymerization process (14):

$$dN_1/dt = k_pM_wM_{jcr-1}^{\bullet} \quad , \tag{8}$$

where k_p is the propagation rate constant, M_w is the monomer concentration in the water phase and $M_{jcr-1}^{\bullet}$ is the concentration of free radicals of a size one repeat unit less than the critical value for self-nucleation. During their growth the oligoradicals may mutually terminate in the aqueous phase or be captured by particles already present. Radicals of any given chain-length j may also disappear by addition of another monomer unit. The primary radicals are formed at a rate equal to that of initiation, R_i. Thus there are three aspects to the kinetics involved:

 a) Radical appearance:

$$dM_o^{\bullet}/dt = R_i \tag{9}$$

 b) Propagation, termination and capture in the water (or continuous) phase:

$$dM_j^{\cdot}/dt = R_{pj-1} - R_{pj} - R_{twj} - R_{cj} \tag{10}$$

c) Particle nucleation:

$$dN_1/dt = R_{p(jcr-1)} \tag{8}$$

Here the symbol R represents a rate, as before, while the subscripts i, p, and tw represent initiation, propagation and water-phase termination respectively. Reactions of the primary radicals, $M_o^{\cdot}$, other than propagation, are assumed to be negligible. The subscript 1 on N_1 signifies that only primary particles are being discussed. No consideration of subsequent coagulation is given yet. This is treated more fully in a subsequent section. Equation 8 is solved upon addition of the series of Equations 9, 10, and 8, expressing them in terms of appropriate constants and concentrations, and for all values of j from 1 to (jcr-1). An exact solution for the number of particles as a function of time, t, involves numerical integration of these equations. However the authors show that by assuming a steady state in $M_j^{\cdot}$, the concentration of oligomers, C_s, is obtained. Further, by assuming that C_s is governed solely by termination in the aqueous phase, and by taking an average value for the rate constant for radical capture by particles, $\overline{k}_c$, an analytical solution is obtained (15) for the particle number:

$$N_{1(t)} = \frac{1}{k_1} \left[\{k_1 R_i t + (k_2+1)^{jcr}\}^{1/jcr} - (k_2+1) \right] , \tag{11}$$

$$k_1 \equiv \overline{k}_c/k_p M_w \qquad (dm^3)$$

$$k_2 \equiv (k_{tw}R_i)^{1/2}/k_p M_w \qquad (-)$$

$$\overline{k}_c = \sum_{j=1}^{jcr-1} \frac{k_{cj}M_j^{\cdot}}{C_s} \qquad (dm^3 s^{-1}) \tag{12}$$

Equation 11 tends to underestimate the number of particles except during the earliest few seconds of reaction, but serves as an extremely useful predictor for assessing the effect of experimental variables on the number of primary particles formed as a function of time.

To see the effects of monomer solubility and other variables on nucleation rate, calculations using Equation 11 were made for a series of monomers with reactivity similar to that for methyl methacrylate. These are shown in Fig. 1; Curve 1 represents the results for methyl methacrylate at a concentration in water close to saturation, i.e. 0.10 molar. The other parameters, given in the caption to Fig. 1, are taken from the literature. To assess the effect of monomer solubility, the parameters j_{cr} and M_w were varied (the values relative to those in Curve 1 are shown on the right side of the figure). In Curve 3 the rate of initiation was also changed. Curves 4, 5 and 6 show the results for a fictitious higher alkyl homolog, less water-soluble than the methyl ester. When j_{cr} is reduced slightly, more particles are nucleated.

When the monomer concentration is sufficiently reduced, however, mutual termination dominates over propagation and fewer particles are formed, as seen in Curves 2 and 6 of Fig. 1 (the calculations assume that particles are not produced upon water-phase termination of two oligoradicals, each with $j<j_{cr}$). Thus no variable has a simple, direct effect on the nucleation rate.

There are optimal levels depending, however, upon the values of the other parameters. These computations, based on Equation 11, have assumed a constant, average value for the specific rate constant for capture:

$$\overline{k}_c = \sum_{j=1}^{j(cr-1)} \frac{k_{cj}M_j^{\cdot}}{C_s} . \tag{12}$$

As mentioned earlier, k_c may be reduced by an electrostatic interaction, whose magnitude may be given the Fuchs symbol W', but also because of reversibility (15). Because the radical, upon collision with a particle, may add monomer units and/or terminate with a radical already present in the

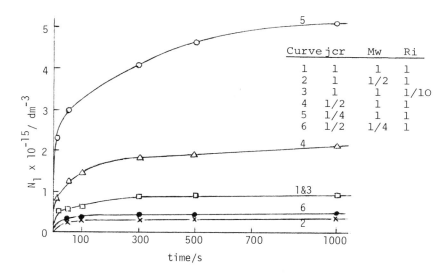

Fig. 1 Kinetics of primary particle nucleation calculated from Eq. 11 for methacrylate-like monomers. Parameters for Curve 1 are k_p = 350 dm^3 mol^{-1} s^{-1}; D = 5x10^{-10} m^{2}s^{-1}; r = 2x10^{-9}m; k_c = 6.3x10^{-18} m^{3}s^{-1}; k_{tw}^* = 10^{-17} dm^3 s^{-1}; M_w = 0.10 mol dm^{-3}; R_i = 10^{20} m^{-3} s^{-1}; j_{cr} = 60. Parameters for Curves 2-6 relative to those for Curve 1 shown at upper right.

particle, Hansen and Ugelstad applied the theory of Dankwerts for diffusion with reaction to determine the overall capture rate. The specific rate constant for capture of a radical of size j then becomes:

$$k_{cj} = 4\pi D_{wj} r F_j,$$ (13)

where F is the rate-reduction factor which takes into account reaction-diffusion into the particle, the degree of reversibility of adsorption and the electrostatic retardation. When the particles and the oligoradicals are both small, the efficiency of capture, F, is very low, on the order of 10^{-6} to 10^{-4}. If the particle contains a radical already, the capture of another from the water phase is much more probable, and nearly all radicals of j-values greater than a few units will be irreversibly captured by the particle. At the outset, all primary particles contain radicals, but obviously this rapidly changes until in a few seconds or minutes the average number of radicals per particle, $\bar{n}$, has declined to 1/2. At any instant there will be a mixture of oligoradicals of all values of j. The largest of these, with j→j_{cr}, will be captured irreversibly regardless of the nature of the particle.

As particles grow, two things occur with opposite effects: (1) they become more highly swollen with monomer (M_p increases), with the result that radicals in temporary contact with particles will increase in j and approach irreversible capture faster; simultaneously the capture rate constant is increasing with r, the radius; and (2) the surface electrical potential is likely to increase as more charged radicals are captured. This gives higher values of the parameters for electrostatic repulsion, which lead to reduced capture efficiency. This electrostatic effect will be much more pronounced in the presence of added surface-active agent or copolymerized ionogenic monomer. Higher rates of nucleation, of course, result from lower rates of radical capture. The former effect, which leads to reduced nucleation, dominates during the early stages when the electrostatic effect is negligible. As the particles become several tens of nanometers in size, electrostatic repulsions will become significant, although the real value may be lower than that calculated because of the "tunneling effect", first proposed by Fitch and Shih (7), in which the hydrophobic part of an oligomeric free radical in solution may "tunnel under" the electrostatic barrier, which affects only the ionic head of the radical, as it approaches a charged particle. Reduction in surface charge may occur by burial of ionic groups during polymerization and by particle coagulation.

EXPERIMENTS ON NUCLEATION KINETICS

Recently attempts have been made to determine experimentally the kinetics of nucleation during the first few tens of seconds of reaction (16). Since the nuclei will scatter light as they are formed, it is possible to study the reaction by following the Rayleigh scattering intensity as a function of time. Since the total volume fraction of polymer produced in this time is on the order of 10^{-9} to 10^{-8} dm^3 polymer/dm^3 aqueous phase, one is pressed to the very limits of detection of this technique. The method employed involves continuous photo-initiation at a short wavelength of light, and scattering at another, longer wavelength. The apparatus is shown schematically in Fig. 2.

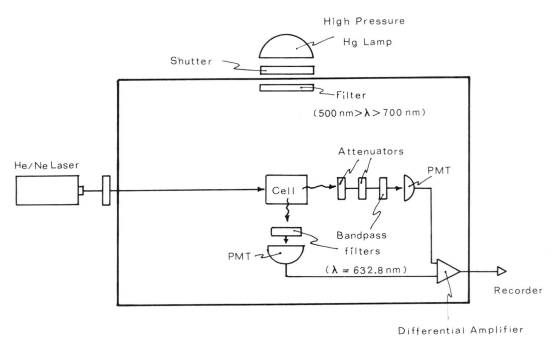

Fig. 2   Light scattering apparatus for nucleation kinetics.

By having two photomultiplier tube (PMT) detectors situated at 0° and 90° to the incident laser beam, and putting their output through a differential amplifier, high sensitivity of detection is obtained (17).

The initiator used was butanedione-2,3,which undergoes homolytic cleavage to produce two acetyl radicals upon radiation with violet-blue light from the mercury lamp. Only the red laser light is detected by the PMT's because of the bandpass filters in front of each. The reaction cell was filled with a solution of methyl methacrylate (MMA) monomer and initiator in water. The chemical system must be scrupulously clean, devoid of dust particles and reactive impurities at concentrations above a few parts per billion. Typical results are shown in Fig. 3. The conditions for these experiments are given below:

Exp.	{MMA}	{SDS}	{butanedione}
1	0.10M	0	0.34 M
2	0.10M	6.0×10^{-4}M	0.16 M

The experiment run in the absence of surfactant, sodium dodecyl sulfate (SDS), has the steeper slope. This pronounced difference due to surfactant concentration, even at levels as low as the 6×10^{-4} molar used here, has always been found in the many experiments which have been run (16).

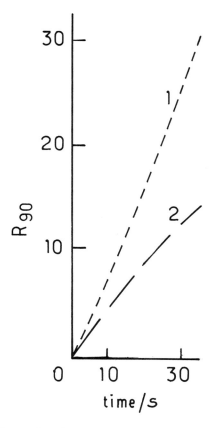

Fig. 3 Particle nucleation and growth kinetics determined by laser light
scattering: Rayleight ratio, R90, versus time. Exp. 1: no surfactant present. Exp. 2: [SDS] = 6 x 10-4 mol dm-3.

Comparison may be made between the experimental curves. The light scattering
intensity, expressed as the Rayleigh ratio, is a function of the size and
number of particles:

$$R_{90} = KV_{T(sw)}^2/N \; , \tag{14}$$

where K contains the wavelength and refractive index parameters; $V_{T(sw)}$ is
the volume fraction of monomer-swollen polymer particles relative to the
volume of the entire system; and N is the number concentration of particles.

Let us assume that both reactions shown in Fig. 3 polymerize at the same
rate, Rp. Then at a given time, say 10 seconds, the total volume of polymer
formed, V_T, will be the same in both. Under these conditions, the scattering
intensity becomes a direct measure of the relative number of particles:

$$R_{90} \propto 1/N \; .$$

Comparison of the two experimental curves at two times, 10s and 30s, is
given in Table 1.

TABLE 1. Relative particle numbers from light scattering

Curve[*]	{SDS}/mol dm^{-3}	time/s	R_{90}(relative)	N_2/N_1
Exp. 1	O	10	6.3	
		30	25	
Exp. 2	6.0 x 10^{-4}	10	5.3	1.2
		30	12	2.1

[*]Ref. to Fig. 3

The results lead to the conclusion that coagulation must be an important
process from the outset, and that it is retarded by the presence of surface
active agent.

Earlier experiments by Fitch and Watson measured coagulation in the absence
of nucleation and growth (18). Using a light scattering apparatus similar to
that of Fitch and Palmgren shown in Fig. 2, but with a single photomultiplier
detector, they followed the increase in R_{90} with time as a function of
surfactant (SDS) concentration in a series of MMA polymerizations. In these
photo-initiation was confined to a single burst of light of about 1 ms
duration. Under these conditions initiation, growth and termination all
occur within a few seconds, so that by the time the first measurement is made
at t = 10s, essentially all polymerization has stopped ($V_{T(sw)}$ is a constant),
and any change observed in R_{90} is due to a change in N (18), according to
Equation 14. Typical results are shown in Fig. 4 in which the variable is

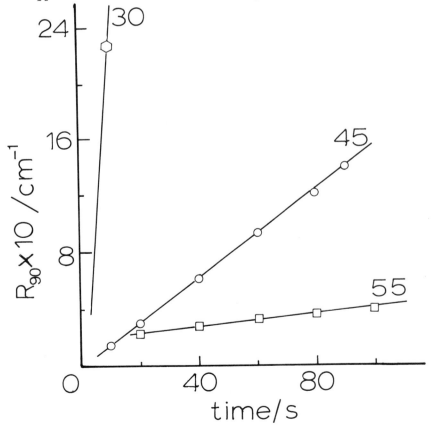

Fig. 4 Particle growth by coagulation. Flash initiation. Rayleigh ratio,
R_{90}, as a function of Time. S represents concentration of sodium dodecyl
sulfate (SDS): Exp. 30, [SDS] = 0; Exp. 45, [SDS] = 7.6 x 10^{-5} mol dm^{-3};
Exp. 55, [SDS] = 88 x 10^{-5} mol dm-3.

SDS concentration. The steepest slope, corresponding to the most rapid coagulation rate, R_f, occurs at {SDS} = O. But even at SDS concentrations on the order of 10^{-4} molar, the surfactant has an enormous effect on R_f. To obtain a quantitative measure, one must apply the coagulation theory of Smoluchowski and Fuchs which states that

$$1/N_{(t)} = (1/N_o)(1 + t/\tau) \quad , \tag{15}$$

where τ is the half-life, or the time required for the number of particles, $N_{(t)}$, to decrease to one-half of its original value, N_o. When this is combined with the light scattering equation of Rayleigh (Equation 14), and allowance made for a distribution in the particle size because of the stochastic nature of the process, one obtains (19)

$$R_{90(t)} = R_{90(O)} (1 + 2t/\tau) \quad . \tag{16}$$

This is an equation for a straight line, the slope of which is $2R_{90}/\tau$, and describes the plots in Fig. 4. From the experimental slopes one may calculate the half-life, τ, and the Fuchs stability factor, W, the latter of which is discussed later. These values are given in Table 2. They confirm

TABLE 2. Half-lives of coagulating primary particles as a function of surfactant concentration

Exp. No.	{SDS}/mol dm^{-3}	slope/ cm^{-1}s^{-1} X 10^8	τ/s	W
30	O	94	0.0043	0.36
45	7.6 X 10^{-5}	1.5	0.30	25
55	88.0 X 10^{-5}	0.21	2.2	180

that coagulation is an important process in determining particle size and number from the outset. With no surfactant present the particles apparently have a net attraction for each other (W <1), and disappear with a half-life on the order of four milliseconds. As surfactant concentration is increased it rapidly will adsorb onto the particles, building up their surface electrical potential, and causing them increasingly to repel each other.

The rate of coagulation **between particles** of size i and size j is given by (19)

$$R_{fij} = 4\pi D_{ij} R_{ij} N_i N_j/W_{ij} \tag{17}$$

where R_{ij} is the distance of closest approach between two particles, and may be taken as approximately equal to $r_i + r_j$; D_{ij} is the mutual diffusion coefficient, and is equal to $D_i + D_j$; and W is the Fuchs stability ratio, which may be expressed as

$$W = R_{f(fast)}/R_{fij} \quad , \tag{18}$$

in which $R_{f(fast)}$ is the collision frequency between i and j-type particles, and measures the coagulation rate when all interparticle collisions are irreversible, and there is no interaction between particles until they touch. The so-called "Smoluchowski fast coagulation rate" may be characterized by a half-life independent of particle size:

$$\tau = 3\eta/4Nk'T \quad , \tag{19}$$

where η is the viscosity of the continuous medium and k' is Boltzmann's constant. If the particle concentration, N, is 10^{16} dm^{-3}, the half-life will be 12 milliseconds. Thus, according to the results in Table 2, the primary particles coagulate almost 3 times as fast as this in the absence of surfactant (due to van der Waals attractions). In the presence of SDS at 7.6 X 10^{-5} molar and at 88 X 10^{-5} molar the half-life is increased 25 and 180 times, respectively. Dunn and Chong estimated lower values of W $\approx$ 5 and 100 at higher SDS concentrations of 5 X 10^{-4} and 2.2 X 10^{-3} mol dm^{-3}, respectively,

in vinyl acetate emulsion polymerizations (10). The difference in stability
or in efficacy of sodium dodecyl sulfate as a stabilizer for the two polymers
may be attributed to the difference in the adsorption isotherms of SDS on
polyvinyl acetate$_2$(PVAc) and PMMA (20). The area occupied per adsorbed SDS
molecule is 1.1nm^2 and 0.79nm^2 on PVAc and on PMMA, respectively (20). The
surface electrical charge densities would be proportional to the reciprocal
of these areas. This, in turn, determines the surface potential, ψ_o (19).
The relationship between W and ψ_o at various ionic strengths is dealt with
by the DLVO theory (9).

It is possible that a steady state may be reached, in which dN/dt = 0, even
though R_f may be positive. Then, according to Equation 1,

$$R_i = R_c + R_f \ ,$$

This is to say that coagulation is occurring throughout the polymerization.
Under these circumstances it is most likely that primary particles are
continuously generated, but quickly coagulate onto larger particles formed
earlier in the reaction, as predicted by Dunn and Chong (10). Thus at very
low surfactant concentrations, the particle size distribution may be
completely determined by coagulation rather than nucleation, in which case
surface electrical potential, particle size and ionic strength become
critical factors. Munro and coworkers have postulated such behavior for the
case of styrene polymerization in the absence of emulsifier (21). They
observed decreases in the number of particles on the order of 10^3 to 10^5
during periods of from 5 to 80 min by quasi-elastic light-scattering. In
nonaqueous systems coagulation has been shown, at least in one case, to be
the principal mode by which particle size is determined. Fitch and Kamath
found that in MMA latex polymerization in alkane media in the presence of an
amphiphilic, polymeric stabilizer that particle size depended on the surface
area occupied by stabilizing moieties, which was constant for all experiments in which it was measured (22).

Further evidence for this was developed by Hansen and Ugelstad in a series of
seeded styrene emulsion polymerizations in which the variables were N_s, r_s,
ψ_o and μ (ionic strength) (23). Here N_s and r_s represent the number and
radius of monomer-swollen polystyrene seed particles. Some results are shown
in Fig. 5. Theoretical curves could only be made to fit experimental points

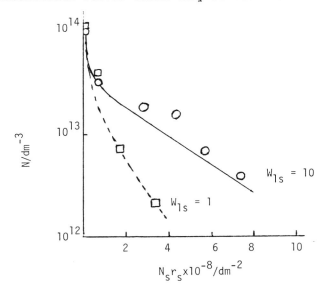

Fig. 5 Nucleation of particles in the presence of seed latex. Points are
experimental: $\bigcirc$ r = 81nm, σ_0 = 9.5μC cm^{-2}; $\square$ r = 40 nm, $\sigma_0 \approx$ 1μC cm^{-2}.
Curves are theoretical: W_{1s} = 1, W_{1s} = 10, and in both F = F_s = 1 and
W_{11} = 0.1. Styrene polymerization (31).

when coagulation among primary particles was calculated to be fast, (W_{11} = 0.1), radical adsorption was irreversible ($F = F_s$ = 1), and coagulation of primary particles with seed particles was more or less rapid (W_{1s} = 1 or 10), depending upon the surface charge density of the seed particles. In the presence of surfactant, however, both radical capture and coagulation rates are reduced so that the tendency towards nucleation is greatly enhanced (24). In terms of Equation 1, emulsifier tends to reduce both R_c and R_f, especially the latter, but the former also as particles become large relative to primary particles. In the presence of increasing amounts of emulsifier, then, the mechanism of particle formation becomes increasingly determined by particle nucleation rather than coagulation. Goebel and coworkers have recently observed behavior in pronounced contrast to these observations, however (25). In a series of emulsion polymerizations of vinyl chloride in the presence of relatively high emulsifier concentrations they observed slow coagulation over periods of up to 250 minutes. They have suggested that the polymerization reaction itself is involved in the coagulation process although the mechanism involved is not yet clear.

Electrostatic effects also can be brought about by copolymerization of ionic monomers in the absence of conventional emulsifiers (26, 27), although it has been shown that these may result in the production of considerable water-soluble polyelectrolyte (28, 29). Not considered above is the likely interaction of growing oligomeric radicals with emulsifier molecules (not micelles) prior to nucleation (30). Much work in this area remains to be done.

EFFECTS OF MONOMER SOLUBILITY; HETEROGENEOUS NUCLEATION

Sütterlin and coworkers have studied the influence of monomer solubility in an homologous series of acrylate and methacrylate monomers (31). For example, in the acrylate series the value of k_p/k_t is relatively independent of the number of carbon atoms in the alkyl group (32), whereas monomer and thus oligomer solubility vary greatly. They studied the effect of SDS concentration on N at constant R_i. Typical results are shown in Fig. 6, in which the

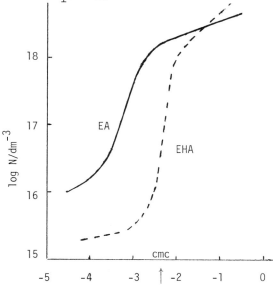

Fig. 6 Dependence of particle number on concentration of sodium dodecyl sulfate (log-log plot) for ethyl acrylate (EA) and 2-ethylhexyl acrylate (EHA) (44).

logarithm of the particle number, N, has been plotted against the logarithm
of the SDS concentration, over approximately four orders of magnitude, for the
emulsion polymerization of ethyl acrylate (EA) and 2-ethylhexyl acrylate (EHA).
The differences in the two curves are representative of the entire series as
well as other monomers differing in water-solubility. In contrast to the EA,
which readily forms many polymer particles at submicellar surfactant concentrations, the poorly soluble monomer, EHA, exhibits relatively low N values
until the CMC is passed, after which it has more particles than the EA
system. Furthermore N increases more rapidly with {SDS} above the CMC in the
case of EHA. The more hydrophobic monomer will be solubilized in surfactant
micelles above the CMC to a greater extent because of more favorable interactions with the hydrocarbon core of the micelles (33). On the other hand,
the solubility of EA is relatively independent of surfactant concentration.
For this monomer N increases with {SDS} primarily because of progressive
reduction in R_f, as observed for MMA in Fig. 3 and 4. Below the CMC, the more
soluble monomer produces more particles primarily because of its greater
propagation rate in the aqueous phase ($k_p M_w$), as predicted by Curves 1 and 6
in Fig. 1. Above the CMC, there is apparently more of the hydrophobic
monomer available for nucleation in the micelles because of its greater degree
of solubilization. The micelles also may be considered energetically favored
sites for particle formation because they will undergo heterogeneous nucleation. Thus EHA produces more particles above the CMC than does EA. This
effect may be enhanced by the fact that SDS is adsorbed more strongly on the
more hydrophobic polymer (20), giving it a higher ψ_o above the CMC and

consequently a lower R_f. Apparently below CMC this effect is insufficient to
overcome the vast differences in M_w between the two monomers (EA is 340 times
more soluble than EHA).

UNIFIED THEORY

The concept of radical capture leads us to a theory which comprehends all
aspects of particle formation in emulsion polymerization. If in the initial
reaction medium there exist no entities capable of radical capture, homogeneous nucleation must occur. The efficiency of homogeneous nucleation will
depend primarily upon monomer solubility. On the other hand if monomer-
swollen surfactant micelles, seed latex particles or emulsified monomer
droplets are present as "nuclei" initially, these may serve as sites for
(energetically more favored) heterogeneous nucleation. The efficiency of
heterogeneous nucleation will depend primarily upon the number-concentration
of the nuclei, since they will compete against homogeneous nucleation more or
less successfully depending upon their ability to capture radicals from the
continuous medium before they reach the critical chain length for phase
separation. The classical theory of Smith and Ewart considered the narrow
case where all radicals are captured by micelles (1). Fitch and Tsai
considered the case where no nuclei were originally present (5). Hansen and
Ugelstad have considered all known cases (15, 23, 24). Thus formerly contending schools of thought have been unified under a single family of concepts.

It should be pointed out that there exist practical differences in emulsion
polymerization technology depending upon the system chosen. A look at
Fig. 6 reveals one example: at 10^{-3}M SDS concentration EHA produces only
about 1/100 the number of particles as EA. Its rate of polymerization would
therefore be proportionately lower, and thus impracticable. Only in the
presence of a more water-soluble comonomer or of surfactant above its
CMC (in the system) would enough particles be formed to achieve reasonable
rates of reaction. Thus although one system may behave in a very different
manner from another, they may both be understood in terms of the overall
kinetic theory presented here.

THE THEORY APPLIED

Practical applications of these concepts can be employed to predetermine or
regulate particle size and size distributions in emulsion polymerizations,
both in the laboratory and the manufacturing plant. These have been discussed
in some detail elsewhere (34). The guiding principles are summarized in
Equation 1, in which it is seen that the three kinetic processes of radical
formation in the continuous phase, R_i, radical capture, R_c, and particle

coagulation, R_f, determine the number-concentration, and therefore the size,
of particles at any instant. By manipulating these processes through
regulation of reactant concentrations, solubilities, temperature and ionic
strength, it is possible to obtain a wide range of sizes and distributions

from monodisperse to broad to polymodal. These, in turn, may directly affect the kinetics of polymerization as well as the molecular weight and its distribution in the polymer produced. Applications of the products of emulsion polymerization are found in almost endless variety: paints, floor finishes, "rubberized" concrete, films, fibers, as well as in clinical diagnostics and as model colloids for fundamental investigations of colloidal behavior.

REFERENCES

1. W.V. Smith and R.H. Ewart, J. Chem. Phys. 16, 592 (1948).
2. J.H. Baxendale, M.G. Evans and J.K. Kilham, Trans. Faraday Soc. 42, 668
      (1946); J.H. Baxendale, S. Bywater, M.G. Evans, ibid. 42, 675 (1946).
3. W.J. Priest, J. Phys. Chem. 56, 1077 (1952).
4. R.M. Fitch and C.H. Tsai, Polymer Letters 8, 703 (1970).
5. R.M. Fitch and C.H. Tsai, Polymer Colloids, R.M. Fitch Ed., p. 73, Plenum,
      New York, (1971).
6. J.L. Gardon, J. Polym. Sci. 6, 643, 665, 687, 2853, 2859 (1968).
7. R.M. Fitch and L.B. Shih, Prog. Colloid Polym. Sci. 56, 1 (1975).
8. K.E.J. Barrett, Ed., Dispersion Polymerization in Organic Media, J. Wiley,
      New York (1975).
9. (a) B.V. Deryagin and L.D. Landau, Acta phys.-chim. USSR 14, 633 (1941).
   (b) E.J.W. Verwey and J.T.G. Overbeek, Theory of the Stability of Lyophobic
      Colloids, Elsevier, Amsterdam (1948).
10. A.S. Dunn and L.C.H. Chong, Brit. Polym. J. 2, 49 (1970).
11. G. Gatta, G. Benetta, G. Talamini and G. Vianello, Adv. in Chem. Ser. 91,
      158 (1969).
12. S.J. Liang and R.M. Fitch, results to be published.
13. M. Morton, S. Kaizerman and M.W. Altier, J. Colloid Sci. 6, 2859 (1968).
14. J. Ugelstad and F.K. Hansen, Rubber Chem. Technol. 49 (3), 536 (1976).
15. F.K. Hansen and J. Ugelstad, J. Polym. Sci.; Polym. Chem. Ed. 16, 1953
      (1978).
16. R.M. Fitch, T.H. Palmgren and T. Aoyagi, unpublished results.
17. G. Kegeles, private communication.
18. R.M. Fitch and R.C. Watson, J. Colloid Interface Sci. 68 (1), 14 (1979).
19. J.T.G. Overbeek, Colloid Science, H.R. Kruyt, Ed., Vol. I; pp 278-298,
      Elsevier, New York/Amsterdam (1952).
20. B.R. Vijayendran, Polymer Colloids II, R.M. Fitch Ed. p. 209, Plenum,
      New York, (1980).
21. D. Munro, A.R. Goodall, M.C. Wilkinson, K. Randle and J. Hearn, J. Colloid
      Interface Sci. 68 (1) 1 (1979).
22. R.M. Fitch and Y.K. Kamath, J. Colloid Interface Sci. 54 (1), 6, (1976).
23. F.K. Hansen and J. Ugelstad, ibid. 17, 3033 (1979).
24. F.K. Hansen and J. Ugelstad, ibid. 17, 3047 (1979).
25. K.H. Goebel, H.J. Schneider, W. Jaeger and G. Reinisch, Acta Polymerica 32
      (2) 117 (1981).
26. H.J. Wright, J.F. Bremmer, N. Bhimani and R.M. Fitch, U.S. Pat. 3,501,432.
27. I.M. Krieger and M.S. Juang, J. Polym. Sci., Polym. Chem. Ed. 14, 2089
      (1976).
28. R.L. Schild, M.S. El-Aasser, G.W. Poehlein and J.W. Vanderhoff, Emulsions,
      Latices and Dispersions, P. Becher, M.Y. Yudenfreund Eds. p. 99, Dekker,
      New York, (1978).
29. Y. Chonde and I.M. Krieger, J. Colloid Interface Sci. 77 (1), 138 (1980).
30. C.Y. Chen and I. Piirma, J. Polym. Sci., Polym. Chem. Ed. 18, 1979 (1980).
31. N. Sütterlin, Polymer Colloids II, R.M. Fitch Ed. p. 583, Plenum, New York,
      (1980).
32. J.A. Brandrup and E.H. Immergut Eds. Polymer Handbook, Wiley, New York,
      (1975).
33. P.H. Elworthy, A.T. Florence and C.B. Macfarlane, Solubilization by
      Surface-active Agents, Chapman and Hall, London, (1968).
34. R.M. Fitch, Proceedings Water-Borne and Higher Solids Coatings Symposium,
      New Orleans, Univ. So. Miss. and Southern Soc. Coatings Tech.,
      Feb. 25-27 (1981).

II. CHEMISTRY AT THE INTERFACE.

INTRODUCTION

It has been known for a long time that latex particles invariably carry
functional groups chemically bound to the interface between the particle and
the surrounding medium (the particle "surface"), and that these are ordinarily
derived from initiator, comonomers bearing such groups or from hydrolysis of
the monomer or polymer during or after polymerization. Recently attention
has been devoted to the synthesis of model polymer colloids, the analysis and
characterization of their surfaces and still more recently, to the investigation of chemical reactions occurring there (1). There are a number of
reasons why we are interested in the surface chemistry of these systems,
some of which are listed below.

Commercial latex systems. Latex paints have often exhibited problems with
"shelf stability", in which coagulation occurs slowly upon storage. This is
often accompanied by a drift to lower pH, followed by corrosion problems in
metal containers. Such behavior is especially characteristic of vinyl
acetate polymers in which hydrolysis of the particle surface is presumed to
occur. The mechanical properties of paint films derived from polymer colloids
have been shown to be dependent upon the distribution of polar and functional
groups on or near the polymer/water interface (2, 3).

Water-borne adhesives are often made from polymer colloids. Their adhesion
to substrates will strongly depend on the nature and concentration of surface
groups which will ultimately be found at the polymer:substrate interface.
The sensitivity of the adhesive bond to moisture is often a problem and
usually due to surface-active substances added to stabilize the system during
and after polymerization. The proper employment of chemically bound
stabilizing groups may largely overcome these problems.

Some polymers used for fibers are formed by emulsion polymerization, most
notably, polyacrylonitrile. Surface properties related to adsorbancy of and
substantivity to dyes may depend upon the surface groups bound to the latex
particles which are not removed during coagulation and washing. Many
plastics and elastomers, not used as latexes, but as molding or extruding
resins are likewise synthesized by emulsion polymerization followed by coagulation and washing. For example ABS (acrylonitrile, butadiene, styrene)
plastics must be made in this way. The disposal to the environment of the
surface-active agents and salts used in the polymerization represent a large
problem, which again may be solved through an understanding of what chemical
groups are present at the interface, how they are introduced, and their
ultimate fate and influence on bulk properties.

Emulsion polymerization. Not only are these functional groups placed at the
particle surface during emulsion polymerization, but they may subsequently
play a role in determining the rates of various kinetic processes which occur.
Among these are the coagulation of primary particles formed by homogeneous
nucleation (4), and the capture of oligomeric free radicals from the aqueous
phase by the polymer particles (5, 6).

Heterogeneous catalysis. It was suggested in 1975 that because of their high
specific surface areas, monodispersity, wide range of particle properties
and variety of synthetic methods available for introducing surface groups,
that polymer colloids were potentially attractive as candidates for heterogeneous catalysts (1). Since that time some work has been done to support
this contention and it will be dealt with in this paper.

Biochemistry, molecular biology and medicine. A recent and exciting development involves the attachment to latex particles of immunospecific reagents
which cause the particles to attach themselves only to certain receptor sites
on the surfaces of living cells (7). Such systems may be used for diagnostic
purposes, for cell sorting by subsequent electrostatic or magnetophoretic
methods and possibly for chemotherapy in which the latex particle releases a
drug to the tissue to which it has become attached.

Model colloids. (a) Determination of forces: interactions among particles
as well as between particles and surfaces may be measured by means of
coagulation kinetics and rates of deposition. From these are derived information on forces arising from electrical double layers, polymer-polymer
interactions, dispersion attractions and hydrodynamic interactions.

(b) Adsorption studies: colloids may be used to study the adsorption behavior of gases, liquids, and solutes of all kinds dissolved in the continuous medium.
(c) Flocculation and coagulation: these processes can only be well understood quantitatively when the colloids involved are fully characterized.

Controlled release systems. Recent studies have shown that certain acrylic polymer colloids undergo autocatalyzed hydrolysis to release low-molecular weight compounds with zero order kinetics (8). Some examples of potential applications are catalysts, corrosion inhibitors, pheromones for insect control, fungicides, plant growth regulators, appetite stimulants and drugs.

ORIGINS OF SURFACE GROUPS

TABLE 1. Comonomers which give surface functional groups

A. Weakly Surface Active

Name	Structure	Abbreviation	Ref.
Acrylic acid	$CH_2 = CH \cdot COOH$	AA	9
Methacrylic acid	$CH_2 = C \overset{CH_3}{\underset{COOH}{}}$	MAA	9
Sulfoethyl methacrylate	$CH_2 = C$, CH_3, $C=O$, $O-CH_2-CH_2SO_3^- M^{+a}$	SEM	10
2-acrylamido-2-methyl propane sulfonate	$CH_2 = CH-\overset{O}{\overset{\|}{C}}-NH-\overset{CH_3}{\underset{CH_3}{\overset{\|}{C}}}-CH_2-SO_3^- M^+$	AMPS	11
Sulfo methyl styrene	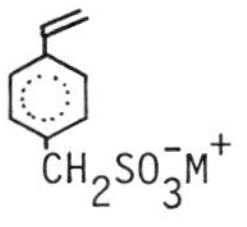 $CH_2SO_3^- M^+$	SMS	12
Styrene sulfonate	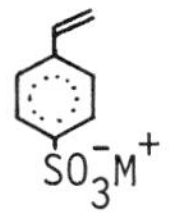 $SO_3^- M^+$	SS	10
Methacroyloxyethyl trimethyl ammonium	$CH_2 = \overset{CH_3}{\underset{COOCH_2-CH_2-N^+-(CH_3)_3}{\overset{\|}{C}}}$ X^- b	META	12

B. Strongly Surface Active

Name			
10-p-Styrylundecanoate	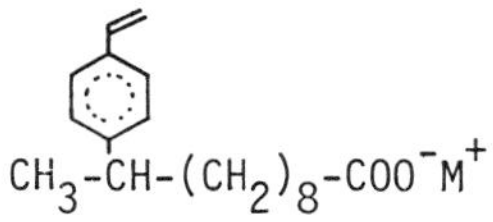 $CH_3-CH-(CH_2)_8-COO^- M^+$	SU	13
9-Acrylamido stearate	$CH_2=CH-\overset{O}{\overset{\|}{C}}-N-H$ $CH_3-(CH_2)_8-CH-(CH_2)_7-COO^- M^+$	AAS	14
Maleoyloxy ethyl dimethyl-p-dodecylbenzyl ammonium	$\overset{H}{\underset{H}{}}C=C \overset{COOH}{\underset{COOCH_2-CH_2-\overset{CH_3}{\underset{CH_3}{\overset{+\|}{N}}}-CH_2-C_{12}H_{25}}{}}$ X^-	MDDA	15

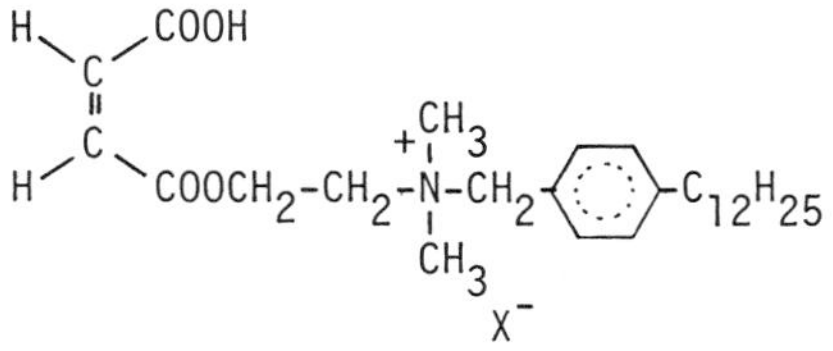

Footnote a: M^+ represents a metal ion or H^+

Footnote b: X^- represents any anion

Comonomers

<u>Weakly surface-active comonomers.</u> The most common method for the introduc-
tion of chemically bound groups at the particle/aqueous solution interface
is by the copolymerization of ionogenic monomers. All commercially available ones are of relatively low molecular weight, are weakly surface active
and as a result are water-soluble. Examples are given in Table 1. One
consequence of this combination of physical properties is that comonomer
tends to be found not only at the interface, but also buried within the
particles and in the aqueous serum as soluble polyelectrolyte.

To optimize placement at the interface, comonomer should be strongly surface
active and as insoluble in both polymer and water as possible. Three such
compounds have been reported, and employed in polymer colloids. They are
listed in Table 1B.

Initiators

In emulsion polymerization the initiator is nearly always water-soluble, so
that radicals are generated in the aqueous phase. As a result they must
enter the polymerizing particles from the surrounding fluid. Often the
monomer is sufficiently soluble in the aqueous phase, and the barrier to
entry of an ionic free radical into an electrostatically charged particle
is sufficiently high, that some polymerization occurs prior to entry (1).
This "capture" or "radical adsorption" process is depicted in Fig. 1. This

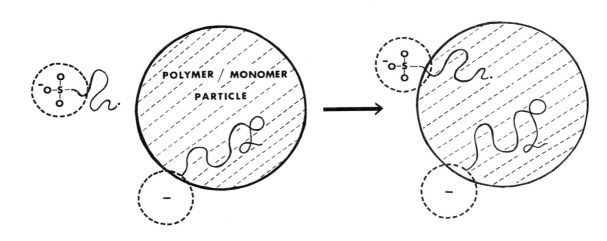

Fig. 1. Schematic representation of introduction of surface functional
groups through capture of oligomeric free radicals from continuous phase
by monomer-swollen polymer particle.

process may be repeated several to several thousand times during the
polymerization of each particle, providing a corresponding number of ionogenic
and/or functional groups. The initiator need not produce ionic radicals,
although if initiator-derived groups are the sole source of colloidal
stability, they must be electrically charged.

A great variety of initiators has been employed for emulsion polymerization.
A list of the more important ones is given in Table 2 along with the surface
groups which are known to be introduced by them. Many other systems have
been shown to initiate emulsion polymerization; in fact nearly any oxidation-
reduction (redox) pair will do so. Generally, however, no careful investigation of the resulting surface groups has been made.

TABLE 2. EMULSION POLYMERIZATION INITIATORS AND DERIVED SURFACE GROUPS

	Name	Symbol	Chemical Formula	Derived Free Radical	Derived Surface Group	Ref.
1.	Persulfate	PS	$S_2O_8^= / M_2^{2+}$	$\cdot OSO_3^-M^+$ $\cdot OH$	$ROSO_3^-M^+$ ROH $RCOOH(?)$	16
2.	Persulfate/bisulfite/ iron	PBI	$S_2O_8^= / HSO_3^- / Fe^{2+,3+}$	$\cdot OSO_3^-M^+ / \cdot SO_3^-M^+ / \cdot OH$	$ROSO_3^-M^+ / RSO_3^-M^+ / ROH$	17
3.	Bisulfite/iron (III)	BI	HSO_3^- / Fe^{3+}	$\cdot SO_3^-M^+$	$RSO_3^-M^+$	18
4.	Bisulfite/silver (I)	BAg	$HSO_3^- / Ag(NH_3)_2^+$	$\cdot SO_3^-M^+$	$RSO_3^-M^+$	19
5.	Bisulfite/copper (II)	BCu	$HSO_3^- / Cu(NH_3)_4^{2+}$	$\cdot SO_3^-M^+$	$RSO_3^-M^+$	19
6.	Azoisobutyronitrile	AIBN	$\begin{array}{c}CH_3\\ \vert\\ -(N-\overset{\vert}{C}-CN)_2\\ \vert\\ CH_3\end{array}$	$\begin{array}{c}CH_3\\ \vert\\ \cdot C-CN\\ \vert\\ CH_3\end{array}$	RCN	12
7.	Azoisobutyramidinium	AIBA	$\begin{array}{c}CH_3 \quad\ NH_2^+X^-\\ \vert \qquad \Vert\\ -(N-C-C-NH_2)_2\\ \vert\\ CH_3\end{array}$	$\begin{array}{c}CH_3 \quad\ NH_2^+X^-\\ \vert \qquad \Vert\\ \cdot C-C-NH_2\\ \vert\\ CH_3\end{array}$	$\begin{array}{c}NH_2^+X^-\\ \Vert\\ R-C\\ \backslash NH_2\end{array}$	12
8.	4,4'-Azobis-4-cyano pentanoic acid	ABCPA	$\begin{array}{c}CH_3\\ \vert\\ -(N-C-CH_2-CH_2-COOH)_2\\ \vert\\ CN\end{array}$	$\begin{array}{c}CH_3\\ \vert\\ \cdot C-CH_2-CH_2-COOH\\ \vert\\ CN\end{array}$	RCOOH	20
9.	Hydrogen peroxide/iron (II) (Fenton's Reagent)		H_2O_2 / Fe^{2+}	$\cdot OH$	ROH	21
10.	Perphosphate	PPh	$H_2P_2O_8^= M_2^{2+}$	$HO\dot PO_3^-M^+$ $\cdot OH (?)$	$ROPO_3^-M^+H^+$ $ROH (?)$	22
11.	p-sulfomethyl benzoyl peroxide	SMBP	$\left(\text{COO}\!-\!\!\bigodot\!\!-\!\text{CH}_2\text{SO}_3^-M^+\right)_2$	$\cdot\!\bigodot\!\!-\!\text{CH}_2\text{SO}_3^-M^+$	$RSO_3^-M^+$	23

M - C

POLYSTYRENE COLLOIDS WITH ONLY SURFACE SULFONATE GROUPS

In spite of the fact that polymer colloids can be synthesized with very narrow particle size distributions and with a variety of surface functional groups, they have not been totally satisfactory for basic studies because of the facts that usually more than one kind of group appears at the surface and that these tend nearly always to be labile, particularly to hydrolysis. Reference to Table 2 reveals that of the common groups which have been studied, sulfate, nitrile, amidine and phosphate are all hydrolyzable; hydroxyl usually results from hydrolysis (except for Fenton's Reagent, No 9), and is itself oxidizable to carboxyl. Sulfonate alone is a strong electrolyte at all pH's and stable. Therefore we have expended considerable effort recently on the synthesis of polystyrene colloids with only sulfonate groups at the surface of the particles. The BI initiator system (No. 3, Table 2) was first mentioned by Duban (24), and is comprised of the reaction between bisulfite and iron (III) ion. It is particularly attractive for this purpose because of its simplicity, so that we have investigated it further (18). Duban used it to polymerize tetrafluoroethylene, but no proof of structure was given. Although this system generates the required surface groups, the stoichiometry of the reaction dictates equimolar amounts of bisulfite and Fe^{3+}:

$$Fe^{3+} + HSO_3^- \rightarrow Fe^{2+} + \cdot SO_3H.$$

In terms of latex stability, this fact probably precludes the use of surfactant-free systems, since it is well known that trivalent cations tend strongly to induce coagulation of electronegative sols. Therefore sodium dodecyl sulfonate (SDSo) was employed in all of our experiments reported below. If any grafting of emulsifier to the polymer particles occurred, no "foreign" functional groups would thereby be introduced.

There has been some controversy over the relative merits of dialysis and ion exchange as methods for purification of polymer colloids (25, 26). Both techniques have been suspected of introducing impurities due to soluble polymer leached from either dialysis membrane or ion exchange resin; dialysis has been thought to be incomplete as well. We believe that difficulties have arisen in the past partly becquse of the lability of the surface groups.

McCarvill and Fitch reported for the first time the synthesis and characterization of all-strong-acid, nonhydrolyzable polystyrene latexes (18). This enabled them to resolve the questions concerning the relative merits of ion exchange over dialysis for the ultimate purification of colloids prepared via emulsion polymerization.

Some examples will illustrate: An all-sulfonate polystyrene colloid ("Latex 64") was ion exchanged five times, each time with fresh, highly purified, mixed bed resin. The titer/gram of polymer remained constand after the second exchange through successive exchanges as shown in Table 3. The initial increase in titer is probably due to the incomplete conversion of all surface groups to the hydrogen form during the first exchange (insufficient resin). The solids content of the latex typically decreases with each successive exchange.

TABLE 3. The effect of repeated ion exchanges on Latex 64[a]

Ion exchanges (MB) 2 hr each time	Wt of polymer/ 10 ml of latex	meq of H^+/g of polymer[b]
1	0.752	0.0612
2	0.718	0.0668
3	0.696	0.0660
4	0.668	0.0658
5	0.630	0.0666

[a]Latex pretreatment: (a) Ion exchange against Na^+-form cation resin; (b) dialysis against distilled H_2O for 24 hr at 55°C.

[b]Determined by conductometric titration against 0.020 N Bq(OH)$_2$. Only strong acid found.

Thus there is no evidence of contamination by highly purified ion exchange resins. On the other hand dialysis, even with extremely pure distilled water (conductivity was 0.8 to 1.0µmho cm^{-1}), invariably caused contamination of highly purified polymer colloids (18). The very rapid hollow-fiber dialysis technique was employed. After dialysis, there was a drop in the titer/gram, which returned to the initial value after ion exchange. Since the latex particle is capable of cation exchange itself, it will selectively exchange the acidic hydrogen ions in its stabilizing double layer for trace multivalent cations in the distilled water. Typically, 100 ml of latex would be dialyzed against 29 liters of water in 12 hr. The extremely small amounts of contaminants in this water will slowly decrease the titer/gram exhibited by the latex. Latex 60, dialyzed for 22 hr, showed the same drop in titer as Latex 65 which was dialyzed for 12 hr. Since Latex 65 had a greater ion-exchange capacity, it was more efficient in scavenging ions than Latex 60. Both latexes regained their initial hydrogen ion concentration after ion exchange, as shown in Table 4.

TABLE 4. The effect of sequential heating, dialysis, and ion exchange on latexes 60 and 65

Latex 60		Latex 65	
Time of heating at 95°C (hr)	meq of H^+/g of polymer[a]	Time of heating at 95°C (hr)	meq of H^+/g of polymer[a]
0	0.0168	0	0.0460
25	0.0182	66	0.0476
37	0.0166	--	--
48	0.0166	--	--
63	0.0166	--	--
Dialyzed 22 hr	0.0078	Dialyzed 12 hr	0.0388
Ion exchanged after dialysis	0.0160	Ion exchanged after dialysis	0.0466

[a]Determined by conductometric titration against 0.020 N Ba(OH)$_2$.
   Only strong acid found. Latexes initially purified by ion exchange.

As further evidence of their stability, these colloids have retained their surface charge densities for over three years in a few cases which have been examined.

SYNTHESIS OF ALL-SULFONATE MONODISPERSE POLYSTYRENE LATEXES

To circumvent the problems involved with the use of Fe^{3+} ions in the BI initiator, we investigated the replacement of this ion with Ag^+. The standard reduction potential of the silver ion is 0.800 V, somewhat higher than the 0.770 V for ferric ion. Silver ion reacts readily with bisulfite to form free radicals which initiate polymerization. Our analysis indicates that the net reaction must be

$$Ag^+ + HSO_3^- \xrightarrow{k_1} Ag^0 + \cdot SO_3H \quad ,$$

and must thus be capable of producing all-sulfonate latexes. Because silver bisulfite is highly insoluble, the Ag^+ must be complexed with NH_3. This forms an equilibrium mixture of several species, all with different reduction potentials:

$$Ag^+ + NH_3 \underset{k_{-2}}{\overset{k_2}{\rightleftarrows}} Ag(NH_3)^+$$

$$Ag(NH_3)^+ + NH_3 \underset{k_{-3}}{\overset{k_3}{\rightleftarrows}} Ag(NH_3)_2^+ \quad .$$

$$2Ag^+ + SO_3^{2-} \underset{k_{-4}}{\overset{k_4}{\rightleftarrows}} Ag_2SO_3$$

Studies on the kinetics of this reaction using an Ag^+ specific ion electrode
indicated rather complex behavior, with the silver ion activity rising
initially in the first few minutes and subsequently falling. Generally, it
can be said that the following variables increase the rate of free radical
generation (27):

    (a) increase in temperature - strong dependence

    (b) decrease in $\{NH_3\}/\{Ag^+\}$ mole ratio, within the range of 2/1 to
        16/1. Silver (I) mono- and di-ammines are known to have lower
        standard reduction potentials than the corresponding aquo ions.

    (c) increase in $\{Ag^+\}/\{SO_3^{2-}\}$ mole ratio, within the range of
        5/1 to 1/5.

In the synthesis of latexes by emulsion polymerization, a steady rate of
initiation over the entire reaction time is desired. Optimization of the
above experimental parameters along with the ionic strength, enabled us to
synthesize polystyrene colloids of various particle sizes and surface charge
densities (28). An example of one such study involved a partial Latin-Square
statistical design in four dimensions, the results of which are displayed
in Fig. 2. The polystyrene latexes were all judged to be monodisperse in
that they displayed Higher Order Tyndall Spectrq (HOTS), from which particle

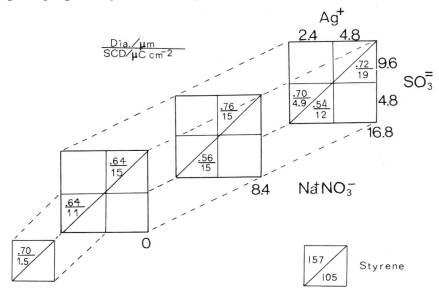

Fig. 2. Effect of experimental variables in BAg initiator system on
particle size and surface charge density (SCD) in polystyrene colloids.
$\{NH_3\}/\{Ag\} = 8/1$, $T = 70°C$. SDSo = 0.21% (by weight/styrene) in one exp't.

sizes were estimated (29). Surface charge densities (SCD) were obtained by
conductometric titration of the ion-exchanged colloids (18). The particle
radii (in micrometers) and SCD's (in microcoubmbs per square centimeter) are
reported in Fig. 2; the quantities of experimental variables are expressed
in millimoles per unit volume of reaction mixture. The surface charge
densities may be converted into areas occupied per group, A_g, as follows:

$SCD/\mu CCM^{-2}$	$A_g/Å^2$
1.5	1060
10	160
20	80

Several conclusions may be reached on the basis of these data:

    (a) changes in ionic strength (NaNO$_3$) have little effect on particle size or SCD;

    (b) an increase in the amount of styrene increases size and decreases surface charge density;

    (c) an increase in amount of initiator (SO$_3^=$ + Ag$^+$)   increases the SCD; and

    (d) addition of small percentages of surface active agent (SDSo - sodium dodecyl sulfonate) dramatically lowers the surface charge density.

Other similar studies have covered a broader range of particle sizes (28). These results generally can be rationalized in terms of the competitive kinetic processes of homogeneous nucleation, radical capture by particles and coagulation (21), although the insensitivity to large changes in ionic strength remain unexplained. What is important here, however, is that we have a means of synthesizing monodisperse polystyrene colloids with controlled particle size and of independently varying surface charge density. The surface ionic groups are all of one kind, hydrolytically and oxidatively stable, and they are strong electrolytes, so that the SCD is independent of pH.

We have chosen this route for synthesis over the apparently simpler one of copolymerizing functional monomers, such as those listed in Table 1, because considerable evidence has been developed that these monomers lead to the formation of water-soluble polyelectrolyte which is difficult to remove and may cause irreproducible results (9, 11, 30).

SYNTHESIS OF BIFUNCTIONAL POLYSTYRENE COLLOIDS

It is possible to synthesize latexes containing both sulfonate and hydroxyl groups at the particle surfaces. This is accomplished by first synthesizing latexes containing both sulfonate and sulfate groups, ion-exchanging all counterions for H$^+$ (the latex is then said to be in the acid form), and subsequently heating to cause the auto-catalyzed hydrolysis of ROSO$_3$ to ROH (31).

The persulfate/bisulfite/iron (PBI) initiating system yields a latex stabilized by a mixture of nonhydrolyzable sulfonic acid groups and hydrolzable sulfate esters (17):

$$HSO_3^- + Fe^{3+} \rightarrow Fe^{2+} + \cdot SO_3H$$

$$S_2O_8^{2} + Fe^{2+} \rightarrow SO_4^{2-} + Fe^{3+} + \cdot OSO_3^- \quad .$$

Since iron is present only in catalytic amounts, it is possible to prepare surfactant-free latexes with this initiator system.

Surfactants stabilize the growing latex particle against coagulation and appear to become irreversibly attached, in part, to the particle surface. Sodium dodecyl sulfate (SDS) will contribute sulfate groups, while the corresponding sulfonate (SDSo) will yield nonhydrolyzable surface acid groups.

The following systems of mixed hydrolyzable and nonhydrolyzable initiators and surfactants were investigated in polystyrene colloids: PBI/no surfactant, PBI/SDS, PBI/SDSo, BI/SDS, and BI/SDSo. The HSO$_3^-$/Fe(III) (BI) initiator system leads solely to sulfonate groups, whereas the S$_2$O$_8^{2-}$/HSO$_3^-$/Fe initiator system gives both sulfate and sulfonate. Grafting of the emulsifier via free radical chain transfer reactions apparently occurs even at room temperature to contribute as many as half of the total number of surface functional groups. So it was found that latexes prepared with the BI/SDSo were stabilized by only sulfonate end-groups. The PBI/SDS system had the largest ratio of sulfate to sulfonate, the PBI/SDSo had a smaller ratio, and PBI/no surfactant had end-groups due solely to the initiator system (31).

Results obtained on four different polystyrene colloids derived from these mixed initiator/surfactant systems are given below:

Latex No. Initiator/Surfactant	Latex 26 PBI/no Surfactant	Latex 24 PBI/SDS	Latex 27 PBI/SDSo	Latex 56 BI/SDS	Latex 65 BI/SDSo
Surface ROH/RSO$_3^-$ mole ratio	3.0	3.7	2.3	1.1	0.0

Further analysis was done on these colloids: the results obtained for
Latex 24, will give some idea of the absolute values of the quantities
involved. It is compared to Latex 65, which had only surface sulfonate
functionality, as shown in Table 5.

TABLE 5. Surface characteristics of selected polystyrene colloids

	Latex 24 (PBI/SDS)	Latex 65 (BI/SDSo)
Particle diameter/nm	44	37
Specific surface area/m^2g^{-1}	134	163
Surface charge density/μCcm^{-2}	1.73	2.73
Surface area per group/$\overset{\circ}{A}^2$		
All groups	200	590
ROH	250	--
RSO$_3^-$	920	590

HETEROGENEOUS CATALYSIS

Chemically functional polystyrene colloids such as those just described have
been investigated as heterogeneous strong-acid catalysts for various
hydrolytic reactions (32). The process of heterogeneous catalysis must involve five fundamental steps: (a) diffusion of "substrate" molecules to the surface, (b) adsorption and
interaction with the catalyst at the surface, (c) reaction within the adsorbed layer, (d)
desorption of the reaction products (to provide a site for subsequent events), and (e)
diffusion of products away from the surface.

As one may see from the data in Table 5, there is a considerable fraction of
the total area on the surface of the latex particles which is "bare" polystyrene and which may provide for the adsorption of hydrophobic molecules
or hydrophobic moieties within molecules. The exact nature of the adsorbed
state is not known because the exact nature of the interface has not yet been
fully explored. Polystyrene is hard and glassy below its glass transition
temperature, Tg, of ∿105°C. This is, of course, a bulk property. Segmental
motions at the interface are presumably greater than in the interior;
segments near a sulfonate group may be drawn further into the aqueous
medium. These effects tend to make the interfacial zone less well defined
than that for, say, a crystalline catalyst. Nevertheless, polystyrene
colloids have been shown to exhibit well characterized adsorption isotherms
for surface active substances (33). Additionally, these colloids will have
particles surrounded by a swarm of counterions whose concentration falls off
exponentially with the radial distance from the interface according to a
Poisson-Boltzmann distribution (34) (Note a). For colloids in the H$^+$ form
this means that the acid concentration can be several molar at the interface
when it is on the order of 10^{-3} molar overall.

Note a. This distribution must break down at small distances from the
interface because the surface ions are discreet, they are spaced
relatively far apart and the surface is non-conducting.

If a substrate molecule, capable of undergoing acid-catalyzed hydrolysis diffuses to the interface, it may be held by adsorption in this zone of high acidity and thus be hydrolyzed much more rapidly than in a homogeneous solution of the same overall acid concentration. The magnitude of the adsorption energy, determined by hydrophobic interactions, will thus influence the rate of the reaction: if there is no net attraction between substrate and particle surface, there will be no concentration effect and no acceleration; if there is too strong an attraction, the products won't desorb and the catalyst becomes "poisoned".

An example of this kind of behavior may be found in the acid-catalyzed hydrolyses of an homologous series of acetate esters (32). An all-sulfonate polystyrene colloid was used as the catalyst. The effects may be expressed in terms of relative rates, i.e. the specific rate constant for the latex-catalyzed reaction (k) relative to that for simple acid catalysis, e.g. that for H_2SO_4 (k_0):

Acetate Ester	k/k_0 at 50°C
Methyl	2.6
n-Butyl	7.7
n-Peutyl	3.7

It appears that the catalytic effect is optimal for a substrate of intermediate hydrophobicity at this particular interface, which had a surface charge density of $4.4 \mu C cm^{-2}$. As an example of the effect of the nature of the catalyst surface, the values of k/k_0 were obtained for a single substrate, n-butyl acetate, on 3 different catalysts. These are shown in Table 6, and indicate that the catalyst with the highest surface acid concentration

TABLE 6. Relative rate constants for hydrolysis of n-butyl acetate

Catalysts				Relative Rate Constants	
Latex No.	Area per Surface Group/$\overset{\circ}{A}^2$		Surface Acid Concentration/$\mu mole \times 10^5 cm^{-2}$	k/k_0	k''/k_0
	RSO_3^-	ROH			
24	925	255	4.60	2.73	5.9
65	585	--	7.26	3.80	5.2
17	363	--	11.7	7.70	6.6

gives the greatest increase in rate. To test this hypothesis we have divided k/k_0 by the relative surface acid concentration to obtain k''/k_0, where k'' is a relative second order rate constant based upon the two-dimensional surface acid concentration. When looked at in this way (last column in Table 6) the differences in rate constants largely disappear, demonstrating that it is the differences in surface acid concentration and not the presence or absence of hydroxyl groups which are responsible for the observed effects. It would be interesting to see how much higher in surface acid concentration one can go before this phenomenon which depends on the hydrophobic effect would be reversed, but such experiments have not yet been done. In any case, it has been shown that these polystyrene colloids can improve overall rates of hydrolysis by almost 8 times over those catalyzed by simple acids such as sulfuric.

We have also observed that in the hydrolysis of sucrose (the "inversion" reaction), catalyzed by polystyrene sulfonic acid colloids, that rate enhancements, k/k_0, of up to about 80% are achievable, and that surface OH groups again seem to have little effect. When compared to the same process catalyzed by ion-exchange resin beads, the latexes were found to be 40 to 60 times more effective, depending on experimental conditions (32). Because the resin bead particles are on the order of 10^{12} times the volume of the latex particles, essentially all of the functional groups (RSO_3H^+) are on the insides of the beads, so that the rather large substrate molecules must diffuse into the cross-linked matrix of the beads in order to react, and the products must diffuse back out (35). On the other hand all of the catalytic groups are on the surface of the latex particles, easily accessible to the substrate molecules in solution.

The reasons for the superiority of the latexes over sulfuric acid probably lie in the fact that there can be some hydrophobic interaction between sucrose molecules and the polystyrene surface in spite of the fact that sucrose is polar and has several -OH groups. In addition, it has been shown that the conjugate anions may participate in the HCl- catalyzed reaction (36). So it is likely that surface RSO_3^- groups may engage in ion-dipole interactions with the sucrose substrate molecules to further enhance the catalytic effect. The k/k_0 values are also about 32% greater than those catalyzed by polystyrene sulfonic acid in homogeneous solution, so that the fact that the catalytically functional groups are at an interface is apparently important.

AUTOCATALYTIC POLYMER HYDROLYSIS

If the polymer comprising the latex particles is itself hydrolyzable and there are surface strong acid groups present, autocatalyzed hydrolysis may occur. For example, polymethyl acrylate (PMA) latexes bearing surface $ROSO_3^-H^+$ and $RSO_3^-H^+$ groups hydrolyzed slowly at rates which were directly proportional to the surface strong acid concentration, and zero order overall (8). When the H^+ ions were exchanged for Na^+, the rates dropped to almost nil. Similar kinetics were observed for latexes of poly methyl methacrylate, and polycyclohexyl-, polybenzyl-, and polynaphthyl- acrylates. Because they can be turned on and off at will by ion-exchange, display constant rates of hydrolysis over long periods of time (more than 6 1/2 weeks at T=90°C, as an example), can have rates regulated at almost any level by variation in the surface acid concentration, and because upon hydrolysis they produce low molecular weight compounds which may diffuse into the surrounding medium, these latexes are promising candidates for controlled release systems. In the case of the acrylates cited above, the corresponding alcohols or phenols are released at constant rates over very long periods of time at room or body temperature. Other hydrolytically cleavable entities may be envisaged. Applications may be found in the fields of drugs, insecticides, corrosion-control, pheromones, and fungicides, among others.

CONCLUSION

From humble beginnings in the fields of synthetic rubber and latex paints, the surface chemistry of polymer colloids is rapidly entering a new era in which the study of chemical reactions at the interface between the polymer particles and the continuous medium is of interest. Methods are being developed for the synthesis of monodisperse colloids with particle sizes anywhere within the colloidal range, and with chemically functional surface groups of various kinds and surface concentrations. This has led, in turn, to the study of the kinetics of various hydrolytic reactions occurring at the interface. Some of these involve the polymer colloids as heterogeneous catalysts in which hydrophobic interactions hold the substrate molecules at the interface, where a relatively high concentration of counterions catalyze the reactions. Ion-dipole interactions with the fixed groups at the interface may also be involved.

Slowly "self-destructing" acrylic latexes have been found which undergo autocatalyzed hydrolysis accompanied by the release of small molecules at constant rates over long periods of time. These suggest potential applications in the burgeoning field of controlled release systems.

Acknowledgements - The author acknowledges with thanks the support of the Délégation Générale à la Recherche Scientifique and the C.N.R.S. Centre de Recherches sur les Macromolécules, Strasbourg during the preparation of this manuscript.

REFERENCES

1. R.M. Fitch, Polyelectrolytes and their applications, A. Rembaum and E. Sélégny, Eds., pp. 51-70, Dordrecht: Reidel Publ. Co. (1975).
2. V.I. Yeliseyeva, Preprints, NATO/ASI on Polymer Colloids, Trondheim (1975).
3. K.L. Hoy, J. Coatings Tech. 51, (651) 28 (1979).
4. R.M. Fitch and R.C. Watson, J. Colloid Interface Sci. 68 (1), 14 (1979).
5. A. Netschey, D.A. Napper and A.E. Alexander, Polym. Letters 7, 829 (1969).
6. F.K. Hansen and J. Ugelstad, J. Polym. Sci., Chem. Ed. 16, 1953 (1978).

7. A. Rembaum, S.P.S. Yen and W. Volksen, Chemtech, March, 1978, 182.
8. R.M. Fitch, C. Gajria and P.F. Tarcha, J. Colloid Interface Sci. 71 (1), 107 (1979).
9. B.W. Greene, J. Colloid Interface Sci. 43 (2), 449 and 462 (1973).
10. M.S.D. Juang and I.M. Krieger, J. Polym. Sci., Chem. Ed. 14, 2089 (1976).
11. R.L. Schild, M.S. ElAasser, G.W. Poehlein and J.W. Vanderhoff, Emulsions, Latices and dispersions, P. Becher and M.N. Yudenfreund, Eds., pp. 99-128, Marcel Dekker, New York (1978).
12. H.J. Wright, J.F. Bremmer, N. Bhimani and R.M. Fitch, U.S. Pat. 3,501,432.
13. H.H. Freedman, J.P. Mason and A.I. Medalia, J. Org. Chem. 23, 76 (1958).
14. B.W. Greene, D.P. Sheetz and T.D. Filer, J. Colloid Interface Sci. 32 (1), 90 (1970).
15. C.M. Samour and M.C. Richards, U.S. Pat. 4,011,259 (1977).
16. H.J. van den Hul and J.W. Vanderhoff, Br. Polym. J. 2, 121 (1970).
17. W.T. McCarvill and R.M. Fitch, J. Colloid Interface Sci. 67 (2), 204 (1978).
18. W.T. McCarvill and R.M. Fitch, J. Colloid Interface Sci. 64 (3), 403 (1978).
19. J. Clarke, G.R. Traut and R.M. Fitch, to be published.
20. D.E. Yates, Nato Advanced Study Institute on Polymer Colloids, Preprints, Univ. of Trondheim - NTH, Trondheim (1975).
21. R.M.Fitch and C.H. Tsai, Polymer Colloids, R.M. Fitch Ed., pp. 103-116, Plenum, New York (1971).
22. (a) H.M. Castrantas and D.G. MacKellar, Polym. Prepr., Amer. Chem. Soc., Div. Polym. Chem. 10 (2), 1381 (1969).
    (b) H.M. Castrantas, P.R. Mucenieks, B. Cohen and D.G. Mackellar, Ger. Offen. 2,002,865 (1970).
23. P.K. Mallya and R.M. Fitch, Polymer Colloids II, R.M. Fitch, Ed., pp. 457-476, Plenum, New York (1980).
24. R.C. Duban, Abstracts of Papers 130th Meeting, Amer. Chem. Soc., Sept. 1956.
25. H.J. van den Hul and J.W. Vanderhoff, Polymer Colloids, R.M. Fitch,Ed., pp. 1-27, Plenum, New York (1971).
26. J. Hearn, M.C. Wilkinson and A.R. Goodall, Adv. Colloid Interface Sci. 14 (2/3), 173 (1981).
27. J. Clarke and R.M. Fitch, unpublished results.
28. G.R. Traut, T.T. Chen, J. Clarke and R.M. Fitch, unpublished results.
29. D. Sinclair and V.K. LaMer, Chem. Rev. 44, 256 (1949).
30. Y. Chonde and I.M. Krieger, J. Colloid Interface Sci. 77 (1), 138 (1980).
31. R.M. Fitch and W. T. McCarvill, J. Colloid Interface Sci. 66 (1), 20 (1978).
32. R.M. Fitch, P.K. Mallya, W.T. McCarvill and R.S. Miller, 27th Int'l. Symp. Macromolecules, Strasbourg, July (1981).
33. P. Connor and R.H. Ottewill, J. Colloid Interface Sci. 37 (3), 642 (1971).
34. E.J.W. Verwey and J.T.G. Overbeek, Theory of the Stability of Lyophobic Colloids, Elsevier, Amsterdam (1948).
35. E.W. Reed and J.S. Dranoff, Ind. and Eng. Chem. Fundamentals, 3 (4) 304, (1964).
36. V.K. Krieble, J. Amer. Chem. Soc., 57, 15 (1935).

M - C*

SOME PROBLEMS OF THE THEORY OF POLYMERANALOGOUS AND INTRA-
MOLECULAR REACTIONS OF MACROMOLECULES

N.A. Platé, O.V. Noah, L.B. Stroganov

M.V. Lomonossov State University of Moscow, Polymer Department, Moscow, USSR

Abstract.- The main results of the theoretical description
of the kinetics and statistics of macromolecular reactions
are reviewed, and some modern problems of the theory of
these reaction (the description of conformational effects,
the solution of reverse problems) are discussed.

The beginning of the development of the general theory of macromolecular
reactions more probably is the beginning of sixties when the first publications concerning the kinetic description of the reactions of functional
groups of polymers with the neighbouring group effect arised. Let's review
what is the situation now, after about 20 years of the research in this
field having in mind that partly these problems were reviewed by us at
Boston Symposium on Macromolecular Chemistry in 1971 /1/.
What are macromolecular reactions? Firstly these are reactions of functional groups of macromolecules with low-molecular reagents, so-called polymeranalogous reactions; secondly these are reactions of functional groups
of one macromolecule with another functional group, intramolecular reactions,
and thirdly these are intermacromolecular reactions. The last ones are of
special interest, and they will not be considered in this lecture. Speaking
about quantitative description of first two types of the reactions we are
thinking about the possibility to calculate their kinetics and statistical
properties of reaction products with the specific effects which are typical
for polymeric reactions and which are determined by the chain-type structure. In general the theory of macromolecular reactions should include the
possibility to estimate the contribution of all these effects (neighboring
group effect, configurational, conformational, supermolecular, concentratio-
nal, electrostatic and so on).However even the detail consideration of
single effect is a rather complicated task. The situation is even more complicated in the cases when the conditions of particular reactions give grounds for several such effects simultaneously, and these effects could be
interralated. Let us consider which specific polymeric effects could be
nowadays quantitatively considered when one deals with a description of macromolecular reactions.

I. POLYMERANALOGOUS REACTIONS

The theory of polymeranalogous reactions includes three main moments: 1) Calculation of the kinetics; 2) Calculation of units sequence distribution in
the intermediate products; 3) Calculation of compositional heterogeneity of
the reaction products.
In the case when the reaction proceeds in the conditions which exclude the
probability for any of polymeric effects, all these three tasks can be solved on a trivial way. The reaction kinetics is described by the first order
equation:

$$P(A) = P(A_o)e^{-kt} \qquad (1)$$

where $P(A)$ is a mole fraction of nonreacted groups, k is a rate constant
of $A \rightarrow B$ substitution. The calculation of probabilities of any units sequences A and B in the chain is extremely simple because the intermediate reaction product at any moment of time is nothing else as a binary copolymer
with Bernullian distribution of units for which the probability to find out
some sequence of units $X_1 X_2 \ldots X_n$ is determined as

$$P(X_1 X_2 \ldots X_n) = \prod_{i=1}^{n} P(X_i) \qquad (2)$$

where X=A or B. Because $P(A) + P(B) = 1$ the solution of kinetic equation
(1) permits to calculate the parameters of units distribution in the chain.
The compositional heterogeneity of reaction products in that case will be
normal with the dispersion D_n which is determined as

$$\lim_{n \to \infty} D_n/n = P(A) P(B) \qquad (3)$$

where n is the length of the macromolecular chain.

1. Neighboring group effect

Kinetics. In the case when polymeranalogous reaction proceeds with the neighboring group effect (for instance hydrolysis and esterification with the
participation of side groups of acrylic polymers, the chlorination and halo-
genation of polyethylene, the quaternization of polyvinylpyridine and so on)
the particular task is nothing else as to find out the expression of mole
fraction of nonreacted units as a function of time and rate constants k_0,
k_1, k_2 for A units which have 0, 1 and 2 reacted neighbors B correspondingly.
The first attempt to solve the kinetic task was done by Fuoss and co. (1960)
who have proposed the scheme of the solution, but were not able to derive
the final kinetic equation /2/. That is while probably the publications of
1962-1965 are usually cited as pioniering works. In these works authors proposed various approaches for the solution of this task which gave however
coinciding results /3/.
The most accurate solution of the kinetic task was done by McQuarrie who had
considered the change in time of the probabilities of two types of the sequences of nonreacted units: j-clusters which correspond to the sequences
from j A units having on both sides units B - $P(BA_jB)$ and j-tuplets, the sequences of A units for which the nature of end units is not determined - A
or B - $P(A_j)$. The main kinetic equation in this case is looked as following:

$$dP(A)/dt = -k_2 P(A_1) + 2(k_2-k_1)P(A_2) + (2k_1-k_0-k_2)P(A_3) \qquad (4)$$

where

$$P(A_j) = e^{-jk_0 t} \exp \left\{ 2(k_0-k_1) \left[t-(1-e^{-k_0 t})/k_0 \right] \right\} \qquad (5)$$

All this is related with the simplest model of polymeranalogous reaction
(irreversible reaction of the first order which proceeds in homogeneous solution at the excess of low-molecular reagent).
In the publications of Silberberg and Simha (1968-1974) and Vainshtein,
Berlin and Entelis (1978) /3/ the model of reversible reaction was considered, and the condition of the equilibrium state was derived:

$$(k_0/k_2')(k_2/k_0') = (k_1/k_1')^2 \qquad (6)$$

where k_0', k_1', k_2' are the rate constants of reverse transformations for
units B having 0, 1 and 2 A neighbors correspondently.
Krishnaswami and Vadav (1976) have considered the case when the reactivity
of the functional groups of macromolecules is dependent on the 4 close
units (two units from each side). The approach which was proposed by Ser-
dyuk (1978) permits to consider the influence of any number of neighboring
units.
Noah and Litmanovich (1977) /4/ taking as an example the reaction of chlorination of polyethylene show the possibility to apply the Keller's approach
for the derivation of the kinetic equation of the reaction which has the
kinetic order equal to 1/2 toward highmolecular reagent.

Units distribution. The quantitative description of the distribution of nonreacted units (it means the calculation of probabilities of the type $P(A_j)$,
$P(A_jB)$ and $P(BA_jB)$ was solved in the same works in which the kinetic
equations were derived /3/. $P(A_j)$ are determined by the relationship (5),
$P(A_jB) = P(A_j) - P(A_{j+1})$, $P(BA_jB) = P(A_j) - 2P(A_{j+1}) + P(A_{j+2})$.

The task of the calculation of parameters of the distribution of reacted
units is essentially different from the viewpoint of irreversibility of the
reaction. The studies carried out by Plate and co. (1971-1974) were devoted

for the solution of this task /3/. In these researches the accurate approach to calculate the parameters of the distribution of A and B units in the chain was proposed, as well as different approximate methods for the calculation of parameters of units distribution. One of such approaches is an approximation by Markov chains of different order and method which was based on the consideration of the kinetics of blocks transformations from already reacted units, so-called "B-approximation".

Another approximate approach to the calculation the parameters of units distribution is a mathematical simulation of polymeranalogous reaction. Firstly this approach was used by Platé and co. (1969) /3/ to calculate the average number of blocks in the chain. Smidsrod and al./5/ (1970) have applied the Monte Carlo method when they have studied the oxidation reaction of amylose by periodate ion. In this work the simulation of the reaction permitted to calculate the rate constants which gave the best coincidence with the experimental data and to calculate also the number-average length of the sequence of nonreacted units. Klesper and al. /3/ have used the Monte Carlo simulation to check the hypothesis about statistical distribution of blocks of reacted and nonreacted units.

Berlin, Vainshtein and Entelis /6/ (1978) having applied the thermodynamic approach have considered the units distribution in the chain of products of reversible polymeranalogous reactions with neighboring group effect.

<u>Composition heterogeneity</u>. Firstly the parameters of composition heterogeneity of the products of polymeranalogous reactions with neighboring group effect were considered by Frensdorf and Ekiner (1967) /3/ for the reaction of chlorination of polyethylene. These authors have applied the Markov approach to describe the statistics of the substitution of hydrogen in the chain. They thought that the probability of the substitution in (n+1)th unit does not depend on the presence of the substituent in (n-1)th unit. Such supposition means that the process can be modelated with Markov chain of the first order.

In 1974 we have estimated the possibility to apply the first order Markov approximation to calculate the composition heterogeneity /3/ when we have compared the results of such approximation with the results of the direct mathematical simulation which we have carried out earlier. It turned out that the modified form of the first order Markov approximation is quite good for the calculation. When this modified form was proposed, it was supposed that the composition heterogeneity is a normal one with a dispersion which is determined through Markovian transitional probabilities $P_{A/A} = P(AA)/P(A)$ and $P_{A/B} = P(BA)/P(B)$ as

$$\lim_{n \to \infty} D_n/n = (1-P_{A/A})P_{A/B}(1-P_{A/B}+P_{A/A})/(1-P_{A/A}+P_{A/B})^3 \qquad (7)$$

where $P(A)$, $P(B)$, $P(AA)$, $P(BA)$ could be found from the solution of accurate kinetic equations (4) and (5).

In other words on building the modified first order Markovian approximation it is considered that at the moment for which all the calculation of the composition heterogeneity is carried out the chain itself is Markovian of the first order, but all its prehistory is described by accurate kinetic equations.

The results of the calculation of the dispersion of composition heterogeneity for various ratios of kinetic rate constants are summarized on the Fig. 1.

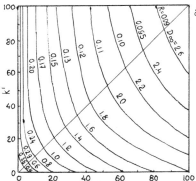 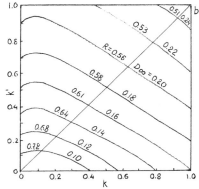

Fig. 1. A dependence of D_∞ and R on $k_0:k_1:k_2$ for accelerating (a) and retarding (b) effects at 50% conversion ($k=k_1/k0$, $k'=k_2/k_0$).

In the frames of above described approximation one can establish also the
immediate relationship between parameters of units distribution and composition heterogeneity. Really it is easy to show that

$$D_\infty = X(4X - R)/R \qquad R = 4X^2/(D_\infty + X) \qquad (8)$$

where $X = P(A) \cdot P(B)$, $R = 2P(AB) = 2[P(A) - P(AA)]$, $D_\infty = \lim_{n \to \infty}(D_n/n)$

In other words the diagrams 1 summarize also the results of the calculation
of the run number R.
There exists also another one approach to calculate the composition heterogeneity of the products of polymeranalogous reactions, and this was proposed by Kutchanov and Brun (1976) /7-9/. These authors have introduced N-me-
ric random vector as a characteristic of statistical properties of polymeric chains of a finite length N. The components of this vector n_j are the
values which are equal to the number of j-clusters in the chain. At the
condition $N \to \infty$ the authors have derived asymptotic formulae which permitted to calculate the composition heterogeneity being described by one-dimensional normal relationship.
For certain polymeranalogous reactions we have succeeded to compare results
of the calculation with the experimental data. For instance, for the quaternization of poly-4-vinylpyridine with benzyl chloride from the kinetical data the ratio of the rate constants $k_0:k_1:k_2 = 1:0.3:0.3$ was calculated, and
the functions of composition heterogeneity based on this ratio of rate
constants were compared with experimental data obtained by GPC method. This
comparison permitted to find out the existence of a certain additional effect beside neighboring group effect which is pronounced at higher degrees
of conversion. Analogously following the rate constants which were found
from the kinetic data the composition heterogeneity of the products of chlorination of polyethylene was calculated and compared with the experimental
data on fractionation. The coincidence of calculated and experimental values has confirmed once more the proposed mechanism of the reaction /10/.
Thus the mathematical approaches for the polymeranalogous reactions which
proceed with the neighboring group effect is now a rather developped field
of the theory of macromolecular reactions and is already successfully used
when the particular chemical reactions of polymers are studied.

2. Configurational effect.

The consideration of configurational effect in polymeranalogous reactions
of vinyl polymers means that the reactivity of nonreacted unit A is depending on the fact at which center of three possible triads - iso, hetero or
syndio - this unit is placed. In the case when configurational effect is
not followed with the neighboring group effect one can consider that the
process includes three parallel reactions with rate constants k^i, k^h and k^s.
The calculation of the kinetics for such reactions is rather simple. But in
the case when you have both effects (which is more probable situation) the
development of the theoretical apparatus for the kinetics and statistics
description means that you should move from the task which includes three
parameters (the rate constants k_0, k_1, k_2) to the task with ten parameters
because in that case the nonreacted unit A can be placed in the center of
one among ten various triads and consequently be transformed to B unit with
one among ten possible rate constants.
The solution of such problem generally speaking is not very different from
the above considered case, but its application to the particular chemical
reactions seems not very probable, which is related with experimental difficulties to determine rate constants. Even in the case of stereoregular
samples, for which you should determine from the experiment only three rate
constants, this problem is rather complicated and needs a special methodic
approach (we shall discuss this question more in details later on). For
atactic polymers the solution of this problem needs a synthesis of model
iso- and syndio-tactic polymers with the high degree of stereoregularity.
It is obvious that the calculation of the kinetics and statistics of the
reactions in atactic chains will be simplified essentially when some of
these ten constants could be equal one to another or certain other limitations will be imposed on these rate constants. Pis'men (1972) /3/ has calculated the kinetics of polymeranalogous reaction in atactic chains for a
particular case of ratio of rate constants, when he had differentiated
three types of "n-clusters" (i.e. sequences of n nonreacted units having
on both sides reacted units) - the open, semiopen and closed clusters. This
was done having in mind whether both end nonreacted units are in iso-posi-
tion toward already reacted neighbors (this gives you the open cluster), is
there one end unit which means semiopen cluster, and is there no one which
corresponds to closed cluster.

II. INTRAMOLECULAR REACTIONS

Intramolecular reactions, i.e. reactions of functional groups of the same macromolecules, which proceed with the participation of low-molecular reagent or without it, could be divided on three classes. The first class includes the reactions which proceed predominantly between <u>neighboring</u> functional groups. These are for instance the elimination of chlorine by zinc from polyvinyl chloride, the aldol condensation of polymethylvinyl ketone and so on /3/. The kinetics and statistics of such reactions are described almost by the same mathematical apparatus as in the case of polymeranalogous reactions. The second class of intramolecular reactions includes the reactions of <u>intramolecular catalysis.</u> These reactions are characterized with the fact that their proceeding is a function of the conformation of a macromolecule, because the interacting units could be very far situated along the chain one from another, and the probability of their get together is determined by the flexibility of the polymeric chain. Beside that this type of reactions is characterized with the fact that independently on the relationship between the reactivity of groups and conformation of the chain this particular conformation practically is not changed in the course of the reaction. The third class of intramolecular reactions includes the reactions of irreversible intramolecular <u>cross-linking.</u> These reactions also proceed as a function of chain conformation, but each act of cross-linking leads to the change of the conformation of macromolecular coil.

1. Reactions of neighboring groups

The first theoretical consideration of such reactions was given by Flory in 1939, who has calculated the average number of groups which remain nonreacted at the end of the reaction, and which was found to be equal to n/e^2, where n is the length of the chain, going to 0.1353 at $n \to \infty$ /11/. Later on some authors gave attention to the kinetics of intramolecular reactions /3/. Cohen and Reiss (1963) have applied the multiplets method to solve the kinetic problem and have obtained the following expression:

$$P(t) = \exp\left[-2(1-e^{-kt})\right] \tag{9}$$

where k is the rate constant of intramolecular reaction. When $t \to \infty$, $P \to 1/e^2$ which coincides with the classical result of Flory. McQuiston and Lichtman proposed that paired interactions in the polymer chain could be modelated by the process of random throwing of a "dumb-bell" consisting of two units on the one-dimensional lattice of N cells and derived the same kinetic equation.
Boucher (1972) has calculated the kinetics of intramolecular reaction with neighboring group effect.
Gonzalez and Hemmer (1976-1977) have considered the particular case of intramolecular reaction, when the reaction is stopped at the certain degree of conversion, and then the chemical bonds which block the part of nonreacted groups are destroyed and then the product is subjected to subsequent reaction.

2. Intramolecular catalysis.

The theoretical consideration of the reactions of that type was done by Morawetz and co. (1966-1973) /3/. The simplest type of the reaction between functional groups of the same macromolecules are the reactions of cyclization, i.e. the reactions of the groups which are on the ends of the chain. The interaction between such groups is possible only if they are close one to another. The probability of such proximity can be calculated using the end-to-end distances distribution function. The same approach can be applied to the case of interaction of catalytic and reactive groups being placed in any n-th and j-th units of a polymeric chain. In studies of Morawetz the mathematical simulation of the kinetics of intramolecular reactions was also done for polymers which contain a certain fraction ω of catalytically active substituents.
To calculate the probability of get together for reactive and catalytically active groups Sisido /12-21/ has applied the thermodynamical approach and calculated the thermodynamical functions by Monte Carlo method for the chains on the lattice having certain energetic limitations. The absolute values of rate constants, conformational energy and activation entropy were obtained.
Later on Sisido and co. have carried out a lot of experimental studies of intramolecular reactions: hydrolysis of end p-nitrophenyl ester group catalyzed by end pyridyl group in saccharide /15/ and oxyethylene /20/ chain,

the formation of charge transfer complex between end p-dimethylaminoanilide and 3,5-dinitrobenzoyl groups /16/, the formation of disulfide bonds during the oxidation of end sulfhydrile groups /19/. In these researches the dependences of rate constants on the length of the chain and its flexibility were obtained, certain thermodynamic parameters were calculated, and the comparison of experimental data with the theoretical calculation and Monte Carlo simulation results was done.

Kozlov /22/ has considered the model of the end groups reaction of one macromolecule including the consideration of excluded volume effect and selective adsorption. The approach which was proposed by Kozlov permitted to treat the experimental results of Morawetz and gave the values of rate constants which turned out to be closer to the experimental ones than the results of the theoretical calculations of the same authors.

3. Intramolecular cross-linking

Dimensions of cross-linked macromolecules. The first publication devoted to the calculation of dimensions of macromolecular coils which contain intramolecular cross-linkages belong to Zimm and Stockmayer who in 1949 have calculated the mean square radius of gyration for the model of a freely jointed cyclic chain. It was shown that

$$R_c^2 = Nb^2/12 \qquad\qquad\qquad (10)$$

while for the linear chain

$$R_l^2 = Nb^2/6 \qquad\qquad\qquad (11)$$

here N is the number of the segments in the chain, b is the segment length. To apply the approach which was proposed by Zimm and Stockmayer to the chains containing more than one cycle (or more than one cross-link) is impossible due to the extremely high increase of the number of possible topological structures and essential complication of the calculation algorithm for each of them. Thus for the chains with two cross-links three different structures are possible, with three cross-links - eight different structures and so on (Fig. 2).

Edwards and co. (1973), Gordon et al. (1977), Ross-Murphy (1978) /3/ have proposed several variations of approximate calculation for chains with m cross-links. Another approach to solve this problem is the simulation of the process by Monte Carlo procedure. Bonetskaya, El'yashevich et al. (1975) have carried out the mathematical simulation of the so-called "momentary cross-linking", when during the time which is necessary for the total completion of the reaction the conformation of the chain has no time to be changed. This corresponds to the case of "fast" reactions for which the time of elementary act is comparable with the time of conformation change. It can be proposed that for usual, so-called "slow" chemical reactions for which the elementary act time is several orders higher than the time of rechange of conformation the effect of densing of macromolecular coil in the course of cross-linking reaction should be expressed in a higher degree. Exactly this case was considered in the works of Platé, Noah. Romantzova et al /23-25/ in which the intramolecular cross-linking reaction was studied by Monte Carlo procedure.

In these researches the chains were simulated on the simple cubic and body-centered lattices with the permission of self-intersection in which cross-linking itself took place. The step-by-step increase of the number of cross-links was followed in each individual chain, and then the parameters obtained were averaged to the chains ensemble with the certain number of cross-links.

Figure 3 represents the relative average dimensions of macromolecular coils as the functions of the degree of cross-linking. It could be seen that the average dimensions are decreased following the degree of cross-linking, and with the increase of the chain length this effect is more pronounced.

Kinetics of the intramolecular cross-linking. Generally speaking the rate of the intramolecular reaction which corresponds to the above described model (relatively slow chemical reaction between any two units of polymeric chain which are in the immediate vicinity one from another due to the flexibility of macromolecular chain) should be proportional to the number of the reactive contacts in partially cross-linked macromolecular coil - Z_j (j is a number of cross-links). This means that the problem to calculate the kinetics of intramolecular cross-linking is nothing else as to find out the equilibrium average values $\bar{Z}_j$. However the analytical calculation of $\bar{Z}_j$ is impossible up today due to the same difficulties as in calculation of

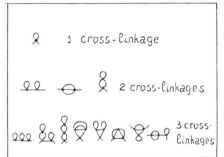

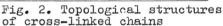

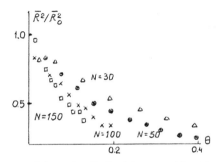

Fig. 2. Topological structures
of cross-linked chains

Fig. 3. Relative coil dimensions
versus the degree of cross-linking

the dimensions of the cross-linked chains. We have studied this problem
using again Monte Carlo procedure.
Figure 4 shows the change of the number of reactive contacts with the increase of the number of cross-links. The increase of the number of contacts
in the course of the cross-linking process is a function of the decrease of
the effective volume of the macromolecular coil, and the decrease of it at
higher degrees of cross-linking is related with the fall of the concentration of free active groups.

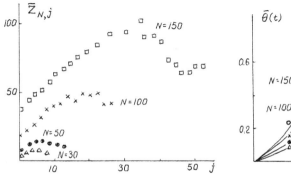

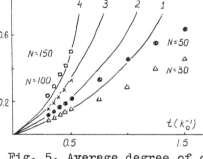

Fig. 4. Average number of reactive contacts versus the number
of cross-links.

Fig. 5. Average degree of cross-linking versus time: curves, linear
approximation; points, simulation.

As it could be seen from Fig. 4 at the initial stage of the reaction the
dependence of the number of contacts on the number of cross-links could be
represented in a linear form:

$$\bar{Z}_j = A_j + B \tag{12}$$

This presentation permits to describe the average number of cross-links by
following way:

$$\bar{n} = B/A\ (e^{k_o At} - 1) \tag{13}$$

The calculation of the kinetics of the intramolecular cross-linking in the
linear approximation with the coefficients A and B which have been found
from Fig. 4 gives a good coincidence with results of the computer experiment
(Fig. 5). It could be seen also from Fig. 5 that intramolecular cross-linking is an autoaccelerated reaction, and that the initial rate and the
degree of the autoacceleration are increased with an increase of the length
of the chain. Besides that the results obtained permit to conclude the existence of an uniform relationship between the kinetics of the reaction and
the equilibrium properties of partially cross-linked chains.

Composition heterogeneity of cross-linking products. The results of the
computer experiment which was carried out by us permitted to calculate the
distribution of the reaction products as a function of the degree of cross-
linking.
At the Fig. 6 the results of the real dispersion of the distribution of the
chains following the number of the cross-links obtained in the computer experiment are plotted. On the same Figure one can see the relationships for

Poisson dispersion and dispersion of the linear approximation. It can be seen that the real distribution is essentially wider than Poisson one, and its width is going up with the increase of the chain length. In the initial stage the dispersion is described well by the linear approximation. For short chains which are subjected to a reaction lasting some time the distribution becomes narrower due to the accumulation of chains with many cross-links (close to the maximum value) and the dispersion tends to a Poisson one.

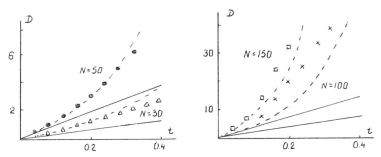

Fig. 6. Dispersion of cross-links number distribution versus time; solid curves, Poisson distribution; broken curves, linear approximation; points, Monte Carlo simulation.

Concluding the consideration of the reactions of intramolecular cross-linking it should be noted that this field is rather poorly investigated in the theoretical aspect as well as in the experimental one. The results of very few experimental works /26-28/ give no grounds unfortunately for the comparison with the results of theoretical predictions. Let's hope that the intramolecular reactions will bring attention of the researchers in coming future and this very interesting and important field of polymer chemistry will be developed more intensively.

As concerned with the experimental study of polymeranalogous reactions one should stress out that now there exists a lot of systems with quantitative data. In the Table 1 which is given below we have summarized the known and published examples of polymeranalogous reactions proceeding with the neighboring group effect for which from the experiment the kinetical rate constants k_0, k_1, k_2 were determined.

However a very important question arises whether the values of these constants are solid ones and could they really serve as a base of the theory of polymeranalogous reactions at the same way as the copolymerization constants are in the Mayo-Lewis theory.

III. REVERSE PROBLEMS OF THE THEORY OF MACROMOLECULAR REACTIONS

The problem of the evaluation of the individual rate constants from the experimental data is not so simple as the calculation from the experimental data of copolymerization constants. In the case of kinetical constants of polymeranalogous reactions the principal problem of the ambiguity of the evaluation of constants arises. The reality of this problem is illustrated by Fig. 7. The kinetical curve for the rate constants k_0 = 10^{-3}, k_1 = 10^{-2}, k_2 = 10^{-1} min^{-1} with the accuracy better than 10^{-5} coincides with the kinetical curve for the constants $9.8 \cdot 10^{-4}$, $1.1 \cdot 10^{-2}$, $1.7 \cdot 10^{-2}$ min^{-1}. To detect experimentally two such sets of the rate constants practically is impossible because the accuracy of the best kinetic studies is about 10^{-2}, but the difference between curves is 10^{-5}. If the method of the evaluation of individual rate constants from the experimental data analogous to the case of Fig. 7 is correctly organized, it should detect both sets of rate constants, i.e. the method should take into account the possible ambiguity of estimation.

It is clear that the development of the unique well grounded method of the evaluation of the individual rate constants which is nothing else as reverse problem of the neighboring group effect is one of the key problems in this field.

To solve this reverse task two different groups of methods exist: linear differential and integral methods.

TABLE 1. Individual rate constants of polymeranalogous reactions

Reaction	Medium	T°	k_0	k_1	k_2	$k_0:k_1:k_2$	Reference
Hydrolysis iso of PMMA	0.2 M KOH	145°	90*	35*	35*	1:0.4:0.4	/29/
syndio	0.2 M KOH	145°	5.8*	1.2*	0.3*	1:0.2:0.05	/29/
syndio	0.05 M KOH	145°	1.9*	1.3*	1.3*	1:0.7:0.7	/29/
syndio	0.05 M KOH	145°				1:0.5:0.25	/30/
iso	0.05 M KOH	145°	5.3*	42*	530*	1:8:100	/29/
syndio	alkali	145°	0.83**	0.42**	0.14**	1:0.5:0.17	/31/
syndio	Py-H$_2$O	145°	1.1*	2.8*	3.7*	1:2.5:3.4	/32/
Hydrolysis of poly-diphe-nylmethyl-methacry-late iso	Py-H$_2$O	145°	0.3*	7*	33*	1:20:100	/29/
syndio			0.5*	0.5*	0.5*	1:1:1	/29/
Hydrolysis of poly-phenyl-methacry-late iso	Py-H$_2$O	145°	6*	240*	6000	1:40:1000	/29/
iso	dioxane buffer	80°	0.15*8.3*	15*		1:55:100	/33/
iso		100°	2*	36*	130*	1:18:65	/33/
radic		80°	5*	10*	50*	1:2:10	/33/
Chlorination of PE (1 order)	chlorben-zene	50°	6.1**	2.1**	0.48**	1:0.35:0.08	/34/
of PE (1/2 order)						1:0.6:0.36	/35/
of n-hexadecane			1.8***	0.68***	0.19***	1:0.38:0.11	/36/
of chlorocyclooctane	CCl$_4$	50°	0.7***	0.245***	0.056***	1:0.35:0.08	/36/
of cyclododecane			0.7***	0.3***	0.13***	1:0.43:0.18	/36/
of cyclooctacozane			0.8***	0.34***	0.14***	1:0.43:0.18	/36/
			1.0***	0.43***	0.18***	1:0.43:0.18	/36/
Quaternization of Poly-4-vinyl-pyridine	nitrome-thane	60°	5.8**	1.7**	1.7**	1:0.3:0.3	/10/
	metha-nol	70°				1:1:0.1	/37/
of Poly-2-methyl--5-vinylpyridine		70°				1:0.6:0.1	/37/
Epoxydation of polyiso-prene	CCl$_4$	25°				1:0.6:0.3	/38/
	benzene	25°				1:0.67:0.42	/38/
Reaction of polymethacryloyl chloride with amines	CH$_2$Cl$_2$--CHCl$_3$					1:0.1:0	/39/
Nucleophylic sub-stitution on PMMA (RCH$_2$Li)						1:0.2:0	/40/
Esterification of PMMA	conc. H$_2$SO$_4$					1:0.1:0	/41/
Reaction of PMMA with methylsul-fonylmethylli-thium iso	DMSO--benze-ne	25°	16.5**	0.9**	0	1:0.055:0	/42/
syndio			81**	81**	0.23**	1:1:0.003	/42/

* const (min^{-1}) x 10^4 *** const (l$^{1/2}$mol$^{-1/2}$sec^{-1}) x 10^3

** const (l mol^{-1}sec^{-1}) x 10^4

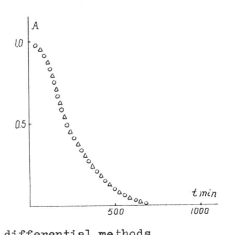

Fig. 7. Kinetics of polymeranalogous reaction for $k_0=10^{-3}$, $k_1=10^{-2}$, $k_2=10^{-1}$ ($\circ$) and $k_0=9.8 \cdot 10^{-4}$, $k_1=1.1 \cdot 10^{-2}$, $k_2=1.7 \cdot 10^{-2}$ min^{-1} ($\triangle$).

Linear differential methods

The main idea of linear differential methods is the search of the values of individual rate constants which are in the best coincidence with the direct differential equations of the model. Because of the fact that the differential equations of the model are linear towards k_0, k_1, k_2 very simple relationships could be derive to evaluate the rate constants. The necessary values of the derivatives one can obtain by the numerical differentiation of corresponding curves.

The simplest illustration of such approach is the method of polymeric models /29/. If initial copolymers can be obtained in conditions which permit to measure or calculate the fractions of different triads $P(AAA)$, $P(AAB)$, $P(BAB)$ after graphical differentiation of the equation (14)

$$-dP(A)/dt = k_0 P(AAA) + k_1 P(AAB) + k_2 P(BAB) \qquad (14)$$

it could be transformed into the system of linear toward k_0, k_1, k_2 algebraic equations

$$\left[-dP(A)/dt\right]_i = k_0 P(AAA)_i + k_1 P(AAB)_i + k_2 P(BAB)_i \qquad (15)$$

($i = 1,2,3,...$) the solution of which is well known.

Doing this way one can obtain only rough values of rate constants. Really the concentrations of triads are known only in very seldom cases, and the operation of graphical differentiation brings an essential non-controlled error. If the experiment itself is the determination and measurement of time dependences of the triads concentrations the fraction of nonreacted units $P(A)$ can be calculated from the data about triads /43/, and after numerical differentiation $dP(A)/dt$ in each moment of time one can come to the excessive system of algebraic equations of the type (15) which contains two equations in each point of time. The total amount of the information here is much more than in the first case which improved the accuracy of rate constant estimation, but the application of the numerical differentiation decreases the valuability of this method.

Klesper /31,44/ has proposed the similar method to evaluate the absolute (k_0, k_1, k_2) and relative (k_0', k_1', k_2') values of the rate constants on the data about triad distribution of units.

The relative rate constants are determined as

$$k_0' = k_0/(k_0 + k_1 + k_2)$$
$$k_1' = k_1/(k_0 + k_1 + k_2) \qquad (16)$$
$$k_2' = k_2/(k_0 + k_1 + k_2)$$

The differential equations of the model with the neighboring group effect in that case could be written as the time dependences or conversion dependences of diads concentrations. For AA diad, for example:

$$dP(AA)/dt = -2k_0 P(AAA) - 2 k_1 (P(AAB) \qquad (17)$$

$$dP(AA)/dP(A) = \left[2k_0 P(AAA)+2k_1 P(AAB)\right] / \left[k_0 P(AAA)+2k_1 P(AAB)+ \atop + k_2 P(BAB)\right] \qquad (18)$$

After numerical differentiation of these dependences obtained from experimental data on triads distribution with the aid of well-known diad-triad relations the equations (17) and (18) are transformed into excessive system of algebraic equations being linear toward the rate constants. This system is solved by graphical intersections method analogous to the well known method of Mayo-Lewis for determination of the copolymerization constants. The utilization of such approach gave the possibility to Klesper and co. to study quantitatively the acidic and alkali hydrolysis of syndiotactic PMMA. The high accuracy of the experiment together with the big total volume of information permitted to authors to get by differential method the solid results. The advantage of differential methods is their simplicity for the realization. Unfortunately nobody has studied yet the limits of the application of the differential approach. We think that mainly they should be used for the calculation of initial approximations for more accurate methods.

Integral methods

The integral methods are the search of such set of individual constants which gives the possibility for the best correspondence of the model with experimentally measured integral characteristics of the process: time dependence of the fraction of nonreacted units, time dependence of the concentrations of diads, triads, tetrades, dependence of the concentrations of diads, triads and tetrades on the composition of the copolymer product and so on. Harwood /45/ was first one who used the integral methods for the evaluation of the relative rate constants k_0', k_1', k_2' from the experimental data on the dependence of the concentration of triads on the fraction of nonreacted units. As a criterium for the confirmity of the model with the experiment the sum of the absolute values of differences in model and experimentally measured concentrations of triads was chosen.

$$\sigma_{mod}(k') = \sum_{i=1}^{m}\{|P(AAA)_i^m \ P(AAA)_i^e| + |P(AAB)_i^m - P(AAB)_i^e| \qquad (19)$$

$$+ |P(BAB)_i^m - P(BAB)_i^e| + |P(ABA)_i^m - P(ABA)_i^e| + |P(BBA)_i^m - P(BBA)_i^e|$$

$$+ |P(BBB)_i^m - P(BBB)_i^e|\}$$

which were calculated using Monte Carlo procedure.
The minimum of σ_{mod} was searched by consequent calculation of the grid with the constant step in very suitable for this case triangular coordinates (Fig. 8). After the calculation of σ_{mod} over all the grid the procedure was repeated on the grid with lesser step built near the point corresponding to the minimal value of σ_{mod} up to getting the reasonable accuracy of constants. The method proposed by Harwood has at the same time the simplicity of the realization and uniformity of approach, but has also some disadvantages. For instance the usage of the Monte Carlo procedure in the internal block of the minimization program results in the serious difficulties in the evaluation of errors. The choice of the sum of modules as a criterium of the proximity of experimental and model dependences is not the best one from the viewpoint of mathematical statistics. This method practically can not be applied, when one try to treat the experimental data on polymeranalogous transformations of copolymers, because in that case each realization of the calculation should be preaveraged using rather big sampling of the initial copolymers with the pre-known distribution of units. In the method proposed by Bauer /46/ the data about the concentrations of triads with central non-reacted units are used. The set of the constants is determined for one point of the time in the way that values of the triads AAA, AAB, BAB concentrations obtained in computer by the numeric calculation of differential equations were in the coincidence with experimentally measured values. But this method is also not the best one because about one half of the experimentally available information containing in the data on B-centered triads distribution is rejected. This method is not good for the treatment of the big volume of the experimental information and can be used only for very rough preliminary evaluation of relative values of rate constants.
In some of ours previous publications /3,29/ we used the method of the evaluation of constants in which the sum of squares of deviations of experimental and model characteristics (triads and conversions) was used as a criterium. In these works we did not study the problem of the unambiguity and accuracy of the constants evaluation.

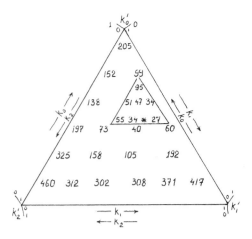

Fig. 8. Triangular diagram; digits, the values of G mod in corresponding points, * - the point of the minimum of G mod, k_0', k_1', k_2' are the relative rate constants.

So it looks like there is no one sufficiently argumented approach to evaluate the individual rate constants from the experiment. There exists some individual methods, but no one could be considered as a satisfactory one. The principal questions related with the unambiguity of the evaluation of constants, with the accuracy, with the correlation characteristics, with the limitation of the applicability of the methods are not studied. During last 2-3 years we have tried to develop a rather general approach for the solution of the reverse problem of the theory of the neighboring group effect which can meet all the above mentioned requests, and to realize it as a rather universal pack of computer programs.

The approach with the minimization of the sum of squares of deviations - SSD($\bar{k}$)

This approach includes the search of a such set of constants k minimizing the sum of squares of deviations between the experimentally measured integral characteristics of the process $f^e(t)$ and model characteristics calculated with a computer following methods which were described in the first part of this lecture /3/:

$$\min_{\bar{k}} SSD(\bar{k}) \; ; \qquad SSD(\bar{k}) = \sum_i \left[f_i^e - f^m(t_i, \bar{k}) \right]^2 \qquad (20)$$

The choice of the sum of squares of deviations as a criterium firstly gives the possibility to use well developped apparatus of the least squares method to calculate errors and correlation characteristics of the constants evaluation /47/, and secondly values themselves obtained have some advantages from the statistical viewpoint.

The pack of computer programs which embraces the reverse problems of the theory of the neighboring groups effect includes a very broad spectrum of problems (Fig. 9), which reflects a difference in the initial polymeric products as well as difference in experimental methods of observation and different realization of the particular method of the evaluation of the rate constants.

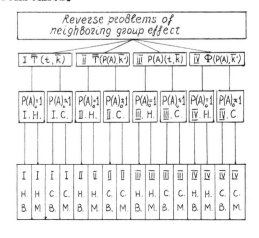

Fig. 9. Classification of reverse problems:
H - initial homopolymer,
C - initial copolymer,
B - problems for big computer,
M - problems for mini-computer.

The reverse problem in fact contains four independent tasks in correspondence with the experimentally measured characteristics of the reaction:
1) The time dependences of the triads concentrations (the longer sequences for the constant evaluation are yet not used) are experimentally measured – $f^e = \bar{T}(t,\bar{k})$; $\bar{T} = \{$ P(AAA), P(AAB), P(BAB), P(ABA), P(BBA), P(BBB)$\}$. 2) The concentrations of triads of units are measured as functions of the fraction of nonreacted units – $f^e = \bar{T}(P(A),\bar{k}')$, $\bar{k}'$ being the relative rate constants determined by Eq. (16). 3) The time dependence of the fraction of nonreacted units is measured – $f^e = P(A)(t,\bar{k})$. 4) The integral function of the composition distribution is analyzed – $f^e = \Phi(P(A),k')$.
For each of these four main tasks one should have two variants of the solution depending on the fact whether the initial product is a homopolymer (H) or a copolymer with the known units distribution (C).
The method of the solution of the reverse problem should have a very high accuracy absolutely necessary when one wants to study the unambiguity of the evaluation of rate constants. On the other hand this method should be simple for the realization, so that it could be used by chemists in their practical work. We develop two different methods for the solution of these tasks: 1) the etalonic method which is oriented on a big computer machine and on the most accurate solution of the direct task of the neighboring groups effect (B); 2) the approximate method which is oriented on a minicomputer and approximate solution of the direct task (M). The etalonic method should serve for the study of the main methodical questions and for the determination of the limits of the applicability of the approximate method, while the approximate method could serve for the practical evaluation of rate constants from the experimental data.

Etalonic method

The etalonic method includes first of all the algorithm which realizes the accurate solution of the direct task $f^m(t,k)$ (Fig. 10,1) as well as the algorithmes of the calculation and minimization of SSD($\bar{k}$) (Fig. 10, 2,3). The calculation of errors of constants evaluation and correlation characteristics in a form of dispersion matrix $\hat{D}\bar{k}$ is accomplished in the block of the statistical treatment of data (Fig. 10, 4).
The main idea used by us studying unambiguity and accuracy of the method is the following: if we introduce into the minimization algorithm the experimental dependence $f^e(t)$ exactly corresponding to the set of rate constants k and obtain the estimate $\bar{k}*$ exactly coinciding with the $\bar{k}$ set independently on the initial approximation k^o we can conclude that k corresponds to the global minimum, and the method is unambiguous in the studied range. For the case which is shown on the Fig. 7 the unambiguity is not fullfilled, and for different $\bar{k}^o$ the algorithm will lead to two different minima. For studying the influence of the experimental error S^2 on the error and correlations of constants estimations Dk* one should introduce into algorithm the experimental dependence $f^e(t)$ corresponding to the constants set $\bar{k}$ and the random experimental error characterized by dispersion S^2. As the dependences with such "dozed" error could not be obtained we simulate the "experimental" dependences $f^{me}(t,\bar{k},S^2)$ with the computer. To obtain the model of the experimental dependence one solves the direct task, and obtained dependence $f^m(t,\bar{k})$ is summed in each point with the model random error being characterized by the O-th expectation and dispersion S^2 calculated with a generator of normally distributed pseudorandom numbers.

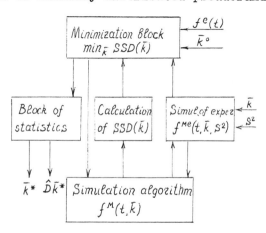

Fig. 10. Block-scheme of the reverse problems algorithm; $\bar{k}$ is the set of constants for the simulation of the experiment, $\bar{k}^o$ is the initial approximation, $\bar{k}*$ are the estimates of constants, $\hat{D}\bar{k}*$ is the dispersion matrix of estimates of constants.

Reverse task of triads distribution

The most detailed information about polymeranalogous reaction is usually obtained when one is studying the triads distribution in the copolymer as a function of the time of the process or as a function of the composition of the copolymer intermediate product (Fig. 9, II) (Fig. 9, I). We have developed the programs for the etalonic methods of solution of both tasks for the cases when initial product is homopolymer (Fig. 9, H) and when it is copolymer with the known units distribution (Fig. 9, C). For all studied range from strong retardation (1:0.1:0.01) to strong acceleration (1:0:100) the reverse problem of triads distribution is unambiguous. For the case which is illustrated by the Fig. 7 the triads distribution gives grounds for detecting and differentiating the sets of constants 10^{-3}, 10^{-2}, 10^{-1} min^{-1} and $9.8 \cdot 10^{-4}$, $1.1 \cdot 10^{-2}$, $1.7 \cdot 10^{-2}$ min^{-1}.

The errors of this method of evaluation are rather essential, and even in favorable conditions they are 20-30% for 10-15 experimental points when the accuracy of the experiment is 3-5%. What does it mean - favorable conditions? This means autoretardation or very slight acceleration in the rather wide range of degrees of conversion. When the acceleration is rather big and the conversion is less than 70% the error in the estimation of k_2 is increased dramatically and can reach 500% and even more with the same accuracy of the experiment 3-5%.

This approach permits to look on the preparation of the experimental work from the absolutely new viewpoint. Really if you imagine even approximately the conditions of your experiment and its accuracy you can calculate and predict the model experimental relationship f^{me} (t, k, S^2), and solve the reverse task. Having this information you can see in advance what will be an accuracy in your estimation of rate constants values.

Reverse task of the kinetics of the process

The study of the kinetics is the most universal and simplest way to have an information about the neighboring groups effect. In the same time the equations which describe the kinetics of the process are formed that way that practically for any set of constants there exists another one which corresponds almost to the same kinetic behaviour. This leads to the appearance of several minima of the $SSD(\bar{k})$ function.

As it is very difficult to study the structure of minima of the complicated function of three variables one can try to go to lesser number of variables by introducing the relationship between k0, k1, k2. As such relationship one can use the dependence of the area below the kinetical curve J on the values of constants which we have obtained in analytical form. Using this approach and relative rate constants k_0', k_1', k_2' (16) one can consider instead of 4-dimensional $SSD(\bar{k})$ the 3-dimensional $SSD(\bar{k}')$ at J=const in a form of level curves on the triangular diagram (Fig. 8). The study of the structure of $SSD(\bar{k}')$ minima shows that for the autoretardation case the local minima are usually in the region of nonmonotonous values of rate constants. For instance the treatment of 20 points of exact kinetical curve for the constants: 0.4, 0.2, 0.1 min^{-1} gives beside the global minimum corresponding to this set of constants the second local minimum corresponding to the set: 0.58, 0.042, 0.29 min^{-1} and having almost the same depth (Fig. 11a, Fig. 7). Introduction of the model experimental error with the dispersion $S^2=2.5 \cdot 10^{-3}$ and $6.5 \cdot 10^{-3}$ corresponding approximately to the experimental accuracy 5 and 10% is illustrated by Fig. 11b and Fig. 11c. It can be seen that at the error being 10% both the minima - global and local are in the region of nonmonotonous values of rate constants. Thus to get from the kinetic data more or less valuable estimates of rate constants one should have a very big experimental set of points which should be measured with the accuracy $\sim 5\%$. If such high accurate experimental data are available one can calculate from the kinetics the rate constants which are almost of the same accuracy as calculated from the triads.

The accuracy of the rate constants estimation is illustrated in Table 2 which summarizes the results for individual constants $(\bar{k})$ of hydrolysis of syndiotactic PMMA and their mean-square errors obtained by various methods from the experimental data of /31/. The linear differential method of Klesper gives the deviations from the estimates corresponding to the minimum of SSD for triads which significantly exceeds the 99% confidence range. The estimates of the meansquare errors are 5-fold higher. The triads distribution is related with the copolymer composition by the simple relationship /43/ and the kinetics of the process can be calculated on the data about triads. The treatment of the kinetics by the minimization of SSD results in two minima of approximately the same depth. Neither accuracy of the estimation nor numerical values of constants in the global minimum do not essentially differ from the data obtained by the treatment of triads distribution, but

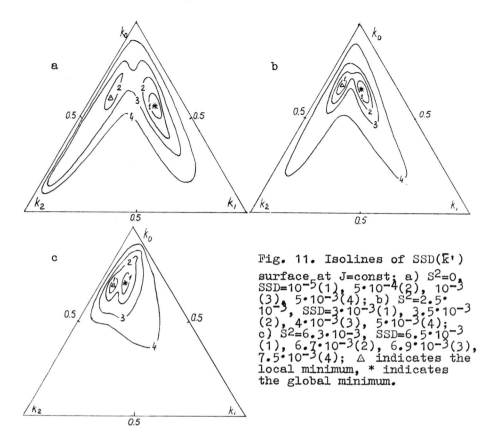

Fig. 11. Isolines of SSD($\bar{k}'$) surface at J=const; a) $S^2=0$, SSD=10^{-5}(1), $5 \cdot 10^{-4}$(2), 10^{-3} (3), $5 \cdot 10^{-3}$(4); b) $S^2=2.5 \cdot 10^{-3}$, SSD=$3 \cdot 10^{-3}$(1), $3.5 \cdot 10^{-3}$ (2), $4 \cdot 10^{-3}$(3), $5 \cdot 10^{-3}$(4); c) $S^2=6.3 \cdot 10^{-3}$, SSD=$6.5 \cdot 10^{-3}$ (1), $6.7 \cdot 10^{-3}$(2), $6.9 \cdot 10^{-3}$(3), $7.5 \cdot 10^{-3}$(4); $\triangle$ indicates the local minimum, * indicates the global minimum.

the choice of the global minimum itself could be changed at the lower experimental accuracy (the problem is solved on the boundary of the unambiguity).

TABLE 2. Estimates of rate constants ($\bar{k}$) and their mean-square errors ($\bar{S}$) calculated by various methods from the data about the triads distribution in products of alkaline hydrolysis of syndiotactic PMMA

	$(k_0 \pm S_0)$ l/mol·h	$(k_1 \pm S_1)$ l/mol·h	$(k_2 \pm S_2)$ l/mol·h	SSD($\bar{k}$) 10^4
Minimization of SSD for triads $T(t,\bar{k})$	0.256 ± 0.006	0.158 ± 0.006	0.0377 ± 0.0006	1.18
Method of Klesper /46/	0.3 ± 0.03**	0.15 ± 0.03**	0.05 ± 0.004**	6.56
Minimization of SSD for conversions* $P(A)(t,\bar{k})$ — Glob. min.	0.244 ± 0.008	0.161 ± 0.0045	0.0366 ± 0.0006	0.153
Loc. min.	0.369 ± 0.028	0.0148 ± 0.0008	0.246 ± 0.03	1.54

* P(A) values are calculated from the triads data /43/
** The mean-square errors are evaluated by division of the confidence range determined in /31/ onto 2.5.

Approximate method for mini-computer

The approximate method is oriented on the low-power computer for instance VSM 15 VSM 5 (USSR) with the storage volume 1 K byte and the speed 300 operations per second.

The program of the simulation of triads distribution uses the approximation by the second order Markov chain and "B-approximation" /3/. The algorithm of the minimization is very simplified coordinate descent. The results which we got by the approximate method practically are not different from the accurate one. The difference is only 10^{-4} from the values of constants for the second order Markov as well as for B-approximation. The difference between these two approximations is quite small (10^{-8} from the values of constants).

It means that in all practical studies which do not need the simultaneous treatment of the big volumes of the information one can very effectively use the approximate method of the evaluation of rate constants from the data on the triads of units distribution using mini-computer.

Concluding the discussion of different methods of the evaluation of individual rate constants of the polymeranalogous reactions with the neighboring groups effect let's look once more on the Fig. 9 illustrating the main problems arising here. Problems I and II which are concerned with the treatment of the experiment on the triads distribution are now solved totally including the development of the not very complicated approximations for mini-computer, the realization of which is possible even using pocket programmed calculators (for instance TI-59). The field of the unambiguous solution of reverse problems of the kinetics of the process (Fig. 9 III) is limited up to now by the experimental data of high accuracy. However if the values of constants can be ordered toward their magnitude from the theoretical viewpoint or from the independent experiments this field can be much wider. Finally if one thinks about the reverse problem of the composition heterogeneity (Fig. 9, IV) this task is not solved at all. The experimental data on the functions of the composition heterogeneity of the macromolecules are hardly available and as usually have a very low accuracy. That is why they probably will not be used in the near future for the evaluation of the rate constants.

IV. CONFORMATIONAL EFFECTS

Speaking about the further development of the general theory of macromolecular reactions the next effect to be studied quantitatively is the <u>conformational</u> one.

From the viewpoint of the development of different mathematical approaches to describe the conformational effect, of the big interest are the reactions of intramolecular cross-linking which always are followed with the change of the conformation of macromolecular coil and change of the reactivity related with this.

As we mentioned before the analytical calculation of the parameters of the reactions of intramolecular cross-linking is not possible due to the big amount of the topological structures which arise in the course of the reaction (Fig. 2). The mathematical simulation of this process is also rather cumbersome. We propose here the approximation to calculate this process not following the evolution of each topological structure, but following only the evolution of some structural elements being common for all structures. One can easily see (Fig. 2) that all topological structures contain elements of three types: "tails", cycles and arcs between two cross-links. The mutual transformations of these elements are presented on the scheme (Fig. 12).

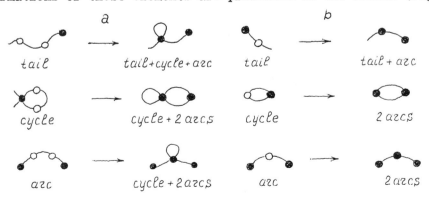

Fig. 12. A scheme of mutual transformations of elements for cross-linking of units of the same element (a) and of two different elements (b).

Using this scheme one can built an algorithm which includes the Monte Carlo procedure: the random choice of cross-linking units, the determination of the type of the elements to which they belong, the rearrangement of elements and the fixation of cross-links.
We have found that the average square of radius of gyration R^2 and the total number of selfintersections in the partially cross-linked chain q (including reactive contacts and "dead" contacts - cross-links) are interrelated one with another by the following relationship:

$$\bar{q}_m (\overline{R_m^2})^{3/2} = const \tag{21}$$

This relationship in the limits of the computer experiment error is valid for the different lengths of the chain and different number of cross-links. Taking into consideration as it was shown above that the kinetics of the intramolecular cross-linking is totally determined by the number of reactive contacts in the macromolecular coil which itself is determined as

$$Z_m = \bar{q}_m - m \tag{22}$$

the relationship (21) brings the problem of the kinetical description of the intramolecular cross-linking only to calculation of the average dimensions.
We have supposed that R^2 of the chain having m cross-links is formed from the R^2 of the linear part which contains "tails" and R^2 of the cross-linked part which contains cycles and arcs as

$$\overline{R_m^2} = \bar{n}_t l^2/6 + (\bar{n}_c + \bar{n}_{ar}) l^2/12 \tag{23}$$

Then basing on the values $\bar{n}_t$, $\bar{n}_c$, $\bar{n}_{ar}$ obtained from the Monte Carlo procedure one can calculate following Eq. (21)-(23) all parameters of the structure of cross-linked coils as well as the kinetics of the process. This approximation gives results which are in good coincidence with the results of mathematical simulation (Fig. 13).

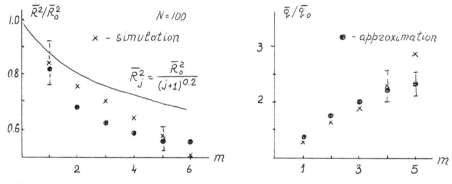

Fig. 13. Relative dimensions and the number of selfintersections calculated by Monte Carlo simulation and by approximation.

This coincidence permits to apply the proposed approximation to more complicated model, i.e. to the reversible reaction of cross-linking. The reactions of such type simulating the polymer-polymer interaction are closely related to the conformational effects in macromolecular reactions.
Figure 14 shows the results of calculation of the kinetics of the reversible intramolecular reaction and of the change in average dimensions of the coil with time in the case of equal rate constants for the formation and rupture of the cross-links. The equilibrium number of cross-links is rather small in this case (5 cross-links for the chain from 100 units). Therefore the main regularities of the process of intramolecular cross-linking calculated in the computer experiment for low degrees of conversion (only such case was studied by Monte Carlo procedure) can be applied to the reversible reaction of intramolecular cross-linking too.
An interesting result of the study of the reversible cross-linking is a linear dependence of the equilibrium number of cross-links on the chain length (Fig. 14, b) obtained for the models with different minimal dimensions of the loop (the minimal distance along the chain between units which can be cross-linked). All these curves are extrapolated into the zero point of coordinates. This regularity permits to extrapolate the obtained results on

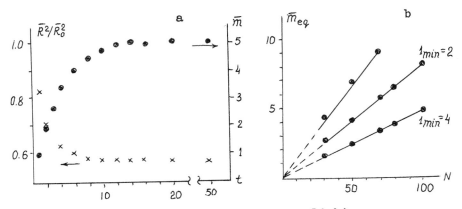

Fig. 14. Reversible intramolecular cross-linking.

the chains of bigger length too.
The processes of reversible and irreversible cross-linking accompany various reactions of chemical modification of polymers, for instance the curing of polymer matrices, coatings, formation of interpenetrating networks, formation of polymer-metal complexes in solution and so on. The comparison of calculated values with valid experimental data will be an important stage in the theory of macromolecular reactions.
The next step of the development of the quantitative approach is the simultaneous simulation of reversible intramolecular interaction and polymeranalogous reaction proceeding with the neighboring groups effect. Such a model will permit to evaluate some parameters of the reaction proceeding with the conformational effect.

V. OTHER EFFECTS

The quantitative description of supermolecular effects should take into account both the nonequivalence of macromolecules participating and non-participating in supermolecular formations and the change of the reactivity due to the chemical transformations of functional groups of different macromolecules.
One of the approaches to the calculation of supermolecular effect was proposed by Litmanovich /48/ who has considered the influence on interchain interaction on the polymer reactivity for the model of the reaction of A into B transformation proceeding in the melt with neighboring groups effect, when the reactivity of each unit depends not only on the state of two nearest neighbors but also on the state of $(Z-2)$ units which belong to other macromolecules (Z is the coordination number of the "lattice"of polymeric melt). Such reaction as in the case of isolated chain can be characterized by three rate coefficients k_0, k_1, k_2 which are here the functions of time, as they depend on the continuously changed composition of the coordination sphere, i.e. on the fraction of A units in the melt.
On the other hand the supermolecular effect can be related with the necessity to take into account the rate of the diffusion of low-molecular reagent into macromolecular accociate. The theory of diffusion-controlled reactions began intensively to be developped in last years.
Up to this moment we have discussed only the reactions of linear macromolecules. The reactions in three-dimensional polymers are also an important and interesting field. The theory of these reactions is now at the beginning of its development. Pakhomova et al. /49/ have considered the features of the kinetics of chemical reactions in network polymers taking into account the diffusion and topology particularly the absence of translational diffusion of macromolecules included into polymeric network.
Leikin, Korshak et al. /50/ have calculated the kinetics of the polymeranalogous reaction in swollen polymer gel with an assumption that the limiting stage is the interaction of functional groups A incorporated into polymeric network with low-molecular reagent B also included into polymer phase (i.e. the reaction is proceeding in kinetical region). These authors proposed that the parts of the chain between two cross-links are independent ones, and their reactivity does not depend on the degree of conversion of other fragments. The total kinetic scheme of such reaction should include beside rate constants k_0,k_1, k_2 also the constants of the transformation of A units situated near the cross-links and end A units. However as the calculation of all seven constants from the experimental data is not possible yet the authors

of /50/ used a semiempirical approach assuming that the kinetics of the reaction is determined only by two rate constants: k_0 corresponding to the reactivity of the particular unit being of the same order as the reactivity of the monomer analog and k_1 corresponding to the reactivity of other units. This approach permitted to treat quantitatively the experimental kinetical data on reactions of phosphorylated macroporous copolymers of styrene with divinyl benzene with the high degree of cross-linking.

Evidently the directions discussed above are not the unique possibilities of further development of the theory of macromolecular reactions. One can firmly predict the appearance of new directions which will lead to new theoretical statements and new possibilities of their applications to particular chemical reactions of polymers.

REFERENCES

1. N.A. Platé, A.D. Litmanovich, Paper presented at 23rd International Congress of Pure and Applied Chemistry, 8, 123-150, Boston (1971).
2. R.M. Fuoss, M. Watanabe, B.D. Coleman, J.Polymer Sci.,48, 5-15 (1960).
3. N.A. Platé, O.V. Noah, Adv.Polymer Sci., 31, 133-173 (1979).
4. O.V. Noah, A.D. Litmanovich, Vysokomolek.Soed., A19, 1211-1217 (1977).
5. O.Smidsrod, B. Larsen, T.J. Painter, Acta Chem. Scand., 24, 3201-3212 (1970).
6. Al.Al. Berlin, E.F. Vainshtein, S.G. Entelis, Vysokomolek.Soed., B20, 275-277 (1978).
7. S. . Kutchanov, Ye.B. Brun, Dokl.Akad.Nauk SSSR, 227, 662-665 (1976).
8. Ye.B. Brun, S.I. Kutchanov, Zh. P.Kh., 50, 1065-1069, (1977).
9. S.I. Kutchanov, Metody kinetitcheskikh Rastchetov v Khimii Polimerov, "Khimiya", Moscow (1978).
10. O.V. Noah, V.P. Tortchilin, A.D. Litmanovich, N.A. Platé, Vysokomolek. Soed., A16, 412-417 (1974).
11. P.J. Flory, J.Am.Chem.Soc., 61, 1518-1521 (1939).
12. M. Sisido, Macromolecules, 4, 737-742 (1971).
13. M. Sisido, Polymer.J., 3, 84-91 (1972); 4, 534 (1973).
14. M. Sisido, Seibutsu Butsuri, 14, 135-147 (1974).
15. M. Sisido , T. Mitamura, Y. Imanishi, T. Higashimura, Macromolecules, 9, 316-319; 320-324 (1976).
16. M. Sisido, Y. Imanishi, T. Higashimura, Macromolecules, 10, 125-130 (1977).
17. H. Tagaki, M. Sisido, Y. Imanishi, T. Higashimura, Bull.Chem.Soc.Jpn., 50, 1807 (1977).
18. M. Sisido, Kobunshi, 26, 260-263 (1977).
19. M. Sisido, F. Tamura, Y. Imanishi, T. Higashimura,, Biopolymers,16, 2723-2738 (1977).
20. M. Sisido, E. Yoshikawa, Y. Imanishi, T. Higashimura, Bull.Chem.Soc.Jpn, 51, 1464-1468 (1978).
21. M. Sisido, Y. Imanishi, T. Higashimura, Bull.Chem.Soc.Jpn. 51, 1469-1477 (1978).
22. S.V. Kozlov, Dokl.Akad.Nauk SSSR, 243, 410-413 (1978).
23. I.I. Romantsova, Yu.A. Taran, O.V. Noah, N.A. Platé, Dokl.Akad.Nauk SSSR, 234, 109-112 (1977).
24. I.I. Romantsova, O.V. Noah, Yu.A. Taran, A.M. Yel'yashevich, Yu.Ya. Gotlib, N.A. Platé, Vysokomolek.Soed., A19, 2800-2807 (1977).
25. I.I. Romantsova, Yu.A. Taran, O.V. Noah, N.A. Platé, Vysokomolek.Soed., A21, 1176-1180 (1979).
26. V.I. Irzhak, L.I. Kuzub, N.S. Yenikolopyan, Dokl.Akad.Nauk SSSR, 214 , 1340-1342 (1974).
27. L.I. Kuzub, V.I. Irzhak, L.M. Bogdanova, N.S. Yenikolopyan, Vysokomolek. soed., B16, 431-433 (1974).
28. Ye.N. Raspopova, L.M. Bogdanova, V.I. Irzhak, N.S. Yenikolopyan, Vysokomolek.Soed., B16, 434-437 (1974).
29. A.D. Litmanovich, N.A. Platé, E. Yun, V.A. Agasandyan, O.V. Noah, V.I. Kryshtob, N.A. Luk'yanova, N.V. Lelyushenko, V.V. Kreshetov, Vysokomolek. Soed., A17, 1112-1122 (1975).
30. E. Klesper, W. Gronski, V. Barth, Makromol.Chem., 139, 1-16 (1970).
31. E. Klesper, V. Barth, Polymer, 17, 787-794 (1976).
32. N.A. Platé, T. Seifert, L.B. Stroganov, O.V. Noah, Dokl.Akad.Nauk SSSR, 223, 396-399 (1975).
33. E. Yun, Thesis, Moscow (1973).
34. L.B. Krentsel', A.D. Litmanovich, I.V. Pastukhova, V.A. Agasandyan, Vysokomolek.soed., A13, 2489-2495 (1971).
35. E.G. Brame, Jr., J. Polymer.Sci., A-1, 9, 2051-2061 (1971).
36. T.I. Usmanov, Thesis, Moscow (1974).

37. J. Morcellet-Sauvage, C. Loucheux, Makromol. Chem., 176, 315-331 (1975).
38. I.A. Tutorskii, I.D. Khodzhaeva, B.A. Dogadkin, Vysokomolek. Soed., A16, 157-168 (1974).
39. A.S. Turaev, Sh. Nadzhimutdinov, Kh.U. Usmanov, Vysokomolek. soed., A19, 1347-1356 (1977).
40. J.J. Bourjuijnon, H. Bellissent, J.C. Galin, Polymer, 18, 937-944 (1977).
41. E. Klesper, D. Strasilla, Maria Christina Berg, European Polymer J., 15, 593-601 (1979).
42. P. Rempp, Plenary Lecture on the 4th Intern. Conf. on Modified Polymers, Bratislava (1975).
43. K. Ito, Y. Yamashita, J. Polymer. Sci., A, 3, 2165-2187 (1965).
44. V. Barth, E. Klesper, Polymer, 17, 777-786 (1976).
45. H.J. Harwood, J. Polymer. Sci., C, 16, 727-731 (1978).
46. B.J. Bauer, Macromolecules, 12, 704-708 (1979).
47. D. Hudson, Statistika dlya Fizikov, "Mir", Moscow (1968).
48. A.D. Litmanovich, Dokl. Akad. Nauk SSSR, 240, 111-113 (1978).
49. L.K. Pakhomova, O.B. Salamatina, S.A. Artemenko, Al.Al. Berlin, Vysokomolek. Soed., B20, 554-558 (1978).
50. Yu.A. Leikin, V.V. Korshak, S.Yu. Gladkov, T.U. Tarasova, A. Khaled, T.A. Tcherkasova, Vysokomolek. Soed., A21, 1220-1228 (1979).

CHEMICAL MODIFICATIONS OF CHLORINATED POLYMERS

Ernest MARECHAL

Laboratoire de Synthèse Macromoléculaire, Université P. et M. Curie,
12 Rue Cuvier - 75005 Paris - FRANCE.

Abstract. Some aspects of the chemical modification of chlorinated polymers
are examined. In the first part the main systems associating an organic halide with a metal or a metal derivative are critically reviewed. In the second part, our works relative to the grafting of chlorinated natural rubber
(C.R.) by metal methacrylate are reported. Metal derivatives are $Ni° P[OC_6H_5)_3]_4$
(NPP) and $Ni° [P(OEt)_3]_4$. Conversion anf grafting efficiency are determined.
From kinetic and NMR studies (^1H and ^{31}p) it can be concluded that the initiation mechanisms reported up to now (20) must be reexamined. Applications of
the grafting of C.R. to the preparation of antifouling paints (7) are described.

INTRODUCTION

The chemical modification of chlorinated polymers has been studied extensively. The ionic
modifications have been used particularly for the synthesis of block and graft copolymers.
Cationic processes have been reviewed recently (1). On the other hand the direct reaction
of chlorinated polymers with amines and alcohols, which has already been widely studied
(2, 3) should undergo new and further development due to phase transfer technics (4, 5).
Radical modifications are highly varied but among the most important are probably those
which associate a chlorinated polymer with a metal or a metal derivative to generate free
radicals. This will be the subject of our lecture. We will examine successively the
following points :

A - Review of the main systems organic halide (polymeric or non) - metal (or metal
derivatives) which have been used to initiate polymerizations.

B - Fundamental study of the grafting of chlorinated natural rubber (Noguès, Dawans and
Maréchal (6)) and application to the synthesis of compounds for antifouling paints (Dawans,
Devaud, Nicolas (7)).

Part A

Critical Review of the initiating systems associating an organic halide with a metal or a
metal derivative.

The initiating processes can be classed under two general schemes.

1 - Scheme I

It includes the systems associating a low valent transition metal derivative with an organic
halide.

$$M^{(n)} + RX \longrightarrow M^{(n+1)} + X^{\ominus} + R^{\cdot} \qquad\qquad I$$

The process can be activated either by heating (8-10) or by photo excitation (11, 12).

2 - Scheme II

It includes the systems associating a transition metallic chelate with an organic halide
(13-15).

$$M^{(n)} L_n \xrightarrow{\text{RX}} M^{(n-1)} L_{n-1} + LX + R^{\cdot} \qquad\qquad\qquad II$$

The polymers formed according to initiation schemes I and II have similar characteristics (microstructure, tacticity and molecular weight distribution) to those obtained in the presence of classical free radical initiators.

When the organic halide is a polymer macroradicals can be prepared by extracting some halogen atom from the polymeric substrate. This method has been successfully applied in preparation of block copolymers from telomers with carbon trichloride end groups (16) and of graft copolymers from polychloroprene, chlorosulfonated polyethylene, polyvinylchloride (17, 18) and chlorinated natural rubber (6, 7).

A' - Initiation according to scheme I

Activation process can be thermal or photochemical.

I/ Thermal processes

I'/ Metal chelates

1) SN_1 processes

a) System $CCl_4 - Mn_2(CO)_{10}$ (19)

Polymerizations initiated by this system are first order in monomer, which shows that the monomer does not contribute to initiation step (or only to a very slight extent). Moreover the presence of carbon monoxide in excess decreases the polymerization rate. From these data Bamford (19) suggested a scheme which can be summarized by :

$$Mn_2^o(CO)_{10} + CCl_4 \longrightarrow Mn^{(I)} + xCO + {}^{\cdot}CCl_3 + \text{inactive compounds} \qquad III$$

b) System $CCl_4 - Ni^o\left[P(OAr)_3\right]_4$ (20)

Bamford (20) suggested the following set of reactions :

$$Ni^o\left[P(OAr)_3\right]_4 \underset{k_2}{\overset{k_1}{\rightleftharpoons}} Ni^o\left[P(OAr)_3\right]_3 + P(OAr)_3 \qquad\qquad IV$$

after which monomer M coordinates with $Ni\left[P(OAr)_3\right]_3$ according to :

$$Ni^o\left[P(OAr)_3\right]_3 + M \rightleftharpoons Ni^o\left[P(OAr)_3\right]_3 M \qquad\qquad V$$

Authors (15) assumed that the equilibrium V is established rapidly and maintains throughout the reaction. Equilibrium IV is the kinetic step.

Then Ni^o isoxydized to Ni^{II} according to :

$$Ni^o \quad \overset{CCl_4}{\nearrow} Ni^I + R^{\cdot} \xrightarrow{CCl_4} Ni^{II} + 2R^{\cdot} \qquad\qquad VI$$

$$\underset{CCl_4}{\searrow} Ni^{II} \text{ (No formation of free radicals)} \qquad\qquad VII$$

From kinetic considerations it appears that steps VI and VII have about the same rate and that the term Ni^o is very probably $Ni\overset{\bullet}{\left[P(OAr)_3\right]}_3 M$.

The order in monomer of the polymerization is 1 which is somewhat different from earlier observations (21). The results found by Noguès, Dawans and Maréchal (6) in their analysis of the initiation step differ to some extent. This will be discussed in part B.

2) SN_2 processes

In these systems the monomer contributes directly to the first step of initiation as shown in the system $CCl_4 - Mo(CO)_6$ (22). The order in monomer is 1.5 and addition of CO decreases reaction rate. From these results the following scheme was proposed.

$$Mo(CO)_6 + M \underset{k_2}{\overset{k_1}{\rightleftharpoons}} \underset{\underline{A}}{Mo(CO)_5 --- M} + CO \qquad\qquad VIII$$

$\underline{A} \longrightarrow$ Inactive compounds.

$$\underline{A} + CCl_4 \longrightarrow M_o(CO)_4 \overset{M}{\underset{CCl_4}{<}} + CO \qquad\qquad IX$$

$$\underline{B}$$

$$\underline{B} \longrightarrow \ ^.CCl_3 + \text{Inactive compounds} \qquad\qquad X$$

$$\underline{A} + \underline{B} \longrightarrow \text{Inactive compounds} \qquad\qquad XI$$

Supposing that species $\underline{A}$ and $\underline{B}$ have constant concentrations it is possible to establish from reactions VIII to XI a rate equation where the order in monomer is 1.5.
In such a system the contribution of the solvent can be very important.

II'/ Metal salts

Some metal salts are able to react with organic halides to give free radicals. This is the case for Cr^{II} derivatives (23).

$$Cr^{II} + RX \longrightarrow R^. + Cr^{III} + X^{\ominus} \qquad\qquad XII$$

Cr^{II} contributes to the termination processes. However when amines, such as ethanolamine, are present the rate of initiation is drastically increased and the contribution of the complex Cr^{II} (ethanolamine) to the termination processes can be neglected (9).
The associations of Cr^{II} derivatives with chloracetic acids are particularly interesting since not only these acids react with Cr^{II} to give free radicals, but are, in addition, inhibitors of termination processes (9).

III'/ Metals

Most transition metals are able to react with organic halides according to processes very close to those described in Ullman, Gomberg and Wurtz-Fittig reactions. Otsu (24, 25) studied several systems and Iwatsuki (26) suggested the following process for the formation of the free radicals :

$$CCl_4 + Me \overset{K}{\underset{\longleftarrow}{\longrightarrow}} \left[Complex\right] \overset{k_i}{\longrightarrow} R^.$$

He established kinetic relations which fit experimental results satisfactorily.

II/ Photochemical Processes

Most studies relative to the initiation by the systems $\left[RX - \text{Metal derivative}\right]$ concern carbonyl derivatives of transition metals. According to literature the first step consists in the dissociation of the complex :

$$M(CO)_n + h\nu \longrightarrow M(CO)_n^* \longrightarrow M(CO)_{n-1} + CO + E \qquad\qquad XIV$$

Then $M(CO)_{n-1}$ can react with an electron donating compound according to :

$$M(CO)_{n-1} + S \longrightarrow M(CO)_{n-1}S \qquad\qquad XV$$

When S is a vinyl monomer $M(CO)_{n-1}S$ reacts rapidly with RX to give a free radical.
Several mechanistic and kinetic studies were carried out. Thus, Bamford (27) studied the irradiation of $M_o(CO)_6$ in the presence of ethylacetate or of methyl methacrylate and pointed out the formation of $(EtOAc)M_o(CO)_5$ or $(MMA)M_o(CO)_5$ and the reaction of this species with the chlorinated compound. However these results are in opposition with Koerner (28) who showed that the irradiation of $Fe(CO)_5$ in the presence of a vinyl monomer M leads to the species $Fe(CO)_4M$ which reacts with the organic halide.
Bamford (29-32) studied extensively the initiation of MMA by $Mn_2(CO)_{10}$ and $Re_2(CO)_{10}$ in the presence of CCl_4. This author suggested the following activation path :

$$Mn_2(CO)_{10} \xrightarrow{h\nu} Mn(CO)_4 + Mn(CO)_6$$

$$Mn(CO)_4 + CCl_4 \longrightarrow Mn(CO)_4Cl + \dot{C}Cl_3$$

$$Mn(CO)_4Cl + CO \longrightarrow Mn(CO)_5Cl$$

$$2Mn(CO)_6 \longrightarrow Mn_2(CO)_{10} + 2CO$$

B' - Initiation according to scheme II
I/ System CCl_4 - Bisephedrine Cu^{II} (A)

The reaction of A with various halides was studied by Barton, (13, 14) who showed that Cu^{II} is reduced to Cu^{I}. However, according to Guyot (16) the oxydation step of Cu is not changed and $CCl_3\dot{}$ is formed.

Amano and Uno (33) showed that $CCl_3\dot{}$ is formed and contributs to initiation. According to Barton (13, 14) Cu^{II} chelate is also a terminating agent.

II/ System CCl_4 - Diaminopropane Cu^{II} (DAPCu^{II}) (15)

Polymerization of acrylonitrile or methyl methacrylate can be initiated by DAPCu^{II} alone or in the presence of CCl_4. In both cases Cu^{II} is changed to Cu^{I}. The order in monomer is between 1.3 and 1.5 which means that monomer takes part in initiation step.

In the presence of CCl_4 the following mechanism has been suggested :

$$Cu^{II}(-NH_2)_2 + CCl_4 \rightleftharpoons \left[\left[Cu^{II}(-NH_2)_2 \right] --- CCl_4 \right] \longrightarrow$$

$$Cu^{I}(-NH_2) + -\dot{N}H_2 \overset{\oplus}{} --- CCl_4 \xrightarrow{Fast} Cu^{I}(-NH_2) + -NHCl + CCl_3\dot{} + H^{\oplus}$$

The resulting polymer does not contain any amino groups which means that the NH_2 group does not initiate. The initiation is due to $CCl_3\dot{}$.

III/ Systems Halogenated polymer - Metal derivative
1) System polychloroprene - Cr^{II} acetate

Lee and Minoura (18) showed that this system behaves like the $CHCl_3$-Cr^{II} acetate system. However at high monomer concentrations the order in monomer is above one. This phenomenon, already mentioned by Smets (34) is due to an increase of viscosity which increases the order in monomer and decreases the order in initiator.

Reactivity analysis shows that initiation takes place on 1, 2 sites :

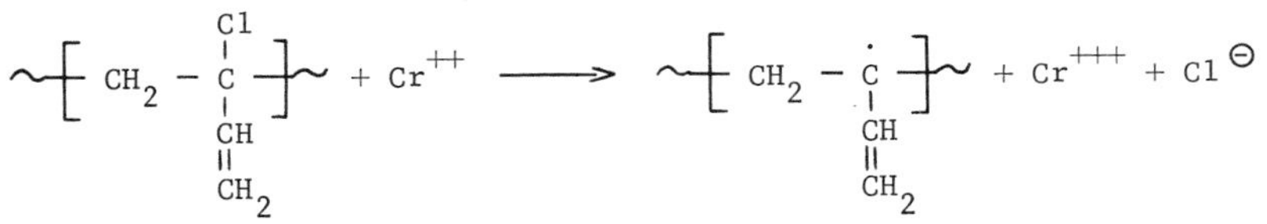

The products of the reaction consist of homopolymer of M, non modified polymer, poly (chloroprene-g-M) and, in some cases crosslinked polymer.

2) Chlorosulfonated polyethylene - Cr^{II} acetate

Initiating sites are the chlorosulfonyl groups ; this fits with the fact that model system : $CH_3PhSO_2Cl + Cr^{II}$ initiates the polymerization of vinyl monomer but that $CH_3PhSO_3H + Cr^{II}$ does not.

3) Chlorinated polypropylene, halogenated polymethylmethacrylate, or halogenated polystyrene associated with Cu^{II}-bis ephedrine

Barton (35) has shown that these systems have the same reactivity as halogenated alkanes. The concentration of chlorine atoms in the medium is less important that their positions in the polymer. From oligostyrene or oligo(vinyl chloride) with CCl_3 end groups Guyot (16) prepared block copolymers PSt-P(MMA) or PVC-P(MMA). PVC without CCl_3 group does not contain a chlorine atom sufficiently reactive to initiate MMA polymerization.

Part B
Grafting of chlorinated rubber (C.R.) by methyl methacrylate (MMA) (6).

A' - Preliminary studies

The chlorine content of C.R. usually ranges from 63 to 68 %. Its molecular structure, which involves both linear and cyclic units, has been described by various authors (36, 37). More recently Wartemberg et al. (38) have begun a ^{13}C NMR study of the compound and shown that several modifications must be made to the structures which are generally accepted.

In order to check its homogeneity we fractionated C.R. by addition of methanol to a CCl_4 solution. The chlorine content ranges from 66.1 to 63.2 % for the 15 first fractions ; the last (16th) has a lower chlorine content (60.2 %).

The addition of a toluene solution of C.R. (% Cl 67.3) to an excess of methanol yields an insoluble fraction A (Cl % 64.3) and a soluble fraction B (Cl % 56.9). The soluble part represents less than 5 % of the initial amount. In order to explain the decrease in the chlorine content, we freeze-dried a benzene solution of the same sample and showed by gas chromatography that the sublimated phase contains 2 to 4 % of CCl_4 remaining from the preparation of C.R. (chlorination of a CCl_4 solution of natural rubber). Preliminary graftings in the presence of tetrakistriarylphosphitenickel (NPP) showed that C.R. is a more efficient halide than CCl_4 and that fraction B contains more reactive chlorine atoms than A.

In the following, C.R. is used without any fractionation. In order to determine the possible contribution of a thermal side-polymerization we heated MMA in the presence of C.R. at various temperatures in the absence of initiator. The PMMA content was determined either by infra-red spectroscopy (ν(C = O), 1730 cm^{-1}) or by elementary analysis ; agreement between the two determinations was satisfactory. It appeared that above 40°C the thermal side reaction cannot be neglected.

The efficiency of 12 metallic compounds towards MMA grafting was determined. It appeared that low valent nickel compounds such as NPP nickel or biscyclooctadiene nickel or even metallic nickel, resulting from the thermal decomposition of nickel formate, are the most effective initiators. In most cases the activity can be increased by adding a small quantity of dimethyl formamide. The other metallic promoters mentioned in literature are less active in the presence of C.R.. Some promoters are almost inactive : $M_o(CO)_6$(DMF), bis ephedrine Cu^{II}, $CuCl_2$ (diaminopropane).

In most cases two quantities were determined :

- The PMMA content τ which corresponds to the global amount of PMMA retained in the copolymer i.e. before any separation of homopolymer.

- The graft efficiency τ_G which is the ratio of the weight of graft (PMMA) on the total weight of PMMA produced. In order to separate homopolymer from graft material we hydrolysed PMMA by concentrated sulfuric acid according to Smets (39) and extracted the resulting polymethacrylic acid (PMA) by methanol or water. After a reesterification of the extracted compound and a second hydrolysis no more PMA could be extracted, which supports the completion of the fractionation after the first hydrolysis and the extraction by alcohol. τ_G values ranging from 0.65 to 0.87 were found for CR-PMMA graft copolymers prepared in the presence of nickel promoters while τ_G does not exceed 0.30 when initiator is a peroxide.

B' - Initiation by tetrakisphenylphosphite nickel (NPP)

The solvent is benzene in all cases. Variations of the conversion and of τ against time are plotted in figure 1. Conversion plots are linear as long as conversions are below 8 %. The plots of log$_e$ (polymerization rate) against log$_e$ M and of polymerization rate against $[NPP]^{1/2}$ are straight lines. The order in monomer is 1 and the order in initiator is 0.5. These results fit literature data and show that the polymerization of MMA initiated by the system CR-NPP is a free radical process.

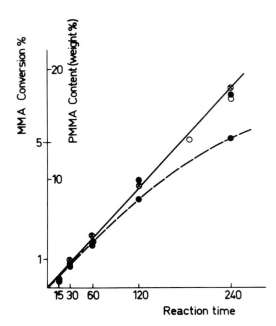

Fig. 1

Polymerization of MMA at 0°C by the system
C.R.-NPP. $[CR]$ = 0.26 mol.1^{-1} ;
$[NPP]$ = 9.67 x 10^{-3} mol.1^{-1}.
—— Conversion ; --- MMA content $\%$.
$[MMA]$ (mol.1^{-1}) :   4.03 ;   1.34 ;
o   0.67

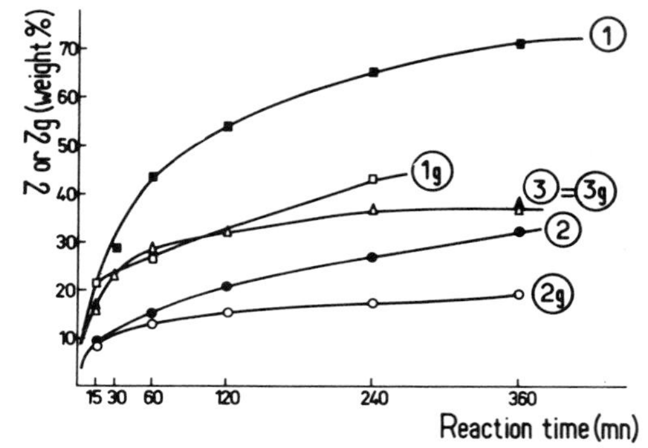

Fig. 2

Polymerization of MMA under various
experimental conditions. Variations with
time of the total amount $\%$ in PMMA and of
content $\%_g$ of grafted PMMA.
Curve 1 : $[MMA]$ = 4.03 ; $[NPP]$ = 9.67 x 10^{-3}
Curve 2 :   1.34 ;   "
Curve 3 :   4.03 ;   0.97 x 10^{-3}

Figure 2 provides a direct comparison of the total amount $\%$ of PMMA retained in the sample
and of the content $\%_g$ of PMMA effectively grafted on CR and determined as reported above
(A'). In all cases index g is relative to the grafted polymer. Examination of figure 2
shows that for low conversions all the PMMA contained in the sample is grafted. Monomer
concentration has only a slight influence on the relative proportions of homopolymer and
of grafted PMMA. On the other hand, the ratio $[Ni^o]/[Cl]$ has a great influence : in the
case of curve 3 nohomopolymer is formed but in the case of curves 1 and 2 (in which the
ratio $[Ni^o]/[Cl]$ is 10 times higher than in experiment 3) a large amount of homopolymer
is produced.

The influence of temperature is very important : at 0°C all the PMMA contained in the
sample is grafted. Homopolymer content increases with increasing temperature.

C' - Initiation by the system C.R.-Ni$^o[P(OEt)_3]_4$ and influence of Ni$^o[P(OEt)_3]_4$ on the
     polymerizations initiated by the system C.R.-Ni$^o[P(OAr)_3]_4$.

The variations of conversion with time are reported in figure 3. It appears that the
activation energy is very high, since at 35°C initial slope is much higher than at -20,
-10 and 0°C. However, in all cases, Ni$^o[P(OEt)_3]_4$ reacts very fast with the chlorine atoms
of C.R.. For experiments carried out at -20, -10 and 0°C there is no limitation of the
conversion. However at 35°C the reaction stops completely when the conversion is above
10%.

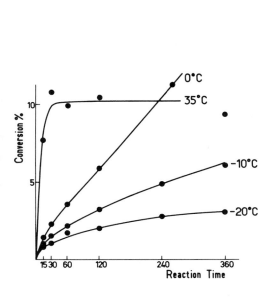

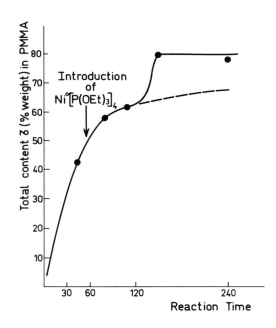

Fig. 3

Polymerization of MMA initiated by the
system CR-Ni^o$\left[P(OEt)_3\right]_4$ at different
temperatures.
Concentrations (mol.l^{-1}) : $\left[CR\right]$ = 0.26 ;
$\left[Ni^o\right]$ = 9.07 x 10^{-3} ; $\left[MMA\right]$ = 4.03.
Variations of the conversion with time.

Fig. 4

Polymerization of MMA initiated by the
system CR-NPP at 35°C.
Concentrations as in fig. 3.
Influence of the introduction of
Ni^o$\left[P(OEt)_3\right]_4$ after one hour of polymeriza-
tion. $\left[Ni^o\left[P(OEt)_3\right]_4\right]$ = 9.07 x 10^{-3} mol.l^{-1}.

In order to see whether Ni^o$\left[P(OEt)_3\right]_4$ has a high reactivity only with labile chlorine atoms
present at the begining of the reaction we introduced it in a polymerization of MMA
initiated by NPP. The introduction was carried out 1 hour after the begining of the
polymerization (fig. 4). One hour after this introduction a net increase of the slope was
observed, then the reaction stopped. This shows that the action of Ni^o$\left[P(OEt)_3\right]_4$ is not
limited to the most labile chlorine atoms, since these have been consumed by the reaction
with NPP. The fact that 1 hour elapsed between the introduction and the change of slope
is probably due to a decrease of the accessibility of chlorine atoms resulting from an
increase of the viscosity of the medium.

D' - Study of the interaction between chlorinated and organometallic compounds

Examination of the literature shows that the mechanism of initiation and the reactivity of
chlorine atoms towards NPP have not really been studied. We tried (6) to obtain some
information about these problems.

I/ Influence of the order of introduction of the reactants

In part A' to C' reactants were introduced in the reactor in the following order : C.R.,
MMA, Ni^o derivative ; moreover it was shown that MMA does not react with the organometallic compound. In what follows MMA was introduced in third position and some time
after the organometallic compound. The study was carried out both on model (CCl$_4$) and on
C.R..

The results relative to CCl$_4$ are reported in figure 5. Curve A corresponds to a reaction
where the reactants were introduced in the same order as in grafting : CCl$_4$, MMA, then
Ni^o derivative. In the case of curve B the order of introduction was CCl$_4$, NPP followed

45 mn later by MMA. Comparison of curves A and B shows unambigously that the activity of
the initiating system CCl_4 + NPP greatly decreases with time. Since, the concentration of
CCl_4 can be considered as constant, due to its high excess, it can be concluded that a non
negligible part of NPP is destroyed by the chlorinated compound.

The same type of experiment was carried out on C.R. (fig. 6). This shows clearly that when
the Ni^o derivative is $Ni^o\left[P(OEt)_3\right]_4$ no polymerization is observed showing that the complex
is almost deactivated after two hours (If introduction order is CR, Ni^o then MMA no polymer
is obtained).

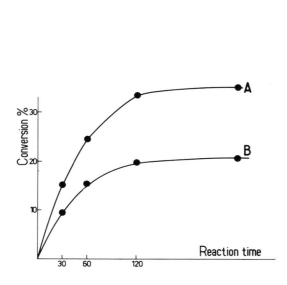

Fig. 5

Polymerization of MMA by the system CCl_4-
NPP at 35°C. Solvent : benzene.
Concentrations (mol.1^{-1}) : $\left[CCl_4\right]$ = 0.49 ;
$\left[MMA\right]$ = 4.03 ; $\left[NPP\right]$ = 9.67 x 10^{-3}.
Order of introduction :
Curve A : CCl_4 ; MMA ; NPP.
Curve B : CCl_4 ; NPP ; 45 mn after MMA.

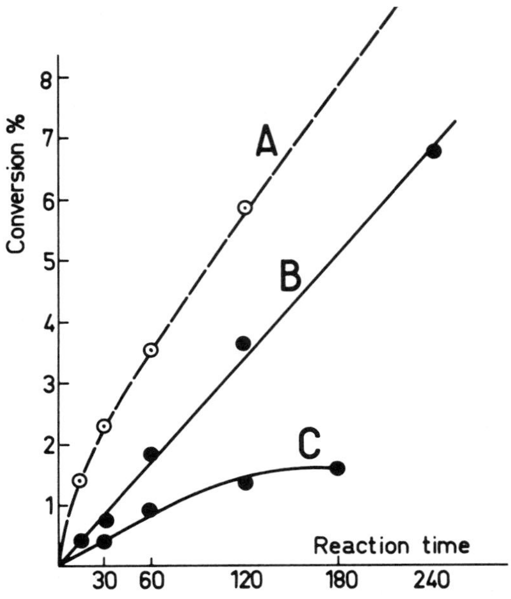

Fig. 6

Polymerization of MMA by the system C.R.-Ni^o
derivative. 0°C. Solvent : benzene.
Concentrations as in figure 5.
Order of introduction :
Curve A : C.R. ; MMA ; $Ni^o\left[P(OEt)_3\right]_4$.
Curve B :   " ; NPP.
Curve C : C.R., NPP ; after 2 hours at 35°C
introduction of MMA in the mixture cooled at
down at 0°C.

In the case of NPP (curve C) the deactivation is not complete. The plot of conversion
against reaction time is a straight line, at least during the two hours following the
introduction of the monomer. The slope of this line is lower than when the order of
introduction is C.R., M.M.A., followed by the Ni^o derivative (Curve B).

2) Study of initiation by ^{31}PNMR

^{31}PNMR gives interesting informations on Ni neighbourhood. The characteristics of the
spectrum depend both on the dissociation of the nickel complex (Ni^oL_4) and on the rate of
exchange between the ligand L and the complex. The characteristics of various Ni^oL_4
complexes are reported in table 1.

Table 1

Behaviour of $Ni^{0}L_{4}$ complexes in $^{31}PNMR$ [a]

Extent of dissociation	Nature of the ligand L	
	Rapid exchange	Slow exchange
low	PMe_3 $P(OMe)Ar_2$	$P(OEt)_3$ $P(OiPr)_3$ $P(OArX)_3$
high	$PMeAr_2$ PEt_3 PAr_3	$P(OArX)_3$ Y

(a) $X = CH_3$, Cl ; $Y = CH_3$, H, Cl

When dissociation is low there is only 1 peak ; when it is high there are 3 peaks corresponding to NiL_4, NiL_3 and L respectively. If the exchange between the complex and the ligand is slow the chemical shifts are not modified. On the other hand these are modified when exchange is rapid. In the case of $Ni^{0}\left[P(OEt)_3\right]_4$ and $Ni^{0}\left[P(OAr)_3\right]_4$ exchanges are slow and dissociations low. The relative $^{31}PNMR$ spectra show the following peaks (Standard 80 % H_3PO_4) (40).

$$Ni^{0}\left[P(Oet)_3\right]_4 = 158.4 \text{ p.p.m.}$$

$$Ni^{0}\left[P(OAr)_3\right]_4 = 128.8 \text{ p.p.m.}$$

The fact that the peak is unique shows that the structure is tetrahedric which is in agreement with literature (40). The chemical shifts relative to various nickel-complexes (NiL_4) and to the corresponding ligands (L) are reported in table 2.

Table 2

^{31}P chemical shifts of various NiL_4 and L

Ligand L	δNiL_4	δL	$\delta NiL_4 - \delta L$
$P(OMe)_3$	-163.3	-140.4	-22.9
$P(OEt)_3$	-158.4	-137.8	-20.6
$P(OAr)_3$	-128.8	-127.3	-1.5

Examination of table 2 shows clearly that the formation of complex shifts ^{31}P resonance towards low fields. This is due to the fact that the establishment of a donnor-acceptor bond between Ni and P decreases electronic density on phosphorus atom. This deshielding makes δNiL_4 lower than δL.

It is interesting to know whether $^{31}PNMR$ results fit reactionnal schemes which were proposed to explain the initiation of monomers by nickel phosphite - RX systems and which have been reported in part AA'II'1b (20).

We studied (6) the behaviour of the system MMA, RCl, $Ni^{0}\left[P(OR')_3\right]_4$ (where R' is Ph or Et) by $^{31}PNMR$. Chlorinated compound is $C_6H_6Cl_6$, CCl_4 or C.R.. From the results reported in table 3 the following conclusions can be drawn.

In so far as the observation takes place just after the mixing of the reactants the addition of a chlorinated compound to $Ni\left[P(OR)_3\right]_4$ where R = Ph or Et does not change the chemical shift of phosphorus on the other hand, the addition of MMA to $Ni\left[P(OR)_3\right]_4$ shifts the peaks towards high fields, with the shift increasing with increasing amounts of MMA. However, when

MMA is present in a very high excess it is difficult to know the cause of the shift as there is a change in dielectric constant.

Table 3

^{31}PNMR study of RX + Ni$^{\circ}$$\left[P(OR)_3\right]_4$ system

Reactant	$\dfrac{[MMA]}{[Ni]}$	$\dfrac{[Cl]}{[Ni]}$	$[Ni^{\circ}]$ mole/l	t (a, b)	Observations (c)
P(OEt)$_3$	$\dfrac{0}{0}$	$\dfrac{0}{0}$	0	10 mn	Single peak (−138.75)
Ni$^{\circ}$$\left[P(OEt)_3\right]_4$	0	0	0.018	10 mn and 3 h	" (−159.6)
Ni$^{\circ}$$\left[P(OEt)_3\right]_4$ − MMA	1	0	0.043	10 mn	" (−159.5)
Ni$^{\circ}$$\left[P(OEt)_3\right]_4$ − MMA	329	0	0.021	(40°C) ; 10 mn and 3 h	Single peak (−159.1) which does not shift
Ni$^{\circ}$$\left[P(OEt)_3\right]_4$ − CR	0	16.3	0.029	10 mn and 3 h	Single peak (−159.6)
$\left[$Ni$^{\circ}$$\left[P(OEt)_3\right]_4$ − C$_6$H$_6$Cl$_6$ − MMA	304	146	0.023	40°C ; 10 mn	Single peak (−159.2)
" "				40°C ; 3 h	Two peaks −159 (70 %) and −105.8 (30 %)
" " $\left.\right]$				40°C ; 24 h	Two peaks −159.2 (10 %) and −105.7 (90 %)
Ni$^{\circ}$$\left[P(OEt)_3\right]_4$ + CCl$_4$	0	(CCl$_4$ solvent)	0.03	120 h and several days	Peak at −102 (10 %) and many secondary peaks between −15 and −10 (90 %)
P(OAr)$_3$	$\dfrac{0}{0}$	$\dfrac{0}{0}$	0	10 mn	Single peak (−128.4)
Ni$^{\circ}$$\left[P(OAr)_3\right]_4$	0	0	0.016	"	" (−129.9)
Ni$^{\circ}$$\left[P(OAr)_3\right]_4$ + CR	0	3.21	0.016	"	" (−130.7)
Ni$^{\circ}$$\left[P(OAr)_3\right]_4$ + MMA	1.9	0	0.037	"	" (−129.8)
Ni$^{\circ}$$\left[P(OAr)_3\right]_4$ + MMA	406	0	0.018	40°C ; 10 mn and 3 h	" (−129.6) does not shift
Ni$^{\circ}$$\left[P(OAr)_3\right]_4$ + C$_6$H$_6$Cl$_6$ + MMA	464	210	0.016	40°C ; 10 mn and 3 h	Single peak does not shift

(a) t is the time between the homogeneisation of the mixture of the reactants and record of the spectrum.

(b) Spectra were registered at 25°C, except when another temperature is indicated.

(c) δ values are relative to H$_3$PO$_4$.

Initiation reaction was studied under the polymerization conditions (B', C'). In the case of Ni$^{\circ}$$\left[P(OAr)_3\right]_4$ no change is observed in the spectrum, even several hours after the mixing of the Ni$^{\circ}$ derivative with the chlorinated compound and MMA. However, when nickel derivative is Ni$^{\circ}$$\left[P(OEt)_3\right]_4$ a peak at −105 p.p.m. appears several hours after mixing. In consequence, two peaks are present at −159 and −105 p.p.m.. The sum of their intensities is constant ; peak at −105 p.p.m. is noted Ni$\left[P(OEt)_3\right]_4^{*}$. After 24 hours nearly all nickel derivative is in the form of Ni$\left[P(OEt)_3\right]_4^{*}$.

All these remarks fit with conclusions relative to kinetic study : the reaction of $Ni[P(OEt)_3]_4$ with a chlorinated compound is faster than that of an aromatic derivative and is quantitative. When $Ni[P(OEt)_3]_4$ reacts with CCl_4 other compounds are formed. Their chemical shifts are in the range -15 to o p.p.m. which corresponds to pentavalent phosphorus. During MMA polymerization no free phosphite is formed which is in opposition with Bamford's mechanism.

Since the species $Ni^o[P(OEt)_3]_4^*$ has the same state of symetry as $Ni[P(OEt)_3]_4$ (only one peak) and as no free phosphite is formed this species could be $Ni^{\oplus}[P(OEt)_3]_4$ or $Ni^{\oplus\oplus}[P(OEt)_3]_4$ since (43) $Ni[P(OCH_3)_3]_5^{\oplus\oplus}$ PNMR spectrum shows a peak close to -105 p.p.m.

3) Study by ^1H NMR

Results relative to $Ni^o[P(OEt)_3]_4$ and its reaction with CCl_4 are reported in table 4.

Table 4

^1H NMR study of the reaction $CCl_4 + Ni[P(OEt)_3]_4$

Compound	CH$_2$		CH$_3$	
	Pattern	δ	Pattern	δ
$P(OEt)_3$	quintuplet	3.80	triplet	1.10
$Ni^o[P(OEt)_3]_4$	unresolved	4.10	triplet	1.25
$CCl_4 + Ni^o[P(OEt)_3]_4$	quadruplet - poor resolution	4.30	triplet - poor resolution	1.38

In the case of free phosphite, protons of methylene group are coupled with phosphorus and methyl protons ; this results in a quintuplet. When the phosphite is coordinate to Ni^o the pattern is unresolved because - due to its paramagnetic character - the nickel atom acts as a screen. The pattern is shifted towards low fields because nickel attracts electrons. After the reaction $CCl_4 + Ni^o[P(OEt)_3]_4$ the CH_2 pattern is a quadruplet; the fact that the pattern is resolved shows that nickel is no longer zerovalent. Moreover, the shifting towards low fields (4.30 instead of 4.10) shows that the electronic density around the protons is lower after the reaction than before. This fits the assumption (see above) that Ni^o is oxydized without destruction of the complex.

The methyl pattern is a triplet. Its chemical shift depends very little on the neighbourhood of phosphite group. However, the changes fit the modifications of CH_2 shift.

Decrease of electronic density around the proton results in a shielding, and protons of both CH_2 and CH_3 are more acidic in $Ni^o[P(OEt)_3]_4$ than in free phosphite. After the reaction $CCl_4 + Ni^o[P(OEt)_3]_4$ has taken place this tendency is even more pronounced.

E' - Application to the preparation of new antifouling coating material (7)

The deposition of marine fouling on the hull of boats and on the various equipments immersed in the sea increases their resistance to friction, their weight and corrosion. For example the deposition of marine organisms on the hull of ships increases their fuel consumption by 20 to 40 %. Efficient protection against fouling necessitates the presence of a homogeneous layer of toxic agents of sufficiently high concentration in the immediate neighbourhood of the surface to be protected.

In order to improve the control of lixiviation, i.e. of the releasing of the toxic compound into the aqueous phase, the use of coatings where the biocid agent is chemically bonded to a polymer have been developped for some years. For example poly(alcoyltinmethacrylate)s were prepared and added to the binder used for the coating. This process is fairly satisfactory when the binder is an epoxy resin but cannot be used when it is C.R. which is not compatible with the polymethacrylates. In consequence Dawans et al. (7) prepared graft copolymers of C.R. where the grafts contain side biocid groups.

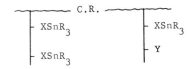

This process allows a better control of the releasing of the toxic agent. The ability of the metallic cation $\oplus SnR_3$ to be released depends both on the nature of the counteranion X and on the hydrophilic or hydrophobic character of the graft which can be changed by the introduction of a comonomer Y.

The following monomers were prepared :

$CH_2 = C(CH_3)COOSnR_3$ (R : nBu, Ph, $-CH_2C(CH_3)_2C_6H_5$)

$R_3SnOOC - CH = CH - COOSnR'_3$ (R : nBu ; R' = nBu, Ph, H)

$CH_2 = CH - CH_2Sn(X)(nBu)_2$ (X : F, Cl, Br, nBu)

$CH_2 = CH(CH_2)_8COOSn(nBu)_3$

(benzene ring with $COOSn(nBu)_3$ and $COOR$) (R : $CH_2 = C(CH_3)COOCH_2CH_2$; $CH_2 = C(CH_3)CH_2$)

(cyclohexene ring with $COOCH_3$ and $COOSn(C_4H_9)_3$) $CH = CH - CH = N - $(benzene)$- Sn(nBu)_3$

Their ability to homo. or to copolymerize was determined and found to be below that of methylmethacrylate.

Grafting was initiated by $Ni^o[P(OR)_3]_4$ where R is an alkyl group. It was shown that the organotin groups do not react with the chlorine atoms of the substrate as is the case when vinyl chloride is copolymerized with trialkyltin methacrylate.

Another technic consisted in the modification of C.R. grafted with acid monomers, by reaction with bis(tributyltin)oxyde.

$$\overline{\quad\underset{COOH}{|}\quad\underset{COOH}{|}\quad} + (Bu_3Sn)_2O \rightleftharpoons \overline{\quad\underset{COOSnBu_3}{|}\quad\underset{COOSnBu_3}{|}\quad} + H_2O$$

Water is eliminated with the help of azeotropic distillation or by addition of a dehydrating agent.

The grafted coatings have the advantage of stabilizing a thin layer of water containing the biocide agent which is maintained in the immediate neighbourhood of the surface and is not immediately dispersed in the sea.

The properties of the antifouling paints prepared with these grafted biocides show a net superioricy compared to those obtained by mixing of a biocide polymer with chlorinated rubber.

REFERENCES

(1) J.P. Kennedy and E. Maréchal, Carbocationic polymerization, Chapt. 8 - J. Wiley and Sons (1981).

(2) H. Kawabe, Bull. Chem. Soc. Japan 49, 2043 (1976) and preceeding papers.

(3) S. Dragan, I. Petrariu and M. Dima, J. Pol. Sci. Polym. Chem. 19, 3077 (1972).

(4) Thanh Dung N'Guyen, A. Deffieux and S. Boileau, Polymer 19, 423 (1978), Tetrah. Lett. 28, 2651 (1979).

(5) J.M.J. Fréchet, J. Macromol. Sci. Chem. A 15, 877 (1981).

(6) P. Noguès, F. Dawans and E. Maréchal, Makromol. Chem., 182, 843 (1981) and to be published.

(7) F. Dawans, M. Devaud and D. Nicolas (Institut Français du Pétrole). French Patent 2, 401, 207 (22-8-1977).

(8) C.H. Bamford, Reactivity, Mechanism and Structure in Polymer Chemistry Edit. A. Ledwith and A.D. Jenkins (1973) p.52 and ref. therein.

(9) M. Lee and Y. Minoura, J. Pol. Sci. 13, 315, 1427 (1975).

(10) T. Otsu and M. Yamaguchi, J. Macromol. Sci. Chem. A3, 177 (1969).

(11) C.H. Bamford, Polymer 17, 321 (1976).

(12) Y. Inaki, M. Takahashi and K. Takemoto, J. Macromol. Sci. Chem. A9, 1133 (1975).

(13) J. Barton, F. Szocs and J. Nemcek, Makromol. Chem. 124, 28 (1969).

(14) J. Barton and M. Lazar, Makromol. Chem. 124, 38 (1969).

(15) K. Kimura, Y. Inaki and K. Takemoto, J. Macromol. Sci. Chem. A9, 1399 (1975).

(16) A. Guyot, M. Ceysson, A. Michel and A. Révillon, Information Chimie, 116, 131 (1973).

(17) M. Lee, H. Nakamura and Y. Minoura, Nippon Kagaku Kaishi 8, 1559 (1974).

(18) M. Lee, H. Nakamura and Y. Minoura, J. Pol. Sci. 14, 961 (1976).

(19) C.H. Bamford and R. Denyer, Trans. Farad. Soc. 62, 1567 (1966).

(20) C.H. Bamford and E.O. Hugues, Proc. Roy. Soc. A, 326, 469 and 489 (1972).

(21) C.H. Bamford and K. Hargreaves, Proc. Roy. Soc. A, 297, 425 (1967).

(22) C.H. Bamford, R. Denyer and G.C. Eastmond, Trans. Farad. Soc. 61, 1459 (1965).

(23) M. Lee and Y. Minoura, Nippon Kagaku Kaishi, 1, 169 (1972).

(24) T. Otsu, M. Yamaguchi, Y. Takemura and Y. Kusuki, Polym. Lett. 5, 697 (1968).

(25) T. Otsu and M. Yamaguchi, J. Pol. Sci. A 6, 3075 (1968).

(26) S. Iwatsuki, H. Kasahara and Y. Yamashita, Makromol. Chem. 104, 254 (1967).

(27) C.H. Bamford, G.C. Eastmond and F.J.T. Fildes, Chem. Comm. 1970 p.144.

(28) E. Koerner Von Gustorf, Z. Naturforsch. 21B, 42 (1966) and 18B, 503 (1963).

(29) C.H. Bamford and S.U. Mullik, J. Chem. Soc. Farad. Trans. 1, 69, 1127 (1973).

(30) C.H. Bamford and S.U. Mullik, J. Chem. Soc. Farad. Trans. 1, 71, 625 (1975).

(31) C.H. Bamford and M.U. Mahmud, J. Chem. Soc. Chem. Comm. 1972 p.762.

(32) C.H. Bamford and S.U. Mullik, J. Chem. Soc. Soc. Farad. Trans. 1, 72, 368 (1976).

(33) Y. Amano and T. Uno, J. Chem. Soc. (Japan), Pure Chem. Sect. 86, 1105 (1965).

(34) F. de Schrijver and G. Smets, J. Pol. Sci. A1, 4, 2201 (1966).

(35) J. Barton, M. Lazar, J. Nemcek and Z. Manacek, Makromol. Chem. 124, 50 (1969).

(36) G.F. Bloomfield, J. Chem. Soc. 1943 p.289.

(37) M. Troussier, Rev. Gén. du caoutchouc 32, 229 (1955).

(38) C. Wartemberg, M. Brigodiot and E. Maréchal (to be published).

(39) G. Smets and W. De Coecker, J. Pol. Sci. 45, 461 (1960).

(40) J.J. Levison and S.D. Robinson, J. Chem. Soc. A, 1970, 96.

(41) K.J. Coskran, R.D. Bertrand and J.C. Verkade, J. Am. Chem. Soc. 89, 4535 (1967).

CATALYTIC ASYMMETRIC SYNTHESIS USING POLYMER SUPPORTED OPTICALLY ACTIVE CATALYSTS

John K. Stille

Department of Chemistry, Colorado State University, Fort Collins, Colorado 80523, USA

Abstract - Several styryl- and acrylate-type monomers bearing chiral phosphines have been synthesized and copolymerized with a crosslinking monomer and a monomer that is compatible with the solvent ultimately to be used in effecting a specific catalytic reaction. The exchange of a transition metal onto the crosslinked polymer containing the chiral phosphine yields catalysts that give chiral products from prochiral substrates in high entantiomeric excess. The synthesis of amino acids in high optical yield from acylaminoacrylic or acylaminocinnamic acids, for example, can be carried out utilizing a catalyst slurry. In this way, the product is easily separated from the catalyst and purified while the catalyst can be recovered and used again.

INTRODUCTION

A practical limit to performing homogeneously catalyzed reactions in the liquid phase is the difficulty in separating the product from the catalyst or in removing the product continuously. Amino acids, and dipeptides, for example, can be synthesized in high optical yields by the asymmetric hydrogenation of a prochiral substrate, α-N-acylaminocinnamic acids with optically active Wilkinson-type catalysts (Ref. 1). In these reactions, neither the rhodium nor the optically active phosphine is readily recovered, however.

To overcome the difficulty of separating product and catalyst, homogeneous catalysts have been attached to a variety of supports, including crosslinked polymers (Ref. 2). In this way, the catalyst acquires the property of insolubility but may retain the same reactivity as exhibited in solution. Most of the synthetic polymer supports for catalysts containing polymer-attached phosphine ligands coordinated to a transition metal, are crosslinked polystyrenes and have been synthesized by running various reactions on crosslinked polystyrene.

Until a few years ago, very few asymmetric polymer-attached ligands have been synthesized, so that the full potential of these polymer catalysts has not been realized. Polystyrenes containing optically active phosphine ligand sites have been prepared (Ref. 3-7), but the results, aimed at effecting asymmetric synthesis, have been somewhat disappointing. A chiral phosphine type ligand, similar to 2,3-O-isopropylidene-2,3-dihydroxy-1,4-bis(diphenylphosphino)butane (DIOP)(1) has been attached to a crosslinked polystyrene bead. The polymer-bound ligand was allowed to react with soluble rhodium(I) complexes to give

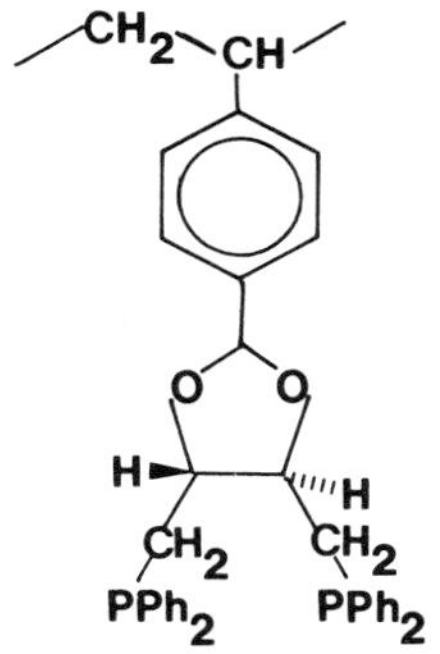

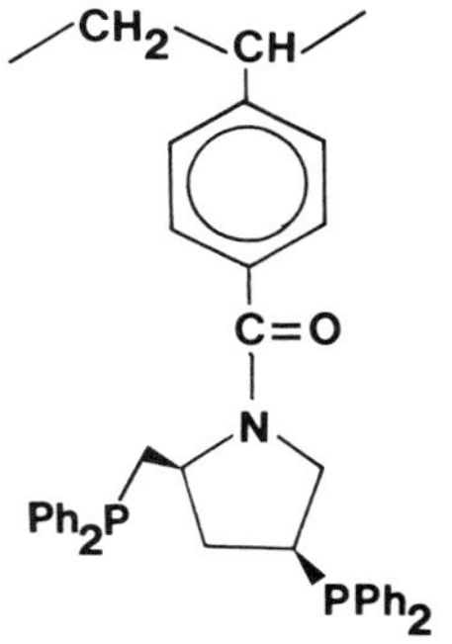

asymmetric hydrogenation (Ref. 3 & 7) or hydroformylation (Ref. 5) catalysts. The hydroformylation of styrene gave predominately 2-phenylpropanal in only 2% enantiomeric excess. Hydrogenation of α-substituted styrenes gave lower optical yields (1.5%) than the homogeneous analogue (15%) (Ref. 3). This is surprising, considering the optical yields that have been achieved with homogeneous DIOP-rhodium catalysts.

Although the crosslinked polystyrene catalyst swells in nonpolar solvents, it collapses in polar solvents, preventing the penetration of the substrate. Acylaminoacrylic acids are hydrogenated to acylamino acids in high optical yields in the presence of a homogeneous rhodium catalyst containing the DIOP ligand. This same catalyst bound to crosslinked polystyrene beads will not hydrogenate acylaminoacrylic acids in nonpolar solvents, since the substrates are not soluble (Ref. 3). In polar solutions of the substrates, the beads collapse, apparently preventing entry of the acylaminoacrylic acid to the catalyst site.

Another polymer supported optically active phosphine catalyst, prepared by the copolymerization of 1-(4-vinylbenzoyl)-2-(S),4(S)-2-diphenylphosphino-4-diphenylphosphino-methylpyrrolidine with hydroxyethyl methacrylate (2) has been described. While this system (containing rhodium) was able to catalyze the reduction of itaconic acid to methyl succinic acid in about the same optical yield as the homogeneous analog (82 vs. 89% o.y.), the reduction of Z-acetamidocinnamic acid to N-acetyl phenylalanine proceeded in far lower optical yield (23 vs. 91% o.y.) with the polymer supported catalyst. When the rhodium(I) species bound to the polymer was changed from a neutral to a cationic species, the optical yield for the cinnamic acid hydrogenation was raised to 70%, still far below the results obtained with the homogeneous catalyst.

The overall objective of this research has been, therefore, to design appropriate polymer-attached chiral catalysts for the catalytic syntheses of chiral products, particularly from the hydrogenation, and hydroformylation, of prochiral substrates. This has been accomplished primarily by attention to the synthesis of the appropriate polymer supports.

RESULTS AND DISCUSSION

The most challenging problems in polymer-supported catalysis are encountered in the choice of the polymer matrix and the synthesis of the catalyst site in the matrix. Most of the reported methods require the introduction of a reactive site on a crosslinked polystyrene bead, followed by the reaction of an optically active phosphine-containing ligand at the site. Our approach, the synthesis of ligand-bearing monomer, followed by its copolymerization with a second monomer, has several advantages. First, the optical purity of the ligand on the monomer can be assured. Second, the concentration of the ligand-bearing monomer in the polymer can be controlled, and polymers containing a wide range of ligand concentrations can be synthesized. Third, depending on the comonomer, and thus the reactivity ratios of the two monomers, isolation of the ligand-bearing monomer can be assured. Fourth, the nature of the polymer backbone, polar or nonpolar, can be varied depending on the selection of the comonomer. Fifth, varying degrees of crosslinking may be introduced.

Chiral monomers

Four different types of monomers have been synthesized, only one of which **R-6** is a monophosphine chiral at phosphorous. The remaining monomers are chiral at a carbon adjacent to phosphorous or one carbon atom removed therefrom. Although monomer **3** does not contain the phosphine ligands, the tosyl groups may be displaced by diphenyl phosphide once the monomer has been copolymerized. Both the R,R enantiomers of monomers **3** and **4** have been synthesized as well as the S,S isomers shown. In addition, monomers **3−5** contain chelating groups, and therefore should not be as susceptable to leaching of the transition metal from the support with repeated use.

SS-3   SS-4   R-5   R-6

Because at this time there is no reliable way to predict which catalyst enantiomer will hydrogenate a prochiral substrate to the desired enantiomeric product it is advantageous to have potential accessability to both enantiomeric phosphines. Further, it is a distinct advantage to utilize available optically active precursors for the syntheses of these phosphines rather than working with racemic mixtures that must be subjected to resolution procedures at some stage during the synthesis. The enantiomers of tartaric acid, both of which are readily available, are the starting compounds for the synthesis of **RR** and **SS-3** (Ref. 8 & 9), while L-hydroxyproline is the natural product available for the synthesis of **SS-4** by a relatively direct pathway (Ref. 10). Its epimer, **RR-4** also can be obtained from L-hydroxyproline, but by a more tedious synthetic scheme (Ref. 10). Monomer **R-5** is derived from D-manitol (Ref. 11). The monophosphine, **R-6**, requires resolution as a menthyl ester of a phosphinic acid in the course of the synthesis (Ref. 12).

Catalytic asymmetic hydrogenations

Of the methods of asymmetric synthesis, the generation of the enantiomeric product from a prochiral reactant by use of an optically active catalyst or enzyme has several advantages, the most important of which is that resolution, if necessary, is achieved with the catalyst, instead of with the product. Thus, large amounts of product of high enantiomeric excess can be obtained from catalytic quantities of optically active catalyst. This is the situation in the hydrogenation of Z-aminocinnamic acids to S-amino acids with a number of homogeneous optically active Wilkinson-type catalysts. Since the Z-acylaminocinnamic acids are not soluble in non-polar solvents, their hydrogenation is effected in ethanol. As a result, it is necessary to prepare a polar polymer support such that the resin will swell in the reaction solvent allowing accessibility to the transition metal sites.

Polymer synthesis. The radical polymerization of **3** with hydroxyethyl methacrylate (HEMA) was effected with azobisiso-butyronitrile (AIBN), using the reactivity ratios of styrene and HEMA to incorporate 8 mol% of **3**. Since commerical hydroxyethylmethacrylate contains ethylene dimethacrylate, the copolymer was obtained as an insoluble, crosslinked material that swelled in polar solvents. Conversion of the tosylated polymer to the phosphinated polymer was accomplished by reaction with sodium diphenylphosphide in THF. Enough sodium diphenylphoshide was used to react with the hydroxyl functions in the HEMA portion plus the tosylate groups. Reaction of this polymer with the bis(ethylene)chlororhodium dimer gave the optically active polymer catalyst **7**, that swelled in alcohol but could be filtered easily (Ref. 8 & 9).

SS-7

Diphosphine monomer **SS-4** was copolymerized with hydroxyethyl methacrylate by free radical initiators using ethylene dimethacrylate as a crosslinking agent. In a similar fashion, a polymer containing the enantiomeric phosphine, **RR-4** was prepared (Ref. 10). Elemental analyses of the resulting copolymers showed incorporation of the diphosphine monomer corresponding to the monomer feed ratio of 3-5%. This relatively low percentage incorporation coupled with a crosslink density of 10% was maintained in order to insure the isolation of the catalyst sites on the polymer backbone. The polymers were white free flowing powders that swelled in polar solvents such as ethanol. The reaction of these polymers with rhodium(II) in the form of the biallylchlororhodium dimer or the cyclooctadienechlororhodium dimer gave the light yellow polymer bound catalysts, **SS-8** and **RR-8**.

Similarly, polar polymer **R-9** (Ref. 11) and **R-10** (Ref. 12) containing phosphine moncmers **R-5** and **R-6**, respectively, were prepared by copolymerization reactions with hydroxyethyl methacrylate and a crosslinking monomer, followed by exchange with a rhodium(I) complex. Initially, polymer **R-10** contains one chiral phosphine unit and diolefin coordinated to rhodium as a result of cleavage of the chlorine bridge by phosphine. On addition of hydrogen and substrate, however, during which the diolefin is removed, possibly two of the chiral phosphines became coordinated to rhodium.

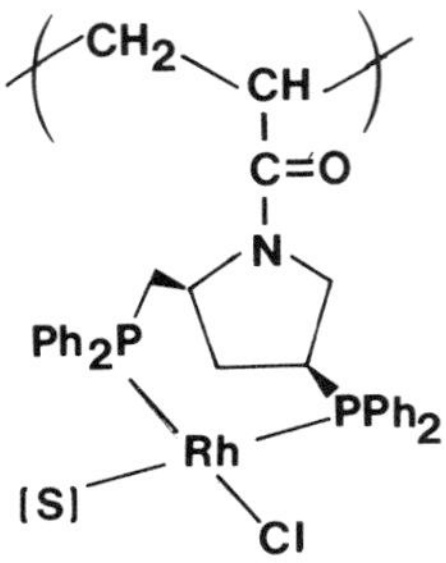

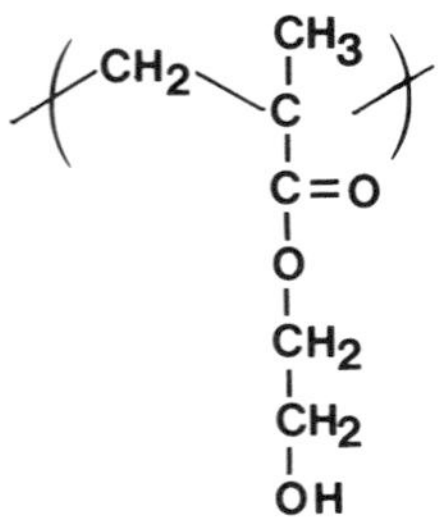

SS-8

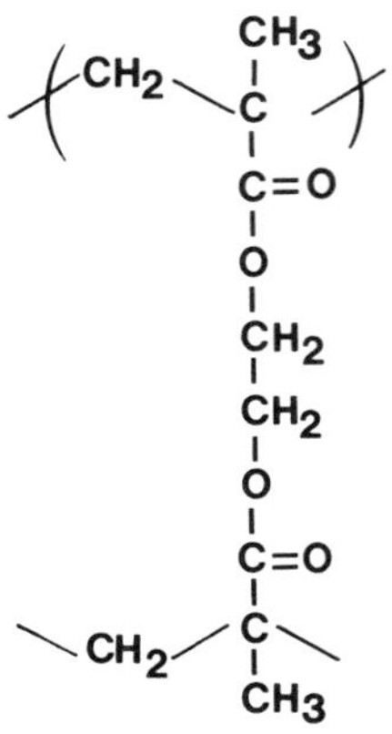

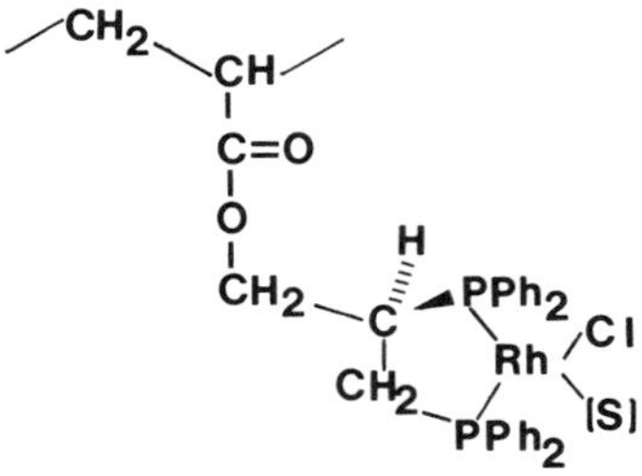

R-9 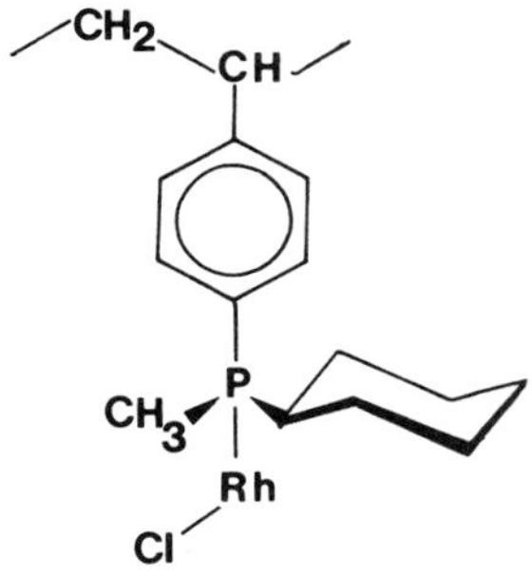　　　　　　　**R-10**

Hydrogenation results.　　In order to compare the % ee and the absolute configuration of the products of hydrogenation obtained from the polymer-attached catalysts with their homogeneous analogs (H-series), parallel reactions were run. Most hydrogenations were run at 25°C at 1-2.5 atm with a substrate to catalyst ratio of 100.

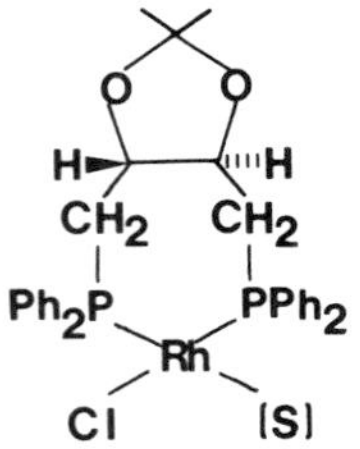

SS-7H 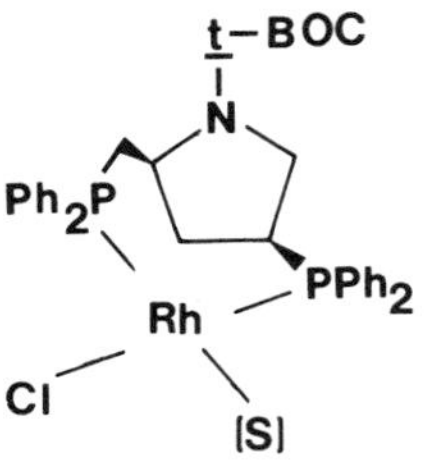　　　　　**SS-8H**

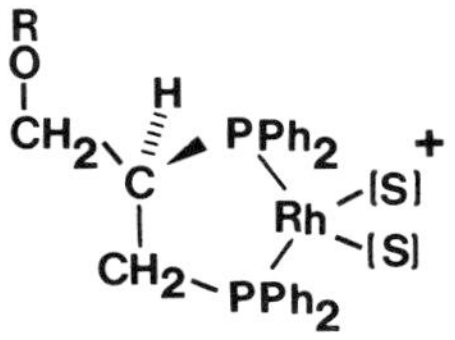

R-9H
R= a But-
　 b PhCH$_2^-$

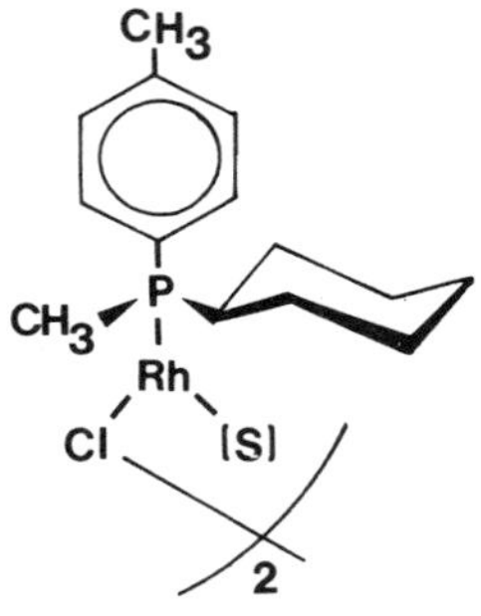

R-10

$$H_2 + \underset{H}{\overset{R}{}}C = C\underset{CO_2H}{\overset{R'}{}} \xrightarrow[\text{EtOH}]{7 \text{ or } 7H} R-CH_2-\underset{CO_2H}{\overset{R'}{\underset{|}{C}}}-H \qquad [1]$$

The high optical yields obtained with polymer-attached catalyst were comparable to those obtained with the homogeneous DIOP catalyst **7H**. The same absolute configurations of the products were observed Table 1. The rates of hydrogenation with polymer supported catalyst were slower than those of the homogeneous catalyst. The hydrogenation of α-acetamidoacrylic acid with **7H** was complete in less than 1 hour, whereas its hydrogenation with **7** required 5 hours to achieve 100% conversion.

Although the polymer-attached catalysts **7** was less sensitive to oxygen than the homogeneous DIOP catalyst, **7H** lower optical yields and slower rates were experienced after the catalyst had come in contact with air. The insoluble catalyst **7** could be reused many times by its filtration from the reaction solution, and this filtration generally was carried out under an inert atmosphere in order to preserve catalyst activity and the optical yield.

TABLE 1. Asymmetric hydrogenation of olefins by polymer-attached (**7**) and homogeneous (**7H**) DIOP-type catalysts (Ref. 8 & 9)

Substrate		Catalyst	Conditions			
R	R'	(SS-Series)	T°C	t(hr)	%ee	Config.
H	NHCOCH$_3$	**7**	28	5	52-60	R
		7H	25	1	73	R
Ph	NHCOCH$_3$	**7**	25	24	86	R
		7H			81	R
H	C$_6$H$_5$	**7**	25	24	58-64	S
		7H			63	S

Hydrogenation with BPPM-type catalysts **8** and **8H** require higher pressures (800 psi). The addition of triethylamine (6 mol % based on substrate) was essential for high optical yields. The reductions catalyzed by both polymers **SS-8** and **RR-8** gave the same optical yields as could be obtained by the homogeneous analogs Table 2.

$$H_2 + \underset{H \quad \quad COOH}{\overset{R \quad \quad NHAc}{\diagup\diagdown}} \xrightarrow[\text{EtOH/Et}_3\text{N}]{\text{8 or 8H}} RCH_2\overset{NHAc}{\underset{COOH}{CH}} \qquad [2]$$

TABLE 2. Hydrogenation of dehydroamino acids with polymer bound (**8**) and homogeneous (**8H**) BPPM-type catalysts (Ref. 10)

R	Homogeneous Catalyst		Polymer Attached Catalyst	
	8H	%ee (Config.)	**8**	%ee (Config.)
(phenyl)	SS	91(R)	SS	91(R)
	RR	90(S)	RR	90(S)
AcO-(phenyl)	SS	87(R)	SS	83(R)
	RR	87(S)	RR	87(S)
MeO, AcO-(phenyl)	SS	86(R)	SS	88(R)
	RR	88(S)	RR	88(S)

Hydrogenations with PROPHOS-type catalysts have been completed for the homogeneous analog, **R – 9H.** The results Table 3 again reveal that high optical yields can be obtained. Particularly significant is the fact that very high enantiomeric excesses can be obtained on the conversion of N-acylaminoacrylic acid to N-acylalanine. This represents the highest enantiomeric excess obtained in the hydrogenation of this substrate with any catalyst, enantiomeric excess in the range 70% being usual for other phosphine-rhodium catalysts. In these examples, the R-phosphine derived from the naturally occurring D-manitol gives the natural S-amino acids.

$$H_2 + \underset{H \quad \quad CO_2H}{\overset{R \quad \quad NHCOR'}{\diagup\diagdown}} \xrightarrow{\text{R-9H}} RCH_2-\overset{NHCOR}{\underset{CO_2H}{C}}-H \qquad [3]$$

TABLE 3. Asymmetric hydrogenation of dehydroamino acids with PROPHOS-type catalysts **R-9Ha** and **b** (Ref. 11)

Substrate R	R'	**R-9H**	%ee (Config.)
H	CH$_3$	a	92(S)
		b	90(S)
i-Pr	Ph	a	88(S)
		b	78(S)
Ph	CH$_3$	a	85(S)
		b	87(S)
Ph	Ph	b	74(S)
AcO—⟨ring⟩—CH$_3$ / MeO		b	91(S)

Hydrogenation of N-acylaminoocrylic acid with rhodium complexes of the type **R-10** at ambient temperature at 2-3 atm of hydrogen provided complete conversion to N-acetylalanine, but only in 13% ee. These low optical yields are not surprising, since N-acylaminocinnamic acid derivatives are hydrogenated by a rhodium catalyst containing optically pure cyclohexylmethylphenylphosphine ligands in less than 32% ee (Ref. 1).

Hydrogenations with polymers containing ancillary chiral alcohol sites. Polymers containing hydroxyethylmethacrylate units have the disadvantage that they could be expected to undergo hydrolysis and alcoholysis in continued use. Thus, copolymerization of phosphine-containing monomers with methyl vinyl ketone was of interest for a number of reasons. The polar character of the resulting copolymer could be expected to provide a suitable catalyst matrix for the asymmetric hydrogenation of α-N-acylaminoacrylic acids. The ketone function could be reduced to a secondary alcohol, another suitable polar group, and, under the appropriate conditions, could be asymmetrically reduced to an alcohol of either configuration. Both ketone and alcohol would be relatively stable under the reaction conditions.

The role of polymer in providing "cooperative effects" in enzymes and in certain other polymer-attached reagents is well documented (Ref. 13). The introduction into the polymer matrix of a second optically active site that does not directly participate in the asymmetric hydrogenation is of interest because it allows the observation of the effect of polymer environment on asymmetric synthesis. It has been demonstrated (Ref. 14) that solvent is involved in the transition state leading to the generation of the asymmetric center in the reduction of N-acyl-α-aminostyrene with the C1(DIOP)Rh(I) catalyst **7H**. Changing the solvent from ethanol to benzene not only affects the rates and optical yields, but, most remarkably, results in reversal in the absolute configuration of the product. The polar environment on the polymer surrounding the catalyst, especially an asymmetric polar environment, could be expected to have an effect on the optical yield. An asymmetric synergism from these ancillary polymer-bound groups could give the polymeric catalyst an important optical yield advantage over the homogeneous species.

The copolymerization of methylvinyl ketone within the presence of 2% divinyl benzene produced copolymer containing the methyl ketone chain. Hydrosilylation of the ketone functions was effected in THF with either **RR-** or **SS-7H**, and diarylsilanes. Phosphination of the polymer containing alcohol functions gave a polymer containing both the optically active phosphine sites and either R or S ancillary secondary alcohols groups on the polymer backbone with a 1.2 ratio of sodium diphenylphosphide to the sum of tosyl and hydroxyl groups, complete conversion of the tosyl groups to diphenylphosphide could be achieved. The rhodium-containing catalyst (**11**) was prepared as before, by the reaction with μ-di-chlorotetraethylene dirhodium(I).

The asymmetric hydrogenation of α-acetamidoacrylic acid with **7H** was first carried out to determine the effect of solvent on the optical yield and hydrogenation rates. In ethanol, the rates and optical yield corresponded to those reported (Ref. 14). However, in THF, slower rates and low optical yields were observed. The addition of ethanol to THF improved both the rates and optical yields.

SS-11

The asymmetric hydrogenation of α-N-acylaminoacrylic acids with various polymer catalysts, **SS – 11**, was carried out under a variety of reaction conditions to afford the amino acid derivatives Table 4.   In alcohol solvents, the high optical yields obtained with polymer-attached catalyst, **SS—11**, were comparable to those obtained with the homogeneous hydrogenation cataysts, and the same absolute configurations of the products were also observed.

TABLE 4.   Some representative asymmetric hydrogenations of N-acylaminoacrylic acids with **SS—11.** *

$-\overset{|}{\underset{|}{C}}-OH$	R[+]	Solvent	% Conv.	%ee[‡]	Config.
R	AcO⟨◯⟩ MeO	C_6H_6/EtOH(1/2)	100	78.5(83)	R
R	HO⟨◯⟩	C_6H_6/EtOH(1/2)	100	70.6(80)	R
R	C_6H_5	C_6H_6/EtOH(1/2)	100	83(81)	R
R	H	C_6H_6/EtOH(1/2)	100(2h)	76(70)	R
S	H	C_6H_6/EtOH(1/2)	100(2h)	74-78(70)	R
R	H	THF	28(24h)	40(6.9)	R
S	H	THF	28(24h)	24(6.9)	R

*Hydrogenations carried out at 25°C/1 atm H_2.

[+]For structure containing R, see Eq. 2.

[‡]Numbers in parentheses are those optical yields with **7H**.

When alcohol was present in the solvent, the chiral alcohol function on the polymer had no effect on the optical yield. As in the case of the soluble catalyst, **7H**, hydrogenations in THF not only reduced the rates but lowered the optical yield. In THF, the optical yields were not diminished to 7%, as in the case of **7H**, probably because of the availability of some of the alcohol functions on the polymer. Involvement of these ancillary alcohol sites in the transition state leading to the generation of the asymmetric center in the product is evident from the fact that a 15% difference in optical yield is obtained with otherwise identical catalysts containing the opposite chirality at the alcohol site. The presence of large amounts of solvent alcohol washed out this effect.   Unfortunately, higher optical yields in THF, where only polymer-attached chiral alcohol is available, were not obtained. This could be due to the inability of some alcohol sites to have enough mobility to become involved in the reaction.

Although an ancillary effect was observed with **SS—11**, there are several drawbacks to this approach. The reduction of the methyl ketones is not completely stereospecific, although nonpolymeric ketones have been reduced in 60-80% ee. In addition, it was possible that the polymer bound alcohol adjacent to the backbone did not have sufficient mobility to interact efficiently with the metal center. In order to overcome these problems, a series of polymers containing optically pure alcohols were prepared. These alcohols were removed from the polymer backbone by several atoms (spacer effect) to ensure greater mobility (Ref. 16).

An optically active comonomer suitable for use in preparing a polymer containing optically active pendent alcohols should be available in both R and S enantiomers so that the chirality of the alcohol and that of the catalyst may be matched to provide any synergistic effect. A suitable starting material for the synthesis of optically active comonomers is 2,3-butanediol, since the R,R isomer is commerically available, and the S,S isomer can be synthesized in a straightforward manner from tartaric acid.

Optically active acrylates were prepared from the diols in three steps to give optically pure monomers. A sample of the racemic monomer was also synthesized.

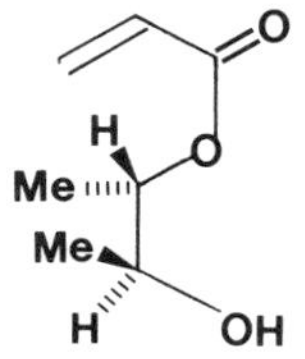

SS-12

Pyrrolidinephosphine containing polymers were prepared by the copolymerization of **SS—4** and **RR—4** with ethylene dimethacrylate and acrylates **12**(SS, RR, and racemic). Copolymerization of **3** with ethylene dimethacrylate and arylates **12** ultimately gave DIOP- containing polymers after phosphination. Polymers containing BPPM and hydroxyethylacrylate (**SS—7**) units were also prepared so that the effects of changing the polymer bound alcohols from primary to secondary could be observed. Rhodium-containing catalysts **13** and **14** were obtained by exchange with μ-dichlorobis—(1,5-cyclooctadiene)dirhodium(I).

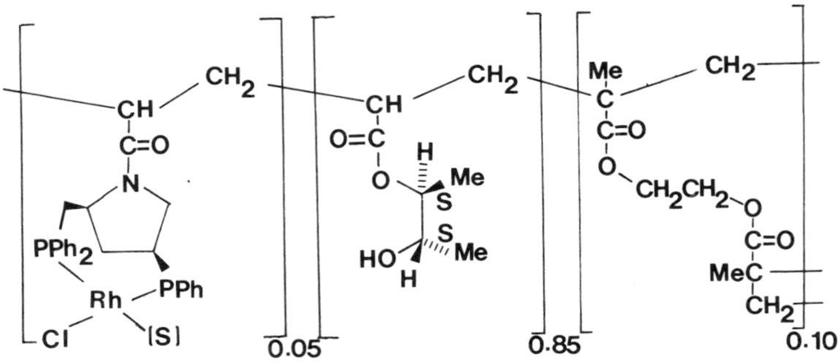

SS-13 [S,S-OH]

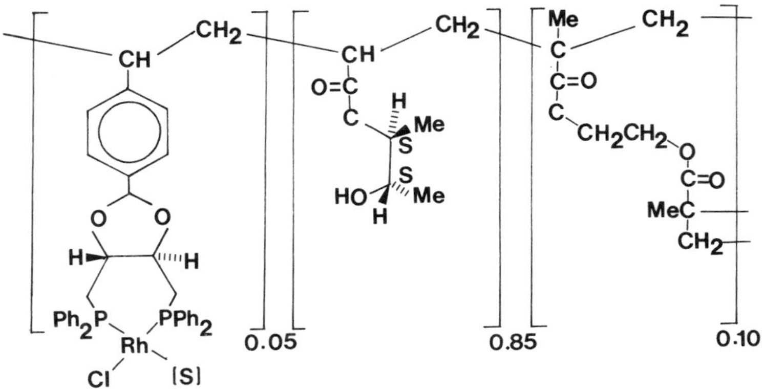

SS-14 [S,S-OH]

The substrate chosen to probe the effects of the optically active alcohols was 2-acetamidoacrylic acid. The hydrogenation of 2-acetamidoacrylic acid with pyrrolidinephosphine containing polymers gave disappointing results. The schematic in Fig. 1 is an interesting illustration of a polymer backbone influencing the behavior of a polymer bound catalyst. The homogeneous hydrogenation in THF with catalyst **SS—8H** proceeded in 15% ee. Switching to ethanol as the solvent led to a reversal of the predominant configuration from R to S. The use of alcohol containing polymers in THF gave values intermediate between the two homogeneous results. A difference of 6% ee was noted when the alcohol structure was changed from secondary alcohols to primary alcohols (**SS-13** racemic OH vs. **SS—8**) with the primary alcohol-containing polymer more closely mimicking the homogeneous reaction in ethanol. A difference of 11% ee was noted between reactions carried out with **SS—13** polymers that possess optically active pendent alcohols of opposite configuration.

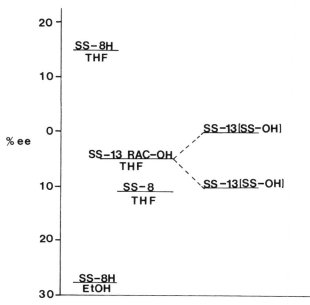

Fig. 1.   Enantiomeric excess obtained with BPPM-asymmetric catalysts.

If ethanol is considered to be a strongly interacting solvent and THF a relatively weakly interacting solvent, then the more bulky secondary alcohol would be expected to interact less efficiently with the catalyst site than primary alcohols as a result of steric interactions. The observed effect in the ee should be the ordering of the polymer catalysts from the weakly interacting pendent alcohols (**13**) to the more strongly interacting alcohols (**8**) with the weakly interacting support more closely resembling the reaction in THF. Polymer containing alcohols of R chirality apparently interacts more efficiently with the catalyst site than polymer which contains S alcohols, leading to an ee which is nearly that found when the primary alcohol containing polymer is used.

Similar results were obtained when DIOP containing polymers were used as catalysts Table 5. Again the gross effects of the alcohol structure were dominant, with a 20% difference in ee noted when polymers **SS—14** (racemic OH) and **SS—7** were used. As expected, the result obtained using polymer **SS—14** containing racemic alcohols, fell between the results obtained with polymers **SS—14** containing optically active pendent alcohols (S,S OH and R,R OH).

Thus, the polymer support can have a profound influence on the performance of polymer supported catalysts. The seemingly minor structural change of replacing a primary with a secondary alcohol as the predominant support structure led to a change of as much as 19% in the observed ee. In addition, the use of the alcohol containing polymer with THF as the reaction solvent, can result in ee's that differ by 42% from the results obtained with their homogeneous analogs. This is clear evidence that the pendent groups can act to provide a solvent environment around the catalyst site which is different from that which would be provided by the solvent alone.

The solvent interactions of chiral alcohols with the catalytic site have yet to be fully exploited. It is apparent that the number of phosphine-rhodium catalysts suitable for such an investigation would be limited to those which are relatively insensitive to changes in solvent from alcohol to THF. Unfortunately, this insensitivity may indicate that few solvent-catalyst interactions occur at critical steps during the reduction, lessening the chance of significant synergistic interactions to be exploited.

TABLE 5. Hydrogenation of 2-acetamidoacrylic acid with polymer bound DIOP

Phosphine	Solvent	ee (Config.)
SS－7H	EtOH	67(R)
SS－7H	THF	59(R)
SS－7	THF	26(R)
SS－14 [SS-OH]	THF	1(R)
SS－14 [RAC-OH]	THF	6(S)
SS－14 [RR-OH]	THF	10(S)

Catalytic asymmetric hydroformylations

Catalytic asymmetric homogeneous hydroformylation reactions with rhodium(I) catalysts are generally carried out in nonpolar solvents, so a crosslinked polymer catalyst containing a nonpolar backbone would be ideally suited for this reaction. The copolymerization of **3** with styrene and varying amounts of divinylbenzene (none to 20%) in a suspension system gave polymer beads 60μ in diameter. Subsequent reaction with sodium diphenylphosphide, or sodium dibenzophosphole, and hydridocarbonyltris(triphenylphosphine)rhodium(I) gave the hydroformylation catalyst **SS－15**. Electron microprobe analysis of these beads showed a uniform distribution of both phosphorus and rhodium atoms throughout the beads.

The hydroformylation of styrene with a catalyst prepared by attaching DIOP to a crosslinked, chloromethylated polystyrene, followed by exchange of rhodium(I) onto the phosphine sites, gave 2-phenylpropanal in only 2% enantiomeric excess (Ref. 5). Hydroformylation of styrene with **SS－15 ab** (both 20% crosslinked) gave 2-phenyl- propanal, generally in lower optical yields than had been obtained with the homogeneous analogues (Ref. 18-21) Table 6. The productaldehyde was very readily racemized under the reaction conditions, so a comparison of optical yields in this case was not too meaningful.

$$RCH = CH_2 + CO + H_2 \longrightarrow R - \overset{\overset{\textstyle CH_3}{|}}{\underset{\underset{\textstyle H}{|}}{C^*}} - CHO \qquad [4]$$

The selectivity to aldehydes was complete, and in no case were alcohols or alkanes detected. The production of branched chain aldehydes was quite pronounced with the polymer-attached catalysts, however. The branched to normal ratio (b/n) of aldehydes obtained from the hydroformylation of styrene with a soluble rhodium-DIOP catalyst was about 2 (Ref. 21), while the b/n ratios obtained with **15a** and **15b** were 5.2 and 21 under comparable reaction conditions.

TABLE 6. Hydroformylation of styrene and vinyl cyclohexane

Olefin	Catalyst	P(psi)	T(°C)	t(hr)	b/n	%ee
Styrene	**Rh**–DIOP*	16	40	192	2.2	25.2
	Rh–DIPHOL*	1500	80	28	8.4	30.9
	SS–15a	400	40	12	5.2	4.25
	SS–15a	45	25	72	2	11.4
	SS–15b	400	40	12	21	5.2
Vinyl Cyclohexane	**Rh**–DIOP*	16	50	168	0.025	–
	SS–15a	400	40	96	0.305	–

*DIOP or DIPHOL plus **Rh[H][CO][PPh₃]₃**

Hydroformylation of vinylcyclohexane with either homogeneous or polymer-attached catalysts gave predominantly linear aldehyde, probably both because of electronic effects and the fact that the vinyl group is more hindered. The amount of branched chain aldehyde was too low to determine the optical yield. However, even in this case, the polymer-attached catalyst produced a b/n ratio that was six times that of the homogeneous catalyst.

Cis-2-butene is hydroformylated by a homogeneous rhodium-DIOP catalyst to give aldehyde in 27% enantiomeric excess (Ref. 18). In this case, the product aldehyde is not racemized under the reaction conditions. Hydroformylation with **SS–15a** gave 2-methylbutanal in a 28% optical yield, demonstrating that the polymer-bound catalysts behaves identically to the homogeneous catalyst, at least in this olefin, where there is no regio choice Table 7. Although optical yields in the hydroformylation of 1-pentene were not determined, it is significant that, again, higher branched to normal ratios of aldehydes were obtained with the polymer-attached catalyst than with the homogeneous analogue.

TABLE 7. Hydroformylation of aliphatic olefins

Olefin	Catalyst	P(psi)	T(°C)	t(days)	b/n	%ee
Z-2-butene	**Rh**–DIOP*	16	20	30	–	27
	SS–15a	45	25	26	–	28.5
E-2-butene	**SS–15a**	45	25	21	–	7.2
1-pentene	**Rh**–DIOP‡	400	75	1	0.43	–
	Rh–DIOP*	16	25	1	0.074	–
	SS–15a	400	75	1	0.95	–
	Rh–DIPHOL‡	400	75	1	0.38	–
	SS–15b	400	75	1	1.03	–

*DIOP plus **Rh[H][CO][PPh₃]₃**

‡**Rh[CO][Cl][**DIOP**]** or DIPHOL

A possible explanation for this behavior lies in the steric hindrance, or lack thereof, presented by the polymer-attached catalyst to the coordinating olefin, although it is difficult to believe that the polymer chain would have this great an influence at the metal site. ^{31}P-T_1 measurements obtained on DIOP and polymer-attached phosphorous ligands show an interesting correlation with the b/n ratio Table 8. The shorter relaxation times and broader line widths are a result of slower molecular motion or longer correlation times, τ_c. Higher normal to branched ratios have been ascribed to increased steric hindrance in the transition state for insertion of the olefin into the meta hydride bond.

Our results indicate, therefore, that the homogeneous DIOP catalyst must present greater steric hindrance to insertion than does **SS-15a**. Certainly this would be the case if the molecular motion associated with the phosphine were greater in the case of DIOP than in that of **SS-15a**, i.e., the volume swept by phosphine per second is greater for DIOP than for polymer-attached DIOP. This enhanced molecular motion in the homogeneous catalyst, as compared with the polymer-attached catalyst, certainly is confirmed by the T_1 measurements.

TABLE 8. $^{31}P - T_1$ Measurements

Phosphine	T_1 (sec)	Line Width (Hz)
DIOP	10.5	1
DIOP on soluble polymer	1.75	22
DIOP on 20% crosslinked polymer	1.27	140
(with **Rh**) **SS-15a**	0.95	250
DIPHOL	10.4	1
DIPHOL on soluble polymer	2.9	20
DIPHOL on 20% crosslinked polymer	1.4	200
(with **Rh**) **SS-15b**	0.95	200

Acknowledgement - This research was supported by the National Science Foundation, Division of Materials Research.

REFERENCES

1. K.E. Koenig, M.H. Sabacky, G.L. Bachman, W.C. Christopfel, H.D. Barnstorff, R.B. Friedman, W.S. Knowles, B.R. Stults, B.D. Vineyard and D.J. Weinkaff, Ann. N.Y. Acad. Sci., 333, 16 (1980), and references therein.
2. C.U. Pittman, Catalysis by Polymer Supported Transition Metal Complexes, P. Hodge and D.C. Sherrington, Eds., Wiley, N.Y. (1980).
3. W. Dumont, J.C. Poulin, T.P. Dang and H.B. Kagan, J. Am. Chem. Soc., 95, 8295 (1973).
4. G. Strukul, M. Bonivento, M. Graziani, E. Cernia and N. Palladino, Inorg. Chim. Acta, 12, 15 (1975).
5. E. Bayer and V. Schurig, Chem. Tech., 212 (1976).
6. K. Achiwa, Chem. Letters, 905 (1978).
7. K. Ohkubo, K. Fujimori and K. Yoshinaga, Inorg. Nucl. Chem. Lett., 15, 231 (1979).
8. N. Takaishi, H. Imai, C.A. Bertelo and J.K. Stille, J. Am. Chem. Soc., 98, 5400 (1976).
9. N. Takaishi, H. Imai, C.A. Bertelo and J.K. Stille, J. Am. Chem. Soc., 100, 264 (1978).

10. G.L. Baker, S.J. Fritschel, J.R. Stille and J.K. Stille, _J. Org. Chem._, _46_, 0000 (1981).
11. P. Amma and J.K. Stille, _J. Org. Chem._, _46_, 0000 (1981).
12. P.D. Sybert, C. Bertelo, W.B. Bigelow, S. Varaprath and J.K. Stille, _Macromolecules_, _13_, 0000 (1981).
13. C.G. Overberger and K.N. Sannes, _Angew. Chem. Int. Ed. Engl._, _13_, 99 (1974).
14. H.B. Kagan, N. Langlois and T.P. Dang, _J. Organomet. Chem._, _90_, 353 (1975).
15. T. Matsuda and J.K. Stille, J. Am. Chem. Soc., _100_, 268 (1978).
16. G.L. Baker, S.J. Fritschel and J.K. Stille, J. Org. Chem., _46_, 0000 (1981).
17. S.J. Fritschel, S.J.H. Ackerman, T. Keyser and J.K. Stille, _J. Org. Chem._, _44_, 3152 (1979).
18. G. Consiglio, C. Botteghi, C. Salomon and P. Pino, _Angew. Chem. Int. Ed. Engl._, _12_, 669 (1973).
19. P. Pino, G. Consiglio, C. Botteghi and C. Salomon, _Adv. Chem. Ser._ _132_, _Homogeneous Catalysis_, 2, 295 (1974).
20. M. Tanaka, Y. Ikeda and I. Ogata, _Chem. Lett_, 1115 (1975).
21. C. Salomon, G. Consiglio, C. Botteghi and P. Pino, _Chimia_, _27_, 215 (1973).

RECENT DEVELOPMENTS IN THE USE OF POLYMERIC REAGENTS

A. Patchornik

Department of Organic Chemistry, Weizmann Institute of Science, Rehovot 76100, Israel.

Abstract - Recent developments in the field of polymeric reagents are reviewed with an emphasis on polymeric transfer reagents and on the exploitation of polymeric reagents for execution of reactions only possible because of the mutual inaccessibility of the reactive groups on the solid matrix. The advantages of insoluble polymeric reagents over their low-molecular-weight counterparts are critically discussed. Examples are given showing how polymeric reagents can be used advantageously to alter the courses of common reactions.

INTRODUCTION

A polymeric reagent is a novel type of substance possessing the physical properties of a high polymer as well as the chemical properties of the attached reagent. The polymer may be either organic or inorganic and the reactive group may be attached via either chemical bonds or physical interactions. These reagents are useful reactants in conventional or unconventional *(vide infra)* chemical or biochemical transformations. Polymeric reagents have, therefore, developed over the past decade from a somewhat exotic and esoteric research area into a fast-growing field, making its valuable contribution to biochemistry, organic synthesis, specific separations, and analysis. For previous reviews see Reference 1.

"Polymeric reagents" are formed either by attaching a low-molecular-weight reagent to a preformed polymer or by polymerizing a monomer, containing an extra functionality which does not interfere with polymerization [eq.(1)].

$$\text{P} + A \longrightarrow \text{P} - A \xrightarrow{B} \text{products} \tag{1}$$
$$M - A$$

P = polymer

A,B = soluble reactants

M-A = polymerizable monomer

The advantages of polymeric reagents over conventional low-molecular-weight reagents have been outlined many times before (1), and are given here in brief. Polymers and polymeric reagents are easily separable from low-molecular-weight compounds; with insoluble polymers, simple filtration or centrifugation is usually sufficient; and with soluble polymers, ultra-filtration or selective precipitation is required. This obvious feature of the polymeric reagent has been its most widely utilized property. It enables one, for example, to use a large excess of either soluble or polymeric reagent in order to increase reaction rates and yields. The easily removable excess starting material may often be reused. Moreover, the polymer can usually be easily cleaned of soluble reactants and side products. This feature makes the polymeric reagent usable either in columns or in batch processes, and it may be recycled many times. This often compensates for the time and capital investment required for preparing the polymeric reagents. The polymers also solve the problems of lability, volatility, toxicity or odor, often experienced with standard reagents.

Because the polymer acts as an immobilizing medium for the attached species, reactions which in solution are possible only at extremely high dilutions may be carried out at relatively high concentrations on polymeric carriers. In addition, the polymeric backbone may be chosen or tailormade to provide a special microenvironment for reactions of the pendant reactive group. Thus, special electronic and steric conditions, significantly different from those existing in the bulk solution may be created in the close vicinity of the reacting species.

Most of the reported chemical uses of polymeric reagents may be classified according to a few general types of processes. Such a general classification is presented in Table 1. This classification, however, is by no means exhaustive or absolute. Some applications of polymeric reagents may properly fall into more than one of the reagent groups listed in Table 1.

TABLE 1. General classification of processes utilizing polymeric reagents

Polymer property	Type of process or reagent
A. Facile separation from low molecular weight compounds	1. Catalyst 2. Specific separation 3. Transfer agent 4. Carrier and blocking group
B. Immobilization of attached species	1. High dilution 2. High concentration
C. Microenvironmental effects	1. Utilizing polar effects 2. Utilizing steric effects
D. Mutual inaccessibility of polymeric reagents	1. Analysis 2. Synthesis

Others may be included only with difficulty in any of these categories:

(A1) *A polymeric catalyst* may contain an enzyme, an inorganic compound, or an organometallic compound as its active center. Its reaction may be represented schematically by eq.(2).

$$\text{(P)}- A + \text{substrate} \longrightarrow \text{(P)}- A + \text{products} \tag{2}$$

Upon completion of the process, the catalyst can be washed and reused without difficulty. The convenience and economy of such catalysts have engendered widespread interest.

(A2) In *specific separation*, polymer-attached reagents are used to selectively bind one or a few species out of a complex mixture [eq.(3)]. Separation of the polymer-bound compound followed by its release from the polymer consitutes a simple highly effective separation method in biochemistry, hydrometallurgy and analysis.

$$\text{(P)}- A + B,C,D,E \longrightarrow \text{(P)}- A \text{------} B + C,D,E \xrightarrow{\text{filter}} \text{(P)}- A \text{----} B \longrightarrow \text{(P)}- A + B \tag{3}$$

(A3) *Polymeric transfer agents* react with low-molecular-weight reactants, transferring to them a functional moiety, thus yielding products which may often be obtained in pure form simply by filtering the polymer and removing the solvent [eq.(4)]. The use of a large excess of polymeric reagent leads, in many cases, to very high yields, and the reagent can often be regenerated by a single synthetic step.

$$\text{(P)}\!\!\begin{array}{l}\alpha \sim A\\ \alpha \sim A\end{array} + B \longrightarrow \text{(P)}\!\!\begin{array}{l}\alpha\\ \alpha \sim A\end{array} + A-B \tag{4}$$

A,B = soluble reactants
α   = attachment site of reactant A
∿   = reactive chemical bond

The use of several transfer polymers in series constitutes a "chemical cascade" which may greatly simplify multistep syntheses.

(A4) When a polymer is used in synthesis as a *carrier or as a blocking group* it acts either to make a chemical reagent easily separable in multistep syntheses or to block selectively certain functional groups in multifunctional compounds. Synthesis may require one or more steps, and the successively modified reactant remains attached to the polymer. The final product is then released from the polymer in a separate reaction [eq.(5)].

$$\text{(P)} + A \longrightarrow \text{(P)}- A \xrightarrow{B} \text{(P)}- A-B \xrightarrow{C} \text{(P)}- A-B-C \longrightarrow \text{(P)} + A-B-C \tag{5}$$

As large excesses of soluble reactants may be used in each step, very high yields of polymeric products are often obtained by this technique.

(B1) A different use of polymeric reagents relies on the fact that a rigid polymer retards ("*immobilizes*") translational motion of reactive species attached to it. When diluted on the polymer these reactive species may participate in highly specific reactions without the self-interactions [eq.(6)] that would occur in solution.

$$\text{Solution:} \quad A + B \longrightarrow A-A + A-B$$

$$\text{Polymer:} \quad (6)$$

(B2) Additionally, when reactants are *highly concentrated* on the polymer, unusual kinetic and thermodynamic effects may be observed, due to the spatial proximity imposed by the polymer backbone [eq.(7)]

$$(7)$$

(C1,2) *Microenvironmental effects* created by the chemical and steric structure of the polymer backbone may also be used to influence the outcome of chemical processes taking place in them. Thus, the polarity of the polymer may influence the reactivity of functional groups attached to it to a greater extent than the polarity of the solvent. The specific steric requirements of the channels and pores of a crosslinked polymer may impart size and structure selectivity on substances attached to the polymer or diffusing into it, thereby affecting reaction outcomes.

(D1) A recent use of polymers takes advantage of the fact that reactive groups on different insoluble polymers are *mutually inaccessible*. One application of such reactions has been the detection of highly reactive, relatively short-lived reaction intermediates [eq.(8)].

$$(8)$$

A* = short-lived species

The free existence of such species is proved by their formation in one insoluble polymer and their being trapped in another.

(D2) Simultaneous application of two or more polymeric reagents has been used in synthesis [eq.(9)] involving reagents which are incompatible with one another in their soluble forms.

$$(9)$$

The highly active species A*, formed by reaction of a precursor A with polymer (P)-C, immediately reacts with the second polymer yielding the product A-B.

While the classification scheme outlined is instructive and useful, most studies to date belong to the category of "facile separation". Historically, the first polymeric reagents prepared (catalysts, polymers used for separations and transfer agents) were of this class due to the early application of the ease of separation of these reagents. It may be assumed that in the future, when polymeric reagents are no longer of such novelty, greater stress will be given to processes where the polymer offers not only obvious technical advantages, but also modifies reactivity of attached species, thereby changing the outcome of common reactions. However, relatively simple, easily reusable polymeric reagents will probably be the first to be used industrially.

In this presentation we will limit our discussion primarily to the use of polymeric reagents as transfer agents, along with some novel applications which take advantage of the mutual inaccessibility of these reagents, which enables simultaneous application of several polymeric reactants which would in their soluble form be incompatible.

TRANSFER AGENTS

A "polymeric transfer agent" is a polymeric reagent that transfers a functional moiety to -- or one that accepts such a group from -- a soluble reactant [eq.(4)]. With such transfer agents, the desired reaction product is usually released into the solution in high yields, an excess of polymeric reagent often being required.

One general advantage of transfer polymers over carrier polymers (e.g. those used in solid-phase peptide synthesis) is that as the reaction proceeds, remaining reactive groups on the transfer polymer become increasingly accessible to the soluble reagent. In carrier polymers, however, accessibility decreases with progress of the reaction. This advantage was first applied to peptide synthesis by Fridkin et. al.(2). We prepared polymeric active esters of N-blocked amino acids which were used to acylate free-amino peptides in solution. An additional advantage of this technique is the fact that the growing peptide remains in solution throughout the multistep synthesis, thus allowing its purification, if necessary, after every step.

An efficient acylating polymer 1 was prepared by Kalir et. al.(3). Using the macroporous polymer which contains 2-nitrophenyl active esters of amino acids, Fridkin et. al.(4) synthesized the decapeptide leutinizing hormone releasing hormone (LHRH) in 40% overall yield. Reaction times, however, are in the range of several hours; hence, a novel polymer 2 carrying 1-hydroxybenzotriazole active esters was introduced (5). This polymer provided

an 100-fold improvement in reaction rates as well as high product yields, and when used to synthesize the tetrapeptide Boc-Leu-Leu-Val-Tyr(Bzl)-OBzl requiring bulky, usually slow amino acids, coupling steps took as little as 20 min each (6). However, because of the high reactivity, the active ester derivatives of this polymer must be protected from moisture and cannot be used with primary alcohols as solvents.

Recently, we have prepared a new, promising nitrophenol-type polymeric reagent by Friedel-Crafts acylation of polystyrene by substituted benzoyl chlorides (7,8)[eq.(10)].

The resulting carbonyl function serves both as an anchor to the polymer and as a *para* activator. Thus, active esters of this polymeric alcohol are some 40 times more reactive than the corresponding esters of the polymeric nitrophenol 1. Synthesis of protected Leu-enkephalin Boc-Tyr(Bzl)-Gly-Gly-Phe-Leu-OBzl in 92% yield was accomplished, with each coupling step requiring about 1 h (7). Active esters of this polymeric alcohol are insensitive to water or alcohols in neutral solutions; this makes for convenient handling.

Many applications of polymeric transfer reagents to other synthetic tasks are described in the literature (Ref. 1). We will discuss herein some of the more recent developments.

Kawabata and his co-workers (9) showed that poly(4-vinylpyridine hypobromide) is a potent reagent for the oxidation of secondary alcohols, yielding the corresponding ketones in good yields and high purity. The polymer may be regenerated electrochemically, *in situ*, while the oxidation is taking place. Thus, only catalytic quantities of polymeric reagent are required (9) [eq.(11)].

electric current

$$(P)-\bigcirc-NH^{\oplus}Br^{\ominus} + H_2O \longrightarrow (P)-\bigcirc-NH^{\oplus}OBr^{\ominus} + H_2$$

(11)

O
‖
R   R'

OH
|
R-CH-R'

Menger et. al. showed that in the presence of $BF_3.Et_2O$, the polymeric reagent poly(vinylpyr-idine,BH_3) reduces ketones, aldehydes, acids and esters (10). The authors show that partial alkylation of the pyridine rings and preswelling substantially improve polymer performance.

A chlorinating polymer containing dichloroiodobenzene moieties was introduced two decades ago by Okaware et. al.(11). Recently, however, Bongini et. al.(12) prepared a similar poly-meric reagent, utilizing a commercially available anion-exchange resin loaded with ICl_2^- anions. The latter polymer apparently is easier to prepare (12). Good yields and select-ivity were obtained when a fourfold excess of this mild chlorinating agent was used. However, when using equimolar amounts of polymeric reagent, iodine containing side-products were formed (12).

Ion-exchange resins carrying cyanide and thiocyanate anions were prepared by Harrison and Hodge (13) for the conversion of halides into the corresponding nitriles, thiocyanates and isothiocyanates.

An interesting silylation reagent was introduced by Murata and Noyori (14). They prepared a stable trimethylsilyl derivative of a commercial perfluorosulphonic acid polymer (Nafion), and used it to silylate alcohols, thiols, acids, amines and phenols.

A novel polymeric fluorinating agent was recently prepared by Banks et al. (15). They used a polymeric form of the Yarovenko-Raksha fluoroakylamine reagent to convert alcohols into their fluoro analogs under mild conditions in good yields. Attempts to prepare this poly-meric reagent by modification of a commercial polystyrene resin failed due to crosslinking of the polymer. The reagent was, however, successfully prepared by polymerizing the corres-ponding monomer (15).

An interesting application of a polymeric Wittig reagent to the selective monoolefination of dialdehydes was devised by Castels and co-workers (16). With respect to selectivity, the polymeric reagent was superior to reaction in solution. Hodge et al.(17) demonstrated the possibility of using aqueous base and phase transfer catalysis in order to perform Wittig re-actions on polymeric supports.

An ionic type of attachment of a Wittig reagent to a solid support is described by Cainelli and co-workers (18). The acidic moieties of cyano- and carbomethoxy-substituted phosphonates were shown to form stable conjugates with anion-exchange resins. High yields of olefins were obtained upon reaction with ketones. Use of these Wittig polymers in *wolf and lamb synthesis* is discussed in the proceeding section.

Manecke and Stärk (19) describe a novel polymer carrying N-acylimidazole moieties which acylates amines in a selective manner.

A solid support bearing light-sensitive anchoring groups (i.e. the 2-nitrobenzyl groups) was introduced by us for the solid-phase synthesis of oligosaccharides (20). Using this and other photosensitive resins, Rich and Gurwara (21) and Wang (22) demonstrated the potential of photoremovable solid supports in peptide synthesis. The 5-bromo-7-nitroindolinyl (Bni) group has been used by us both to protect the carboxylic function and to activate it by light irradiation (23). It was successfully used to condense various peptide segments in solution. We are currently examining the applicability of Bni-containing supports to solid-phase condensation of peptide segments.

MUTUAL INACCESSIBILITY OF POLYMERIC REAGENTS

The fact that insoluble polymeric reagents do not react with one another can be utilized profitably for mechanistic studies as well as for synthesis.

Mechanistic studies

The existence of free highly reactive species as reaction intermediates has been demonstrated for several chemical transformations using an original method termed the "three-phase test" (24). Here two insoluble polymeric reagents are mixed in a common vessel. One reagent is a precursor of the active species in question; the other is a trapping agent for this species. If, after reaction of the polymers with an appropriate initiator, a reaction product of the active species is found on the "trapping polymer", this constitutes proof of the existence of a free active species which had to diffuse from one polymer to the other.

Many exciting examples of such studies are described in a recent review by Rebek (24). Tests for determining reaction intermediates, including cyclobutadiene, acylimidazole, and ketene are discussed.

A similar approach was used by us to prove the existence of free oxocarbonium ion in the Fries rearrangement (25). O-acylation of insoluble hydroxynitrobenzyl polystyrene in the presence of aluminum chloride by a second insoluble polymeric reagent - O-acyl nitrophenyl polystyrene - provided such proof.

Synthetic applications

Multipolymer systems of insoluble polymeric reagents are unique in that reactions can only take place within the matrix of the polymeric reagent. This spacial localization of various reaction steps to their respective polymeric phases can lead to advantages not found in conventional reactions in solution, where all components can interact simultaneously.

First, we will describe two-stage reactions in which a starting material is modified successively by two polymeric transfer reagents. The analogous soluble reagents react with each other rapidly in solution, but they are rendered mutually inactive upon their attachment to the respective polymeric phases. These reactions were accordingly termed "Wolf and Lamb" reactions (after Isaiah 11, 6). Then we will discuss consecutive type reactions in which a soluble starting material is passed through a series of polymeric reagents, each causing a modification of the compound and leading to a new soluble product. These were termed "Cascade" reactions.

"Wolf and Lamb" reactions

A general scheme for these reactions is as follows:

$$(12)$$

A and B react with each other in solution, (P)-A and (P)-B are the corresponding insoluble polymeric reagents which cannot interact when mixed. An indirect reaction between (P)-A and (P)-B is possible only through the mediation of a third, soluble "messenger" reagent S: it will react first with (P)-A leading to a soluble product SA, the latter moving to (P)-B and reacting with it to give the end product SAB. It is impossible to carry out the corresponding reaction with soluble reactants because A and B will interact, and the overall reaction must be separated into two steps: first, a reaction of S with A, followed by a reaction of SA with B.

We examined the advantages of the polymeric reagent approach in reactions, both at equilibrium conditions and under kinetic control. In either case, the two-polymer system proved to be superior to a two-stage conventional reaction carried out with soluble reagents.

At equilibrium, the system in solution may be described as follows [eq.(13)].

$$S \underset{}{\overset{A}{\rightleftharpoons}} SA \underset{}{\overset{B}{\rightleftharpoons}} SAB \qquad (13)$$

The reaction of one equivalent of A with one equivalent of S will not lead quantitatively to SA. Equilibrium concentrations of the three components S, A and SA will depend on the equilibrium constant. The same is true for the reaction of SA with B. If an excess of A is used to promote SA formation, this excess should be removed prior to the introduction of B, otherwise an ensuing reaction between A and B will lower the yields and complicate the separation process. With polymeric reagents, on the other hand, excesses of (P)-A and (P)-B may be used simultaneously, without polymer interaction. S will react with the excess of (P)-A and will lead to quantitative formation of SA, which will further react with the excess of (P)-B, yielding the desired final product. Excess (P)-A and (P)-B are easily separated by filtration, eliminating laborious work-up procedures.

Under kinetic control the following scheme may be presented [eq.(14)]:

$$
\begin{array}{ccc}
& A & B \\
S & \longrightarrow SA^* & \longrightarrow SAB
\end{array}
$$

side products   (14)

The reaction of S and A leads to an active species SA* whose reaction with B gives the
desired product SAB. However, SA* may also react with unchanged starting material S or
undergo spontaneous decomposition, to give side products. In solution, a certain reaction
time is needed to complete the transformation of S to SA*. If B is introduced before
quantitative formation of SA* has taken place, an undesired reaction will occur between B
and unreacted A. During this time, however, side reactions involving SA* molecules already
formed may occur. The extent of these will depend on the various reaction rate constants.
Here again, utilization of a two-polymer system with excesses of ℗-A and ℗-B may be helpful: any SA* molecule formed can immediately undergo a reaction with ℗-B, already available in the reaction mixture, and unlike the reaction with soluble reagents, there is no
need to wait for quantitative formation of SA*.

In the following section these two schemes are illustrated with acylations of carbanions, and
the advantages in the simultaneous use of two polymeric reagents are confirmed. In these
examples, the "A"'s are strong bases, "B"'s are acylation reagents of the active ester type
and "S"'s are ketones, nitriles, esters, etc., possessing α-hydrogens abstractable by A.

Reactions at equilibrium. When acylating carbon acids such as nitriles, esters or ketones,
having two or three α-hydrogens, the mono-acylated product has α-hydrogens more acidic than
those in the starting material. In the benzoylation of phenylacetonitrile, for example,
treating an equivalent of it (S) with one equivalent of a strong base such as trityllithium
(A) will produce one equivalent of the anion (SA). The anion will be formed quantitatively
due to the large difference between the pKa values of triphenylmethane and phenylacetonitrile.
However, addition of an equivalent of an acylation reagent such as an active ester of benzoic
acid (B) will not lead to the quantitative formation of the acylation product (SAB) because
of proton exchange between the product and anion molecules not yet acylated [eq.(15)]:

$$
C_6H_5CH_2CN + (C_6H_5)_3C^- \rightleftharpoons C_6H_5CH^-CN \xrightarrow{\quad\text{NO}_2\text{-}\bigcirc\text{-OCOC}_6H_5\quad} C_6H_5\underset{\underset{COC_6H_5}{|}}{C}HCN
$$

(15)

$$
\xrightarrow[\quad\quad]{C_6H_5CH^-CN} C_6H_5CH_2CN + C_6H_5\underset{\underset{OCOC_6H_5}{|}}{C}^-CN
$$

The result is an incomplete reaction and formation of mixtures. The concentrations of the
various species at equilibrium depend on the pKa values of A, SA and SAB. The increased
acidity of the acylation product vs. that of the starting material makes proton interchange
a common problem in acylations. It is impossible to add excess of trityllithium to promote
the formation of the product since this excess will react competitively with the acylation
reagent.

According to our suggested two-polymer approach, a strong polymeric base and a polymeric
acylation reagent could be used to obtain the desired product in high yield.

To accomplish the "wolf and lamb" acylations we developed (26) a strong polymeric base,
polymeric tryllithium 4 [eq.(16)]:

(16)

$$
℗\text{-}\bigcirc + (C_6H_5)_2CHX \xrightarrow{AlCl_3} ℗\text{-}\bigcirc\text{-}\underset{\bigcirc}{\overset{\bigcirc}{CH}} \xrightarrow{RLi} ℗\text{-}\bigcirc\text{-}\underset{\bigcirc}{\overset{\bigcirc}{C^{\ominus}}}\ Li^{\oplus}
$$

X = Cl, OH

R = CH₃, n-C₄H₉   3   4

Polymer 3, carrying triphenylmethane moieties was prepared by reacting polystyrene with benzhydrol or benzhydryl chloride in a Friedel-Crafts reaction. With benzhydrol, lower loadings (0.7 mmol/g) of trityl groups than with benzhydryl chloride (2.3 mmol/g) were obtained. The latter reaction also required less catalyst. Therefore, polymer 3 obtained from the chloride was further used as our polymeric reagent.

The blood-red polymeric anion 4 was produced by reacting polymer 3 with excess methyllithium or n-butyllithium in THF or in 1,2-dimethoxyethane. In ordinary use, polymer 4 with a loading of ca. 1.5 mmol/g of tityllithium groups, was prepared by applying a 100% excess of n-butyllithium in THF for 2 h at 0°C.

After polymer 4 was prepared, excess n-butyllithium was washed out and the side arm turned so as to add the acylating polymer -- an active ester of polymeric o-nitrophenol 1 -- into the reaction chamber. The "wolf and lamb" effect was clearly demonstrated, as the blood-red polymeric tityllithium and the white polymeric active ester did not change their colors upon mixing, whereas their soluble analogs reacted immediately in solution. Color changes did occur, however, when molecules carrying an acidic hydrogen were introduced. The red polymer 4 was converted to its white protonated form 3, and the acylation of the carbanion produced the orange polymeric o-nitrophenol 1 as its o-nitrophenolate salt [eq.(17)]:

$$(17)$$

In the benzoylation of phenylacetonitrile, the use of the two-polymer system offers a substantial advantage over the reaction in solution. Molecules of phenylacetonitrile, reformed at the acylating polymer as a result of proton interchange between the product and anion molecules, can move back to the polymeric base (present in excess), become deprotonated, acylated, etc., until product formation is complete [eq.(17)]. The stoichiometry of the reaction shows that two equivalents of base are required per one equivalent of starting carbon acid (both polymer 1 and the product will be ionized). We used a 150% excess of polymer 4 over that quantity and a 100% excess of the acylating polymer. Under these conditions, the reactions were complete within less than 15 minutes.

Table 2 demonstrates the results of the "wolf and lamb" reactions vs. control reactions in solution.

The above described reactions result in an ionized product. No second acylation was evident although excesses of the basic and acylating polymers were still present. This was further checked with two of the products. They were reacted in a separate experiment with one equivalent of base, then with an excess of the soluble acylation reagent. No further acylation was detected.

Polymers carrying para-substituted trityl groups were also prepared, and these provide stronger basic groups, higher loadings and reduced steric crowding (26).

Reactions under kinetic control. Ester enolates can undergo self-condensation, a reaction which competes with the desired acylation. When acylating an ester in solution using a strong base (such as tityllithium) and an acylation reagent, quantitative enolate formation must be attained (as evidenced by the disappearance of the red color of the base) before the acylation reagent can be introduced, otherwise it will react with the remaining base. During this waiting period, some self-condensation may occur, depending on the reactivity of the enolate, its concentration, etc. For example, when γ-butyrolactone is benzoylated, two products, α-benzoyl-γ-butyrolactone and α-(1-benzoyloxy-1-tetrahydrofuryl)-γ-butyrolactone are produced, the latter resulting from self-condensation [eq.(18)]:

TABLE 2. "Wolf and lamb" reactions at equilibrium *vs.* control reactions

Starting material	Acylation reagent	Product	Yield	Yield of reaction in solution
$C_6H_5CH_2CN$	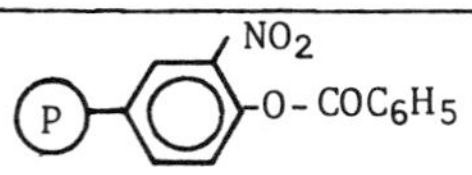	$C_6H_5CH(CN)COC_6H_5$	94%	45%
$C_6H_5CH_2CN$	$-CO_2Et$	$C_6H_5CH(CN)CO_2Et$	91%	37%
CH_3CN	$-COC_6H_5$	$C_6H_5COCH_2CN$	90%	27%
$C_6H_5COCH_3$	$-COC_6H_5$	$C_6H_5COCH_2COC_6H_5$	96%	48%
$C_6H_5COCH_3$	$-CO_2Et$	$C_6H_5COCH_2CO_2Et$	92%	40%
$CH_3CON(CH_3)_2$	$-COC_6H_5$	$C_6H_5COCH_2CO(CH_3)_2$	92%	42%

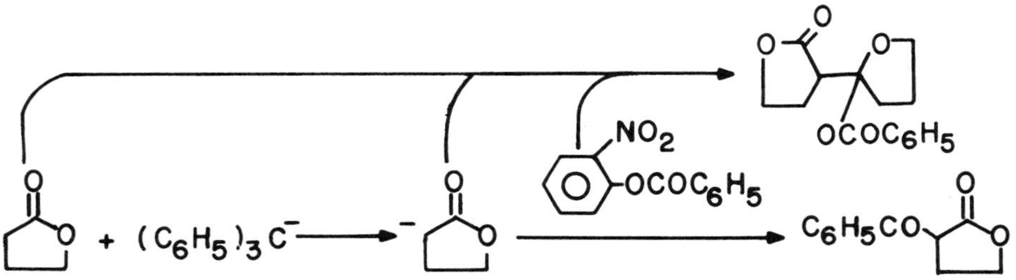

(18)

When the same reaction is carried out as a "wolf and lamb" reaction, using polymeric reagents (26), no waiting for complete enolate formation is necessary. The acylating polymer, present from the start in excess, can immediately react with enolate molecules formed at the basic polymer, without reaction of the two polymers. Thus, the concentration of enolate molecules can be kept low and side products are not detected [eq.(19)]:

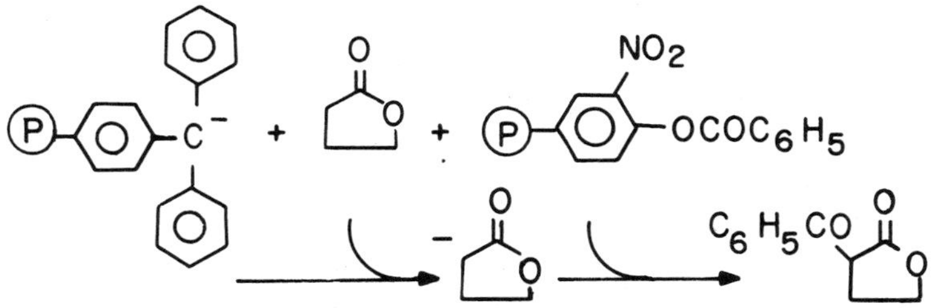

(19)

The results of several "wolf and lamb" acylations of esters and yields of the control reactions in solution are shown in Table 3.

The solution reactions produced various side products attributed to self-condensation. These impurities were avoided with the polymeric approach. In addition to the kinetic advantages, the "wolf and lamb" reactions also proved to be superior by providing high yields of

TABLE 3.   "Wolf and lamb" reactions under kintic control *vs.* control reactions

Starting material	Polymeric acylating agent	Product	Yield	Yield of reaction in solution
CH_3CO_2Et	(P)—⟨NO₂⟩—$O-COC_6H_5$	$C_6H_5COCH_2CO_2Et$	92%	38%
CH_3CO_2Et	$-CO_2Et$	$CH_2(CO_2Et)_2$	88%	35%
$C_6H_5CH_2CO_2Et$	$-COC_6H_5$	$C_6H_5CHCO_2Et$ $\overset{|}{C}OC_6H_5$	98%	47%
$C_6H_5CH_2CO_2Et$	$-CO_2Et$	$C_6H_5CH(CO_2Et)_2$	92%	35%
$C_6H_5CH_2CH_2CO_2Et$	$-COC_6H_5$	$C_6H_5CH_2\overset{COC_6H_5}{\underset{|}{C}}HCO_2Et$	92%	40%
(lactone structure)	$-COC_6H_5$	(lactone with COC_6H_5)	95%	31%

acylation, a result of overcoming the proton exchange phenomenon in the particular esters being reacted, which had at least two α-hydrogens.

A recent example of "wolf and lamb" reaction for preparing olefins from ketals is reported by Cainelli et al. (18).   They introduced the novel polymeric Wittig reagent described above, together with an acidic polymer into the reaction mixture. The ketal is cleaved by the acidic polymer and the resulting ketone is able to react with the Wittig reagent. While the two polymers do not interact, their analogs will neutralize each other in solution.

Cascade reactions
A starting material S may be passed through a series of polymeric reagents, each one causing a modification of the compound, thus leading to a "chemical cascade".

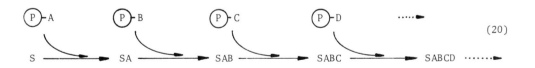

$$(20)$$

The utilization of a large excess of the polymeric reagents (P)-A, (P)-B etc., contained in suitable, interconnected vessels may lead to a pure product obtained in a fast and easy to control way. Multi-step processes of this kind may be suitable for large-scale automatic synthesis.

Stern et al. (27) applied groups of three different types of polymeric reagents to the stepwise synthesis of peptides. Polymeric active esters were used for chain elongation, polymeric 2-thiopyridine for N^α-Nps protecting group removal (28), and a basic polymer for neutralization of the N-terminal amino acid. The scheme was repeated for each coupling step.

A similar approach was studied by us (7). Polymer 2 carrying active esters of different Fmoc-protected amino acids (28) were used consecutively to prepare Leu-enkephalin. Since polymeric reagents for removal of the Fmoc group were found to be too slow, we used dimethyamine for this purpose, which due to its high volatility can be easily removed following the deprotecting step.

Pittman and Smith previously described (29) a reaction involving the simultaneous use of two polymeric organometallic catalysts for consecutive isomerization and hydrogenation of olefins. In this case, however, the use of polymeric reagents provides no protective advantage; non-polymeric forms of these catalysts would not interact in solution either.

CONCLUDING REMARKS

We have presented in this review an update of the latest advances of two aspects of the chemistry of polymeric reagents: the applications of the mutual inaccessibility of these reagents and the use of polymeric transfer agents. We have also included developments made by our own group, such as new acylating polymers, "cascade" reactions and initial examples of reactions using the "wolf and lamb" approach.

We have shown in this review that polymeric reagents, in many instances, exhibit advantages over their low molecular weight counterparts. Advantages such as unique reaction paths, possibility of reuse, ease of workup, stability and safety of these reagents result from the reactive moieties being bound to an insoluble polymeric matrix.

Limitations do however exist, such as extra cost and labor, the cumbersome analysis of polymers that is sometimes required, and the slower reaction rates. These are discussed in our previous review (1c). The use of polymeric reagents, therefore, requires careful considerations and weighing up of the possible advantages and shortcomings of the technique in any particular application.

REFERENCES

1. a. N.K. Mathur, C.K. Narang and R.E. Williams, Polymers as Aids in Organic Chemistry, Academic Press, New York (1980).
    b. P. Hodge and D.C. Sherrington, Eds., Polymer Supported Reactions in Organic Synthesis, Wiley, New York (1980).
    c. M.A. Kraus and A. Patchornik, Macromol. Rev. 15, 55-106 (1980).
2. M. Fridkin, A. Patchornik and E. Katchalski, J. Am. Chem. Soc. 88, 3164-3165 (1966).
3. R. Kalir, M. Fridkin and A. Patchornik, Eur. J. Biochem. 42, 151-156 (1974).
4. M. Fridkin, E. Hazum, R. Kalir, M. Rotman and Y. Koch, J. Solid-Phase Biochem. 2, 175-182 (1977).
5. R. Kalir, A. Warshawsky, M. Fridkin and A. Patchornik, Eur. J. Biochem. 59, 55-61 (1975).
6. M. Fridkin, Y. Stabinsky, V. Zakuth and Z. Spirer, Biochim. Biophys. Acta 496, 203-211 (1977).
7. B.J. Cohen, Ph.D. thesis, Weizmann Institute, Rehovot, Israel (1979).
8. A. Patchornik and B.J. Cohen, in A. Eberle, R. Geiger and T. Weiland, Eds., Perspectives in Peptide Chemistry, pp. 118-128, Karger, Basel (1981).
9. J.-i. Yoshida, R. Nakai and N. Kawabata, J. Org. Chem. 45, 5269-5273 (1980).
10. F.M. Menger, H. Shinozaki and H.-C. Lee, J. Org. Chem. 45, 2724-2725 (1980).
11. M. Okawara, Y. Kurusu and E. Imoto, Kogyo Kagaku Zasshi 65, 1647-1652 (1962).
12. A. Bongini, G. Cainelli, M. Contento and F. Manescalchi, J. Chem. Soc., Chem. Commun. 1278-1279 (1980).
13. C.R. Harrison and P. Hodge, Synthesis 299-301 (1980).
14. S. Murata and R. Noyori, Tetrahedron Lett. 21, 767-768 (1980).
15. R.E. Banks, A.-K. Barrage and E. Khoshdel, J. Fluorine Chem. 17, 93-98 (1981).
16. J. Castells, J. Font and A. Virgili, J. Chem. Soc., Perkin Trans. 1 1-6 (1979).
17. S.D. Clarke, C.R. Harrison and P. Hodge, Tetrahedron Lett. 21, 1375-1378 (1980).
18. G. Cainelli, M. Contento, F. Manescalchi and R. Regnoli, J. Chem. Soc., Perkin Trans. 1 2516-2519 (1980).
19. G. Manecke and M. Stärk, described in G. Manecke and P. Reuter, Pure Appl. Chem. 79, 2313-2330 (1979).
20. U. Zehavi and A. Patchornik, J. Am. Chem. Soc. 95, 5673-5677 (1973).
21. D.H. Rich and S.K. Gurwara, J. Am. Chem. Soc. 97, 1575-1579 (1975).
22. S.-S. Wang, J. Org. Chem. 41, 3258-3261 (1976).
23. Sh. Pass, B. Amit and A. Patchornik, submitted for publication.
24. J. Rebek Jr., Tetrahedron 35, 723-731 (1979).
25. A. Warshawsky, R. Kalir and A. Patchornik, J. Am. Chem. Soc. 100, 4544-4550 (1978).
26. B.J. Cohen, M.A. Kraus and A. Patchornik, J. Am. Chem. Soc. in press.
27. M. Stern, A. Warshawsky and M. Fridkin, Int. J. Peptide Protein 17, 531-538 (1981).
28. Nps - 2-nitrophenylsulfenyl, Fmoc - 9-fluorenylmethyloxycarbonyl.
29. C.U. Pittman and L.R. Smith, J. Am. Chem. Soc. 97, 1749-1754 (1975).

THE EFFECT OF MACROMOLECULAR STRUCTURE ON THE BEHAVIOUR OF POLYMER CATALYSTS

Francesco Ciardelli, Mauro Aglietto, Carlo Carlini, Salvatore D'Antone, Giacomo Ruggeri, Roberto Solaro

Centro di Studio del C.N.R. per le Macromolecole Stereordinate ed Otticamente Attive, Istituti di Chimica Organica e Chimica Organica Industriale, University of Pisa(Italy)

Abstract - In synthetic polymer catalysts the macromolecular structure plays a substantially different role than in enzymes even if some analogies can be evidenced. Apart from the effect of possible heterogenization, polymer attachment gives systems containing several active species per macromolecule which can display a different catalytic behaviour from monomeric analogs because of cooperative interactions. The presence of other groups in the macromolecule can also be important for hydrophobic interactions with the substrate and for selectivity if groups with particular steric requirements are used. In case of transition metal - polymer complexes the differences with respect to monomeric analogs are in general determined by the influence of the macromolecular structure on the environment of the individual metal atoms. Examples are reported to stress influence of polymer structure on catalytic activity, selectivity and stereoselectivity.

1. INTRODUCTION

It is well known that enzymes are macromolecules thus when talking about polymer catalysts it is straightforward thinking to them (1). With enzymes in mind, polymer chemists have started investigating polymer catalysts deriving from synthetic macromolecules. Particularly in the last two decads a very large number of papers appeared dealing with the use of polymer bound catalysts which indeed had very little to do with enzymes, the only analogy being the macromolecular structure. The aim of this paper is not to discuss differences between enzymes and synthetic polymer catalysts, rather we will limit ourselves to an analysis of the latter to show how their features can be affected by a proper selection of the polymer matrix. Several good reviews (2-6) appeared recently on the argument showing the very broad spectrum covered by polymer catalysts, where these last are intended as systems involving a polymer either as active species or ligand or support for species of different type. Even in the first case however the whole macromolecule cannot be considered a single species but it contains many catalytic sites depending on the number of attached reactive groups. On the other side we wish to stress here that the macromolecular chain exerts in many cases, if not in all cases, marked effects on catalytic properties of attached functional groups. Accordingly the binding of an active species to a polymer chain results in general in a new catalyst with different structure, reactivity and selectivity. The rationalization of these effects is the valuable aim of research in synthetic polymer catalysis. In such connection concepts derived from enzymes can help both for understanding of results and designing of new systems, clearly remembering the basic differences between the two types of polymer catalysts (7). The important factors to keep in mind are:
- nature and structure of the polymer matrix
- type of active species and their environment
- number and distribution of active species in each macromolecule
- type of catalyzed reaction
- interactions with substrate and reaction medium

If the catalyst acts as a heterogeneous system additional effects may arise from the macroscopic structure of the supporting resin. In general however it is not easy to distinguish between real homogeneous and heterogeneous polymer catalysts which are better considered as "hybrid catalysts". Thus rather than discussing separately homogeneous and heterogeneous systems it seems more convenient to discuss in two distinct sessions metal free polymer catalysts and transition metals containing polymer catalysts. In our opinion the two groups present different characteristics also as far as the effect of the macromolecular chain is concerned. The discussion will be based on same examples selected on arbitrary basis, no attempt to be exhaustive being made, the interesting papers appeared in the field being really too many for the room available.

2. METAL FREE POLYMER CATALYSTS

As already mentioned synthetic polymer catalysts can be either soluble or insoluble in the reaction medium. Moreover polymers bound onium salts are used in liquid-solid-liquid "phase-transfer" processes (8). In these three different situations the polymer can exert different functions and its role in the reaction can be markedly different.
In the case of the soluble systems the macromolecule can act only at the molecular level thus resulting in i)cooperative effect between equal or different groups attached to the same macromolecule, ii)cocatalytic effect by groups present in the same chain together with the reactive functional group, iii) steric effects controlling selectivity and stereoselectivity. In the second case, in addition to the above effects, the polymer matrix can exert a macroscopic effect depending on the pores sizes,sites distribution on the resin and interaction with the liquid medium. In the third case finally the polymer acts as a phase insoluble both in the organic solvent and in water and contains ionogenic species active in transferring anions between the two liquid fases.

2.1. POLYMERIC CATALYSTS FOR HYDROLYTIC REACTIONS

Even if hydrolytic reactions have been also carried out in the presence of heterogeneous catalysts (2), the main part of work concerns with soluble systems for which the examples reported are really conspicous. An exhaustive review has been reported by Kunitake and Okahata (2) in 1976 and before analysing more recent results it is convenient to summarize some examples already considered by the above authors relevant to the present paper.
Overberger, who has been a real pionier in the field, firstly used poly-4(5)-vinylimidazole (9) and poly-5-vinylbenzoimidazole (10) in the hydrolysis of activated esters such as p-nitrophenyl-acetate (PNPA). This last substrate was chosen as the imidazole group is a too weak base for displacing from normal esters the more basic alkoxide anion, and has been later on generally used. The polymer showed greater efficiency for the solvolysis than imidazole, and this result was attributed to the cooperative action of anionic and neutral imidazole species coexisting in the macromolecular chain (11).
A further effect associated with the polymeric structure was evidenced by Overberger et al (12) who observed a considerable rate enhancement when using poly-4(5)-vinylimidazole for the hydrolysis of a long chain phenylester such as 3-nitro-4-dodecanoylbenzoic acid (NDBA) in water/ethanol mixture. The catalytic activity dependence on solvent composition was associated with conformational variations; indeed at 40-60 % of ethanol the maximum value of solution viscosity was observed paralleled by the lowest efficiency with PNPA as the substrate. The rate enhancement for NDBA was thus attributed to a local high concentration of active groups due to polymer shrinkage. Moreover the catalytic behaviour of the imidazole containing polymers can be furtherly controlled by proper comonomers inserted in the same macromolecule with 4(5)-vinylimidazole units. This can both affect the microenvironment around the imidazole groups and also provide cooperative interactions among different groups.
As far as the former point is concerned Overberger and Guterl (13) reported the synthesis of terpolymers with hydrophobic side chains which give rate enhancements on hydrolysis of phenyl esters. The polymeric catalysts were based on terpolymers of 4(5)-vinylimidazole, acrylamide and either 3-buten-2-one, 1-penten-3-one, 1-undecen-3-one or 1-hexadecen-3-one and 3-nitro-4--acyloxy-benzoic acids with the acyloxy group containing 2,7,12 and 18 carbon atoms were used as substrates, their concentrations being below the critical micelle concentration.
The keto groups in the terpolymer were reduced to alcohols in order to prevent intramolecular reactions. The main conclusions of this study have been that increasing the size of the terpolymer side chains increases hydrolysis rate faster than the corresponding increase in apolar weight of terpolymer, and an effective rate increase is also observed by increasing the number of carbon atoms, and then the hydrophobicity, in the acyloxy group of the substrate. Moreover the same weight percent of apolarity appears more effective on fewer but longer side chains

than on a larger number of shorter side chains.

$$—CH_2 -CH -CH_2 -CH -CH_2 -CH—$$
with side chains: CH-OH, (CH_2)_m, CH_3 ; CO, NH_2 ; imidazole ring

m = 0, 1, 7, 12

with COOH, NO_2, O-CO-(CH_2)_n-CH_3   n = 0, 5, 10, 16

Another approach made possible by the molecular structure and suggested by the knowledge of enzymes actions was based on the introduction in the same macromolecule of imidazole and hydroxyl
groups, which are known to interact cooperatively in the active site of serine esterase.
Overberger et al. found that the copolymer of 4(5)-vinylimidazole with p.vinylphenol is a better catalyst at high pH values than its monomeric or polymeric analogs toward neutral, anionic
and cationic substrates (14).
Other examples of cooperative interaction between imidazolyl group with carboxylic groups and
with hydroxyl plus carboxylic groups have been discussed by Shimidzu in a review paper (15).
Polymers containing hydroxamate, oximate and thiolate groups have been also used as macromolecular nucleophilic catalysts. In the esterolysis of activated esters they give however stable
acyl intermediates, thus not allowing a real catalytic process. It is therefore necessary to
introduce in the polymer chain additional groups capable of decomposing the acyl derivative
formed in the first step of the reaction, as shown for the poly(4-vinylpyridine) with attached
oxime groups used by Kabanov et al. (16).

$$—CH_2 -CH -CH_2 -CH— + CH_3COO-\langle\bigcirc\rangle-NO_2 \longrightarrow {}^-O-\langle\bigcirc\rangle-NO_2 +$$

with pyridine and quaternized pyridinium-N-CH_2-C=N-O^- side group

$$—CH_2 -CH -CH_2 -CH—$$
with pyridine and pyridinium-N-CH_2 side group, C = N-O-\overset{O}{\overset{\|}{C}}-CH_3

Kunitake and Okahata reported (17) that the problem can be solved by incorporating in the same
macromolecule N-phenylhydroxamate and methylimidazole units together with acrylamide units for
solubility in polar solvents.

$$—CH_2 - CH - CH_2 - CH - CH_2 - CH—$$
with CO-N(Ph)(OH), methylimidazole-CH_3, CO-NH_2 side groups

This terpolymer indeed showed acylation rate substantially similar to that of the copolymer
containing only N-phenylhydroxamate and acrylamide units, whereas the rate of hydrolysis of the
acetyl-hydroxamate intermediate was 60-80 times larger. As the authors stated, hydroxamate and
imidazole functions act complementarily in the nucleophilic process, acylation being faster for
the former and deacylation for the latter.
A similar cooperative effect seems to occur in case of catalyst developed in our laboratory
following the same lines of Kabanov and coworkers who prepared and used polymers with oxime
groups attached to partially quaternized poly(4-vinylpyridine) (16,18). Poly(4-vinylpyridine)
was partially alkylated(20 mol%) first with (S)-1-bromo-2-methylbutane and then(30 mol%) with
phenacylbromide followed by reaction with hydroxylamine to give the following terpolymer.

$$-CH_2 - CH - CH_2 - CH - CH_2 - CH-$$

This system in the presence of an excess of PNPA gives, at first, a rapid hydrolysis followed by the decrease of the reaction rate up to a stationary state. This behaviour can be explained assuming that the first step corresponds to the acylation of the oxime and the second to the hydrolysis of the N-acylated derivative assisted by free pyridine nuclei (19). The use of this catalyst with p.nitrophenolates of branched chiral acids show a remarkable steric effect on reaction rate but no chiral discrimination (19,20). A moderate discrimination between the two antipodes of racemic (2-methylbutyl)p.nitrophenolate was however observed in case of the terpolymer containing units from 4-vinylpyridine, N-methyl-4-vinylpyridinium halide and oxime of (S)-5-methyl-1-hepten-3-one.

$$-CH_2 - CH - CH_2 - CH - CH_2 - CH- \qquad -CH_2 - CH - CH_2 - CH - CH_2 - CH-$$

The stereoselectivity must probably be assigned to a local dissymmetric effect rather than to a chiral conformation of the macromolecule.

A potential catalyst where dissymmetric perturbation of reactive groups arises from the chiral macromolecular structure is that obtained by introducing oxime groups by quaternization of pyridine nuclei in the copolymers of 4-vinylpyridine with (S)-4-methyl-1-hexen-3-one (21). In this last copolymer indeed CD studies indicate that electronic transitions of pyridine chromophore are optically active, thus it could be expected that the active sites were located in a dissymmetric environment provided by the macromolecular conformation.

The contemporary presence of hydrophobic and electrostatic interactions for substrate attraction gives further increase of catalyst efficiency. Indeed poly(4-vinylpyridine) quaternized with short and long chain alkyls showed according to Okubo and Ise the maximum efficiency for p.nitrophenylpalmitate used in the presence of the most hydrophobic polymer catalyst. This result was attributed to the attraction of the hydroxide ion to the polycation by electrostatic forces and the accumulation of the hydrophobic ester around the polymer with hydrophobic side chains (22). Again the occurrence of the two effects would not be possible without the macromolecular structure. In the same direction more recent studies of Ise et al. (23) have shown that poly(4-vinyl-N-alkylpyridinium) salts, such as poly(4-vinyl-N-alkylpyridinium bromide), poly(4-vinyl-N-benzylpyridinium chloride) and copolymers of 4-vinyl-N-benzylpyridinium chloride with 4-vinyl-N-cetylpyridinium bromide exert pronounced catalytic effect on the hydrolysis of ester and amide containing indolyl groups, the observed rate enhancement being maximum for the copolymer followed by the benzyl containing homopolymer. This result suggested a possible contribution of charge transfer interactions in addition to electrostatic and hydrophobic interactions between the polymers and the substrates.

Klotz, stating again that enzymes are macromolecules and that their catalysis is based on i) binding of the substrate and ii) provision of a molecular environment conducive to the chemical reaction, indicated lines for the preparation of synthetic enzymelike polymer catalysts (*synzymes*) (6). The polymer backbone was provided by highly branched polyethylenimine to which apolar groups were attached to obtain the binding properties. Truly catalytic functional groups were then attached, mainly based on imidazole moiety. Such structures display catalytic activity under ambient conditions, that is at room temperature and pressure in aqueous media near physiological pH (6).

In the more recent years several other papers have appeared extending the concepts of cooperative interactions between different groups and of polymer substrate interactions to additional polymer catalysts such as for instance rigid copolymers containing 3,3'-substituted biphenyls and vinylimidazole (24), poly[*p*.vinyl(thiophenol)-*co*-acrylic acid] (25), and polyampholytes [polypropylenglycine, polyethylenalanine, polystyryl-*p*-(2-carboxybutylamine)] (26). As clearly stated by Kunitake (2), Ise (4) and Klotz (6) these studies are very limited as far as selectivity and stereospecificity are concerned.

Overberger and coworkers first attempted to obtain stereoselectivity by using optically active polymers containing imidazole, but no particular stereochemical effect was observed (27,28). We have already reported that a moderate enantiomeric discrimination has been observed with optically active oxime containing terpolymers (19). Stereoselective   hydrolysis of amino acid *p*.nitrophenyl esters has been also obtained in the presence of poly(ethylenimine)s containing covalently linked optically active L-histidine moieties. These last appear to be stereoselective both in the preequilibrium step(K_M) and in the kinetic activation(k_2) of the antipodes of N-carbobenzoxy-alanine-*p*.nitrophenolate, N-carbobenzoxy-phenylalanine-*p*.nitrophenolate and N-*t*.butyloxycarbonylglutamine-*p*.nitrophenolate. The constants ratios K_M (D)/K_M (L) and k_2 (D)/ k_2 (L) were in the range 0.49 -1.04 and 1.01 - 1.33, respectively (29). However there is still a long way to go before synthetic enzymes can approach stereoselectivity of natural polymeric catalysts.

2.2 POLYMERIC ONIUM SALTS IN LIQUID-SOLID-LIQUID PROCESSES

A comprehensive review has been very recently reported by Chiellini et al. (8) to which reference must be made for detailed information. The catalytic behaviour of these systems has been analyzed in terms of degree of crosslinking of the polymer matrix, content of active groups and spacing between chain and active functional group.

Regen observed that displacement of bromine in *n*.octylbromide by cyanide anion in the presence of polystyrene resins containing $-CH_2N^+(CH_3)_2C_4H_9$ cations decreases of one order of magnitude by increasing the starting divinylbenzene content from 0.5 to 8.0% (30). A similar, but more remarkable, effect has been observed by Montanari et al. (31) in the Br/I exchange of the same substrate with polystyrene containing $-CH_2P^+(n.C_4H_9)_3$. In this case an increase of crosslinking from 1.0 to 4.5% produces a 12-fold decrease of the rate constant.

The ability of anion exchange resins to swell in water can be modulated by varying the number of ionogenic groups, a reduction of these last producing a catalyst with a greater hydrophobic character, with consequent increase of the interactions with low-polarity organic substrates and decrease of those with water-affine anions. The data available up to now concern only the displacement of bromine by cyanide in *n*.octylbromide catalyzed by microporous trimethylamine quaternized styrene resins. These data indicate that the optimal concentration of ammonium groups is between 1 and 30% (8).

It has been in general observed that resins anchored onium salts display lower efficiency than the monomeric analogs probably because of steric hindrance and diffusion control. These negative effects can be released by moving the cation further from polymer backbone with a proper spacer group. The use of $-NH-CO-(CH_2)_{10}-$ as spacer between the *para* position of phenyl groups in the resins and the quaternary ammonium cation gives a substantial increase of the rate constant for Br^- displacement by I^- on *n*.octylbromide (31).

An anlogous trend has been observed for the *o*-alkylation of β-naphthol in the presence of similar catalysts (32). It is of interest to note that even with very long spacing groups the catalytic activity remains lower than for monomeric analogs (31).

No clear selectivity effect by the polymer matrix has been reported up to now in spite of few attempts which gave very modest selectivity if any (8,33).

More interesting results concern with the use of optically active materials for asymmetric synthesis. This topic has been discussed in detail in the already mentioned review (8). We would like to stress here that copolymers of optically active monomers with functional comonomers bearing onium groups could be conveniently used to demonstrate participation of the polymer to the reaction process. Indeed it has been clearly shown that transmission of chiral perturbation from optically active monomer units to functional comonomer units is determined by the secondary structure of the macromolecules (34). Copolymers of optically active α-olefins, alkyl acrylates or methacrylates with styrene have been conveniently functionalized to insert onium groups in the *para* position of aromatic nuclei (20). A large set of linear and crosslinked optically active onium salts with a different hydrophylic/hydrophobic balance can be prepared by a proper selection of comonomers mixtures. Up to now however no appreciable asymmetric induction was observed in the presence of these systems in heterophase reactions involving prochiral substrates, such as alkylation of phenyl-acetonitrile (33,36), dichlorocarbenation of styrene and benzaldehyde (35) and oxidation of racemic 2-octanol (36).

Copolymers of cinchona alkaloids with acrylonitrile (37,38) give asymmetric Michael reaction with optical yield between 24 and 42% that depends on the type and frequency of the optically active alkaloid units in the polymer thus suggesting a participation of the polymer matrix to the stereochemical control.

This was furtherly confirmed by the better optical yield obtained with the polymeric catalyst with respect to monomeric analogs.

2.3. POLYMERS CONTAINING PHOTOREACTIVE GROUPS

Straightforward influence of the polymer anchoring on properties of immobilized photosensitizing dye have been shown in case of Rose Bengal covalently anchored to an insoluble poly(styrene-*co*-divinylbenzene) matrix (39). This last provides for instance 1) increased photostability of the dye 2) possibility of isolating unstable primary photooxidation products 3) utility in solvents where the free dye is insoluble.

Thus the same photosensitizer was made usable in water by attachment to a hydrophylic ter-polymer from chloromethylstyrene, ethylene glycol monomethacrylate and bis methacrylate as crosslinking agent (40). The effectiveness of this system to sensitize singlet oxygen formation in water was proved by photooxidation of several substrates known to react with singlet oxygen in water.

A direct effect of the macromolecular structure on the photochemical behaviour of benzophenone chromophore has been shown by using poly(4-acryloxybenzophenone) as polymerization photoinitiator of acrylic derivatives (41).

The mentioned polymer showed upon UV irradiation in absence of air, a greater initiation efficiency than the monomeric analog. Addition of a tertiary amine improves the efficiency of both
polymer and monomer, the former being again more active. Finally a copolymer of 4-acryloxy-
benzophenone with 4-N,N-dimethylaminostyrene shows the lowest activity. These data suggest
that the high efficiency of the systems based on poly(4-acryloxybenzophenone) can be associated with exciton migration of energy along the chain through intramolecular interaction between excited and groud state benzophenone groups. The same explanation was put forward to explain the better efficiency of poly(4-vinylbenzophenone) with respect to benzophenone in hydrogen extraction from THF and polymerization of methylmethacrylate in THF (42).

3. POLYMER-METAL CATALYSTS

The basic principles governing the catalytic behaviour of polymer-metal complexes are substantially similar as for the already discussed metal free polymer catalysts. However the presence
of the metal species offers a very sensitive probe for investigating the effect of the polymer,
in this case more properly macromolecular ligand, on the catalysis. This clearly holds both
for metal-enzymes and synthetic polymer-metal catalysts, but as for the preceding section we
will limit the discussion to the latter field.
Examples are available of both homogeneous and heterogeneous transition metals-polymer catalysts, the two groups having been originated by different approaches. Soluble polymer-metal
complexes clearly indicate the aim of obtaining synthetic systems with some analogies with
metal-enzymes whereas the corresponding heterogeneous systems have been investigated with the
aim of obtaining stable catalysts with good reproducibility, activity and selectivity as homogeneous systems, accompanied by easy separation from the reaction products. Again the role of
the macromolecular chain can be different in the two cases as the use of an insoluble resin gives
macroscopic effects in addition to the expected influence at the molecular level.
We have recently reviewed the basic aspects concerning polymer supported transition metal catalysts (43) and indicated that the macromolecular ligand can affect catalyst activity, selectivity, and type of catalysed process of the attached transition metal complexes.

3.1. HETEROGENEOUS SYSTEMS

As already mentioned transition metal complexes have been in general supported on organic
polymers in order only to obtain heterogeneous catalysts maintaining the typical features of
the homogeneous analogs (3). However it soon appeared that the polymer ligand could not be
simply considered as an inert support and it became more and more interesting to look for differences rather than analogies between heterogeneous polymer supported and homogeneous monomeric transition metal complexes.
Polymer swelling by an appropriate liquid phase is of primary importance for catalytic activity;
a high degree of swelling can allow to eliminate possible diffusion problems and make all
active species available for the substrate as occurs for homogeneous catalysts.
It is not easy to exemplify this aspect in a really clean way. In our opinion a clear example
is offered by complexes obtained by reacting $RuBr(C_3H_5)(CO)_3$ with phosphinated polystyrene
samples having different molecular weight and stereoregularity (44). In the different complexes the metal surroundings are practically the same as shown by spectroscopic data and elemental analysis, this last indicating a P/Ru ratio near to one. The proposed structure of the
metal containing units is depitched below together with that of the monomeric analog in which
phosphinated polystyrene has been replaced by triphenylphosphine.

These systems are all active for catalytic isomerisation of olefins. The catalytic activity decreases when going from the homogeneous monomeric complex to the polymer supported catalysts. Moreover these last show decreasing isomerisation rate with increasing molecular weight and stereoregularity of the polymer, that is with decreasing swelling in toluene (44). A similar effect has been observed for the isomerisation of 1-pentene in the presence of polycarboxylate ruthenium halocarbonyl catalysts obtained from the sodium salts of different polymers with carboxylic side chains and $[RuCl_2(CO)_3]_2$ (45). In the case of the poly(sodium acrylate) system the catalyst with high content of polar -COONa groups ($-COO^-/Na$ = 1.5) swells very little in the substantially apolar liquid phase (toluene/ethanol = 25/1) and shows 25 times lower isomerisation rate than the analogous catalyst with much smaller content of such ionizable species ($-COO^-/Na$ = 28.3). Similar results were observed (45) with the terpolymer vinyl alcohol/benzyl vinyl ether/Na-maleate which, under the same conditions, shows decreasing activity with increasing the content of polar vinyl alcohol units.

It may be concluded from these data that the swelling of the polymer in the liquid reaction medium can control the accessibility of the substrate to the active metal sites. Thus, provided the metal environment is the same, the homogeneous system displays the maximum activity, while insoluble systems will show decreasing activity with decreasing swelling in the reaction medium.

The metal environment of the monomeric system is not always strictly maintained in the corresponding polymeric complexes, thus active species with modified catalytic activity can be formed by attachment to a polymer ligand. This is in general due to the polydentate nature of the macromolecules but also to the steric constrains imposed to the metal complex, which can also affect the reaction selectivity.

A first evidence of the modified metal environment is given by the presence, observed in several cases, of "coordinatively unsaturated" metal atoms ("spoiled species"). Their presence previously postulated by few authors (46,47) has been clearly demonstrated by Braca et al.(48) who assigned the appearance of e.s.r. activity in hydrido- and carboxylato-triphenylphosphine-Rh(I) complexes after air exposure to the formation of paramagnetic $Rh(II)O_2^-$ species through coordination of oxygen to "coordinatively unsaturated" diamagnetic Rh(I) species. In the polymer anchored complexes isomerisation rate of 1-pentene was found to increase for systems showing an increasing concentration of e.s.r. species (48). On this basis it is possible to interpretate the catalytic behaviour of polycarboxylate-Ru(II) complexes having a P/Ru ratio lower than the monomeric analog. Polymeric catalysts obtained by reacting $RuH_2(CO)(PPh_3)_3$ with polyacrylic acid (45) and with copolymers of maleic acid with several vinyl ethers (49) show increasing hydrogenation and isomerisation efficiency with decreasing the P/Ru ratio under

the stoichiometric value (x = 2) and in some cases the catalytic activity is even larger than for the homogeneous monomeric analogs where the P/Ru ratio is equal to 2.

The formation of species with a new unexpected catalytic activity is becoming more frequent in polymer anchored metal complexes. Thus Pd(II) species attached *via* anthranilic acid groups to a polymer matrix are able to hydrogenate benzene and nitrobenzene, whereas the N-benzylanthranilic acid complex is not active in such reaction (50,51). Also, polystyrene-supported η^5-cyclopentadienyl cobalt derivatives have been reported to be catalytically active in the hydrogenation of carbon monoxide to give hydrocarbons, whereas soluble η^5-cyclopentadienyl cobalt dicarbonyl is inactive and decomposes under the same conditions (52). A typical example of the different structure attainable in polymer-metal complexes with respect to monomeric analogs is given by the reaction of (η^6-cyclo-octa-1,3,5-triene)(η^4-cyclo-octa-1,5-diene)ruthenium(0) [(COT)(COD)Ru] with polystyrene compared to benzene (53). Whereas with this last under H_2 and in THF solution only COT is displaced by the arene to give (η^6-benzene)(COD)Ru, both olefinic ligands are displaced under the same conditions by the phenyl groups of polystyrene. This different behaviour can be clearly associated with the particular relative geometry of the aromatic groups provided by the macromolecular structure of polystyrene. This different structure gives origin to more versatile catalytic behaviour of the polymer catalyst, which not only hydrogenates and isomerises olefins as (η^6-arene)(COD)Ru, but is also active for the hydrogenation under relatively mild conditions of benzene and other aromatic hydrocarbons, ketones, nitrocompounds, oximes and nitriles (53).

The possible occurrence of steric effects due to the macromolecular nature of the ligand is documented by several examples indicating that in heterogeneous systems both porous structure of the resin and molecular structure of the chain can be effective in a number of ways. Grubbs and Kroll observed a dramatic decrease in hydrogenation rate of Δ^2-cholestene with respect to cyclohexene in the presence of the Wilkinson catalyst attached to phosphinated polystyrene beads that was attributed to the restriction of the size of the solvent·channels by the random cross-links (54). A similar effect was also observed by Innorta et al. (55) investigating the kinetics of hydrogenation of the olefins by catalytic systems prepared by anchoring the Wilkinson catalyst to 1.2 and 4% crosslinked resins from diphenylphosphinated styrene/divinylbenzene copolymers. They observed that hydrogenation rate of 1-hexene in a poorly swelling liquid medium as ethanol is the same as in benzene that has a greater swelling power. By contrast for more hindered cyclohexene hydrogenation rate is almost twice in benzene with respect to ethanol.

Selectivity can be also affected by combined steric effects from the macromolecular chain and other low molecular weight ligands around the metal atom. Catalysts derived from $[RuCl_2(CO)_3]_2$ and sodium salt of several polycarboxylates isomerise 1-pentene to a *trans/cis* pent-2-ene ratio between 2.7 and 3.4 practically identical to that (2.9) obtained in the presence of $[Ru(OCOCH_3)(CO)_2]_n$ prepared from sodium acetate and $[RuCl_2(CO)_3]_2$.

By contrast markedly different values for the above *trans/cis* ratio are obtained in the presence of polymeric and monomeric catalysts derived from $RuH_2(PPh_3)_3L$ (L = PPh_3 or CO) and polyacrylic acid or isobutyric acid, respectively. Indeed the former catalysts give *trans/cis*-pent-2-ene ratios 0.9 - 1.2 for conversion from 10 to 50%, whereas $RuH[OCOCH(CH_3)_2](PPh_3)_3$ and $RuH[OCOCH(CH_3)_2](CO)(PPh_3)_2$ give values around 5.0 at 10% conversion and 3.4 - 4.3 at 50% conversion (45,56).

The use of optically active polymer ligands can provide transition metal polymer complexes able to display stereoselectivity. A dissymmetric effect by the macromolecular chain could be unequivocally identified only if asymmetric carbon atoms are not present in the immediate vicinity of the metal as in the case of poly-α-amino acids-metal complexes (57). A first approach, not followed by successive data up to now, was based on copolymers between optically active α-olefins and phosphinated styrene to which $RhCl_3$ was anchored. This complex displayed a moderate enantiomeric discrimination in the hydrosylylation of prochiral ketones which could only be attributed to a macromolecular chain effect, the asymmetric carbon atoms of the chiral comonomer being to far from the metal to exert an appreciable dissymmetric perturbation (20,58).

The limits of this approach and the possible developments must be referred to further experiments with properly designed systems (59).

An interesting aspect of selectivity control by the polymer matrix in free-radical reaction has been observed by Drago et al. (60) in case of oxidation of 2,6-dimethylphenol to 2,6-dimethyl-1,4-benzoquinone(BQ) and 3,3',5,5'-tetramethyl diphenoquinone(DPQ) in the presence of cobalt diacetate complexes with polymers containing salicylidenimine groups.

The polymeric catalyst(P-SalDPT)Co, obtained by reacting $Co(CH_3COO)_2$ with P-SalDPT in DMF, increases its BQ selectivity 25 times by doubling the resin metal loading; a similar duobling

for (SalDPT)Co increases BQ selectivity only 1.5 times. This result has been interpreted by suggesting that phenoxy radicals in the polymer by low cobalt concentration have high probability to combine for forming DPQ. This probability is low in solution and in the polymer at high cobalt concentration and more BQ is formed by encounter of Co(II) with organic radicals. Attachment to a polymeric matrix can also give steric hindrance with undesidered side effects which can be eliminated by a proper spacer. An example is given by Tsuchida and Nishide for the oxidative polymerization of 2,6-dimethylphenol in the presence of polyamine/copper complexes (61). Copper attached to 5% divinylbenzene/4-vinylpyridine resin, where the metal is relatively close to the main chain, shows low polymerization rate with only oligomers formation.

A rate increase is observed with 5% divinylbenzene/styrene/4-vinylpyridine resins, but the molecular weight remains low. Good rate and higher molecular weight are obtained by using a polymer ligand where the complexing pyridine groups are far away from the main chain (61).

We cannot close this section without mentioning the stabilization of active intermediates by the attachment to the polymer matrix. This was firstly proposed by Grubbs et al. who reported that polystyrene attached titanocene gives by reaction with $n.C_4H_9Li$ a more active hydrogenation catalyst than titanocenedichloride and benzyltitanocenedichloride (62). This result was attributed to the fact that fixation on the polymer hinders the coupling of the reduced titanocene species to inactive species. Similarly Pittman et al. observed activation of polymer

anchored $Ir(CO)Cl(PPh_3)_2$ species during hydrogenation cycles and attributed this to the formation of Ir-hydrides moieties stabilized by the polymeric support (63).
Evidence of long living active intermediates in polymer-metal catalysts has been obtained in our laboratory in the case of the already mentioned polymer catalyst prepared from poly-4-vinylpyridine or phosphinated polystyrene and $Ru(C_3H_5)Cl(CO)_3$ (44). Homogeneous analogs of these systems catalyze isomerisation of α-olefins with the proposed intermediate formation of a labile Ru-alkyl. Accordingly the basic steps of the reaction with the polymeric catalysts should be the following:

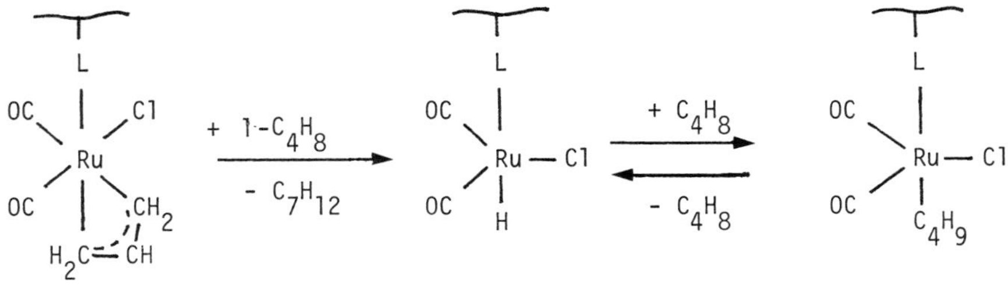

where C_4H_8 indicates either 1-butene, *cis*- or *trans*-2-butene, and C_4H_9 either *n*.butyl or *sec*. butyl. The polymeric catalyst after isomerisation reaction and removal of the olefin shows in its mass spectrum the presence of peaks unequivocally attributable to the butyl group and consequently can be reused with comparable efficiency several times (44).
However, as shown by Valentini et al. (49), for complexes obtained by reacting $RuH_2CO(PPh_3)_3$ with maleic acid/vinyl ethers alternating copolymers the activity decreases with successive cycles more rapidly in benzene than in *n*.octane. As the former solvent has a much better swelling power, the deactivation has been attributed to the saturation of initially formed "coordinatively unsaturated" species by free carboxylic groups made mobile by chain swelling.

3.2. SOLUBLE SYSTEMS

The main portion of work concerning soluble polymer metal complexes has been performed in water with poly-α-amino acids and polypeptides as ligands, with the evident aim to simulate metal containing enzymes. As already mentioned we have not considered this type of polymer catalysts as we have centered our discussion on more typical synthetic polymers, and in particular polymers with hydrocarbon backbone.
Certainly soluble systems can be very convenient to identify macromolecular effects on catalysis at the molecular level, possible porosity effects of reticular matrices being not present. However soluble systems do not appear to have been investigated in such a great extent as heterogeneous systems.
Soluble complexes of copper with polymeric tertiary amines are active catalysts for the oxidative coupling of phenols, as already reported for heterogeneous systems. Soluble complexes of this type have been prepared by Challa with several polymers and low molecular weight analogs (64, 65).

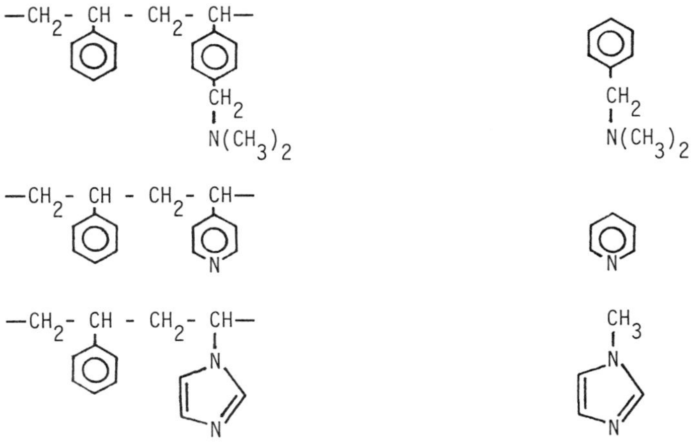

It was observed that the rate of oxidative coupling increases up to an almost asymptotic value if the same amount of catalytic centers is concentrated in a smaller number of polymer coils. It was then concluded that a decreasing intermediate chain length between successive ligand groups enhances the catalytic activity of the metal anchored species. When the concentration in the copolymer of amino units (*p*.dimethylaminostyrene , 4-vinylpyridine or N-vinylimidazole) reaches a certain value (0.4 molar), the intermediate chain segments become so short that complexation between two adjacent amino groups of both Cu(II) ions in the same binuclear complex becomes impossible. As a consequence an adjacent amine unit must be skipped and the intermediate chain length betweeen N-atoms in the complex will be increased with reduction of strain and of the accelerating polymer chain effect.

4. FINAL REMARKS

In this review paper we have consciously not considered polypeptides (66) and immobilized enzymes (67) in order to be concerned with systems very familiar to synthetic polymer chemists and also to avoid a too broad presentation. Clearly some aspects have been lost but we had not the presumption to fully cover polymer catalysts and we have to admit to have not been exhaustive even in the topics treated. We have also not considered extensively polyelectrolytes which can act as very particular polymeric catalysts in several cases (68).

The examples discussed however suffice in our opinion to give a general picture of the possibility offered by synthetic macromolecules in controlling catalytic properties of functional groups attached to them and also to describe the many complex aspects of the field.

It is at present clearly demonstrated that the macromolecule can exert its action on the catalytic species in a number of more or less direct ways. An evidence of direct participation of the polymer to the catalytic process is given by the cooperative effects involving equal groups or different groups with complementary properties. This aspect is well documented by the several types of hydrolytic catalysts acting as nucleophiles (2), but systems of this type can be designed also for metal containing polymer catalysts as the solid bifunctional catalyst for methanol carbonylation (69) and the cobalt-phthalocyanine-poly(vinylamine) complexes acting as bifunctional catalysts in the autoxidation of thiols (70). In this contest designing polymers where the cooperatively interacting groups are not distributed at random but according to a definite mutual situation would be very useful, and in this connection the use of structurally ordered functional polymers and copolymers would be auspicable.

A second effect which has been clearly shown concerns with the presence of hydrophobic groups (13) which provide proper interactions with the substrate thus affecting the chemical reactivity and introducing kinetics of Michaelis-Menten type (2, 6). These characteristics derive from the information provided by enzymes, whose knowledge is of great help in designing synthetic polymer catalysts.

Enzymes possess also a great selectivity and stereoselectivity which is far to be even approached in synthetic systems. It may be argued that these properties are less necessary in synthetic catalysts as they work in a chemically more simple environment, however investigation of structurally better defined polymer catalysts capable of controlled chemical and stereochemical activity would be of great interest. We have only mentioned polyelectrolytes which can in principle act as micellar catalysts; this last area is of good potential development as micelles can affect the catalytic process in considerable way (71). A recent example of studies with microgels as matrices for molecular receptors and catalytic sites have indeed indicated that reactivity of nucleophils can be exceptional in these systems (72, 73).

ACKNOWLEDGEMENT

The authors wish to express their thanks to Prof. Emo Chiellini for many useful discussions.

REFERENCES

1. P.L. Luisi, Naturwissenschaften. 66, 498-504 (1979).
2. T. Kunitake and Y. Okahata, Adv. Polymer Sci. 20, 159-221 (1976).
3. Y. Chauvin, D. Commerenc and F. Dawans, Prog. Polymer Sci. 5, 95-226 (1977).
4. T. Okubo and N. Ise, Adv. Polymer Sci. 25, 136-181 (1977).
5. G. Manecke and W. Storck, Angew. Chem. Int. Ed. 17, 657-670 (1978).
6. I.M. Klotz, Adv. Chem. Phys. 39, 109-176 (1979).
7. P.D. Boyer Ed., The Enzymes, 3rd ed, Vol I and II, Academic Press, New York (1970).
8. E. Chiellini, R. Solaro and S. D'Antone, Makromol. Chem. suppl. 5, 82-106 (1981).
9. C.G. Overberger, T. St.Pierre, N. Vorhheimer and S. Yaroslavsky, J. Am. Chem. Soc. 85, 3513-3515 (1963).
10. C.G. Overberger, T. St.Pierre, N. Vorhheimer, J. Lee and S. Yaroslavsky, J. Am. Chem. Soc. 87, 296-301 (1965).
11. C.G. Overberger, T. St.Pierre, C. Yaroslavsky and S. Yaroslavsky, J. Am. Chem. Soc. 88, 1184-1188 (1966).
12. C.G. Overberger, M. Morimoto, I. Cho and J.C. Salamone, Macromolecules 2, 553-554 (1969).
13. C.G. Overberger and A.C. Guterl Jr., J. Polymer Sci., Polymer Simposia 62, 13-28 (1978).
14. C.G. Overberger, J.C. Salamone and S. Yaroslavsky, J. Am. Chem. Soc. 89, 6231-6236 (1967).
15. T. Shimidzu, Adv. Polymer Sci. 23, 55-102 (1977).
16. Yu.E. Kirsh, A.A. Rahnanskaya, G.M. Lukovkin and V.A. Kabanov, Eur. Polym. J. 10, 393-399 (1974).
17. T. Kunitake and Y. Okahata, Bioorg. Chem. 4, 136-148 (1975).
18. T.S. Lebedeva, Yu.E. Kirsh and V.A. Kabanov, Vysokomol. Soedin. A 19, 1973-1981 (1977).
19. M. Aglietto, G. Ruggeri, B. Tarquini, F. Ciardelli and P. Gianni, Polymer 21, 541-544 (1980).
20. F. Ciardelli, E. Chiellini, C. Carlini and M. Aglietto, Pure & Appl. Chem. 52, 1857-1864 (1980).
21. M. Aglietto, G. Ruggeri, B. Tarquini, P. Gianni and F. Ciardelli, IUPAC Symposium on Macromolecules, Mainz, Preprints 1, 273-276 (1979).
22. T. Okubo and N. Ise, J. Org. Chem. 38, 3120-3122 (1973).
23. T. Ishiwatari, T. Okubo and N. Ise, J. Polymer Sci., Polymer Chem. Ed. 18, 1807-1813 (1980).
24. S. Hayama, M. Takeishi, S. Niino and T. Okamura, J. Polymer Sci., Polymer Chem. Ed. 18, 2941-2948 (1980).
25. S. Hayama, M. Takeishi, K. Takahashi and S. Niino, Makromol. Chem. 181, 1889-1896 (1980).
26. A. Everaerts, C. Samyn and G. Smets, IUPAC Symposium on Macromolecules, Florence, Preprints 4, 167-169 (1980).
27. C.G. Overberger and I. Cho, J. Polymer Sci. A-1, 6, 2741-2754 (1968).
28. C.G. Overberger and K.W. Dixon, J. Polymer Sci., Polymer Chem. Ed. 15, 1863-1868 (1977).
29. M. Nango, H. Kozuka, Y. Kimura, N. Kuroki, Y. Ihara and I.M. Klotz, J. Polymer Sci., Polymer Letters Ed. 18, 647-651 (1980).
30. S.L. Regen, J. Am. Chem. Soc., 98, 6270-6274 (1976).
31. H. Molinari, F. Montanari, S. Quici and P. Tundo, J. Am. Chem. Soc. 101, 3920-3927 (1979).
32. J.M. Brown and J.A. Jenkins, J. Chem. Soc., Chem. Comm., 458-459 (1976).
33. D.C. Sherrington and W.M. MacKenzie, IUPAC Symposium on Macromolecules, Florence, Preprints 4, 178-181 (1980).
34. F. Ciardelli, E. Chiellini, C. Carlini, O. Pieroni, P. Salvadori and R. Menicagli, J. Polymer Sci., Polymer Symposia 62, 143-171 (1978).
35. E. Chiellini and R. Solaro, J. Chem. Soc., Chem. Comm., 231-232 (1977).
36. E. Chiellini, R. Solaro, S. D'Antone and R. Ricci, Chim. Ind. (Milan) 62, 717 (1980).
37. N. Kobayashi and K. Iwai, J. Polymer Sci., Polymer Chem. Ed. 18, 923-932 (1980).
38. N. Kobayashi and K. Iwai, Macromolecules 13, 31-34 (1980).
39. A.P. Schaap, A.L. Thayer, E.C. Bloney and D.C. Neckers, J. Am. Chem. Soc. 97, 3741-3745 (1975).
40. A.P. Schaap, A.L. Thayer, K.A. Zaklika and P.C. Valenti, J. Am. Chem. Soc. 101, 4016-4017 (1979).
41. F. Gurzoni and C. Carlini, paper submitted to the 5° Meeting of Italian Macromolecular Association, Milan, october 1981.
42. M. Kamachi, Y. Kikuta and S. Nozakura, Polymer J. 11, 273-277 (1979).
43. F. Ciardelli, G. Braca, C. Carlini, G. Sbrana and G. Valentini, J. Molecular Cat., in press (1981).

44. C. Carlini, G. Braca, F. Ciardelli and G. Sbrana, J. Molecular Cat. 2, 379-397 (1977).
45. G. Braca, F. Ciardelli, G. Sbrana and G. Valentini, Chim. Ind. (Milan) 59, 766-768 (1977).
46. J.P. Collman, L.S. Hegedus, M.P. Cooke, J.R. Norton, G. Dolcetti and D.N. Marquardt, J. Am. Chem. Soc. 94, 1789-1790 (1972).
47. C.U. Pittman Jr., L.R. Smith and R.M. Hanes, J. Am. Chem. Soc. 97, 1742-1754 (1975).
48. G. Braca, G. Sbrana, G. Valentini, A. Colligiani and C. Pinzino, J. Molecular Cat. 7, 457-468 (1980).
49. G. Valentini, G. Sbrana and G. Braca, J. Molecular Cat. 11, 383-395 (1981).
50. N. Holy, J. Org. Chem. 43, 4686-4688 (1978).
51. M. Terasawa, K. Kaneda, T. Imanaka and S. Teranishi, J. Catalysis 51, 406-421 (1978).
52. P. Perkins and P.C. Vollhardt, J. Am. Chem. Soc. 101, 3985-3987 (1979).
53. P. Pertici, G. Vitulli, C. Carlini and F. Ciardelli, J. Molecular Cat. 11, 353-364 (1981).
54. R.H. Grubbs and L.C. Kroll, J. Am. Chem. Soc. 93, 3062-3063 (1971).
55. G. Innorta, A. Madelli, F. Scagnolari and A. Foffani, J. Organometal. Chem. 185, 403-412 (1980).
56. G. Braca, C. Carlini, F. Ciardelli and G. Sbrana, Proc. 6° Internat. Congress on Catalysis, G.C. Bond, P.B. Wells, F.C. Tampkins Eds., vol. 1, 528-539, The Chemical Society, London (1977).
57. C. Carlini and G. Sbrana, J. Macromol. Sci., Chem., in press (1981).
58. F. Ciardelli, E. Chiellini, C. Carlini and R. Nocci, ACS Polymer Prep. 17, 188-193 (1976).
59. P.L. Luisi in "Optically Active Polymers", E. Selegny Ed., pp. 357-401, Reidel, Dordrecht (1979).
60. R.S. Drago, J. Garl, A. Zombeck and D.K. Straub, J. Am. Chem. Soc. 102, 1033-1038 (1980).
61. E. Tsuchida and H. Nishida, IUPAC Symposium on Macromolecules, Florence, Preprints 4, 147-150 (1980).
62. R.H. Grubbs, C. Gibbons, L.C. Kroll, W.D. Bonds Jr. and C.H. Brubaker Jr., J. Am. Chem. Soc. 95, 2373-2375 (1973).
63. C.U. Pittman Jr., S.E. Jacobson and H. Hiramoto, J. Am. Chem. Soc. 97, 4774-4775 (1975).
64. A.J. Schouten, G. Challa and J. Reedijk, J. Molecular Cat. 9, 41-50 (1980).
65. W. Breemhaar, H.C. Meinders and G. Challa, J. Molecular Cat. 10, 33-42 (1981).
66. S. Inoue, Adv. Polymer Sci. 21, 77-106 (1976).
67. K. Mosbach Ed., Methods in Enzymology, vol. 44, Academic Press, New York (1976).
68. N. Ise, J. Polymer Sci., Polymer Symposia 62, 205-226 (1978).
69. K.M. Webber, B.C. Gates and W. Drenth, J. Molecular Cat. 3, 1-9 (1977).
70. P. Piet, J.H. Schutten and A.L. German, IUPAC Symposium on Macromolecules, Florence, Preprints 4, 151-154 (1980).
71. T. Kunitake and T. Sakamoto, Polymer J. 11, 871-877 (1979).
72. K.A. Stacey, R.H. Weatherhead and A. Williams, Makromol. Chem. 181, 2517-2528 (1980).
73. R.H. Weatherhead, K.A. Stacey and A. Williams, Makromol. Chem. 181, 2529-2540 (1980).

INTERMOLECULAR EXCIMER INTERACTIONS IN POLYMERS

Qian Renyuan

Institute of Chemistry, Academia Sinica, Beijing 100080, P.R. China

Abstract - The intermolecular excimer interactions in polymers refer to the excimer formation between intrachain non-adjacent chromophores and interchain chromophores. The difference in the formation of excimer sites and their relaxation between intermolecular excimers and that of intrachain excimers from adjacent chromophores is pointed out. The exsistence of intermolecular excimers has been experimentally demonstrated without any ambiguity by polymers with chromophores widely separated in the back bone chain. For PS in a good solvent critical concentrations C_s and C^+ have been found from the concentration dependence of I_E/I_M. C_s is interpreted as the concentration at which the expanded polymer coil in a good solvent starts to feel the presence of neighboring coils in the solution with consequent contraction in the coil dimension. C^+ is interpreted as the concentration at which the spatial distribution of polymer segment density becomes continuous and roughly uniform throughout the solution. C_s has been found to be weakly MW dependent while C^+ is almost MW independent. Intermolecular excimer interactions exist also in the solid polymer films. Excimer fluorescence of PET, in which the chromophores are situated in the main chain, was observed down to 77K which demonstrate without any doubt that intermolecular excimer sites are preformed. Steric factors are important for the formation of intermolecular excimer sites. During the crystallization process of poly(polytetramethylene ether glycol 2,6-naphthalene dicarboxylate) direct excitation of excimer sites in the crystalline region of the polymer has been observed for the first time.

INTRODUCTION

Intrachain excimer formation between two adjacent pendant aromatic chromophore groups in aryl vinyl polymers has been widely studied (1-3). However the study and the significance of intermolecular excimer interactions in polymers have been rather overlooked. In recent years we have been looking into such possibility of using the excimer interaction of very short range, i.e. 3.3-3.7 Å, and the excitation energy migration of longer range, i.e. 30-100 Å, still of short range in polymer sense, as a probe to see what information could we get from the excimer fluorescence about the state of molecular aggregation of polymer

chains in concentrated solutions and in solid films. Fig. 1 shows the range of interaction

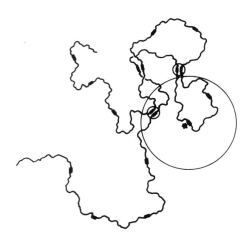

Fig. 1 Schematic showing the range of excimer interaction and energy migration.

for excimers and energy migration. From the view point of the physical nature of the formation

of an excimer site and its relaxation we shall consider the intrachain excimer formation

between nonadjacent chromophores situated at remote distances along the chain by back

coiling and the interchain excimer formation between chromophores situated on two chains

as both belong to the catagory of intermolecular excimer interactions.

The potential energy diagram of two interacting chromophores is schematically shown in

Fig. 2. For an intermolecular excimer site formed by collisional diffusion or micro-

brownian motion of the chain the interaction energy between two highly polarizable π-elec-

tron systems must be attractive to show some energy of stabilization of the excimer site

(A$\cdots$A) in the ground state of chromophores, although this energy of stabilization might be

rather small. When one of the chromophores of the site is excited much higher energy of

stabilization of the excimer ($\overset{*}{A}\cdots$A) will result from charge transfer interactions between

$\overset{*}{A}$ and A, and the equilibrium interchromophore distance will be shortened. For an excimer

site of intrachain adjacent chromophores this is not the conformation of lowest energy, it

has to be formed through internal rotation of the back bone C–C bonds overcoming the

potential barrier of internal rotations involved by a thermal activation process as has been

enumerated by Frank and Harrah (4). Because of longer range of interaction of the excitation

energy migration through dipolar mechanism of Forster (5) the excited state of the chromo-

phore in the excimer site will predominately come from single step energy migration from an

excited state outside of the excimer site. This idea has also been put forward by Johnson

(6) in his study of the excimer formation of pyrene (being a small molecule, necessarily

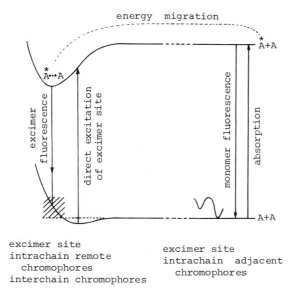

Fig. 2 Schematic diagram of the potential energy curve for two interacting chromophores.

intermolecular) dispersed in polystyrene matrix. No difference could be detected in the absorption spetra or the excitation spetra to give the intramolecular and intermolecular excimer fluorescence in polymers in usual circumstances, since the concentration of the excimer site is usually low and the excimer fluorescence is not the result of direct excitation of the chromophore in the excimer site. More direct evidence for the exsistence of stable ground state dimer (the excimer site) in concentrated solutions of pyrene in a solid polystyrene matrix has been revealed recently (6,7). Thus the factors controlling the intensity of intermolecular excimer fluorescence involve the life of the excited state, the range and efficiency of excitation energy migration, the concentration of excimer sites and the rates of competing nonradiative deactivation processes of the excited state.

Chromophore containing polymers can be classified into three classes: (a) chromophore on the pendant group attached directly to the main chain, (b) chromophore on the side group connected to the main chain through flexible bonds and (c) chromophore in the back bone chain. Trival cases of chromophores as end groups or single mid-chain group will not be discussed in the present paper. Aryl vinyl polymers belong to the class (a) polymers and have been most extensively studied but scarcely from the view point of intermolecular excimer interactions. Examples of class (b) and class (c) polymers are

$$-\text{C}\langle\bigcirc\rangle\text{C-O-(CH}_2)_2\text{-O-}\qquad\text{(PET)}\qquad\qquad\text{(Ref. 13)}$$

$$-\text{C}\langle\bigcirc\bigcirc\rangle\text{C-O-(CH}_2)_2\text{-O-}\qquad\text{(PEN)}\qquad\qquad\text{(14)}$$

$$-\text{N-CH}_2\langle\bigcirc\bigcirc\rangle\text{CH}_2\text{-N-C-(CH}_2)_n\text{-C-}\qquad\qquad\text{(15)}$$

$$-\text{C}\langle\bigcirc\rangle\text{C-O-(CH}_2\text{CH}_2\text{CH}_2\text{CH}_2\text{-O-)}_n\text{-}\qquad\text{(PPTMGTP)}\qquad\text{(16)}$$

$$-\text{C}\langle\bigcirc\bigcirc\rangle\text{C-O-(CH}_2\text{CH}_2\text{CH}_2\text{CH}_2\text{-O-)}_n\text{-}\qquad\text{(PPTMGNDC)}\qquad\text{(17)}$$

EXPERIMENTAL DEMONSTRATION OF THE EXSISTENCE OF INTERMOLECULAR EXCIMERS

Although there have been evidences of increased intensity ratio of excimer to monomer fluorescence, I_E/I_M, for polystyrene in a θ- or poor solvent as compared to that in a good solvent (18, 19), the presence of adjacent chromophores three atoms apart complicates the situation to give a clear cut demonstration of the formation of excimers from remote chromophores along the chain by back coiling. Similar results were reported in the case of polynaphthylmethacrylate (8) where chromophores are separated by 7 atoms but every other atoms on the main chain are connected to the side group containing these chromophores. A clear cut demonstration of the intrachain excimer formation between remote chromophores and interchain excimer formation comes from the solvent dependence of I_E/I_M of a dilute solution and the concentration dependence of I_E/I_M of concentrated solutions of the following polymers with chromophores widely separated in the main chain:

$$-\text{BD}\cdots\text{BD-S-BD}\cdots\text{BD-}\qquad\qquad-\text{C}\langle\bigcirc\rangle\text{C-O-[(CH}_2)_4\text{-O]}_n\text{-}$$

$$\text{(I)}\qquad\qquad\qquad\qquad\qquad\qquad\text{(II)}$$

$$-\text{C}\langle\bigcirc\bigcirc\rangle\text{C-O-[(CH}_2)_4\text{-O]}_n\text{-}\qquad\qquad[(\text{CH}_2)_4\text{-O]}_n\text{: MW}\approx2000$$

$$\text{(III)}$$

The styrene butadiene copolymer (I) was synthesized with low content of styrene (5% by wt. from I.R.) so that there are only isolated styrene units in the chain as proved by proton NMR. In fact, because of the values of reactivity ratios of styrene butadiene copolymerization even in ordinary SBR the styrene units are predominately isolated. The average spacing between two styrene units in the chain of our sample is arround 150 atoms. Any excimer fluorescence observed in a dilute solution must come from remote styrene units along the chain and coiled back to close distances of 3.5Å or so in stacked configuration for the phenyl rings to form excimer sites. In a dilute solution in good solvent such as 1,2-dichlor-ethane (DCE) the polymer showed the monomer fluorescence only which is identical to that of ethylbenzene. As the expanded polymer coil in a good solvent possesses low spatial density of segments the concentration of intrachain excimer site from remote chromophores is negligible. When a non-solvent, methanol, was added to the solution an excimer band in the fluorescence spetrum appeared (20). Maintaining the polymer concentration the same ($7.5 \cdot 10^{-3}$ g/ml) while varying the solvent-nonsolvent ratio until close to the θ-point we observed a steady increase of I_E/I_M as shown in Fig. 3a. It is evident that as the polymer coil shrinks

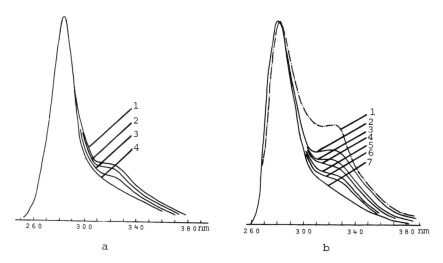

Fig. 3a   Fluorescence spetra of SBD copolymer in the mixed solvent 1,2-dichloro-ethane-methanol. Conc. $7.5 \cdot 10^{-3}$ g/ml, λ260nm excitation.

Curve	MeOH/DCE v/v	$[\eta]$, dl/g
1	0.28	0.65
2	0.20	0.74
3	0.12	0.93
4	0	1.04

3b   Fluorescence spetra of concentrated solutions of SBD copolymer in DCE. λ260 nm excitation. 1-solid film; 2,3,4,5,6,7-conc. 0.17, 0.11, 0.085, 0.057, 0.043, 0.0004-0.01 M/l respectively.

in dimension in a poor solvent the spatial density of segments inside the coil increases so that the excimer site concentration will be increased.

So long as the concentration of the SBD copolymer in DCE is smaller than 10^{-2} M/l chromo-phore (or $4.2 \cdot 10^{-2}$ g/ml), only monomer fluorescence was observed. When the concentration was increased beyond this a concentration quenching of the monomer fluorescence with the appear-ance of an excimer peak was observed as shown in Fig. 3b. The ratio I_E/I_M increased as the first power of concentration indicating the interchain excimer formation, as shown in Fig. 4.

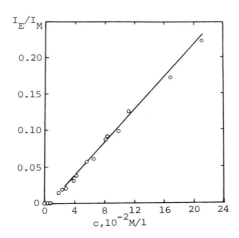

Fig. 4 I_E/I_M of concentrated solutions of SBD copolymer in DCE.

In the cases of polymer (II) and (III) the chromophores are in the main chain with an
average spacing of arround 140 atoms. Entirely similar results were obtained as the polymer
(I) in respect to the increase of I_E/I_M for polymer (II) in good solvent DCE to mixed
solvents DCE-MeOH and from dilute solution to concentrated solutions (16), and for polymer
(III) in $CHCl_3$ from dilute to concentrated solutions (17) as shown in Fig. 5. In contrast to

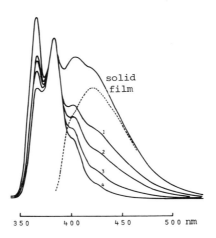

Fig. 5   Fluorescence spetra of concentrated solutions of PPTMGNDC (III) in $CHCl_3$.
λ330nm excitation.   1,2,3,4-Conc. 43,28,16,0.024 g/dl.

the polymers containing phenyl chromophores the polymer (III) showed weak intrachain excimer
fluorescence even in very dilute solutions probably due to longer life time of the excited
state of the naphthyl chromophore. This is also true for other naphthyl containing polymers
reported in literature (8,12,14).

The above described experiments have clearly demonstrated without any ambiguity the existence
of the two types of intermolecular excimer formation in polymer solutions, i.e. intrachain
excimer from chromophores remotely located in the chain by back coiling and interchain

excimer from chromophores located on two chains.

CONCENTRATED SOLUTIONS OF POLYSTYRENE IN 1,2-DICHLOROETHANE

We have examined the transition from dilute to semi-dilute to concentrated solutions of polystyrene in DCE by excimer fluorescence. In a dilute solution PS showed the intrachain excimer fluorescence from adjacent chromophores which was concentration independent. On increasing the concentration the ratio I_E/I_M started to increase according to a fractional power of c and then increased linearly and superlinearly with c above certain value of the concentration. A typical curve is shown in Fig. 6. The experiments were done with monodisperse PS samples in solutions prepared at 60°C, and for every step of concentration increase or decrease the solution was put two hours at 60°C and then cooled to room temperature in

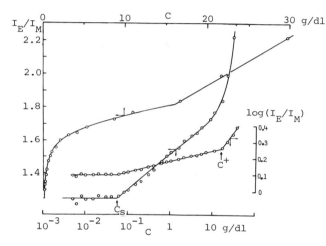

Fig. 6   Concentration dependence of I_E/I_M for PS ($M_W=2.5\cdot10^5$) in DCE at 25°C.

10 hrs and then thermostated at 25°C before a fluorescence spetrum was taken to assure uniformity of the solution. This appeared to be important in dealing with concentrated polymer solutions. Surface fluorescence was recorded in all cases.

It was originally thought that the transition from intrachain excimer to interchain excimer formation would show up in the I_E/I_M vs c plot as a transition from a starting portion of concentration independence to a first power dependence. The transition point would be the overlap concentration $\overset{*}{c}$ of the polymer coils in solution. This idea has been used by Roots and Nystrom (21) in his interpretation of excimer fluorescence of concentrated solutions of the same polymer-solvent system as we used here. Our experimental results are in disagreement with those of Roots and Nystrom and will be interpreted differently. A flexible polymer coil in a good solvent in dilute regime is expanded from the unperturbed dimension due to repulsive interaction between the segments because of polymer-solvent interactions. The polymer coil will shrink with increasing concentration when going from dilute to semi-dilute regime until the polymer coils are highly interpenetrated in the concentrated regime leading

to a coil dimension equal to the unperturbed dimension in a θ-solvent (22). The concentration of the transition from dilute to semi-dilute regime is usually denoted by $\overset{*}{c}$ and that
from semi-dilute to concentrated regime by c^{+}. As will be explained in the following the
usual designation of $\overset{*}{c}$ as the concentration at which two polymer coils in solution will be
in contact and to be defined as $\overset{*}{c} = M/(\tilde{N}\cdot 4\sqrt{2}\cdot R_{o}^{2})$ or $\overset{*}{c} = 1/[\eta]$ (23) is logically not
correct, where $R_{o} = R(c=0)$, the root mean square radius of gyration of the coil in dilute
solution, $[\eta]$ the intrinsic viscosity of the solution and $\tilde{N}$ the Avogadro number. An expanded
flexible coil in dilute solution will start to feel the repulsive interaction of segments of
a neighboring coil resulting in a shrinkage of the coil dimension when the concentration of
the solution gets high enough. This has been demonstrated experimentally by SANS (24) and
X-ray scattering (25). As a manifastation of repulsive interaction between interchain segments to compensate partially the repulsion between intrachain segments, the presence of
increasing number of neighboring coils and decreasing intercoil distances will lead to a continuous shrinking of the coil dimension as the concentration is increased. It is conceivable
that the expanded coils will be highly compressed before they could overlap. So it is not
correct to define $\overset{*}{c}$ from the coil dimension measured in dilute regime, R_{o} or $[\eta]$. The change
of coil dimension and spatial distribution of segment density with increasing solution concentration is illustrated schematically in Fig. 7. At $c < \overset{*}{c}$, the spatial distribution of

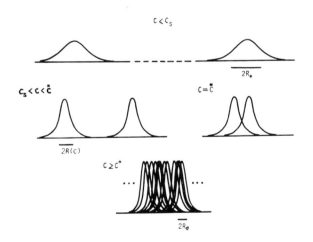

Fig. 7 Spatial distribution of segment density of a polymer coil in good solvent
from a dilute to a concentrated solution (schematic).

segment density is discontinuous throughout the solution. At $c \geq \overset{*}{c}$, the spatial distribution of segment density becomes continuous but non-uniform and at $c > c^{+}$, the segment distribution will be roughly uniform. Then the segment density will increase linearly with c,
as the polymer coils will simply pack into the space occupied already by other coils, with
ever increasing degree of interpenetration of the coils. In highly interpenetrated coils the
interplay of intrachain segmental repulsion and interchain segmental repulsion will balance

the forces exerted on each segment so that the coil dimension will reach the unperturbed dimension R_θ in a very concentrated solution and in the amorphous solid polymer. It is easy to calculate that there will be hundreds of interpenetrating coils within the space of a coil of dimension R_θ in the solid state (26). It is because of the continuing shrinkage of the coil dimension in the semi-dilute regime and in this sense that "the concentration $\overset{*}{c}$ is not a quantity that can be accurately determined either from the theoretical nor from the experimental point of view" as expressed by Daoud *et al.* (24).

Now, using excimer formation as a molecular probe we could determine the concentration at which the polymer coil in a good solvent begins to feel the presence of neighboring coils in the solution. We shall define this concentration as c_s. We treated our experimental data by plotting I_E/I_M vs log c which shows that for the initial part of $c > c_s$ the plot was quite linear so that the intersaction point with the horizontal line of constant I_E/I_M for $c < c_s$ could be conveniently located to give the value of c_s as shown in Fig. 8. The increase of

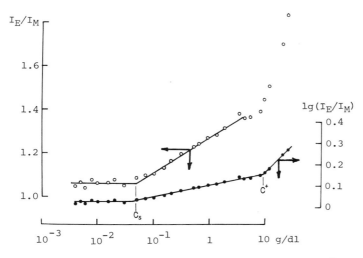

Fig. 8   Concentration dependence of I_E/I_M for PS (M_w=6.1·10^6) in DCE, 25°C.

I_E/I_M of intrachain excimers in this concentration region is a manifestation of coil shrinkage analogous to the case of adding non-solvent to a polymer solution as discussed earlier. On further increasing the concentration the coils begin to overlap at $\overset{*}{c}$. The concentration at which two polymer coils in geometrical contact as originally defined as $\overset{*}{c}$ appeared to be not accessible experimentally. It appears to me that this geometrical contact concentration is perhaps just trival and irrelevant to the understanding of the solution properties. It is the continuous shrinkage of coil dimension with increasing concentration due to the repulsive interactions between interchain segments which governs the behavior of polymer solution in a good solvent in the concentration region between c_s and c^+. It seems to be more logical to define c_s as the boundary between the dilute and semi-dilute regimes.

The transition from a fractional power dependence of I_E/I_M on c to a higher power dependence in the region $c > c_S$ will give us the value of c^+ (Fig. 8). In the region $c > c^+$, the I_E/I_M- c curve should show an initial part of linear dependence on c and then increase superlinearly. Thus both c_S and c^+ as we defined here could be located experimentally by excimer fluorescence. The effect of dissolved oxygen in the solution will change the value of the fractional power dependence but will not alter the intersaction points c_S and c^+ significantly. As it was very difficult to bubble nitrogen gas through a concentrated polymer solution, we worked with solutions as prepared without further treatment. The values of c_S and c^+ so obtained for six monodisperse PS samples in the molecular weight range of 10^4- $6 \cdot 10^6$ are shown in Fig. 9. It is seen that c_S is only weakly dependent on M, being $c_S \propto M^{-0.10}$ while

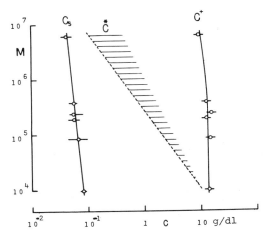

Fig. 9 Molecular weight dependences of c_S and c^+.

c^+ is almost independent on M. This is quite unexpected. Assuming hexagonal close packing of spheres we can calculate the intercoil distances from the value of c_S for $M = 10^4$, 10^5 and $6 \cdot 10^6$ to be 300, 720 and 3200Å respectively. This seems to be not entirely unreasonable as the coil dimensions $2R_0$ are 60, 220 and 2350Å respectively. The rather long range repulsive interactions between interchain segments in a good solvent could only be understood as volume exclusion due to rapid microbrownian diffusion of the segments. Thus this intercoil distance should reflect the instantaneous intercoil distance in dynamical contact. The almost molecular weight independence of c^+ was first suggested by Graessley (27) from arguments based on screening principle. However, his derivation seems to be unsound in two respects, one is the incorrect definition of $\overset{*}{c}$ and the other is the use of $c^{-1/4}$ dependence of $R^2(c)$ dependence in the whole semi-dilute regime. Although there is experimental evidence of such dependence for a single case (24) of PS of $M=1.14 \cdot 10^5$ in CS_2, it is difficult to see that solutions of different molecular weights having different values of $R_0 (c \leq c_S)/R_\theta (c \geq c^+)$ will show the same slope on the log R vs log c plot in the region $c_S \leq c \geq c^+$. The constancy of c^+ for different molecular weights might be understood as a state of roughly the same uniform spatial density of segments at $c = c^+$ irrespective of M, as the polymer in amorphous solid state has the same density irrespective of molecular weight.

EXCIMERS IN SOLID POLYMER FILMS

In solid polymer films of aryl vinyl polymers it is usually found that only excimer fluorescence is observed. For polymers with chromophores connected through flexible linkages either in the side group or in the main chain we found high monomer fluorescence intensity in solid films, for examples $-CH_2-CH(CH_2-O-Np$ or $Ph)-O-$ (12), the polymers I (20), II (16) and III (17). It appears to us that the more flexible are the chromophores linked together in the polymer chain, the closer will the behavior of excimer fluorescence resemble to that of chromophores in small molecules. We believe that this is due to more chance of dissociation of the excimer sites through microbrownian motions of the chain during the film formation processes. So far we have performed only scattered experiments on the study of excimers in solid polymer films. We shall mention a few interesting results obtained.

It is obviously more appropriate to study the intermolecular excimers in the solid films of polymers with chromophores in the main chain. The gradual change of the fluorescence spetra of PET in trifluoroacetic acid (TFA) from concentrated to very concentrated solutions and to a solid film (13) is shown in Fig. 10. For concentrations up to ca. 60 wt% I_E/I_M was found

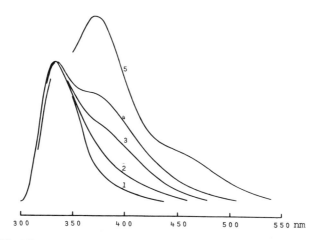

Fig. 10 Fluorescence spetra of PET-TFA concentrated solutions.
λ 285nm excitation. 1,2,3,4,5-conc. 9.3, 28, 45, 64, 93 wt% respectively.

to be proportional to c, so that it is no doubt that in the solid film the excimer fluorescence will come from interchain excimers. It is very interesting to see that both PET and PEN film showed only the excimer fluorescence without monomer fluorescence even at 77K as shown in Fig. 11. For PET and PEN the chromophores are in the main chain, even the local

motions of the chain will not be possible at this temperature (the β-relaxation of PET is

arround 240K). The necessary conclusion is that interchain excimer sites in solid polymers

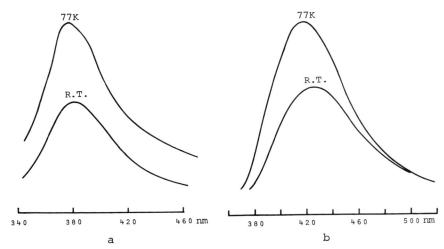

Fig. 11 Excimer fluorescence spetra of (a) PET and (b) PEN films at 77K.

are preformed and the concentration of these sites will be governed by the monomer-dimer

chromophore equilibrium at the temperature of film formation as was first pointed out by

Frank and Harrah (4) for intrachain excimer sites.

We observed that some polymers containing phenyl groups in the main chain showed no excimer

fluorescence in solid films (28). Examples of these are

In these polymers the phenyl groups are bonded through O, S, $C(CH_3)_2$ or SO_2

linkages which are at an angle to each other and with two bulky CH_3 groups

attached close to the chromophores so that close approaching of two phenyl

chromophores to form an excimer site is hindered. It is apparent therefore that steric factors are important for intermolecular excimer formation in polymers.

Uniaxial or biaxial stretching of PET and PEN films usually lead to an enhancement of excimer fluorescence (14). This should be interpreted as an increase of the concentration of excimer sites by stretching, as biaxial stretching is known to bring the planar chromophores to lie parellel to the film surfaces (29), and there seems to be no reason to expect an increase of energy migration efficiency in a biaxially stretched film as compared to the unstretched one.

Poly(polytetramethylene ether glycol terephthalate) (II) and poly(polytetramethylene ether glycol 2,6-naphthalene dicarboxylate) (III) films cast from solution are amorphous. On standing the soft segments first crystallize which melt arround 30°C and the chromophore-containing hard segments will slowly crystallize in the course of a few months at room temperature (30). We found in the case of the solid film of polymer III the excitation spetrum of the excimer emission showed two peaks (17): one at λ362nm which is the normal excitation peak, the other at λ379nm which increased in intensity with increasing degree of crystallization on standing as shown in Fig. 12. Heating the crystallized film under nitrogen to ca. 200°C for 20 min. and then cooling down

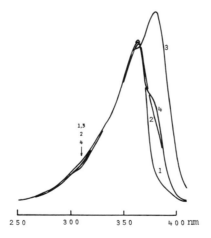

Fig. 12   Effect of crystallization on the excitation spectrum of
          λ420nm excimer emission of PPTMGNDC film.
          1-freshly precipitated, 2-after 25 days, 3-after 9.5 months,
          4-after 9.5 months then heated to 200°C for 20 min. under N_2
          and cooled to room temperature.

to room temperature reduced greatly the intensity of the second excitation

peak at λ379nm. This fact leads us to consider the second excitation peak as
the absorption peak of the chromophores in the crystalline region of the polymer. The excimer spectra excited by λ360 and by λ380nm are entirely the same.
Films of the polymer II behaved similarly. It is very likely that in the crystalline region of the polymer the planar chromophores will be stacked parallel to each other and this will resemble the excimer sites. We would inter-
prete the second excitation peak as the direct excitation of the excimer site.
The energy diagram of Fig. 1 indicates that this direct excitation wave length
should be red shifted as compared to the excitation of a free chromophore with
subsequent energy migration to an excimer site. This interpretation is sub-
stantiated by the fact that by λ370nm excitation only the excimer fluorescence
was observed with no λ380nm monomer fluorescence as shown in Fig. 13b.

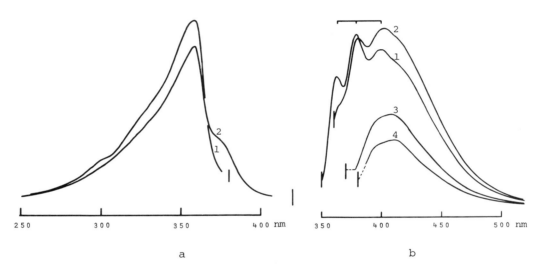

a   b

Fig. 13a   Excitation spectra of PPTMGNDC film.
           1,2-λ emission 380, 420nm.
     13b   Fluorescence spectra of PPTMGNDC film with different exci-
           tation wave lengths.
           1,2,3,4-excitation by λ350, 360, 370,380nm.

Furthermore, for the monomer emission at λ380 the excitation spectrum did not
show the long wave length tail, while the excitation spectrum for the excimer
emission at λ420nm a long wave length tail showed up clearly for the same film
sample studied as shown in Fig. 13a. So we feel confident that we have found
the first example of the direct excitation of excimer sites which are in the
crystalline regions of the polymer II and III.

REFERENCES

1. A.C. Somersall and J.E. Guillet, *J. Macomol. Sci. Revs.* C13, 135 (1975).

2. S.W. Beavan, J.S. Hargreaves and D. Phillips, in J.N. Pitts, Jr., G.S. Hammond, K. Gollnick and D. Grosjean, Ed., *Advances in Photochemistry*, Vol. 11, p. 236, Interscience, New York (1979).

3. H. Morawetz, *Science* 203, 405 (1979).

4. C.W. Frank and L.A. Harrah, *J. Chem. Phys.* 61, 1526 (1974).

5. Th. Förster, *Disc. Faraday Soc.* 27, 7 (1959).

6. G.E. Johnson, *Macromol.* 13, 839 (1980).

7. B. Polacka and I. Weyna, *Bull. l'acad. polon. sci. Ser. sci. math. astr. phys.* 26, 935 (1978).

8. A.C. Somersall and J.E. Guillet, *Macromol.* 6, 218 (1973).

9. T. Nakahira, I. Maruyama, S. Iwabuchi and K. Kojima, *Makromol. Chem.* 180, 1853 (1979).

10. M. Keyanpour-Rad, A. Ledwith, A. Hallam, A.M. North, M. Breton, C. Hoyle and J.E. Guillet, *Macromol.* 11, 1114 (1978).

11. A. Itaya, K. Okamoto and S. Kusabayashi, *Preprint, 22nd Polymer Symposium*, The Society of Polymer Science, Japan, Vol. III, p. 1 (1973).

12. X. Jin, H. Li, X. Fu and R. Qian, to appear in *Scientia Sinica*.

13. R. Qian, F. Bai and S. Chen, to appear in *Kexue Tongbao*.

14. S. Chen, F. Bai and R. Qian, *Scientia Sinica* 24, 639 (1981).

15. J.A. Ibemesi, J.B. Kinsinger and M.A. El-Bayonmi, *J. Polym. Sci. Polym. Chem. Ed.* 18, 879 (1980).

16. R. Qian, T. Cao, F. Bai and S. Chen, to be published.

17. T. Cao and R. Qian, to be published.

18. T. Nishihara, M. Kaneko, *Makromol. Chem.* 124, 84 (1969).

19. L. Gargallo, E.B. Abuin and E.A. Lissi, *Scientia* (Valparaiso) 42, 11 (1977).

20. R. Qian, X. Jin and H. Li, to appear in *Gaofenzi Tongxun*.

21. J. Roots and B. Nyström, *Euop. Polym. J.* 15, 1127 (1979).

22. J.P. Cotton, D. Decker, H. Benoit, B. Farnoux, J.S. Higgins, G. Jannink, C. Picot and J. des Cloizeaux, *Macromol.* 7, 863 (1974).

23. R. Simha and L.A. Utracki, *Rheol. Acta* 12, 455 (1973).

24. M. Daoud, J.P. Cotton, B. Farnoux, G. Jannink, G. Sarma, H. Benoit, R. Duplessix, C. Picot and P.G. de Gennes, *Macromol.* 8, 804 (1975).

25. H. Hayashi, R. Hamada and A. Nakajima, *Makromol. Chem.* 178, 827 (1977).

26. P.J. Flory and D.Y. Yoon, *Nature* 272, 226 (1978).

27. W.W. Graessley, _Polymer_ 21, 258 (1980).

28. R. Qian and X. Jin, unpublished work.

29. C.J. Heffelfinger, R.L. Burton, _J. Polym. Sci_. 47, 289 (1960).

30. R. Qian and M. Wu, to be published in _Kexue Tongbao_.

RELAXATION METHODS FOR STUDYING MACROMOLECULAR MOTION IN THE BULK

H. Sillescu

Institut für Physikal.Chemie der Universität Mainz, Sonderforschungsbereich
Chemie und Physik der Makromoleküle, Mainz/Darmstadt, D-6500 Mainz,W.Germany

Abstract - Macromolecular motion in amorphous and partially crystalline
polymers is discussed in the light of recent relaxation experiments with
particular emphasis on NMR methods. Polystyrene and polyethylene serve as
pertinent examples where a considerable amount of new experimental data
provides a bysis for better understanding molecular processes below and
above the glass transition, and in the melt.

INTRODUCTION

A system that approaches equilibrium from a nearby non-equilibrium state is said to undergo
relaxation. In this narrow sense, relaxation methods are restricted to experiments where
the recovery of the system is observed after an external perturbation. However, the same
information with respect to molecular motion is obtained from stationary experiments where
the response to a periodic perturbation is observed. Stationary absorption and dispersion
methods are therefore also denoted as relaxation methods in a wider sense (e.g., dielectric
relaxation). Recently, inelastic or quasielastic scattering of neutrons and optical photons
has been applied very successfully to studies of macromolecular motion. As a matter of fact,
rather similar molecular correlation functions are used in order to interpret relaxation,
absorption, dispersion, and quasielastic scattering experiments. We have attempted to quan-
tify the above remarks in Table 1 that may be used as a guide to reviews and to some recent
literature.

The polymer scientist interested in macromolecular motion should be aware of the potential
of all relaxation methods in the widest sense although it is impossible to be familiar with
all details that must be mastered by the experimentalist who wishes to apply a particular
method. Thus, any review on relaxation methods given by an experimentalist is lopsided
since he is most familiar with just one or two methods. The present article is written from
the NMR spectroscopist's point of view, and concentrates on recent NMR results that are
compared with results from other methods.

Spin relaxation is peculiar in that the spin relaxation times are usually much longer than
the corresponding motional correlation times. This is because the motional degrees of freedom (termed "lattice" in the NMR literature) are only weakly coupled to the spin system.
Furthermore, the spin energies are usually negligible in comparison with kT. These properties have allowed the development of a wide variety of specialized pulse techniques
generating rather different modes of spin motion tailored to probe particular "lattice"
motions. Fortunately, the intricacies of spin dynamics do not enter the motional correlation
functions resulting from NMR applications. Nevertheless, some familiarity with the possibilities of the different NMR methods is useful in order to know what kind of problems can be

TABLE 1. Experimental methods for studying macromolecular motion

Method	correlation time range (in sec)	remarks	references
mechanical relaxation	10^{-8}- 10^{6}	creep, stress relaxation, forced and free vibrations, sound and ultrasound propagation	(1-3)
Brillouin scattering	10^{-10}- 10^{-9}	hypersound propagation	(4)
differential thermal analysis		C_p changes at the onset of molecular motions	(5)
dielectric relaxation	10^{-10}- 10^{5}	see also: Thermally stimulated discharge (TDS) or -current (TSC); (8)	(1,6,7)
Kerr effect relaxation	10^{-8}- 10^{4}	relaxation of electric birefingence	(9,10)
quasielastic light scattering	10^{-4}- 10^{2}	polarized and depolarized Rayleigh spectroscopy (photon correlation spectroscopy)	(11)
quasielastic neutron scattering	10^{-12}- 10^{-8}		(12,13)
nuclear magnetic resonance	10^{-12}- 1	spin-lattice relaxation: 10^{-12}- 10^{-5} s	(14,15)
		NMR line shapes: 10^{-6}- 10^{-3}	(15,16)
		dipole and quadrupole order relaxation: 10^{-3}- 1	(16-18)
electron spin resonance	10^{-5}- 10^{-10}	ESR line shapes: 10^{-10}- 10^{-7}	(19,20)
		ESR saturation transfer: 10^{-7}- 10^{-5}	(21)

solved. In the following section, we review briefly those NMR methods that have been successfully applied to polymer problems in recent years.

NMR METHODS

Spin-lattice relaxation

Proton spin-lattice relaxation times T_1 have been investigated in all common polymers during
the period of 1950-70 (15 ,22). The essential information is obtained from plots of T_1
versus temperature T where at each T_1-minimum the correlation time τ_c of a particular process is of the order of the reciprocal Larmor frequency (more precisely: $\omega_0 \tau_c \approx 0.62$). In
attempts to fit the whole T-dependence, one has mostly used a phenomenological description
by a distribution of correlation times that can easily be compared with other relaxation
results (22,23,24). This procedure implies the assumption that the shape of the τ_c distribution is independent of T (see Note a).

In recent years, Kimmich and collaborators (25) have applied a field cycling technique that
allows for T_1 measurements as a function of the Larmor frequency in a range of $\omega_0/2\pi \approx$
$10^4 - 10^8$ Hz. In this method, the system is first polarized in a high field B_0 that is
switched to a different field B_0' during the spin relaxation period, and switched back to B_0
for signal registration. The authors have achieved very short field cycle times where relaxation times T_1 down to 5 ms can be determined.

In proton T_1 experiments, the resulting correlation times τ_c are usually related with the
motion of many intra- and intermolecular proton-proton vectors, and an unambiguous assignment of molecular relaxation processes is difficult. In order to separate the different
relaxation contributions, one can systematically replace protons by deuterons thereby replacing particular $^1H - {}^1H$ dipolar couplings by the much weaker $^1H - {}^2H$ couplings (26,27).
Furthermore, one can investigate T_1 of the ^2H-nuclei that only depend upon reorientation
of the ^2H-C bonds through ^2H quadrupole relaxation (26). A further advantage of ^2H relaxation measurements is the strong reduction of "spin diffusion" effects (14,28) that limit
the applicability of ^1H relaxation in solid polymers. Spin diffusion is caused by the dipolar coupling between all protons of the sample, and has the effect that T_1 in a solid is
essentially determined by the rapid motion of relaxation centers, say, mobile side or end
groups. The main potential of ^2H and ^{13}C T_1-measurements seems to be studies of molecular
motion in semicrystalline polymers where the motions in amorphous and crystalline regions
can be probed separately (see below). Recently, it has also been possible to determine
separately T_1 of chemically shifted ^{13}C nuclei in solid polymers by the technique of
CP-MAS-NMR (29,30).

Whereas spin-lattice relaxation times T_1 determined in the laboratory frame have minima for
τ_c values close to the reciprocal Larmor frequency ω_0^{-1}, there is a "spin-locking" technique
(31) providing <u>rotating-frame relaxation times</u> $T_{1\rho}$ that have their maxima at much longer
correlation times close to ω_1^{-1} where $\omega_1 = \gamma B_1$ is determined by intensity of the RF-field
B_1 during the "spin-locking" pulse. We refer to the literature for a description of this
technique (31), and only mention that $T_{1\rho}$ can be used as a probe for studying slow macromolecular motions down to $\tau_c \sim 10^{-3}$s. The influence of spin diffusion is smaller in $T_{1\rho}$-

Note a. The validity of a corresponding assumption is discussed in Chapter 11 of Ferry's
book (2) where the "method of reduced variables" (time temperature equivalence) is developed.

than in T_1-experiments but still limits investigations of different motional degrees of freedom in the same polymer. This is also true for ^{13}C where $T_{1\rho}$ of chemically shifted ^{13}C nuclei can be determined in solid polymers by the CP-MAS technique (32,33).

NMR-line shapes

The advent of cryomagnets and advanced pulse techniques has opened up new possibilities for NMR-line shape studies in solid polymers. In particular, one can avoid the many-spin problem of dipolar coupling that has limited former ^{1}H-NMR studies (15). In ^{13}C NMR experiments (34,35), dipolar decoupling of the protons leads to line shapes that are governed by the ^{13}C chemical shift anisotropy (see Note b). In solid polymers, there are anisotropic molecular motions that can average out part of the chemical shift anisotropy thus causing line shape changes. Since the line shape of a one-spin system can be calculated for any well defined motional process, line shape studies can give valuable information upon the type or mechanism of motion in solid polymers. Besides ^{13}C, motional averaging of chemical shift anisotropies can also be investigated using ^{15}N or ^{19}F nuclei where homonuclear dipolar decoupling by multiple pulse techniques is applied in the case of ^{19}F NMR (36). Even more promising is the ^{2}H nucleus that that has a spin I = 1, and a quadrupole coupling much larger than the dipolar couplings and the chemical shift anisotropy. The experimental difficulties due to the large spectral width ($\sim$ 250 kHz) of ^{2}H spectra in solids have recently been overcome, and applications to molecular motion in polymers are in progress in our laboratory (16). Since we discuss applications of ^{2}H spectroscopy to some extent below, we give a few details of the physical basis in the following paragraph.

There are 2 transitions ($0 \leftrightarrow \pm 1$) of the ^{2}H spin I = 1 that give rise to a NMR doublet with a frequency splitting $\Delta\omega$ that can be written as

$$\Delta\omega = \delta \ (3 \ \cos^2\theta - 1) \tag{1}$$

if the field gradient tensor at the ^{2}H nucleus has axial symmetry. This condition applies with good approximation in C-^{2}H bonds of polymers where θ is simply the angle of C-^{2}H with respect to the external magnetic field. In Equation (1), $\delta = 3 \ e^2Qq/4\hbar$ is related with the known quadrupole coupling constant e^2Qq/h (164 kHz in polyethylene). In a solid powder or glass, the ^{2}H line shape consists of a weighted superposition of doublets the weights being proportional to the number of C-^{2}H bonds between θ and $\theta + d\theta$. In Fig. 1a, the theoretical ^{2}H powder line shape is shown for an isotropic distribution of C-^{2}H bonds (see Note c). The influence of molecular motions is shown in Fig. 1b for the example of rapid exchange between

───────────────────────

Note b. This anisotropy can be averaged out by magic angle spinning (MAS) of the sample in order to obtain high resolution NMR spectra in solids. The coherent sample rotation should be compared with the stochastic molecular reorientation in liquids that is also causing motionally averaged high resolution NMR spectra.

Note c. The same line shape (Pake spectrum) is obtained for the effective I = 1 system of a [1]H-^{1}H pair, and for the S = 1 triplet ESR spectrum. In oriented polymers, the ^{2}H line shape depends upon the orientational distribution which can thus be determined (38).

2 orientations separated by the tetrahedral angle of 109.5^o, and in Fig. 1c for rapid exchange between 3 orientations separated by the same angle (e.g., CH$_3$-rotation). "Rapid" is defined by $\Delta\omega\tau_c \ll 1$ which in our context is $\tau_c < 10^{-6}$s. In the "slow motion region" $\Delta\omega\tau_c \sim 1$, the correlation time τ_c can be obtained by fitting experimental line shapes with model calculations (39). The solid echo technique that is used to determine ^{2}H NMR spectra (37) yields distorted line shapes if τ_c is of the order of the distance ($\sim$ 50 μs) between the 2 RF pulses applied to the sample in order to produce a solid echo (40). A new method for probing "ultraslow" motions with correlation times that are only limited by the ^{2}H spin lattice relaxation time T_1 (typically $\sim$ 1 s) has been developed (18) where a Jeener 3 pulse sequence is applied to the sample. We refer to Ref. 18 for details of this "spin alignment" technique which provides not only the correlation time but also information upon the type of ultra slow motions in polymers.

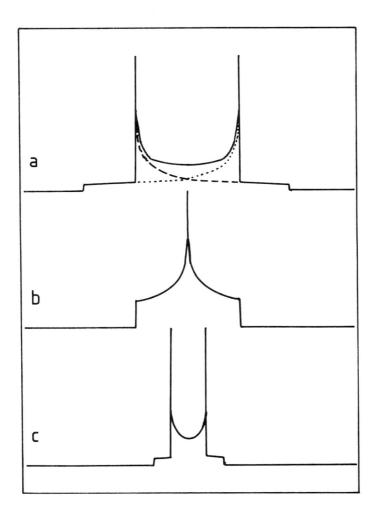

Fig. 1. Theoretical ^{2}H NMR line shapes for a solid glass (a), and for rapid anisotropic molecular motions (b,c) specified in the text. The dashed and dotted lines in (a) correspond to the $0 \leftrightarrow 1$ and $0 \leftrightarrow -1$ transitions, respectively.

MACROMOLECULAR MOTION

Amorphous polymers

Motional processes in polymers that are detected by relaxation methods are usually labeled by Greek letters α, β, γ, δ, ... where α denotes the glass transition in amorphous polymers, and the other transitions occur at lower temperatures. The relation of these transitions with particular molecular motions has been and still is being tackled by the methods listed in Table 1. Rather than attempting a general review, we shall treat polystyrene (PS) as a "typical" example where recent experiments have been rather successful.

The glassy state. Relaxations well below the glass transition temperature T_g are believed to originate from local motions of mobile side or end groups. In PS, the δ-process detected in mechanical and dielectric relaxation at very low temperatures (46 K at 1 kHz) has been related (41,42) with oscillatory motions of the phenyl groups by comparison with experiments in poly-chloro-styrenes (PCS) where relaxations in PS and p-PCS occur at the same temperature whereas those in o- and m-PCS are shifted to lower temperatures thus indicating the higher mobility of phenyl motions around the axis to the backbone chain. A δ'-process ($\sim$ 100 K at 10 kHz) could be assigned to motion of the n-butyl endgroups through proton T_1- and $T_{1\rho}$-measurements in polystyrenes differing in M_w (43,44). The assignment of the γ- and β-processes has been somewhat controversial in the literature (14,42,43). Yano and Wada (42) attribute the γ-process to rotation of the phenyl groups around the bond to the backbone since the γ-peak does not appear in dielectric loss whereas the β-process is believed to originate from backbone chain oscillations. McBrierty and Douglass (14) argue that only small amplitude motions of the phenyl rings are possible since $T_{1\rho} \ll T_1$ over the temperature range of 150 - 350 K. Large scale phenyl rotations are also at odds with calculations of Tonelli (45) which indicate a small probability for rotation. Illers and Jenckel (46), among others, suggest that the molecular motion responsible for β relaxation is the rotation of some phenyl groups possessing less steric hindrance than the majority. The β-transition is further complicated by an influence of the cooling rate of the sample. Thus, Petrie (47) has obeserved that a process seen in the mechanical loss tangent at 300 - 373 K vanishes after the PS sample was annealed for 70 h just below T_g at 365 K. Apparently, there is no unique structure of vitreous PS, and it can be shown by P-V-T-measurements that at least one order parameter must be introduced in order to describe the state of PS below T_g (48). Ballard et al. (49) have measured by neutron scattering the radius of gyration R_z of tagged molecules in PS samples that were cooled from the melt under constant high pressure (6 kbar) resulting in a reduction of R_z by $\sim$ 20%. Annealing at a temperature below T_g led to relaxation of R_z to its equilibrium value thus proving the existence of slow large scale motions in the glassy state.

Spin relaxation experiments at and below T_g are complicated by spin-diffusion (14,28,43) that prohibits a unique assignment of the β- and γ-processes in ordinary PS. This difficulty can be partly circumvented by investigating partially deuterated polystyrenes. In phenyl deuterated PS-d_5, the proton relaxation times are essentially determined by chain motions whereas the phenyl motions are seen in chain deuterated PS-d_3 (27). The difference in the spin relaxation rates, already apparent in T_1^{-1}, is most pronounced in the rotating frame relaxation results (27) shown in Fig. 2. If analysed in terms of a "strong collision" model (17, 50), $T_{1\rho}^{-1}$ is of the order of $(1 - p) \tau_c^{-1}$ where the correlation time τ_c is defined as the mean time between "collisions", and p is a measure of how much the dipolar coupling changes in a collision. In the limit of complete randomization, p = 0 and $T_{1\rho}^{-1} \approx \tau_c$. Values of

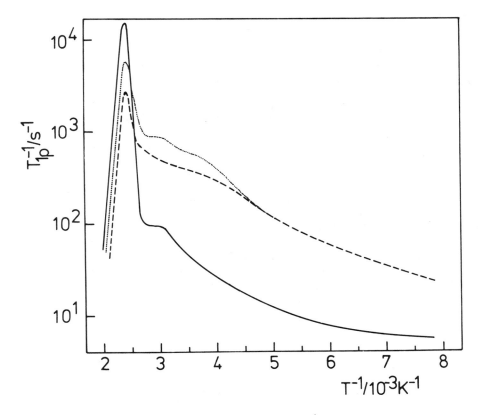

Fig. 2. Rotating frame relaxation rates $T_{1\rho}^{-1}$ of polystyrenes. Dotted line: PS-d_0; dashed line: PS-d_3; full line: PS-d_5.

$1 > p > 0$ are related with differing motional amplitudes of the H-H vectors. Very probably (50), the much larger $T_{1\rho}^{-1}$ in PS-d_3 originates from a small p-value whence the phenyl group motions should have much larger amplitudes than possible chain oscillations. The shoulder seen in the PS-d_5 curve of Fig. 2 can be attributed to a relaxation process with a $T_{1\rho}^{-1}$ maximum at ~ 340 K which is also visible in the PS-d_0 curve. We believe that it belongs to librational chain motions that have a smaller amplitude, but are in the same correlation time region as the phenyl group motions. Since $T_{1\rho}$ is also influenced to some extent by spin diffusion effects, we should not conclude from Fig. 2 that all phenyl groups perform large amplitude librations. As a matter of fact, we can conclude from the ^{2}H NMR line shapes of Fig. 3 and from preliminary solid echo FT-NMR spectra at 55 MHz, that only a fraction of the phenyl groups have motionally narrowed spectra at room temperature (51) thus supporting the suggestion of Illers and Jenckel (46). End group effects should not be responsible for this high mobility. In this case, each end group would influence the mobility of an unreasonably large number of phenyl groups in our PS-d_5 sample having M_w = 225 000 and M_w/M_n = 0.2. However, density fluctuations in conjunction with favourable conformations of adjacent molecules should allow for large amplitude motions within a small fraction of the sample, that are possibly related with the β-process seen in most amorphous polymers below T_g. By future annealing (47) or P-V-T-experiments (48) one might be able to change the fraction of mobile phenyl groups though it should be noted that thermal density fluctuations investigated by small angle X-ray scattering are apparently frozen in at T_g, and do not relax toward their equilibrium value during a time comparable with enthalpy or volume relaxation times (52).

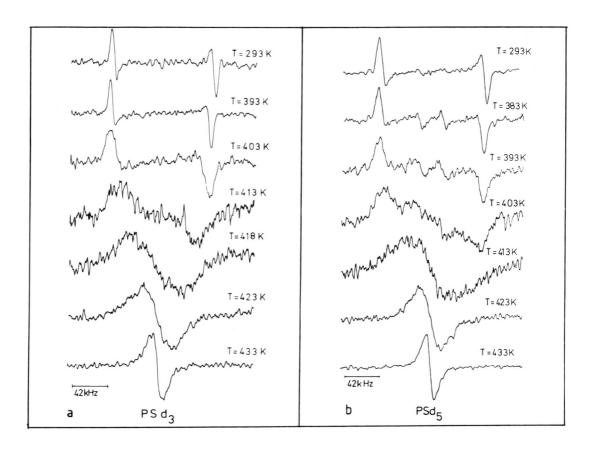

Fig. 3. ^{2}H-NMR spectra of PS-d$_3$ and PS-d$_5$ showing the derivative of the absorption line shape as determined at 13.8 MHz with a Bruker wide line spectrometer.

The glass transition region. In the temperature region between T$_g$ and a temperature of ~1.2 T$_g$ many properties of amorphous polymers (and even semicrystalline polymers, see below) can be described quite satisfactorily within a free volume picture expressed in terms of the WLF equation (2)

$$\log_{10} a_T = c_1(T - T_g)/(c_s + T - T_g) \tag{2}$$

where $a_T = \tau(T)/\tau(T_g)$ is the ratio of the relevant relaxation times at T and T$_g$, respectively. The absolute values of $\tau(T_g)$ differ considerably for different relaxation experiments, and they depend critically upon the shape and width of the distribution of correlation times chosen in order to interpret the experimental data. Thus, the average correlation time $\langle\tau(T_g)\rangle$ obtained from a fit of photon correlation spectra in PS (53) to a Williams-Watts (WW)-distribution (24) has a value of 32 s as compared with 0.32 s obtained from dielectric relaxation (54). On the other hand, the apparent activation energy around T$_g$ has the same value of ~600 kJ/mole for both methods. The WW-distribution is similar to the Cole-Davidson (CD)-distribution of correlation times both being asymmetric on a log τ scale and showing a small τ_c-tail (see Note d). Dielectric relaxation times have been described in the glass transition region of many polymers by asymmetric correlation time distributions (24) their widths extending over $\sim$ 4 decades. In the light of these experiments it appears very remarkable that the motional narrowing behaviour of the ^{2}H-NMR spectra shown in Fig. 3a for PS-d$_3$ can be explained by a single correlation time or by a distribution that extends over not more than about one decade. Although we have not yet performed a quantitative line shape analysis, it can be seen by inspection that a broad distribution of correlation times τ_c should result in superpositions of "rigid" and motionally narrowed line shapes. In particular, the small τ_c-tail of the WW- or CD-distributions should produce a motionally narrowed portion in the center of the spectrum at temperatures around 400 K which is absent in Fig. 3a. In search of a possible explanation we should realize that motional narrowing of the ^{2}H-spectra in PS-d$_3$ is related with the rotational autocorrelation function of single C-^{2}H vectors at the backbone chain. Thus, we see primarily the elementary conformational transitions producing rotational jumps of the order of the tetrahedral angle. Photon correlation spectra as well as dielectric and mechanical relaxation are related with fluctuations of larger volume elements where the cooperativity caused by free volume redistribution is much more pronounced. We expect that ^{2}H-spin-alignment (18) measurements around T_g will provide further sensible information and lead to a better understanding of the molecular processes involved. Preliminary spin-alignment experiments in PS-d$_3$ indicate that the rotational correlation function consists of a rapid initial decay (perhaps insufficient for producing observable line shape changes) followed by a slower decay on a time scale typical for results from other relaxation experiments. It should be noted that a rapid initial decay has also been observed by photon correlation spectroscopy (56). The additional rotational degree of freedom of the phenyl groups causes a motionally narrowed fraction of the ^{2}H-NMR spectra of PS-d$_5$ at temperatures up to $\sim$ 400 K where the whole spectrum starts to merge into the motionally narrowed Lorentzian shape seen at higher temperatures (see Note e). Since the line shape at 413 K (corresponding to $\tau_c \sim 10^{-5}$ s) is almost identical with the PS-d$_3$ shape at 418 K, it is clear that the C-^{2}H bonds of all phenyl groups reorient more rapidly than those of the backbone chain. If an activation energy of $\sim$ 300 kJ/mole is assumed, the temperature difference of $\sim$ 5 K corresponds to a τ_c-change by a factor of $\sim$ 3. The same factor has been obtained from proton T_1-measurements in PS-d$_5$ and PS-d$_3$ where the finite intermolecular contribution in PS-d$_3$ has been eliminated by the deuteron dilution technique (26,27,57).

Note d. The CD-distribution omits all correlation times $\tau > \tau_0$ and decreases at smaller τ_c-values in proportion to $(\tau_0/\tau_c - 1)^{-\beta}$ where β is a width parameter. A detailed comparison of the WW- and CD-distributions is given in Ref.(55).

Note e. The derivative of the absorption line shape is shown in Fig. 3 due to the usual detection technique of wide line spectroscopy. The solid-echo FT-technique applied for the spectra of Fig. 5 is not applicable in PS at temperatures $\geq$ 400 K since the C-^{2}H vectors reorient by isotropic rotational processes where no solid echo is observable if the distance τ between the RF-pulses of the solid echo sequence becomes of the order of the rotational correlation time (40).

In the temperature region of 420 - 500 K, we have used a Fuoss-Kirkwood (FK)-distribution (23,43) of correlation times in order to describe proton spin-lattice relaxation times T_1 measured at 13.8, 55.2, and 90.0 MHz in PS-d$_5$ and PS-d$_3$ (17,57). The frequency dependence is described equally well by a Cole-Cole-distribution which is also symmetrical on a logarithmic τ-scale, but less satisfactorily by the non-symmetrical CD-distribution (23,27). The width parameter β of the FK-distribution was determined as β = 0.44, 0.37, and 0.23 for PS-d$_5$, PS-d$_3$(intra), and PS-d$_3$(inter), respectively. This reflects the increasingly complex motion of H-H chain vectors (PS-d$_5$), H-H vectors at phenyl groups of the same chain (PS-d$_3$; intra), and at neighboring chains (PS-d$_3$; inter), respectively. β = 0.23 corresponds to a very broad correlation time distribution that extends over $\sim$ 4 decades whereas β = 0.44 corresponds to only $\sim$2 decades which is, however, still broader than $\beta \gtrsim 0.6$ estimated from the ^{2}H line shapes of Fig. 3. Since the proton T_1-values of PS-d$_5$ are influenced by the motion of all H-H-vectors within $r \lesssim 2.5$ Å along the chain, the smaller β-value appears plausible. The average correlation times τ_0 obtained by fitting the T_1-values measured at 13.8 MHz in PS-d$_5$ to a FK-distribution, are compared in Fig. 4 with the ratio $a_T = \tau(T)/\tau(T_g)$ for 2 different sets of WLF parameters where the values of c_1 and c_2 given in Ref. (59) are closer to other published WLF parameters (2). It is remarkable that the change in slope of the semilogarithmic plot of τ_0 observed at $\sim$ 470 K is also detectable in the WLF curves although the latter have been obtained from experimental data at lower temperatures, and should not apply in the high temperature region (2). The temperature of 470 K corresponds to T/T_g = 1,24, and is close to the ratio T_{11}/T_g found for a "liquid-liquid transition" at T_{11} that has been investigated in great detail by Boyer and collaborators (60). The authors have also noted (61) that T_{11} agrees with a temperature T_c defined by the experience (34,62) that ^{13}C-NMR line shapes cannot be studied with high resolution techniques at temperatures below T_c since the lines become too broad. Recently, Mandelkern (63) has drawn correlation times obtained from ^{13}C - T_1 values as a function of T/T_g for a number of solid polymers, and has found that they are in harmony with the WLF equation in the whole temperature range $T \gtrsim T_c$ where ^{13}C spectra could be observed. In particular, he finds that the correlation time is $\sim 10^{-8}$ s at $T_c \sim T_{11}$ which agrees with our findings (see Fig. 4). Neumann and Mac Knight (64,65) argue that the T_{11} relaxation is caused by the temperature variation of the Newtonian viscosity of the material, and can be described by the WLF equation. Since Eq.(2) produces very small apparent activation energies at $T \gg c_2$ it appears difficult to interpret the WLF results above T_{11} where further intramolecular processes should be considered (2). However, there is probably agreement about the usefulness of the WLF equation in the region $T_g \lesssim T \lesssim T_{11}$ covering a correlation time range of roughly 8 decades.

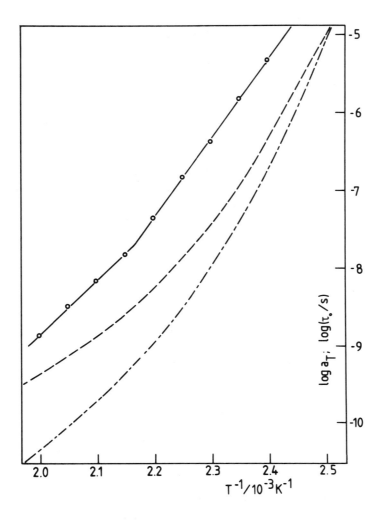

Fig. 4. Comparison of rotational correlation times τ_0 (in s) obtained from ^{1}H spin-lattice relaxation times T_1 in PS-d_5 with the ratio $a_T = \tau(T)/\tau(T_g)$ for 2 sets of WLF parameters:

— — — — —: $c_1 = 12.4$; $c_2 = 41.0$ K (Ref. 58).

——·——·——: $c_2 = 14.5$; $c_2 = 50.4$ K (Ref. 59).

<u>Semicrystalline polymers</u>

Molecular motion in partially crystalline polymers depends considerably upon the crystallization conditions that determine the crystallinity and morphology of the sample under investigation (34,66). Nevertheless, many experiments are interpreted in terms of a 2 phase model
where the properties of the "amorphous" regions are compared with those of truly amorphous
polymers. This conception has been remarkably successful, but it has also led to difficulties
of which the controversy about the glass transition temperature in polyethylene (PE) provides
ample evidence (67). Davis and Eby (68) have even published a histogram showing the distribu-

tion of T_g values of 50 publications on PE with maxima between 140 and 260 K. Since there is presently no general agreement on structure and dynamics of crystalline polymers we restrict our discussion to some recent NMR results that may contribute to better understanding of the molecular processes involved.

The motional narrowing of ^{1}H NMR spectra in semicrystalline polymers has been investigated since 1950 and comprehensive reviews have been given (15,34,69). Attempts to decompose the line shapes into crystalline and amorphous components have revealed that more than one amorphous component is necessary (70). However, the analysis in terms of particular motional models is difficult since it entails the problem of dipolar interaction between many protons. Furthermore, there is no unambiguous procedure for separating the contribution from protons in the crystalline regions since the crystalline line shape is only approximately known. This necessitates additional assumptions in order to simplify the line shape problem. Zachmann and collaborators (71) have calculated the ^{1}H second moments and line shapes for CH_2 groups in a chain with fixed ends as a function of the chain length, and the end to end distance where the possible chain conformations are restricted to the positions of a diamond lattice. As we have noted above, one can readily calculate the ^{2}H-NMR line shape for any well defined motional model. Thus we have also calculated line shapes for this model of chains between fixed ends where any chain motion in the diamond lattice can now be viewed as a special mechanism for producing tetrahedral jumps of the C-^{2}H vectors since all possible C-^{2}H directions coincide with the 4 tetrahedral axes of the lattice (72). We have compared these calculations with experimental ^{2}H line shapes measured in a sample of melt crystallized deuterated PE having an X-ray crystallinity of 74% (16,73). Typical line shapes shown in Fig. 5 can easily be decomposed into the Pake spectrum (see Note f) for rigid C-^{2}H vectors (see Fig. 1), and a motionally narrowed spectrum. By comparing the latter with our model calculations we can infer that the average length of the flexible chains between fixed ends increases from 3-5 bonds at room temperature to more than 10 bonds at 380 K. On melting, the experimental line width of the mobile deuterons decreases further by an order of magnitude thus indicating that the motion in the amorphous regions is still constrained at temperatures just below the melting point.

The nature of the constraints that are described by fixed carbon atoms in our model calculations has been investigated by the ^{2}H spin-alignment technique (18) mentioned above. The results can be summarized as follows: At each temperature, the time averaged quadrupole coupling (responsible for the line shape) remains constant for times that are equal or longer than the spin-lattice relaxation time of about 50 ms. Thus we conclude that independent of any particular model chain motion is subject to long lived constraints. The ^{2}H line shapes provide a measure as to what extent the number of conformations accessible for a given segment increases with increasing temperature. The segmental motion itself is very rapid with correlation times below 10^{-8} at temperatures above room temperature.

We have been able to investigate the NMR-line shape of mobile deuterons in a temperature range of 125 - 390 K (16,74) by application of a pulse technique where the total magnetization is saturated first by a series of 90^0 pulses and then the solid echo is created after a waiting period between 20 and 200 ms. This assures that the signal is essentially due to

---

Note f. Small deviations from the Pake spectrum at higher temperatures can be explained by oscillations around the chain axes the root-mean square angle being 11^0 at 373 K (73). The 180^0 jumps of the α-process in the crystalline regions are not observable in the ^{2}H spectrum since they do not change the quadrupole coupling tensor.

deuterons in the amorphous regions those in the crystalline regions still being saturated because of their much longer spin-lattice relaxation time. It is concluded from an analysis of
the experimental data that only highly localized motions are possible at low temperatures
above $\sim$ 140 K. Around 230 K, the line shape is similar to the $\eta = 1$ spectrum shown in Fig.1b
which is consistent with a 3-bond motion (75) where each C-^{2}H bond performs jumps between
2 orientations. These are augmented by 5-bond motions (Schatzki crankshaft) at room temperature, and by motions involving more than 10 bonds below the melting point. It should be emphasized that these findings apply to a PE sample of high crystallinity, and a particular
morphology resulting from the crystallization conditions applied. Higher mobility may be
found in samples of branched or less crystalline PE. Finally, we should contrast our findings in PE with those in amorphous chain deuterated polystyrene (see above). Here, the
spin-alignment time is drastically reduced above the glass transition, and neither spin-alignment nor solid echoes are found above 400 K. The ^{2}H line shape changes within $\sim$ 50 K from the
rigid solid limit at 390 K to the motionally narrowed Lorentzian at 440 K, and no indications
of "constrained" motions are seen that characterize the spectrum in PE between the γ-transi-
tion at $\sim$ 140 K up to the melting point.

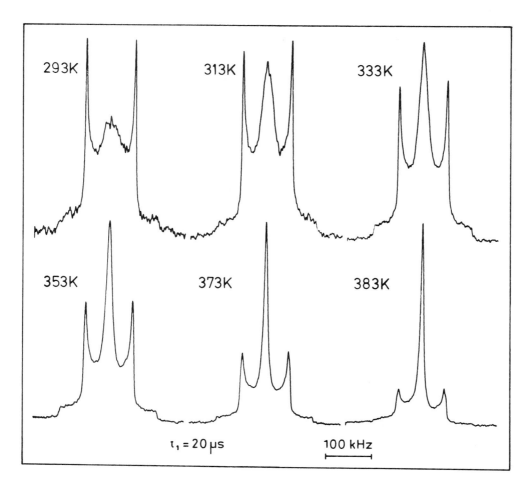

Fig. 5. ^{2}H NMR spectra of deuterated polyethylene.

Polymer melts

Molecular motion in polymer melts can be investigated on different time and length scales. In shear viscosity and self diffusion experiments, the time scale is determined by the motion of the center of gravity of a macromolecule over roughly the end to end distance. This global motion is caused by an extremely large number of elementary steps performed by the monomeric units on a time scale that is typical for the motion of small molecules in liquids. NMR relaxation methods are usually most sensitive to the high frequency segmental motions with correlation times in the ns- to μs-region (see Fig. 4). However, Kimmich and collaborators (25) have been able to investigate spin-lattice relaxation times T_1 at Larmor frequencies between 10^4 and 10^8 Hz where the results contain information upon fast and slow motions. The data are interpreted in terms of a 3-component model the components being anisotropic segment reorientation via defect diffusion, longitudinal chain diffusion caused by the same defect process, and the conformational fluctuation of the environmental tube incorporating the reference chain. The model parameters determined from the frequency dependence of T_1 are a short correlation time τ_s for segmental motion, a medium correlation time τ_1 that measures chain diffusion over a correlation length l of the order of the persistence length of the chain, and a very long time constant τ_r that is the mean life-time of the local tube orientation in the range of the correlation length. In polyethylene, it was possible to fit T_1 values at different frequencies (10^4 - 10^8 hz), different temperatures (423, 473 K) and different molecular weights (M = 2 200, 9 600, and 100 000) in terms of this model where τ_1 and τ_r are proportional to M and M^3, respectively (76). By comparison with the diffusion experiments of Klein (77), the correlation length is estimated to ~ 8 Å at 423 K. Although, the model applied by Kimmich and collaborators is similar in spirit to the reptation model of De Gennes, Doi, and Edwards (78), the tube constraint is treated in a different manner since the tube dimensions are essentially determined by the adjacent molecules whereas in the Doi and Edwards theory, the tube diameter is of the order of the root-mean square distance between entanglements (~ 60 Å in PE). The exciting idea of chain reptation has provided a number of predictions that will, in part, be checked by current NMR and self diffusion experiments. Still unsolved questions are related with the state between T_g and T_{11} where it is debated whether this "supercooled" melt is a "fixed fluid" (60), a region of largely cooperative motion (79), or an ordinary melt with all motional degrees of freedom just becoming slower by lowering the temperature (65).

REFERENCES

1. N.G. McCrum, B.E. Read, and G. Williams, Anelastic and Dielectric Effects in Polymeric Solids, Wiley, New York 1967.
2. J.D. Ferry, Viscoelastic Properties of Polymers, 3rd Ed., Wiley, New York 1980.
3. I.M. Ward, Mechanical Properties of Solid Polymers, Wiley, New York 1971.
4. G.D. Patterson, CRC Critical Reviews in Solid State and Materials Sciences 1980, 373.
5. W. Wrasidlo, Adv. Polym. Sci. 13, 1 (1974).
6. P. Hedvig, Dielectric Spectroscopy of Polymers, Adam Hilger, Bristol 1977.
7. G. Williams, Adv. Polym. Sci. 33, 59 (1979).
8. H. Block, Adv. Polym. Sci. 33, 93 (1979).
9. E. Fredericq and C. Houssier, Electric Dicroism abd Electric Birefringence, Clarendon Press, Oxford 1973.
10. M.S. Beevers, J. Crossley, D.C. Garrington, and G. Williams, J. Chem. Soc. Faraday Trans. 72, 1482 (1976).
11. G.D. Patterson, in Methods of Experimental Physics: Polymer Physics, R. Fava, Ed., Academic Press, New York 1980.
12. A. Machonnachie and R.W. Richards, Polymer 19, 739 (1978).
13. J.S. Higgins, L.K. Nicholson, and J.B. Hayter, Polymer 22, 163 (1981).
14. V.J. McBrierty and D. Douglass, Macromol. Revs., in press.
15. I.Ya. Slonim and A.N. Lynbimov, The NMR of Polymers, Plenum Press, New York 1970.
16. D. Hentschel, H. Sillescu, and H.W. Spiess, Macromolecules, in press.

17. D. Wolf, Spin-temperature, and Nuclear Spin Relaxation in Matter, Clarendon Press, Oxford 1979.
18. H.W. Spiess, J. Chem. Phys. 72, 6755 (1980).
19. B. Ranby and J.F. Rabek, ESR Spectroscopy in Polymer Research, Springer, Berlin 1977.
20. L.J. Berliner, Ed., Spin Labeling, Acad. Press, New York 1976.
21. L.R. Dalton, B.H. Robinson,L.A. Dalton, and P. Coffey, Adv. Magn. Resonance 8, 149 (1976).
22. W.P. Slichter, in NMR Basic Principles and Progress, Eds.: P. Diehl, E. Fluck, and R. Kosfeld, Springer, Berlin 1971, Vol. 4, p. 208.
23. F. Noack, ibid., Vol. 3, p. 83.
24. C.J.F. Böttcher and P. Bordewijk, Theory of Electric Polarization, 2nd Ed., Vol. II, Elsevier Sci. Publ. Co., Amsterdam 1978.
25. H. Koch, R. Bachus, and R. Kimmich, Polymer 21, 1009 (1980), and references therein.
26. B. Willenberg and H. Sillescu, Makromol. Chem. 178, 2401 (1977).
27. P. Lindner, E. Rössler, and H. Sillescu, submitted to Makromol. Chem.
28. D.C. Douglass and G.P. Jones, J. Chem. Phys. 45, 956 (1966).
29. J. Schaefer, in Structural Studies of Macromolecules by Spectroscopic Methods, K. Ivin, Ed., Wiley, New York 1976, p. 201.
30. J.R. Lyerla, in Contemporary Topics in Polymer Science, M. Shen, Ed., Plenum Press, New York 1979, Vol. 3, p. 143.
31. Th.C. Farrar and E.D. Becker, Pulse and Fourier Transform NMR, Academic Press, New York 1971.
32. J. Schaefer, E.O. Stejskal, and R. Buchdahl, Macromolecules 10, 384 (1977).
33. A.N. Garroway, W.B. Monitz, and H.A. Resing, Faraday Symposia Chem. Soc. 13, 63 (1978).
34. J.J. Dechter, R.A. Komoroski, D.E. Axelson, and L. Mandelkern, J. Polym. Sci. (Polym. Phys. Ed.) 19, 631 (1981), and references therein.
35. D.L. VanderHart, Macromolecules 12, 1232 (1979).
36. A.J. Vega and A.D. English, Macromolecules 13, 1635 (1980).
37. R. Hentschel and H.W. Spiess, J. Magn. Res. 35, 157 (1979).
38. R. Hentschel, H.W. Spiess, and H. Sillescu, Polymer, in press.
39. H.W. Spiess, in NMR Basic Principles and Progress, Eds.: P. Diehl, E. Fluck, and R. Kosfeld, Springer, Berlin 1978, Vol. 15.
40. H.W. Spiess and H. Sillescu, J. Magn. Resonance 42, 381 (1981).
41. R.D. McCammon, R.G. Saba, and R.N. Work, J. Polym. Sci. A-2 7, 1721 (1969).
42. O. Yano and Y. Wada, J. Polym. Sci. A-2 9, 669 (1971)
43. T.M. Connor, J. Polym. Sci. A-2 8, 191 (1970).
44. B. Christ Jr., J. Polym. Sci. A-2 9, 1719 (1971).
45. A. Tonelli, Macromolecules 6, 682 (1973).
46. K.H. Illers and E. Jenckel, Rheol. Acta 1, 322 (1958).
47. S.E.B. Petrie, A.C.S. Polymer Preprints 15/2, 336 (1974).
48. H.-J. Oels and G. Rehage, Macromolecules 10, 1036 (1977), and references therein.
49. D.G.H. Ballard, A. Cunningham, G.W. Longman, A. Nevin, and J. Schelten, Polymer 20, 1443 (1979).
50. C.P.Slichter and D.C. Ailion, Phys. Rev. A 135, 1098 (1964).
51. U. Pschorn, H.W. Spiess, H. Sillescu, and M. Wehrle, to be published.
52. J.H. Wendorff, J. Polym. Sci. (Polym. Lett. Ed.) 17, 765 (1979).
53. C.P. Lindsey, G.D. Patterson, and J.R. Stevens, J. Polym. Sci. (Polym. Phys. Ed.) 17, 1547 (1979)
54. S. Kastner, E. Schlosser, and G. Pohl, Kolloid Z. Z. Polym. 192, 21 (1963).
55. C.P. Lindsey and G.D. Patterson, J. Chem. Phys. 73, 3348 (1980).
56. H. Lee, A.M. Jamieson, and R. Simha, Macromolecules 12, 329 (1979).
57. H. Sillescu, E. Rössler, and P. Lindner,A.C.S.Polymer Preprints 22, 107 (1981).
58. K.C. Rush, J. Macromol. Sci. - Phys. B-2, 179 (1968).
59. A.V. Tobolsky, J.J. Aklolonis, and G. Akovali, J. Chem. Phys. 42, 723 (1965).
60. S.J. Stadnicki, J.K. Gillham, and R.F. Boyer, J. Appl. Polym. Sci. 20, 1245 (1976).
61. R.F. Boyer, J.P. Heeschen, and J.K. Gillham, J. Polym. Sci. (Polym. Phys. Ed.) 19, 13 (1981).
62. D.E. Axelson and L. Mandelkern, J. Polym. Sci. (Polym. Phys. Ed.) 16, 1135 (1978).
63. L. Mandelkern, Communicated at the A.C.S. Spring Meeting, Atlanta 1981.
64. R.M. Neumann and W.J. Mac Knight, J. Polym. Sci. (Polym. Phys. Ed.) 19, 369 (1981).
65. R.M. Neumann, G.A. Senich, and W.J. Mac Knight, Polym. Eng. Sci. 18, 624 (1978).
66. I.G. Voigt-Martin, E.W. Fischer, and L. Mandelkern, J. Polym. Sci. (Polym. Phys. Ed.) 18, 2347 (1980).
67. U. Gaur and B. Wunderlich, Macromolecules 13, 445 (1980), and references therein.
68. G.T. Davis and R.K. Eby, J. Appl. Phys. 44, 4274 (1973).
69. R. Kitamaru and F. Horii, Adv. Polym. Sci. 26, 139 (1978).
70. K. Bergmann, J. Polym. Sci. (Polym. Phys. Ed.) 16, 1611 (1978), and references therein.
71. K. Rosenke and H.G. Zachmann, Progr. Colloid & Polym. Sci. 64, 245 (1978), and references therein.
72. K. Rosenke, H. Sillescu, and H.W. Spiess, Polymer 21, 757 (1980).
73. D. Hentschel, H. Sillescu, and H.W. Spiess, Makromol. Chem. 180, 241 (1979).
74. D. Hentschel, H. Sillescu, and H.W. Spiess, to be published.

75. W. Wunderlich, J. Chem. Phys. 37, 2429 (1962).
76. R. Kimmich and H. Koch, Colloid & Polymer Sci. 258, 261 (1980).
77. J. Klein, Nature (London) 271, 143 (1978).
78. P.G. De Gennes, Scaling Concepts in Polymer Physics, Cornell University Press, Ithaca
                                                                              1979.
79. A.M. Lobanov and S.Ya. Frenkel, Vysokomol. Soedin. 12, 1045 (1980).

SOME NEW ASPECTS OF CRYSTALLIZATION MODES IN POLYMERS

A. Keller

H.H. Wills Physics Laboratory, University of Bristol, Bristol BS8 1TL. U.K.

Abstract - The present article covers some recent experimental developments in two aspects of polymer crystallization. The first, (A), deals
with melt crystallization from the random state, the second, (B), with
crystallization from the orientated state in the case of solutions. The
purpose of (A) was to establish the primary lamellar thickness (ℓ^*) over
as wide a temperature range as possible (an information not available so
far owing to the intervention of isothermal thickening). The effect of
isothermal thickening was circumvented through the newly discovered 'enhanced self nucleation' method and at the same time the achievable supercooling
range could be significantly extended towards low crystallization temperatures and the corresponding ℓ^* values determined. In addition to providing
primary information for future theoretical work the result has led to the
recognition of astonishingly high crystal growth rates (more than $2m\ sec^{-1}$)
which nevertheless still yield well defined lamellae raising some renewed
fundamental issues as regards the existence and nature of chain folding and
the ability of chains to form organized structures. The results in part B,
amongst others, point to the predominant importance of agitation in influencing the state of the solution above the temperature where it normally crystallizes, as diagnosed by its ability to gel on cooling after all agitation
has stopped. The formation of temporary, yet surprisingly long lived, associations is inferred with far reaching implications for shish-kebab formation,
for the creation of high modulus fibres and for flow induced phase segrega-
tions in general, beyond the subject of crystallization, linking up with
long standing but unexplained observations in solution rheology.

INTRODUCTION

The present paper will cover two separate topics in the area of polymer crystallization,both
relying on polyethylene as a model substance. The first, (A), will deal with crystallization
from the random melt, the second,(B),with crystallization from the oriented state in the case
of solutions. While I consider the subject matter in both parts central to the general issue
of polymer crystallization, the scope and style of the two parts are different. (A) will be
a summarising account of some recent experimental work, and while containing a fairly extensive introduction to set the scene, it is homogeneous and sharply focussed in its content.
(B) on the other hand is a very broad, even if brief survey of a range of interconnected subjects with an unexpected widening of the issues involved at the end.

A) CRYSTALLIZATION FROM THE RANDOM STATE

General Background and Objectives
My main concern will relate to straightforward observational facts, such as are beyond any
dispute. First of these is the lamellar nature of crystallization, the second is chain folding and the third the dependence of the fold length on crystallization temperature. Let us
take these in turn.

The first and most unambiguous observations were made on crystals grown from solution a fact
which still holds to the present day. As well known, here the primary products of crystallization could be identified as isolated lamellae. A simplest monolayer is illustrated by
Fig. 1. It is an indusputable fact that the molecules are perpendicular (or at a specific
large angle) to the basal plane of the lamellae. By the familiar common sense argument the
chain can only be accommodated by an isolated lamella if it folds back on itself. Here the
lamellar thickness, a clearly measurable quantity (ℓ),corresponds to the fold length, or conversely the fold length, hence a molecular feature, determines a morphological feature the
layer thickness,hence chain folding(without explicit committal at this point on how the chain
folds,adjacently, regularly etc. which are the much contested issues). The fold length (ℓ)

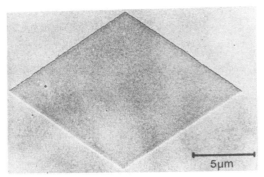
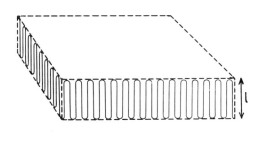

Fig. 1.   a) Solution grown single crystal lamella of polyethylene. Electron micrograph (Ms. S. Organ, Bristol) b) Schematic representation of the chain folded structure in a). The folds are drawn   sharp and adjacently reentrant. This is a contested issue, not to affect the content of this article: for the present purpose the essential point is that the lamellae have a well defined thickness (ℓ) which is determined by the fold length irrespective of the precise nature of the fold.

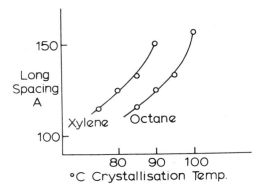

Fig. 2. Lamellar thickness (ℓ, as determined by low angle X-ray scattering) as a function of crystallization temperature for polyethylene crystallized from two solvents. The relative shifts of the curves indicates that the supercooling (ΔT) is the fold length determining factor (39).

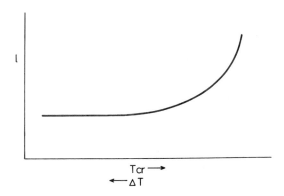

Fig. 3. The generalised ℓ v. T_{cr} (or ΔT) relationship for solution crystallized polymers. The curve is schematic as no given single polymer can be crystallized over a wide enough temperature range to reveal the full trend represented in the figure; namely the horizontal plateau at high and the significant upswing of the curve towards low supercoolings (ΔT). Poorly crystallizable polymers display the former and readily crystallizable ones (as polyethylene in Fig. 2) the latter behaviour.

is a unique function of the crystallization temperature or rather supercooling (ΔT) in the manner expressed by Fig. 2. Fig.3 gives the more general shape of the ℓ v.T_{cr} or ΔT curve covering a much larger supercooling range than normally achievable by a given polymer. E.g.ℓ in polyethylene (Fig.2), which is not highly supercoolable, is confined to the upswinging right hand side and, while in most nylons which crystallize at high supercoolings, only the horizontal plateau is dominant.

Now let us consider the situation for crystallization from the melt. It is by now generally recognized and accepted as a fact that the basic morphology is lamellar also in this case. However, as here the lamellae are contiguous, the same argument for chain folding as in the case of an isolated crystal does not carry the same weight because here chains can pass from one lamella to the next. Nevertheless, there are several cogent reasons why chain folding should also apply in this case, which are as follows: i) The analogy with the isolated solution grown crystal layer. Indeed,the comparative separateness of layers, hence the absolute necessity of chain folding, can be asserted at least as revealed by certain preparations (Fig. 4). ii) Why should crystal growth stop along the chain direction and this always at the same regular interval ? Clearly chain folding provides an answer. iii) Irrespective of ii) what should be the chain structure between the layers ? If the chain carried on straight there is nothing to distinguish the lamella. If it becomes random before carrying on to the next layer the density in the laterally confined space becomes unacceptably high. It follows that at least a portion of the chains must return to the same lamella from which they have emerged, and this in a tight and adjacently reentrant fashion so as to provide room for the rest of the interlamellar material to randomise and give rise to amorphous regions; hence there must be chain folding with at least a portion adjacently reentrant as an a priori geometric necessity. So far everybody concurs. The arguments in this field relate to the amount of folding, and in particular to the minimum amount of adjacent reentry that is required by a priori geometric arguments; namely whether the latter represented a minority or a majority component. Apart from giving some references (1,2) I shall not be concerned with this disputed question any further at this place. In what follows I shall concentrate on the value of ℓ in the case of the bulk. Clearly ℓ is a measurable quantity irrespective of the various arguments on the fold structure, and its knowledge is a basic requirement for the understanding of how the crystals grow.

However, in the case of melt crystallization the determination of ℓ corresponding to that of primary crystal growth (to be denoted ℓ^*) is not generally possible owing to the phenomenon of isothermal thickening discovered by Hoffman and Weeks (3). It is well known from solution grown crystals that ℓ can increase on heat annealing subsequent to crystallization - an effect associated with chain refolding. In the case of crystallization from the melt ℓ increases while the crystal is still growing. This isothermal thickening is strikingly indicated by Fig. 5. It follows that the crystal thickness we are actually recording does not correspond to the one which had formed originally, but to a thickened version, depriving us from the knowledge of the primary crystal thickness and of its variation with supercooling. It was one of the aims of our latest works to arrive at ℓ^*, i.e. at the unthickened primary value of the layer thickness, also in the case of the crystallization from the melt. This we shall term as our Objective I.

At this point a very pertinent interjection could be made: how do we know that the ℓ values measured for solution crystallization are the primary ℓ^* values and do not correspond to crystals which have thickened subsequently ? The answer is that we cannot be sure, although this has always been implicitly assumed, and in fact, as will be referred to later, formed the basis of crystallization theories. In fact there is a recent departure from this implicit assumption by Rault (4) who takes the constant flat portion of the ℓ v ΔT curve (Fig.3) as the only one relevant to primary crystallization and associates the upswing at low ΔT values with thickening. In itself this is a perfectly legitimate suggestion which,however, would necessitate a radical departure in the interpretation of crystallization, as is in fact expressed by Rault's new approach.

In this latter context all I can do is to express my personal view and working hypothesis which is the current guiding principle in our Bristol laboratory. This is as follows. The primary lamellar thickness (ℓ^*) is determined by the supercooling (ΔT), as expressed by Figs. 2,3, while subsequent thickening is essentially influenced by the absolute temperature (T) which governs the mobility of the chain in the lattice. As crystallization in solution occurs at a lower temperature than in the melt I assume that in the temperature range of the upswing in Figs. 2,3 (which is 70-90°C in polyethylene) isothermal thickening does not yet operate. In fact it is known that in order to produce such thickening by means of heat annealing, such crystals have to be heated to 110°C or beyond (in the dry state,otherwise they dissolve). All melt crystallization in polyethylene in fact occurs above 110°C (see later), hence in the range where thickening is expected by the annealing experiments on solution grown crystals.

The above working hypothesis, if correct, is central to polymer crystallization and would need confirming. This is our current endeavour. It could be done in two ways 1) to crystal -lize at lower temperatures from the melt, in order to demonstrate that isothermal thickening

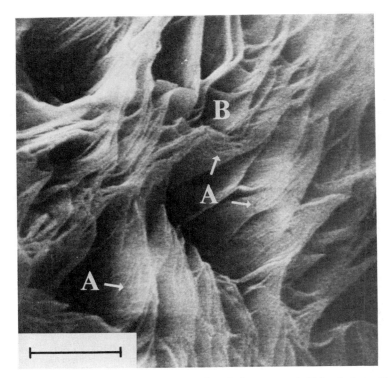

Fig. 4. Interior of a spherulitic polyethylene crystallized from the melt
at a low undercooling (128.8°C). Scanning electron micrograph. The three-
dimensional nature of the detail revealing the lamellae, in places singly,
has been obtained by extracting the low molecular weight which had crystallized at a lower temperature (40).

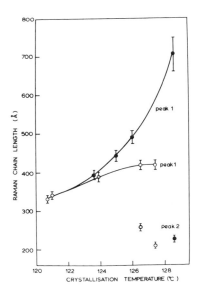

Fig. 5. Prominent illustration of lamellar thickening in the course of
isothermal crystallization of polyethylene. The curves represent the lamellar thickness, as assessed by the Raman LAM technique. The upper curve
corresponds to long crystallization times during which the primary crystallization has reached completion at the corresponding temperatures. The
lower curve corresponds to a crystallization time of 1 hr during which,
particularly at the higher crystallization temperatures, the primary crystallization would not yet be complete. The difference in lengths (i.e. ℓ
values) represented by the two curves corresponds to isothermal thickening
(for other symbols see original source) (41)

can be avoided in this way also in the case of the melt, and 2) to crystallize at increasing temperatures from appropriate solutions in order to demonstrate that such isothermal thickening can in fact be produced also in solutions according to the above hypothesis. This plan, presently in progress, would also establish automatically the continuity (so far unachieved) between solution and melt crystallization as regards ℓ^*. So far we have realized 1) i.e. our programme involving the melt. It is a summary of a far ranging activity by a team of colleagues to be reported in detail elsewhere (5,6,7).

To appreciate the background to 1) above, let us look at the temperature scale in Fig. 6. The normal temperature range of melt crystallization, is broadly between $129^{\circ}C$ and $115^{\circ}C$ as indicated in Fig. 6. With polyethylene, as normally crystallized, crystallization becomes unmeasurably slow above $129^{\circ}C$ and uncontrollably fast below $\sim 115^{\circ}C$, so that the sample becomes fully crystallized before a temperature lower than $\sim 115^{\circ}C$ can be attained. Our further aim therefore was to find means of extending the temperature of melt crystallization as far below $115^{\circ}C$ as possible and measure the corresponding ℓ values. This we shall call our Objective II.

The combined Objectives I and II are therefore in service of the following purposes. First and chiefly, to obtain ℓ^* values for melt crystallization (a), and this over as wide a temperature range as possible (b). As ℓ^* is a basic parameter of the morphology and of the underlying molecular conformation (fold length) both a) and b) represent a search for basic facts outside and beyond existing disputes. c) To obtain information on isothermal refolding, in particular to identify the temperature below which such refolding ceases to play a part. d) ℓ^* is the test stone of theories, hence the knowledge of ℓ^* over a wide temperature range should be of major consequence in the reassessment of existing theories and in their extension, and possibly in the construction of new ones.

In the context of d) above, a brief interjection will be made relating to theories, and it is on this point that we may step on disputed ground. The theories of crystallization were set up in the first place to account for the ℓ v. T_{cr} (or ΔT) curves in solution grown crystals (Fig. 2). As well known the kinetic theories in various versions were highly successful in establishing such a relation with very realistic input parameters, the resulting expression being

$$\ell^* = \frac{2\sigma_e T_m^{\,o}}{\Delta H \Delta T} + \delta\ell \qquad\qquad 1)$$

where ΔH is the heat of fusion, $T_m^{\,o}$ the melting point (or dissolution point, $T_d^{\,o}$, for solutions) of the infinitely extended crystal and σ_e the free energy of the fold surface. For low and moderate supercoolings $\delta\ell$ is a small quantity, hence the first term on the right hand side dominates ℓ^* and describes curves as in Fig. 2. In the early versions of the theory, however, $\delta\ell$ (as expressed explicitly in terms of the relevant variables including ΔT) increases beyond bounds (the so called $\delta\ell$ catastrophe) at very high supercoolings, in obvious contradiction with experience (curves such as in Fig. 3 became available later). This $\delta\ell$ catastrophy however, could be avoided by later adaptations of the theories by which the validity of the whole approach was safeguarded, there being two ways in which this was achieved (8,9). The theories in all their various versions relied on the chain folded deposition of the molecules along a growing prism face, the nucleation of a new chain folded ribbon along an initially smooth face being the rate determining step. The flux of the net attachment of the chain is taken as a function of ℓ and the mean ℓ, i.e. for which the total net flux is the greatest, is calculated. This ℓ value will be the dominant fold length, i.e. ℓ^*. In brief, the chains will fold with a particular ℓ^* because it is in this manner that the crystal grows fastest.

For the above calculations a specific path of deposition, hence a model had to be chosen. In all the theories this is taken as adjacently reentrant folding. It is on this structural concept on which most current disputes are centred. To this the following comment is appropriate. Irrespective of how closely the model of adjacent reentry is obeyed it is the simplest pathway to describe mathematically. There is nothing to say that other more complex pathways (e.g. where the depositing fold stems are one stem separation removed, or various mixed stem separations) could not be introduced and similar theories derived along such models. Such treatments would be much more complicated and the plain fact is that such have not been attempted. Thus if the present theories were totally dismissed on grounds of objections to the underlying model (this is not to say that I concur with most of the objections), we would be left even without an attempt to account for the most basic facts of observation. This is as far as I shall refer to the fold surface and stem reentry dispute in this article.

Whatever the case, the fact remains that we have a successful class of theory to account for the main observations amongst which the foremost and most quantitative one is the ℓ^* v. ΔT relation. Originally the theories have been set up for solution crystallization where ℓ^* was directly assessable. With the recognition of the lamellar morphology also in the bulk material they were increasingly applied also to melt crystallization. However, as here ℓ^* could not be assessed (because of isothermal thickening) testing of the theories relied not on the lamellar thickness but on the lateral growth rate (G) of the crystals (e.g. 10).

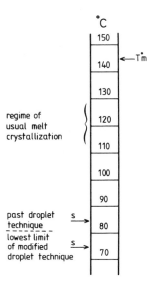

Fig. 6. Temperature scale relevant to crystallization of polyethylene from the melt. (see also later in text).

At this stage a few words about growth rates. The same theory which defines ℓ^* as a function of ΔT also leads to an expression for the linear crystal growth rate (G), as both are determined by the fastest net flux of chain deposition. In its first established form it yields

$$G = G_o \exp\left(-\frac{\Delta F}{KT}\right) . \exp\left(-\frac{4\sigma_e \sigma T_m^o}{\Delta H(\Delta T)T}\right) \qquad 2)$$

Here G_o is a constant, the first exponential accounts for the interfacial transport and the second exponential contains the most relevant relationship (for the present paper) that between G and ΔT. The validity of eq. 2 has been well tested on solution grown crystals giving further support for the underlying theory (e.g. 11). Now, it has been known for long (12) that growth of spherulites in the melt obeys a relation of the form

$$G \propto \exp\left(-\frac{K}{T(\Delta T)}\right) \qquad 3)$$

which conforms to equation 2, derived and observed for single crystals at a much later date. In fact, later work established close identity between the constants (i.e. K and the parameters in the second exponential of eq. 2). In brief, transfer of the theories from solution to melt crystallization proved fruitful and gave confidence to a common model underlying both. Further elaborations of the theory, as applied to the melt on the basis of growth rates, scored a further significant success by predicting, and in fact observing a transition with increasing ΔT. This consists of the number 4 in eq. 1 changing to 2, i.e. a sudden increase in G at a particular ΔT range, a transiton in fact observed experimentally (10). (This is the transition from Regime I to Regime II crystallization where in the latter the rate of nucleation of a new growth strip along a given prism face begins to compete with the spreading rate of a chain folded strip; in brief, multiple nucleation along a given growth face will set in).

Nevertheless, in spite of it all the basic fact remains: namely, that the crystallization theories were set out to account for ℓ^*, and in melt crystallization ℓ^* has been so far inaccessible. Consequently in the melt case all testing of theories had to fall back on the measurement of a derivative quantity, for the present purpose at least, namely the growth rate (G). Provision of a measure of ℓ^*, also in the case of melt crystallization, should therefore be of first order theoretical interest. This diversion into theory, with all its problems, should not, however, obscure our purpose: assessment of the primary structural parameter ℓ^*.

For appreciation of the practical measures required by our Objectives I and II the following facts need recalling. The overall crystallization rate (R) is determined by a combination of nucleation and crystal growth rate (G). As the latter is determined by the supercooling alone according to eq. 2, at a particular crystallization temperature we can only influence R through primary nucleation. As in polymers nucleation is always from predetermined centres we would need to be able to influence the number of these centres. This is a common requirement both for Objectives I and II but in different ways in each case.

For Objective I (i.e. avoidance of isothermal thickening in the conventional temperature range of crystallization) we need to keep the crystals at the intended crystallization temperature for as short a time as possible, and during such a short time to obtain adequate amount of crystal to enable determination of their ℓ value. Clearly, for this to be achievable, the number of nucleating centres has to be sufficiently high. For Objective II (i.e. to extend the crystallization temperature range downwards) we need to avoid substantial amount of crystallization during cooling, which means we have to reduce, if possible eliminate, the number of preexisting nucleating centres. Thus the realization of our Objectives I and II require means which are diametrically opposed, namely increase and decrease of the preexisting nucleation centres respectively.

The predetermined nucleating centres can be of two different origins α) extraneous heterogeneities β) residual polymer crystals or fragments thereof (self seeds). Nuclei of class α) are impurities, mostly catalyst residues, which unless fully removed, are intrinsic to a given synthetic process. In commercial material these catalyst residues are never fully removed, in more recent products they are in fact left in altogether. Even if there are variations in this respect (as manifest e.g. by the different spherulite sizes - the spherulite size is inversely proportional to the number of nuclei - in the different technical products, usually an intrinsic property of a given product) the practicable temperature range for crystallization, is not basically affected by these differences and remains as indicated by Fig. 6. In contrast, nuclei under β) are affected by the morphology the sample possessed before melting and by the rate of heating to the melt, and the maximum temperature reached by the melt before cooling down to achieve its final solidification. Crystal fragments of sufficiently high molecular weight possessing sufficiently high fold lengths may survive when the overwhelming portion of the sample is already molten and is registered as a true melt by conventional tests. (The highest surviving fold lengths can be the result of refolding during the heating up process, hence the heating rate dependence). It follows that this 'self seeding' will always raise the number of nuclei beyond those in category α), which thus represents the limiting minimum value for a given material, achieved when the melt is heated to temperatures which are sufficiently high to melt all crystal fragments which could act as self seeds.

In view of the above our action is therefore clear: For Objective I we need to enhance β) (the self seeds) as much as possible for Objective II to eliminate all nucleation sources including α) (extraneous heterogeneities).

Objective I: obtaining ℓ^* at low supercoolings

The enhanced self nucleation phenomenon. While totally general, the effect was discovered(6) and is best illustrated in a material with intrinsically large spherulites, which means few heterogeneous nuclei of class (α). Linear high density polyethylene, Sclair 2907, proved to be an exceptionally good source. It gave large spherulites and enabled spherulitic growth to be followed conveniently on a microscope hot stage. Fig. 7 represents the usual spherulitic development under isothermal conditions. Here spherulites start to grow from predetermined centres and expand until they impinge. The number of centres can be increased by the self seeding process outlined above, but the increase achievable proved to be insufficient for our ultimate purpose. It was an attempt to maximise self seeding that a totally new effect to be termed 'enhanced self nucleation' was discovered. Its essentials are as follows.

A spherulitic film is heated very slowly to and beyond the melting point. Melting is signalled by the gradual disappearance of the spherulites as shown by Fig. 8. To note, the spherulites do not shrink in the way they had grown originally (Figs. 7 and 8 show the same field) but gradually fade out, the overall birefringence decreasing until at a particular temperature (optical melting point, T_{om}) the field of view becomes totally dark. We found that there is a narrow temperature interval just beyond T_{om} from which, if not exceeded, the spherulites reappear on cooling along the same route as they had melted. I.E. they do not grow out from a central nucleus as in Fig. 7 but the whole spherulite brightens up more or less uniformly until the appearance of the original spherulite is regenerated. This process is shown in Fig. 9. As noted, even the original concentric banded structure reappears, even if in a slightly less regular form. This means that in the narrow temperature interval in question self seeds are retained throughout the entire spherulite which regenerate the spherulite on cooling, (the banding becomes increasingly irregular with the increasing holding time in the isotropic molten state, but dependent on the melt temperature may still reappear after many hours).

For our purposes two features are of overriding importance. First, that the above method of totally regenerating the spherulites proceeds by a magnitude faster rate than it would by normal radial growth as in Fig. 7 at the same temperature. Thus the residence time of the crystals, and with it that of isothermal thickening at the crystallization temperature, will be much reduced. Secondly, the lamellar thickness in the regenerated spherulite bears no relation to that of the initial spherulite but depends on the temperature of recrystallization (augmented by isothermal thickening such as may still occur). This signifies that while the residual nuclei beyond T_{om} are sufficiently numerous to regenerate the initial spherulitic architecture, the initial lamellar structure is not preserved,which means that the bulk of the material on the level of the lamellae is truly molten. It will be evident that even the novelty of this remarkable self nucleation phenomenon apart (itself deserving a study of its own) the associated acceleration of the crystallization enables us to 'catch' the lamellae younger, hence less affected by isothermal thickening, and thus approach ℓ^*. Inspection of Fig. 5 will reveal that the method becomes increasingly useful and important at the higher crystallization temperatures.

Determination of ℓ^* at low supercoolings; kinetics of isothermal thickening. ℓ itself was determined by two methods. First and chiefly by low frequency Raman spectroscopy (LAM). For extrapolation to ℓ^* the sample such as in Fig. 9 was cooled to room temperature during crystallization when ℓ was determined. ℓ decreased with decreasing crystallization time but so did of course the area of the LAM peak corresponding to the smaller amount of material which has crystallized at that temperature. It could be clearly distinguished from the Raman signal due to material crystallized during cooling, setting a limit to the shortest crystallization times for which ℓ could be measured at any given crystallization temperature. The second method relied on the determination of the melting point. Crystallization (apparent as in Fig. 9) was interrupted at an early stage and the sample rapidly heated up again and the corresponding optical melting point (T_{om}) determined. T_{om} could be identified for much lower amounts of crystal, hence for shorter crystallization times than the LAM peak. ℓ is then obtained from the relation

$$\ell = \frac{CT_o}{T_o - T_m} \qquad\qquad 4)$$

where T_o and C in the first instance are treated as fitting parameters. They were assigned values from samples on which T_{om} and LAM ℓ values could both be determined (they gave 146^oC

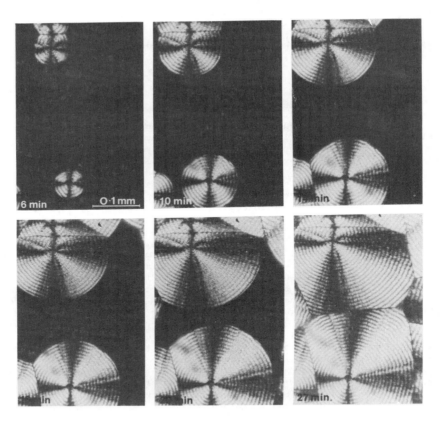

Fig. 7. Stages in the isothermal crystallization of polyethylene from the melt at 123^oC in the form of a thin film as observed under the polarising microscope. Exposures were taken after the marked times. Spherulites are seen to grow from a fixed number of centres (6).

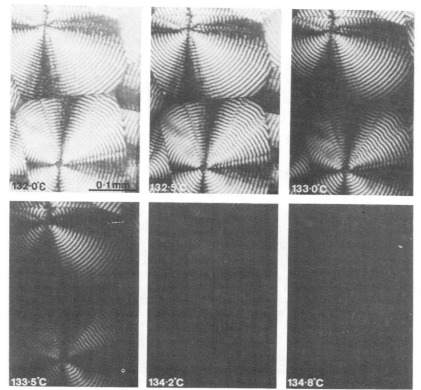

Fig. 8. Stages in the melting of the sample in Fig. 7 observed immediately after crystallization at 123°C(last stage in Fig. 7). The temperature was raised gradually and held constant at the marked values during each exposure (6).

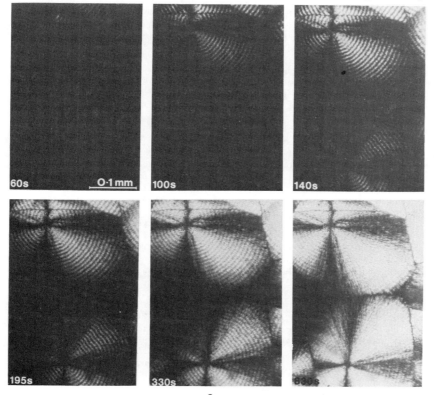

Fig. 9. The crystallization at 126°C of the sample in Fig. 8 immediately after its melting at 134.8°C (last stage in Fig. 8). Photographs were taken after the times marked under conditions identical to Fig. 8 (6).

for T_m^o (thermodynamic melting point) and 130 erg/cm^2 for σ_e (fold surface energy) if T and C are expressed in terms of these quantities in the familiar way. (The above values seem certainly very reasonable for such an identification, nevertheless, not having calibrated for thermal lag we would not claim this to represent an accurate determination of T_m^o and σ_e itself). With C and T_o thus established the determination of ℓ could then be carried out down to very short crystallization times (t) allowing an extrapolation to t = 0, hence to ℓ^*. (ℓ^* values thus obtained are represented by the high temperature end of Fig. 11).

The ℓ v. t plots at a given temperature in themselves represent a study of the kinetics of isothermal thickening which was in fact part of the project. It will not be given here in detail beyond stating that an incubation period (τ) and a rate constant (B) was identified, both τ and B increasing for crystals formed at the higher temperature which had the result that ℓ v. t curves for crystals formed at different temperatures could cross over. The latter means that when comparing samples with different ℓ values those with higher ℓ-s could have started life with lamellae having lower ℓ-s than those displaying lower ℓ values in the fully crystalline state. Thus isothermal thickening may not only shift the numerical values of ℓ but would actually reverse the sequence order in lamellar thickness.

Objective II

Modified droplet method. At low crystallization temperatures the rate of growth is very fast and in the presence of a few nuclei crystallization is complete before the intended crystallization temperature is attained. As already stated in order to prevent this, pre-existing nuclei need to be removed. This in fact has already been achieved in the past by the classical method of Turnbull applied to polymers by several authors (13, 14, 15). The method consists of dispersing the melt into tiny droplets within an inert medium and then cool. Only few of the droplets contain the heterogeneities which act as nuclei, and these will crystallize at the usual temperatures while the droplets which are free of such nuclei can then be supercooled further until the supercooling range, where homogeneous nucleation sets in, is attained. In brief, this method isolates the preexisting nuclei and confines their effect to the individual drops containing them. In past works, this method as applied to polyethylene, enabled about 84% of the droplets to be supercooled to $\sim 85^\circ$C (see scale Fig. 6) before they crystallized. However, in these studies, the droplets were suspended in an inert medium in minute quantities. The kinetics of nucleation was studied, however the droplets could not be isolated and the properties and structure of material crystallized at such remarkably high supercoolings were not examinable. This is what has been currently achieved (7) to be described briefly below.

The underlying idea was to prepare solution grown single crystals via conventional self seeding (16,17). It was anticipated that here most of the crystals will nucleate on material of its own kind and only a minority fraction on nuclei of extraneous origin. As the latter are likely to be catalyst residues of high density they may be expected to impart a slightly higher overall density to the whole crystal which they have initiated, than possessed by the self seeded crystals consisting of polyethylene alone. If so, the former should be elimin-able by gravity separation.

This expectation was successfully realized. A self seeded crystal suspension was prepared and the density of the suspending medium adjusted so that after centrifugation there was a wide spread of suspension along the centrifuge cuvette. The suspension was then sampled from different levels and sprayed on to hot slides where the sprayed droplets become molten blobs of $\sim 3\mu$ diameter. The solidification of these droplets on cooling could then be observed under the microscope. It was found that droplets taken from the bottom of the suspension crystallized around 120-125°C, as normal polyethylene does under similar cooling rates. However, the drops originating from the top of the suspension could be highly supercooled to the region of 80°C as in the original Turnbull type suspended droplet experiment. However, the present dry drops could be readily scraped off the slide and collected for other investigations which was the purpose of the work. In particular, ℓ values could be determined by the Raman LAM method, the principal objective of our work. It will be stated at this point that a sharply defined LAM peak was always obtained indicating a narrow distribution of stem lengths.

In particular detail, the droplets even from the same upper suspension level formed two populations. On cooling a small fraction crystallized at $\sim 125^\circ$C, these obviously still contained the original extraneous heterogeneities. The majority, however, at the lower temperatures referred to above. Fig. 10 shows such an essentially two step histogram.

The nature of the nucleation at high supercoolings. Before, however, commenting on the ℓ values obtained for such material the following important observation has to be reported. T_n as defined in Fig. 10 was found to depend strongly on the substrate, whether coated or uncoated (with carbon), freshly cleaned or wiped with detergent or oil (lanolin). Detergent gave a T_n value of 86°C (interestingly the value reported in the previous suspended droplet technique with detergent as a suspending medium (14) while lanolin gave the lowest value yet,

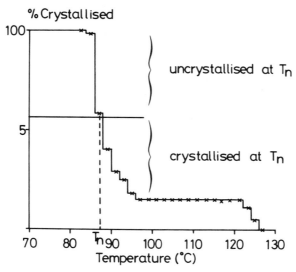

Fig. 10. Typical histogram showing the percentage of droplets which have crystallized during cooling at 1°C/min from the melt. T_n is the temperature at which one half of the droplets, which have remained unfrozen at $\sim 120^{\circ}$C, have solidified.

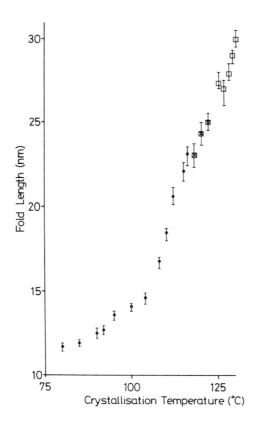

Fig. 11. Graph showing the dependence of the initial, i.e. unthickened, fold length ℓ^* on crystallization temperature (the different symbols refer to the different preparations and routes of determination described in the original text) (5).

75°C.

In the first instance all this means that the nucleation, even at these low temperatures, is still not truly homogeneous but is brought about by the substrate. In fact each drop could be seen to display a spherulitic cross between crossed polaroids, hence must have been nucleated at a particular site on the surface (a fact to be utilized below). Secondly, the different nucleating power of different surfaces enables a whole T_n range to be realized. As a consequence ℓ could be determined over a wide temperature interval.

ℓ^* at high supercoolings; the full ℓ^* v T_{cr} curve for the melt. Examining samples for ℓ as a function of time it was observed that the ℓ values were invariant with time, hence corresponded to the true ℓ^*-s up to a crystallization temperature of 112°C where a very small amount of thickening, 1-2A, could first be detected over a period of tens of minutes. This sets the upper limit of isothermal thickening as ∿112A in good agreement with long standing experience on annealing dry solution grown crystals. Conversely, from the point of view of melt crystallization this means that down to 112°C the effect of isothermal thickening needs reckoning with when measuring ℓ, but not below this temperature.

Combining with the results of the preceding section, the full ℓ^* v. T_{cr} (Fig. 11) can now be presented for a melt crystallized polymer, to our knowledge for the first time in polymer crystallization studies. While the overall trend is not very dissimilar from that anticipated from solution crystallization (Figs. 2,3) some notable discontinuities are discernible. We shall not comment on these, or on any other features further, beyond stating that a reliable base has now been established for theoretical work on melt crystallization. Evaluation is presently in progress. Attention will merely be drawn to the fact that the lower values of ℓ^* are now in the range of those familiar from solution crystallization (it will be recalled that up till now ℓ values for melt crystallization were always much higher).

It will be clear that the extension of the crystallization range of polyethylene to such low temperatures, and the accessibility of the material obtained in this way, widens the scope of the investigation of many important additional issues. Such are e.g. the effect of nucleating agents and the phenomenon of refolding for which it sets new base lines. Two points, however, will be separately mentioned: growth rate and morphology.

Growth rates. When watching in the polarising microscope the crystallization of a supercooled drop is registered as a sudden, instantaneous brightening in a dark field. The spherulitic cross (even if sometimes distorted, and/or off central) indicates that crystallization has started from one centre and has proceeded radially outwards. If the time taken by the 'lighting up' of the drop to its maximum brightness could be assessed, this, in the knowledge of the diameter of the drop would give the radial growth rate at the temperature in question. This could in fact be achieved in the following way. The image of a drop was focussed onto a photodiode and the build up of the brightness followed by the output of the photo current displayed on an oscilloscope. It was found that the rise of the signal to its maximum value, as produced by crystallization, was indistinguishable from that given by the natural rise time of the optoelectronic system measured to be 10^{-6} sec. For a typical droplet of 2μm this corresponds to a radial growth rate of 2msec^{-1}. Accordingly we arrive at the astonishing conclusion that the minimum radial growth rate is 2msec^{-1} and it could clearly be faster than that.

Morphology. Some droplets grown at 85°C were prepared specially for examination under the electron microscope. Fig. 12 shows an example. The result was surprising. The surprise was not the novelty of the morphological features observed, but the fact that the basic morphology does not differ in any significant way from what we are accustomed to in samples crystallized in the conventional lower supercooling range. The constituent elements are lamellae. In the thicker central portion they aggregated in a way broadly familiar in spherulites while at the thinner periphery (Fig. 12b) the lamellae are seen in isolation, in places displaying clearly defined facets, some of which are marked on the print.

Combining the observations on rates and morphology the following remarks will be made. The very high rates at the low temperatures is in itself not unsuspected, nevertheless we consider it significant that we are now able to place at least a lower bound to it, something which prior to this work has been beyond the scope of possibility. While at 85°C very high rates were indeed expected, the rate now found as a lower limit is nevertheless surprising. Even more surprising is this high growth rate when considered in conjunction with the morphology. The unexpectedness of the clearly defined lamellar morphology at such high supercoolings has already been commented on; it is even more unexpected for the extremely high growth rates in question. As stated earlier the existence of lamellae implies substantial amount of chain folding, and for individual, separate lamellae such as in Fig. 12, exclusive chain folding (the molecules have nowhere to go but to return into the same lamella), in both cases at least a certain amount of sharpness of the fold structure and adjacency in the fold reentry being implied, (the amounts needed being a subject of argument). One of the points for and against the various views, as debated in the field of polymer crystallization, is the

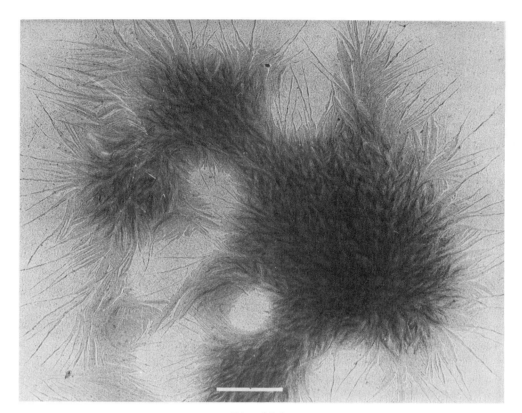

Fig. 12a)

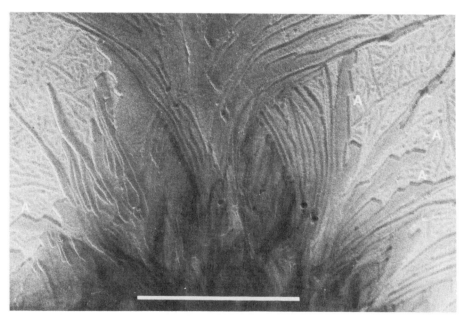

Fig. 12b)

Fig. 12. Transmission electron micrographs of droplets crystallized at 85°C
a) Overall view   b) Higher magnification detail showing the faceting on
the lamellae (marked with A). The white bars correspond to 1μm.

M - G

intrinsic ability or otherwise of the chains to organize themselves on laying down more or less neatly along the crystal face, as opposed to merely freezing in into a state resembling their configuration in the melt. Typical linear growth rate values of $\sim 1\mu m$, already regarded as very high, have been quoted in past argumentations (1) against which various models of chain mobility are being tested. The growth rate of $2 m sec^{-1}$ now obtained is by six magnitudes higher, and hence (even when admitting the possibility that the rates at the single lamellar edges of the droplets may not be quite as high as in the rest) creates a situation which lies totally outside the values underlying past arguments. At this stage I leave the issue open with the remark that in the light of the new findings the regularity of the morphology and the organizational ability of the chains seems to exceed anything that has been foreseen a priori. Clearly more experimentation is needed on both, growth rates and on the morphology, including chain inclination, but so may be some rethinking of the molecular dynamics in the light of the new facts which have now emerged."(Note a)".

In conclusion, whatever the final model which will emerge the experimental studies just outlined, we think, have significantly increased our factual knowledge about polymer crystallization and, what is more significant, have raised some new unforeseen specific issues with corresponding widening of our whole approach to the subject.

Note a: Perhaps not quite coincidentally the $2 m sec^{-1}$ growth rate corresponds to a recent prediction by Hoffman for the same undercooling based on a growth mechanism he denotes as regime III. This regime III should set in at very high supercoolings, where the nucleation rate along the growth face is very much faster than the spreading rate along that face (18).

B)   ORIENTATION INDUCED CRYSTALLIZATION

General Principles

In this second part of my report my main concern will be to convey some recent developments on orientation induced crystallization, only very briefly for the record. The account will include two aspects: 1) The structure of shish-kebab crystals and 2) the origin of the under-lying chain extension with some comments on high modulus fibre formation. Both of these aspects provide surprises and 2) in particular is open ended.

First, the present picture will be recapitulated. As a broad schematization, if the chain molecule is extended from its random coil configuration into its stretched out state, and is allowed to crystallize, it will form fibrous crystals; in fact only then will it give rise to fibres otherwise it crystallizes in the form of chain folded platelets. As in a given system of flexible chains, the chains are never all completely extended (this is because the extensional flow field is usually not completely uniform, hence equally effective everywhere; also there is a critical threshold molecular weight beyond which the chains will extend, see e.g. review (19)), the unextended chains (or chain portions) will crystallize later on cooling in a chain folded form onto the central thread already formed, thus giving rise to the platelet component (the kebabs) of the shish-kebab in the familiar fashion. As well known, this whole subject acquired special significance as it provided one route towards the achievement of ultra-high modulus and strength attributed to the fully extended component of the structure (see review 19).

Structure of Shish-Kebabs

Here, first we have to distinguish between the kebabs which are detachable and those which are molecularly connected to the central thread (macro amd micro-shish-kebabs see Fig. 13). Our further concern will be only the latter. It has been observed by Pennings et al. (20) that in one and the same preparation the appearance of the kebabs (their size and frequency along the shish) can be altered by the way the shish-kebab containing solution or suspension is being cooled down from its formation temperature, or alternately is stored at a particular temperature, this alteration being reversible on appropriate treatments. (E.g. consider that the fibrous crystallization occurs, say at $120^{o}C$ (see temperature scale Fig. 14): store the suspension at 95° and then cool to room temperature for examination. The kebab structure will be different, the kebabs are larger and more widely spaced, than that resulting from direct cooling to room temperature). The authors inferred that the fibres as formed are 'hairy' and it is the loose hairs which nucleate and form the kebabs on subsequent lowering of the temperature with a frequency determined by the nucleation temperature, hence cooling and/or storage conditions. According to all foregoing experience the kebab content is being reduced with achievement of higher formation temperatures (T_f), hence reduction of the amount of hairs is being inferred. In fact overgrowth free, smooth fibres have been obtained in this way, which possessed the highest moduli achieved so far.

Our own contribution to this subject, in addition to the above, was the recognition that even smooth, kebab-free fibres are convertible into kebab containing ones, and this in a reversible manner by appropriate storage (T_s) below the original formation temperature (T_f) (Fig.15).

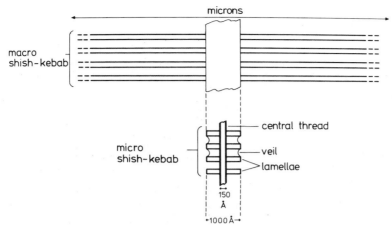

Fig. 13. Sketch illustrating the architecture and classification of shish-kebabs. In the macro shish-kebabs the large platelets can be removed (in case of solution crystallization) to reveal the central micro shish-kebab whose platelets are molecularly connected to the central thread.

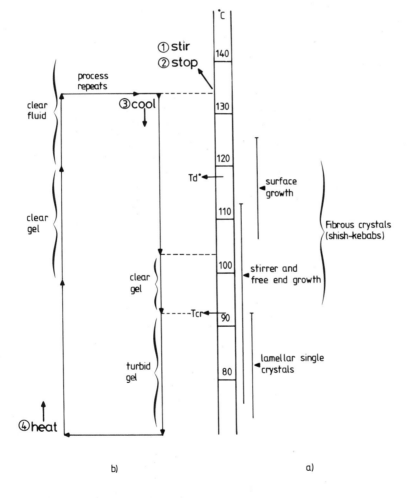

Fig. 14. Temperature scale relevant for a) shish-kebab formation and high modulus fibre preparation b) stirring induced thermoreversible gelation, as described in the text.

It follows that hairiness is never fully avoided,and as a consequence the mere appearance
(which we termed 'hairdressing'(21))is no reliable indicator of the formation temperature,
neither that of the resulting stiffness and strength. The latter, accordingly is determined
not only by the shish-to-kebab ratio but chiefly by the 'quality' of the shish which in turn
is affected by T_f.

As regards the last point, separate studies in our laboratory have revealed that the shish,
i.e. the fibrous core,does not consist of fully defect-free crystals, but is rather segmented in nature (22, 23). The defects can range from lattice imperfections to fully, even
if constrained, amorphous regions totally interrupting crystal continuity in a probably very
narrow and tapering fringed micellar fashion   (these imperfections are revealed, distinguished and quantified by thermal studies and dark field electron microscopy (22,23)). It was
established that the defect distribution along the shish is exponential. It is the mean
crystal length in this distribution which increases with T_f, and it seems that it is this
crystal length which is the main factor determining the modulus, longer defect-free crystals
yielding stronger and stiffer fibres (for latest analysis see Barham (24)). This is clearly
a new approach to the whole problem of what determines the modulus of a fibre, a composite -
mechanical issue we are not pursuing here further. It will merely be stated that the recognition of this segmented nature of the 'shish' has stimulated some new theoretical activities
as regards the genesis of such fibrous crystals. The model recently proposed (25) envisages
multiple crystal nucleation along a bundle of extended, but yet uncrystallized chains.

The origin of chain extension
This subject has a rather tortuous history. At this place I shall merely spotlight the main
historical landmarks; for a survey with more explanations I refer the reader to our recent
review and to the bibliography contained therein (26).

The first stage in the history of the subject was the recognition of the need for elongational flow, i.e. a flow field with a 'sufficiently' (for what is 'sufficient' see original papers
referred to in the various reviews on the subject e.g. (27)) large component of the velocity
gradient along the flow direction, an accelerating flow being the simplest example. (This
is in contrast with the more familiar simple shear flow with the velocity gradient perpendicular to the flow direction as in the familiar capilliary flow). In later developments,
while the need for an accelerating flow field was retained as regards nucleating the fibre
formation, it was recognized, both theoretically and experimentally, that this was not needed
for continuing growth of the fibres, which can take place in simple shear flow or in any
other flow. The reason for this lies in the fact that the tip of the fibre, once the fibre
has formed, creates a localised elongational flow field downstream in its immediate vicinity
which ensures continuous extension of the newly attaching molecules. Zwijnenburg and Pennings
recognized the potential of this effect for continuous fibre production which they put in
effect in their so called 'free end growth' method (28). The next stage was the recognition,
again by Zwijnenburg and Pennings (29),that growth of fibres becomes greatly accelerated if
in contact with a moving solid surface immersed in the solution (in practice a rotating cylinder in a Couette type apparatus) which they attributed to a thick adsorbtion layer ('adsorb
-tion entanglement' layer) forming along the solid surface in question. Accordingly, an
immersed seed fibre would hook on to this 'adsorbtion entanglement layer' and stretch the
chains by means of which the fibre tip becomes the site of continuous growth. By this method
the temperature at which the fibres could be formed , T_f,was raised (see Fig. 14) with associated gain in modulus and strength. It was at this point that a further, totally new and
unexpected development occurred which I shall proceed to describe.

The key word in which follows is gel formation. A gel is normally understood to be a swollen
network,continuous throughout the sample. The junctions in the networks are physical (as
opposed to chemical as in the familiar elastomers) in nature, local association of chains,
which in the simplest case we understand to be small crystals tying together two or more molecules. Small micellar crystals in the traditional sense, but too small to produce overcrow-
ding at their end surfaces where the amorphous chain portions (the actual network elements)
emerge, can serve as such physical junctions. Gels arising in this way are thermoreversible:
they form and dissolve on lowering and raising of the solution temperature respectively. Such
gel forming crystallization was found to be common in poorly crystallizable polymers but not
originally in such a highly crystallizable material as polyethylene which, as familiar, tends
to form particulate structures on crystallization such as the well known suspension of chain
folded platelets. The essential step at this stage was the recognition that flow, agitation
in particular, greatly promotes thermoreversible gel formation even in a system such as polyethylene and that at least some (if not all, this would need to be ascertained retrospective-
ly) of the past methods leading to shish-kebabs and used for fibre production consist of the
stretching of such gels. The most effective (as far as modulus and strength is concerned)
surface growth method, referred to above (29),consists of stretching out gel particles which
adhere to the solid rotor surface and has been attributed to an entanglement adsorbtion layer
previously.

At this stage the subject divides into two aspects according to the interest of the parties
concerned: a) production method for high modulus, high strength fibres,b) the origin of gel

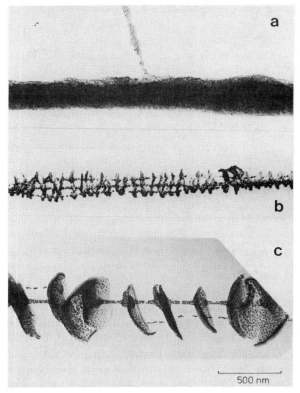

Fig. 15. Electron micrographs demonstrating the three principal appearances (including smooth fibres) which may be adopted by any shish-kebab on the way in which it is treated after its formation. These and other similar appearances are interconvertible by appropriate heat treatments in solution (hairdressing) (21).

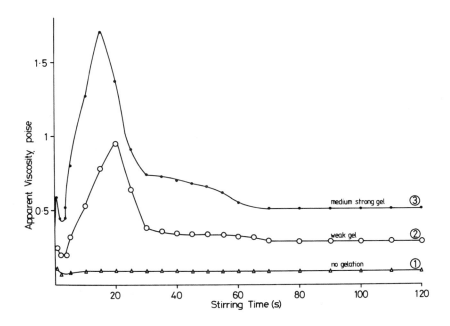

Fig. 16. Apparent viscosity as measured by a Couette type viscometer as a function of stirring time for a solution of high molecular weight polyethylene (M_w = 1.5 x 10^6) in decalin at 125°C for a series of concentrations ● conc. 0.4% w/w; o conc. 0.2% w/w; Δ conc. 0.1% w/w (36).

formation, and in particular the part played in it by chain elongation.

a) On the attainment of high modulus fibres It is now well recognized that moduli and strengths achieved by the conventional drawing process are by up to 2 magnitudes lower than expected from the properties of the fully extended chain. (The theoretical modulus of polyethylene should be ∿300GPa,that of a conventionally drawn fibre is around 5GPa). It is also appreciated that this is a consequence of the fact that in an initially chain folded structure the chains, while oriented, may remain partially urfolded. In order to achieve full chain extension higher draw ratios than conventionally attained (4-6x) are required. The main obstacle for achieving this is the limiting strain a sample can support before breaking. Indeed, once higher draw ratios could be attained (up to 40x) the resulting moduli came within a factor of 2-3x of the theoretical. Conditions by which this could be realized in the case of drawing the solid, crystalline material (route 1) have remained largely empirical,although elevated temperature and moderate (i.e. not too high) molecular weight have emerged as general requirements.

Another method (route 2) by which to achieve the same goal was by extending the random chains, usually in solution, and allowing them to crystallize in the extended state subsequently.This is the method which ties up with shish-kebab formation and has been our concern in the foregoings. As we have seen however, it has emerged that this method, at least as realized in present practice, corresponds to drawing out of a gel. This means, as a broad statement, that in principle routes 1 and 2 have become identical: both correspond to drawing out of a structure already formed, except that in the case of route 2, this structure is a very loose network where to begin with, the polymer occupies only a fraction of the sample volume.

The above developments have led to the recognition of the following overriding qualitative principle. High modulus requires high chain extension. The realization of such high chain extension before the sample breaks is facilitated by a porous structure where the number of junctions per unit volume is low (we have been considering such junctions as being crystals; entanglements would form the same role in as far as they would induce or promote the formation of junction forming crystals). Conversely, entanglements such as present in the usual melt crystallized material are prone to lead to failure at moderate extension, particularly when the molecular weight, hence the intrinsic entanglement concentration is high. Gel structures, irrespective of how the gel has been achieved, obviously satisfy the requirement of porous structures with just enough stable junction points to convert the material into a network, but not too many to impair its extensibility. As a consequence, some recent works in the high modulus fibre field are aimed at producing gels directly through crystallization (most of which, but not all, involve flow at some state), and the gels are stretched subsequently either in the wet, or partially of fully dried state (where the latter still preserves the initial gel porosity) (30,31,32). Even if this term 'gel' does not feature in the reporting of all these methods, we judge from the description that in all cases at some stage of the process gelation is likely to have been involved (e.g. in method of ref. 33). Direct gel production (for details see review (26)) thus seems to replace the previously applied ingenious, more specific methods (surface growth in particular) which nevertheless retain their interest for their own sake.

b) Flow induced gelation Gel formation of high molecular weight polyethylene in the course of stirring has been reported by Pennings earlier, but this in the range where fibrous crystallization normally occurs (34). To this we have the following to add. Consider the temperature scale in Fig. 14. If we stir at a temperature above that of fibrous crystal formation, say at 135°C, absolutely nothing visible happens. If we now stop stirring and proceed to cool the quiescent solution we find that gelation may set in (35). The gel is transparent and forms at a temperature above 90°C,i.e. above that of normal single crystal formation. If such a transparent gel is cooled further it becomes turbid as the rest of the polyethylene will precipitate in the usual platelet form, either as overgrowth onto shish-kebab fibres, or as isolated crystal platelets. Our interest in what follows will be centred on what happens during the initial stirring (135°C in the above case) and on the structure of the transparent gel which arises first on cooling. While we do not know the true answers to either,the first issue is the most puzzling and possibly most basic.

It is apparent that the initial stirring at the high temperature induces an incipient aggregation which is the source of the subsequent gelation, as without this stirring only the usual single crystal suspension (without any gel) results on subsequent cooling. It was observed that the memory of this stirring, as diagnosed by the above gelation test on cooling in the quiescence condition, persisted long after it has been stopped; dependent on temperature and stirring conditions it could persist for hours. This is far too long to be attributable to molecular processes characterising solutions, while not long enough to allow its definitive attribution to stable crystals. The most remarkable feature of this stirring induced state is the asymmetry of the time scale of its formation and disappearance: stirring times of tens of seconds suffices for inducing the gel on subsequent cooling, while the effect, as diagnosed by the gelation it gives rise to, decays over a period of hours. It was found that only material capable of undergoing this gelation induced by stirring, but itself occurring under quiescent conditions, yields fibres by Pennings' surface growth method, supporting the contention that in the latter method gel particles adhering to the rollers are being stretched.

While nothing apparent is happening during the initial high temperature stirring as registered by eye, pronounced viscosity effects can be recorded. Fig. 16 shows the shear stress as a function of time in a Ferranti type viscometer (two concentric cylinders with the outer one rotated and torque measured on inner cylinder). After the familiar initial overshoot effect a further,major peak is procuded in the shear stress which then decays to a constant level. Now it was observed that the gelation effect in question,i.e. gelation setting in on cooling after the stirring has been stopped, only occurs when the peak in the shear stress v. time curve has started to develop. It is more pronounced with higher peaks,and for a given run the resulting gel is strongest when the interruption of stirring occurs at the peak value itself, nevertheless it does occur after interruption at all times beyond the peak, i.e. where the stress v. time curve has already become level. We interpret the peak as a sign of gel formation spreading across the gap between the cylinders. The gel adsorbs along the cylinder surface which reduces the gap. The peak then represents the maximum shear stress which the gel can withstand. Beyond this the gel is broken up into particles,but never fully dissolved, a steady state between the new kind of gel formation and disruption being reached. On cessation of the rotation these gel particles can then be the source of the transparent macroscopic gel formation on subsequent cooling with the long lived memory already stated (36).

Thus curves such as in Fig. 16 serve as an objective documentation of the initial aggregation, which is the source of gelation, and consequent shish-kebab formation and of the fibre production methods arising therefrom. Such curves are not new in themselves. Pennings (34) reported such a curve earlier, and indeed attributed it to transient gelation, but apparently did not recognize the fact that beyond the peak the gel does not disappear but merely breaks up and remains a source of gelation in the quiescent state for hours after stirring has ceased and thus the nature of the whole system has changed in a transient yet long lived manner. Indeed, and most significantly, there are even much earlier precedents. Shear stress v. time curves of the kind as in Fig. 16 feature in the literature of the 1960-s (e.g. 37,38) associated with intrinsically amorphous materials (atactic polystyrene, although one crystallizable polymer, polyethylene oxide also features), high molecular weight, viscous and poor solvents being common features. Observation of gel particles is frequently mentioned, but the effect was never fully explained. It appears to us that we are dealing with a totally general phenomenon which transcends the present issue of crystallization. Most likely orientation induced phase separation is at play in the most general sense, of which the present polyethylene, being highly crystallizable, presents a specially clear cut example, where in contrast to the early works we have well defined reference points as regards the temperature scale where happenings are expected (Fig. 14). Further, our own experiments are providing a definite novel diagnostic effect in the form of gelation in the ensuing quiescent state in addition to the rheological observation itself. Little can be said at this early state of all the implications regarding the structures forming, the conditions of phase segregation which are involved, and the relaxation of these effects on the scale of the molecular and that of the larger scale structures. That much, however, should be apparent that the study of orientation induced crystallization seems to have opened wider horizons than initially envisaged marrying new discoveries with hitherto unexplained old experiences. These seem to embrace such widely apparently disparate areas of polymer science as crystallization, chain orientation, formation and stability of new phases, and that of solution rheology an open ended note on which I am terminating this survey.

Acknowledgement - I wish to express my appreciation of the salient contributions by my collaborators,past and present, to the material forming the basis of this article. I am particularly indebted to Drs. Barham, Martinez,Jarvis, Messrs Chivers and Narh, for their consent to quote their works(referenced as 5,6,7 and 36) before their appearance in the open press.

REFERENCES

1. Faraday Discussion No. 68 (Organization of Macromolecules in the Condensed Phases)(1979).
2. E.A. DiMarzio and C.M. Guttman, Polymer, 21, 733 (1980).
3. J.D. Hoffman and J.J. Weeks, J. Chem. Phys.,42, 4301 (1965).
4. J. Rault, J. Physique Lettres, 39, L-441 (1978).
5. P.J. Barham, R.A.Chivers, D. A. Jarvis, J.Martinez-Salazar and A. Keller, J.Polymer Sci. Lett. Ed. in the press.
6. R. A. Chivers, P.J. Barham, J. Martinez-Salazar and A. Keller, J. Polymer Sci. Phyd.Ed. in the press.
7. P.J. Barham, D.A. Jarvis and A. Keller, J. Polymer Sci. Phys. Ed. in the press.
8. J.D. Hoffman, G.T. Davies and J.I. Lauritzen in Treatise on Solid State Chemistry, ed. N.B. Hannay, (Plenum Press, New York)3, Chapt.7 (1976).
9. J.J. Point, Macromolecules, 12, 770 (1979).
10. J.D. Hoffman, L.J. Frolen, G.S. Ross and J.I. Lauritzen, J. Res. Nat. Bur. Stand.79A, 671 (1975).

11. A. Keller and E. Pedemonte, J. Crystal Growth,18, 111 (1973).
12. L. Mandelkern in Growth and Perfection of Crystals, Proceedings of International Confer-
     ence on Crystal Growth, Cooperstown eds: R.H. Doremus, B.W. Roberts, D. Turnbull,
     John Wiley & Sons, Inc. New York, 467 (1958).
13. R.L. Cormia, F.P. Price and D. Turnbull, J. Chem. Phys., 37, 1333 (1962).
14. F. Gornick, G.S. Ross and L.J. Frolen, J. Polymer Sci.C18, 79 (1967).
15. J.A. Khoutsky, A. Walton and E. Baer, J. Appl. Phys., 38, 1831 (1967).
16. D.J. Blundell, A. Keller and A.J. Kovacs, J. Polymer Sci. B. (Letters) 4, 481 (1966).
17. A. Keller and F.M. Willmouth, J. Polymer Sci. A-2, 8, 1443 (1970).
18. J.D. Hoffman, private communication.
19. A. Keller, J. Polymer Sci. Polymer Symposia, 58, 395 (1977)).
20. A.J. Pennings, R. Lagaveen and R.S. de Vries, Colloid and Polymer Sci., 255, 532 (1977).
21. M.J. Hill, P.J. Barham and A. Keller, Colloid and Polymer Sci. 258, 1023 (1980).
22. D.T. Grubb and A. Keller, Colloid and Polymer Sci.,256, 218 (1978).
23. D.T. Grubb and M.J. Hill, J. Crystal Growth, 48, 321 (1980).
24. P.J. Barham, to be submitted.
25. J.D. Hoffman, Polymer, 20, 1071 (1979).
26. A. Keller and P.J. Barham, Plastics and Rubber International, 6, 19 (1981).
27. M.R. Mackley, J. Non-Newtonian Fluid Mech. 4, 11 (1978).
28. A. Zwijnenburg and A.J. Pennings, Colloid and Polymer Sci., 253, 452 (1975).
29. A. Zwijnenburg and A.J. Pennings, Colloid and Polymer Sci. 259, 868 (1978).
30. P. Smith, P.J. Lemstra, B. Kalb and A.J. Pennings, Polymer Bull., 1, 733 (1979).
31. P. Smith and P.J. Lemstra, J. Material Sci., 15, 505 (1980).
32. P. Smith ,P.J. Lemstra and H.C. Booij, J. Polymer Sci. Phys. Ed., 19, 877 (1981).
33. B. Kalb and A.J. Pennings, Polymer, 21, 3 (1980).
34. A.J. Pennings, J. Polymer Sci. Polymer Symposium, 59, 55 (1977).
35. P.J. Barham, M.J. Hill and A. Keller, Colloid and Polymer Sci.258, 59 (1980).
36. K.A. Narh, P.J. Barham and A. Keller, Macromolecules, submitted.
37. T. Matsuo, A. Pavan, A. Peterlin and D.T. Turner J. Colloid Interface Sci., 24, 241 (1967).
38. A. S. Lodge, Polymer, 2, 195 (1961).
39. T. Kawai and A. Keller, Phil. Mag., 11, 1165 (1965).
40. M.M. Winram, D.T. Grubb and A. Keller, J. Material Sci., 13, 791 (1978).
41. J. Dlugosz, G.V. Fraser, D.T. Grubb, A. Keller, J.A. Odell and P. L. Goggin, Polymer. 19,
     361 (1978).

NEUTRON SCATTERING STUDIES ON THE CRYSTALLIZATION OF POLYMERS

E.W. Fischer

Institut für Physikalische Chemie der Universität Mainz, D - 6500 Mainz

Abstract - Neutron scattering on mixtures of deuterated and undeuterated molecules offer a new approach for solving long standing problems of the structure of semicrystalline polymers. The various ranges of the amount of the scattering vector q yield different informations. From small angle studies the radius of gyration R_g can be obtained, which does not change markedly by crystallization. The scattering intensity in the intermediate angle range yields mainly the average distance $<x>$ of the crystalline stems. Experimental results show that not only R_g remains almost constant, but that also the pair distribution function $g(r)$ of the monomer units for $r \ll R_g$ does not change appreciably during crystallization, since the average distance $<x>$ is controlled by the inherent statistical properties of the chain. There is no indication of privileged adjacent re-entry for melt crystallized polymers.
The problems of "intramolecular clustering" due to crystallization can be treated by an Ornstein-Zernike approach. The direct correlation function $c(x)$ is related to the pseudopotential of the interaction between the crystalline stems, which is caused by the amorphous loops. The intermolecular clustering can be described in a similar way, it leads to an enhancement of the intramolecular clustering.

INTRODUCTION

The problem of the structure of the amorphous material between neighbouring crystalline regions in a semicrystalline polymer of lamellar morphology is not yet solved. This question is important for two reasons: i) In order to develop correct theoretical models for the melt-crystal phase transition the nature of the amorphous regions and of the interface between the crystalline and the amorphous regions must be known. ii) For the interpretation of the physical properties of semi-crystalline polymers a rather detailed picture of the structure is required. An integral number of the degree of order is not sufficient in most cases.

It is quite clear that there exists no unique and simple answer to this question. Otherwise it should have been found after about 25 years of intensive studies carried out in many laboratories. It is also obvious from the experimental results that the nature of the amorphous regions depends on the mode of crystallization, on the molecular weight and on the chemical structure of the macromolecules. From one polymer different samples can be prepared with about the same degree of crystallinity but with very different mechanical behaviour as it is nicely demonstrated by the so called "hard elastic fibres" (1). Arguments about the various models in many cases neglect those variations of sample properties.

The most direct informations about structures can be obtained from scattering studies. The relatively new technique of neutron scattering on mixtures of deuterated and undeuterated molecules allows the evaluation of the single-chain structure factor $P(q)$ ($q = (4\pi/\lambda)\sin \theta/2$). This method has been proved to be very successful in the study of amorphous polymers (2) and its application to semicrystalline polymers enables at least in principle the verification of the various models which have been proposed. For that purpose the measured structure factor $P(q)$ has to be compared with calculated scattering curves.

The main aim of the following paper is the discussion of the problem, what kind of informations can be obtained from the various ranges of q and what difficulties may arise both from an experimental and from a theoretical point of view. Some present results are discussed in the light of these more general considerations.

PRINCIPLES OF EVALUATION OF NEUTRON SCATTERING DATA

The various models for the conformation of macromolecules in the semicrystalline state are well known. The two extreme cases are the regular folding model with adjacent re-entry and the "switchboard" or random re-entry model. Induced by the neutron scattering experiments an intensive discussion started about two years ago on the space requirement of the non-crystallized chain sequences (3)-(5) and on the kinetic control of chain folding (6)-(8). We will not be concerned with those questions at all, but we will only discuss the results of scattering experiments and the relevance of these data.

The principle of the method is well known. For a mixture of H- and D-molecules without any specific interaction the differential scattering cross section per unit volume is given by

$$\frac{d\sigma}{d\Omega}(q) = c_D(1-c_D) \; n_w \; K \; P_n(q) \tag{1}$$

where c_D is the concentration of deuterated molecules and K is the contrast factor given by

$$K = \frac{\rho N_L}{M_O} (m|b_D - b_H|)^2 \tag{2}$$

with ρ = density, N_L = Avagadro number, M_O = monomer molecular weight, b_D, b_H = scattering lengths of deuterium or hydrogen, m = number of exchanged hydrogen nuclei per monomer. In eq.(1) n_w means the degree of polymerization of the macromolecules and is supposed to be equal for both kinds of molecules. $P_n(q)$ is the formfactor of the polymer molecule

$$P(q) = \frac{1}{n_w^2} < \sum_{i,j} \exp\left[i\underline{q}(\underline{R}_i - \underline{R}_j)\right] > \tag{3}$$

which may be approximated for small q by

$$P(q) \approx \exp\left(-\frac{1}{3} q^2 R_g^2\right) \tag{4}$$

or by

$$\frac{1}{P(q)} \approx 1 + \frac{1}{3} q^2 R_g^2 \tag{5}$$

where R_g is the radius of gyration.

For the evaluation of the data it is useful to introduce the reduced scattering intensity

$$J(q) = \frac{d\sigma/d\Omega}{c_D(1-c_D)K} = n_w \; P(q) \tag{6}$$

or a "scattering function"

$$F_n(q) = n_w \; P(q) \; q^2 \tag{7}$$

The eq.(1) needs some more discussion. It can be derived rigorously (9), (10), (11) from

$$\frac{d\sigma}{d\Omega}(q) = \frac{1}{V} \sum_{\alpha,\beta,i,j} < b_i^{\alpha s \sigma} \; b_j^{\beta s'\sigma'} \exp\left[i\underline{q} \; (\underline{R}_i^{\alpha s} \; \underline{R}_j^{\beta s'})\right] > \tag{8}$$

where $b_i^{\alpha s \sigma}$ is the scattering length of the nuclei i; belonging to the chain α and characterized by isotope s and spin state σ. The averaging has to be done over the spin states, the isotope species and over the configurations of all molecules. Under the assumption, that there are no specific H-H, H-D or D-D interactions, eq.(1) is obtained. If such interactions are present the reduced intensity defined by eq.(6) will depend on concentration, as it has been observed in crystalline polymers. Those clustering effects will be considered later on in more detail.

In the past the formfactor $P_n(q)$ of the single chain in bulk polymeric materials has been evaluated mostly from measurements of rather diluted mixtures of protonated and deuterated macromolecules because it was assumed that the interchain interferences of the tagged molecules will disturb the evaluation for higher concentrations. It has been demonstrated experimentally, however, that in the case of semicrystalline polymers the single-chain conformation can also be evaluated from measurements on concentrated mixtures (12), (13). In general the eq.(1) can be used for any concentration c_D of deuterated molecules.

We have developed (10,11) an evaluation method based on eq.(8), which has the great advantage that no background scattering of the "zero-sample" must be substracted. In addition besides the structure factor of the single chain also the overall density fluctuation is obtained. The method was first tested with amorphous polymers (11) and then applied to crystalline samples. In the case of semicrystalline polyethylene we made use of eq.(1) in earlier work as already mentioned.

The use of high concentrations enabled us to measure the scattering intensity over a very wide range of the scattering vector q. In Table 1 the various informations are summarized (13,14) which can be obtained in 3 different q-ranges characterized very roughly by the numbers given. In the small angle range (SANS) the radius of gyration R_g of the molecules in the semicrystalline state can be measured. The apparent molecular weight obtained by extrapolation q → o can be used for control of clustering or segregation. In the intermediate angle range (IANS) informations are obtained about the spatial correlation of crystalline stems. As it will be shown later mainly the average density of stems can be measured.

Table 1. *Informations obtainable from neutron scattering experiments in various angular ranges.*

Range	Informations
SANS $0.005 < q/\text{Å}^{-1} < 0.03$	Molecular weight Radius of gyration R_g Clustering and segregation
IANS $0.03 < q/\text{Å}^{-1} < 0.5$	Space correlation of crystalline stems Average density of stems
WANS $0.05 < q/\text{Å}^{-1} < 5$	Direct correlation function of stems Formfactor of crystallized sequences Phase separation (from Bragg intensities)

Most valuable results have to be expected from the upper range of IANS and from the wide angle scattering (WANS), that means in the range of about $0.1 < q < 1\text{Å}^{-1}$. The reasons are quite obvious if one looks at models for the crystalline amorphous surfaces as drawn in Fig. 1 (15). The various models differ with regard to the spatial correlation of the crystalline stems of one and the same molecule in the range of $5 < x < 20\text{Å}$, let say. The drawing also shows that some of the different structure may lead to the same scattering curves, for example model d) and f).

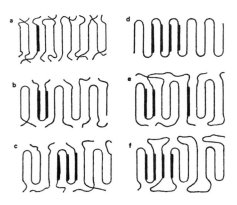

*Fig. 1. Models for crystalline-amorphous surfaces according to Peterlin (15).
The correlation of crystalline stems belonging to the same molecule depends
on the nature of chain re-entry.*

THE SMALL ANGLE RANGE

In the range $q R_g < 1$ the scattering is due to the overall mass distribution of a polymer chain
and yields valuable information about the "dilution" of a single chain in the crystalline
state. The radius of gyration R_g can be evaluated from eq. (4) or (5) without any specific
model assumptions. On the other hand only this one integral number can be obtained from SANS
measurements.

The evaluation of R_g is distorted by phase separation between H- and D-molecules or by
clustering effects, which are considered in more detail later. In so far as reliable information could be obtained it turned out that the gyration radius in the crystalline state does
not differ markedly from that in the melt or in dilute solution. This was found by Schelten
et al. (16) for melt-quenched linear polyethylene and by Ballard et al. (17) for isotactic
polypropylene crystallized and annealed under various conditions. In the case of polyethylene-
oxide we found a small increase (14,18). Repeated experiments using a PEO-sample with a more
narrow molecular weight distribution (19) revealed also an increase of about 20%. Some deviations from this rule of almost constant R_g are observed with isotactic polystyrene (20,21)
especially if the deuterated molecules are dispersed in a lower molecular weight matrix.

Clearly more experiments should be carried out with regard to the molecular weight dependence
of R_g under various crystallization conditions. The remarkable result so far obtained shows
however, that during crystallization no large reorganization of the macromolecule takes
place.

The most simple explanation for the approximate invariance of R_g may be based on a model,
which we call "Erstarrungsmodell" (solidification model) (12-14, 22). In this rather naive
picture, see Fig. 2, it is assumed that crystallization occurs only by straightening suitably
oriented sequences of the coil without a long range diffusional process. The model is supported by considerations on the crystallization rate (6), which are still a matter of arguments as already mentioned. We are here only concerned with the results of neutron scattering and we will show in the next section that the "Erstarrungsmodell" is strongly supported
by the data obtained in the intermediate angle range.

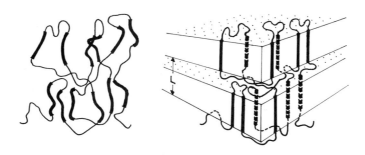

Fig. 2. "Erstarrungsmodell" (solidification model) of the crystallization process of chain molecules. The fully drawn sequences of the coil are incorporated into the growing lamellae without long range diffusion or major reorganization of the chain conformation (13, 14).

THE INTERMEDIATE AND WIDE ANGLE RANGES

In contrast to the small angle range where only one integral quantity (R_g) can be measured one may expect to obtain detailed informations from measurements in a larger q-range in which the scattering intensity depends on the correlation of monomer units in the range of about $5 < x < 30$Å. In the following we will concentrate on the discussion of neutron scattering by highly crystalline polymers, so that in first approximation we can neglect the contribution of the amorphous regions, that means of the chain folds or loops. This contributions can be taken into account by Monte-Carlo calculations (23-26), but at the moment we prefer the discussion of analytically expressed scattering curves since some more general conclusions can be got out of these considerations.

Under the assumptions made in connection with eq. (1) one obtains for the coherent differential scattering cross section per unit volume for an isotopic sample consisting of a mixture of H- and D-stems

$$\frac{d\sigma}{d\Omega}(q) = c_D(1-c_D)\frac{K}{n_{st}} < |F_D(\underline{q}) - F_H(\underline{q})|^2 > \{1 + \frac{1}{N_s}\int n^{(2)}(r)\frac{\sin qr}{qr} 4\pi r^2 dr\} \qquad (9)$$

where $n^{(2)}(r)$ is the usual density correlation function for the spatial arrangement of N_s stems belonging to one and the same molecule. $|F_D(\underline{q}) - F_H(q)|$ is the difference of the scattering amplitudes of the deuterated and protonated crystalline sequences divided by $(n|b_D-b_H|)$. With regard to the orientational averaging <> the assumption was used that all sequences of one molecule are parallel to each other, which is certainly a good approximation for a lamellar morphology (see Note a). Then we may write

$$< |F_D(q) - F_H(\underline{q})^2> = \sum_{ij}\frac{\sin q\ Rij}{q\ Rij} = n_{st}^2\ P_s(q) \qquad (10)$$

where n_{st} is the number of monomer units per stem and Rij is the distance between the i-th and j-th hydrogen site of one crystalline sequence. By eq. (10) a formfactor $P_s(q)$ of the stems was defined. It may be useful to consider the case $q \rightarrow o$. Then the integral in eq. (9)

Note a. For a random orientation of the crystalline stems one would obtain

$$d\sigma/d\Omega \sim (<F^2> - <F>^2) + <F>^2\{1 + \frac{1}{N_g}\int n^{(2)}(r)\frac{\sin qr}{qr} 4\pi r^2 dr\}$$

see for example Ref. (27).

tends towards $N_s(N_s-1)$, $P_s(q) \rightarrow 1$ and accordingly

$$\frac{d\sigma}{d\Omega}(q \rightarrow o) = n_{st}N_s \approx n_w \tag{11}$$

Now we consider eq. (9) for the case that all the N_s stems form a "bundle" of stems within one lamella, see Fig. 3. (The spreading of the stems to different lamellae is not important at the moment, we will regard this later on). The differential scattering cross section is then given by (13, 14, 30)

$$\frac{d\sigma}{d\Omega}(q) = c_D (1-c_D) \frac{K}{n_{st}} \cdot \frac{1}{N_s} < <|F_D(\underline{q}) - F_H(\underline{q})|^2>_{rot} \sum_{m,n} J_0(Rx_{mn})>_{sp} \tag{12}$$

where J_0 is the Bessel function of order zero, x_{mn} is the distance between the stem centers in the lamella, R is the projection of $\underline{q}$ on the equator plane of the reciprocal space. $<>_{rot}$ means rotational averaging (see Note b) and $<>_{sp}$ means spatial averaging.

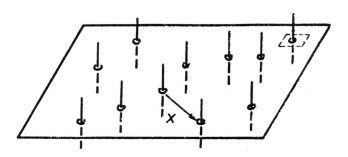

Fig. 3. *Distribution of crystalline stems of a tagged molecule in one lamella.*

For long stems L, that means for $1/L < 0.01$Å which is the range of interest here, the averaging of eq. (12) leads to the reduced intensity (Ref. 28)

$$J(q) = \frac{\pi F_Q^2(q)}{q\ell_0} \{1 + \rho_s \int h(x) J_0(qx) 2\pi x dx\} \tag{13}$$

where $F_Q^2(q)$ is the formfactor of the cross section of the crystalline sequence, ℓ_0 is the length of the monomer unit. Here we introduced the pair correlation function $h(x)$, so that $\rho_s h(x)$ gives the probability to find another stem belonging to the same molecule at the extremity of the vector $\underline{x}$ in the plane of the lamella.

Since

$$(1 + \frac{1}{N_s} \sum_{m \neq n} J_0(qx_{mn})) = 1 + \rho_s \int h(x) J_0(qx) 2\pi x dx \tag{14}$$

the function $h(x)$ is normalized by the condition

$$\rho_s \int h(x) 2\pi x dx = \rho_s H(q=0) = N_s - 1 \tag{15}$$

Note b. For reasons discussed in connection with eq. (10) the averaging $<>_{rot}$ requires that all the crystalline sequences are parallel with regard to their planes.

For the interpretation of the measured scattering curves in the IANS range, which we will discuss first, the eq. (13) can be used to calculate the reduced intensity for certain models. As an example in Fig. 4 (Ref. 28) the scattering function defined by eq. (7) is plotted in dependence on q for scattering models with various mean square average distances $< x^2 >^{1/2}$ of the stems. For that case the pair distribution function was constructed by repeated convolution of a Gaussian distance distribution between consecutive stems (28).

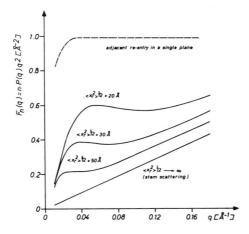

Fig. 4. *Analytically calculated scattering functions for a Gaussian distance distribution between consecutive stems (28). h(x) was constructed by repeated convolution.* $N_s=10$, $\ell_o=1.27Å$, $F_Q^2(q)=1$, $L=160Å$

The scattering functions $F_n(q) = n\, P(q)q^2$ so far measured from melt-crystallized polymers can be described qualitatively by two features. There exist a constant (or almost constant) plateau in the region of about $0.02<q<0.1Å^{-1}$ and a maximum at about $0.3 - 0.6Å^{-1}$. We want to discuss this characteristic behaviour in connection with eq. (13).

As an example the q^2-plateau for isotactic polypropylen isothermally crystallized from the melt (crystallinity ~ 0.70) is shown in Fig. 5 (Ref. 17). The level does not depend significantly on the crystallization conditions and agrees rather well with that found by neutron scattering from the melt, that means with the well-known plateau calculated from Debye's scattering law for a Gaussian coil. Similar results were obtained for isotactic polystyrene (21) (crystallinity ~ 0.35) with deviations of $\pm$ 15% from the melt depending on the molecular weight of the matrix. In the case of polyethylene-oxide a q^2-plateau with a level about 10% lower as in the melt was found (29). For melt-quenched linear polyethylene either a q^2-plateau was observed (30) or a slow increase with q (16, 13) and the $F_n(q)$ value of the crystalline sample at $q \sim 0.1Å^{-1}$ agreed very well with that of the melt (16).

As it can be seen from Fig. 4, the plateau also appears in the calculated scattering functions for this special model. Obviously the level of the plateau depends on the average distance $<x^2>^{1/2}$ of the stems. This dependence is a rather general feature of all scattering curves so far calculated. As an example Fig. 6 shows the results of Monte-Carlo calculations for isotactic polypropylen chains assuming various values of the a-priori probability P_{ar} of adjacent re-entry (25). The q^2-level of $F_n(q)$ is the higher the larger P_{ar}, that means the smaller the average distance of the stems belonging to the same molecule. It can be noticed that the best agreement can be reached for $P_{ar} = 0$.

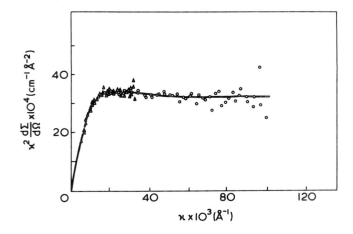

Fig. 5. Kratky plot of scattering intensity for isothermally crystallized polypropylene (Ref. 17).

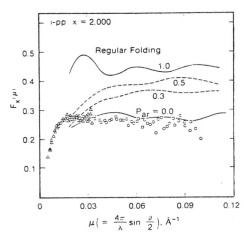

Fig. 6. The scattering functions $F_n(q)$ for isotactic polypropylene computed for various values of the a-priori probability of adjacent re-entry P_{ar} (25). Experimental points are from Ref. (17).

Both analytical and Monte-Carlo calculations give some indication what kind of information can be obtained from neutron scattering in the IANS range up to about q = 0.1Å^{-1}. Let us idealize the experimental observations and compare them with conclusions obtainable from eq. (13). We assume that in a certain x-range, which is specified in more detail in the following section, the pair distribution function is given by

$$\rho_s h(x) = \frac{c}{2\pi x} \qquad\qquad 0 < x < R_o \qquad\qquad (16)$$

then the Fourier-Bessel transformation yields approximately (see Note c)

$$\rho_s H(q) = \frac{c}{q} \qquad\qquad\qquad (17)$$

Note c. Approximately means that we neglect the convolution of H(q) with a rather sharp declining Fourier-transform of the shape function given by R_o.

and since in this q-range the scattering intensity is much higher than that of a single stem, one obtains from eq. (13) with $F_Q^2(q) = 1$

$$J(q) \approx \frac{\pi}{\ell_0} \cdot \frac{c}{q^2} \tag{18}$$

The simple pair correlation function of eq. (16), which is just an heuristic approach for describing the experimental results, has an interesting property which may offer an additional insight of the crystallization process.

Firstly we may state that eq. (16) represents the pair correlation function for a linear arrangement of stem centers in the lamella, that means for crystallization along one net plane. For that case the normalization condition of eq. (15) yields

$$c = 2/a \tag{19}$$

where a is the average distance of stems, and one obtains according to eq. (18) the reduced intensity

$$J(q) = \frac{2\pi}{\ell_0 a} \cdot \frac{1}{q^2} \tag{20}$$

which is the well known scattering law of a sheet.

For linear polyethylene it is $F_n (q = 0.1) \sim 0.5\text{\AA}^{-2}$ resulting in a $\approx 10\text{\AA}$ and indicating a "dilution" of tagged stems along a sheet, which can be modeled in various ways (26). It is very important to notice, however, that the correlation function of eq. (16) can be realized also by a two-dimensional arrangement of stem centers in the lamella as it is clearly demonstrated by the results of the Monte-Carlo calculations in Fig. 6. We will show in the next section by introducing a direct correlation function that a relationship as approximated by eq. (17) is just the consequence of the fact that there are only two consecutive stems connected by a loop, that means the connectivity of the stems gives rise to a pair distribution function approximated by eq. (16) in a certain q-range.

Before we do that we look at another interesting generalization of the experimental results. As mentioned already there is no significant change of the q^2-level during crystallization observed. For an ideal coil we have a (three dimensional) density correlation function

$$g(r) = \frac{3}{\pi b^2} \cdot \frac{1}{r} \qquad\qquad r \ll R_g \tag{21}$$

where b = length of the statistical segment. The 3-dimensional Fourier-transformation yields the reduced intensity per monomer unit

$$J(q) = \frac{12}{b^2} \cdot \frac{1}{q^2} \tag{22}$$

If the q^2-level of the scattering function $F_n = q^2 n_w P(q)$ does not change as a result of crystallization it follows that

$$c \approx \frac{12\ell_0}{\pi b^2} \tag{23}$$

which means that the average distance of the stems is controlled by the length of the statistical segment or by the characteristic ratio of the chain molecule.

The experimental result described approximately by eq. (18) has another interesting feature. As a result of the normalization condition eq. (15) one gets

$$c R_0 = N_s - 1 \tag{24}$$

where R_0 is the "radius" of the bundle of stems in the lamella belonging to one molecule.

(It may be noted that this in contrast to the assumption of a constant density, where $R_o^2 \sim N_s$). As a consequence from eq. (24) (or from eq. (16)) the number density of monomer units in the volume occupied by a tagged molecule is

$$\rho_{mon} \approx \frac{c}{\pi \ell_o} \cdot \frac{1}{R_o} \tag{25}$$

On the other hand the density of monomer units in a coil scales like

$$\rho_{mon,coil} \sim \frac{n_w}{R_g^3} = \frac{R_g^2/b^2}{R_g^3} = \frac{1}{b^2} \cdot \frac{1}{R_g} \tag{26}$$

This scaling argument shows that the pair distribution function of the stem centers of the type $h(x) \sim c/x$ takes care for maintaining the overall density of the tagged monomer units in the volume of the coil.

One may think that this result follows only from the assumption that all N_s stems crystallize in one lamella. The same conclusion can be drawn, however, if the 3-dimensional density distribution function $n^{(2)}(r)$ of eq. (9) is considered. In order to obtain a scattering law

$$J(q) = \frac{K}{q^2} \tag{27}$$

the density correlation function of the monomer units must read

$$g(r) = \frac{K}{4\pi} \cdot \frac{1}{r} \tag{28}$$

$g(r)$ has to be generated by

$$g(r) = n^{(2)}(r) \otimes g_{st}(r) \tag{28}$$

where

$$g_{st}(r) = \frac{1}{4\pi r^2 \ell_o} \tag{30}$$

is the density distribution function of a stem.

The convolution theorem leads to

$$n^{(2)}(r) = \frac{K\ell_o}{4\pi^3} \cdot \frac{1}{r^2} \tag{31}$$

which is the threedimensional analogon to the distribution function of eq. (16).

Without specifying a certain model of chain re-entry we can draw the following conclusion. Most experimental results show that not only the radius of gyration R_g remains almost constant but also the monomer density distribution function $g(r)$ in a range $10 < r < 50 \text{Å}$ does not change markedly. In order to establish this structure the crystalline sequences arrange themselves according to a distribution function like eq. (16) or eq. (31). As a consequence the volume of the coil is maintained and the molecular weight dependence is the same as in the melt. The evidence of invariance of the monomer distribution function $g(r)$ with respect to crystallization strongly supports the solidification model. Minor changes of $g(r)$ are probably due to the fact that the sequences have to fit into the crystalline lattice. We have already shown by Monte-Carlo calculations that taking into account the lattice structure shifts the q^2-level to some extent (see Fig. 6 of Ref. 13).

The level of the q^2-plateau is determined only by the average distance of the stems and there is no way to prove or to reject a specified model on the basis of neutron measurements in the range 0.01 to 0.2$\mathring{A}^{-1}$ alone.

We will show in the next section that a distribution function $h(x) \sim (1/x)$ in a range of about $0.02 < q < 0.1\mathring{A}^{-1}$ is the natural consequence of the connectivity of the chain.

THE DIRECT CORRELATION FUNCTION

It is well known that during crystallization phase separation or "paraclustering" (31) may occur due to small differences in the chemical potential of the H- and D-molecules. The question what "segregation" means is not easy to answer, since besides the inter-molecular clustering also an intramolecular clustering exists which is caused by the loops or folds connecting the crystalline stems. The problem is quite clear if one thinks about a regular folded chain in a net plane. This chain is clearly segregated from the other chains without any special H-D-interactions. Deuteration only detects the "intramolecular" clustering.

Both types of clustering can be described by applying the well known Ornstein-Zernicke theory (32) of critical opalescence to our problem in so far as the scattering law is concerned. We start from eq. (12) which we rewrite in the form

$$\frac{d\sigma}{d\Omega} (q) = c_D(1-c_D)K \, n_{st} < P_s(R,Z) \, \{1 + \rho_o \int h(x) \, J_o \, (Rx) \, d^2x\} > \tag{32}$$

where the formfactor of the stem (eq. (10)) was used. ρ_o is the number density (per unit area) of the crystalline sequences. If there is no correlation between the stems belonging to the same kind, the reduced intensity in the limit $q \to o$ becomes $J(q=o) = n_{st}$, if N_s stems belong to the same molecule the effect of "intramolecular" clustering leads to $J(q=o) = n_{st}N_s$, that means $\{ \} \to N_s$ and finally if there is an additional correlation between stems of the same kind but belonging to different molecules the result will be $J(q=o) = \alpha \, n_{st}N_s$, where α depends on the concentration c_D.

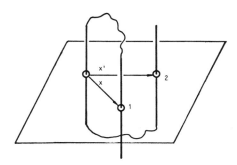

Fig. 7. Application of the direct correlation function to the crystallization of polymers.

As usual in the O.-Z.-theory we introduce a direct correlation function $c(x)$, the meaning of which is demonstrated in Fig. 7. The stem at 0 is directly correlated with the stems at 1 and 2 by means of the amorphous loops. But there is also a probability to meet another stem at point 1 because of a direct correlation between 2 and 1:

$$h(01) = c(0,1) + c(0,2) \otimes c(2,1) \tag{33}$$

The convolution has to be repeated and leads finally to O.-Z. integral equation

$$h(x) = c(x) + \rho_o \int c(x') \, h(x'-x)d^2x' \tag{34}$$

By Fourier-transformation one obtains

$$H(q) = \frac{C(q)}{1-\rho_o C(q)} \tag{35}$$

Now we have to take into account that the direct correlation functions consists of two distinct terms

$$c_{tot}(x) = c(x) + \phi(x) \tag{36}$$

Where $c(x)$ is the chain correlation function determined by the loops or folds as shown in Fig. 7 and $\phi(x)$ is a "thermodynamical" correlation function taking into account the small intermolecular interactions between D-D-crystalline stems. We assume that the pseudopotential responsible for $c(x)$ is much stronger than the potential causing $\phi(x)$, so that $c(x)$ is not disturbed by the thermodynamic interaction. For the following discussion we switch off $\phi(x)$ and consider the intermolecular clustering later.

For small q we obtain by serie expansion

$$C(q) = F.B.\{c(x)\} = \int c(x) \left[1 - \frac{1}{4}(qx)^2 \pm \ldots\right] d^2x \tag{37}$$

$$\approx c_o - c_1 q^2$$

with

$$c_o = \int c(x)d^2x = F$$

$$c_1 = \frac{1}{4}\int x^2 c(x)d^2x = \frac{1}{4}\varepsilon^2 \tag{38}$$

and finally from eq. (35)

$$1 + \rho_o H(q) = \frac{\ell^{-2}/4F\rho_o}{\kappa^2 + q^2} \tag{39}$$

Here we defined a direct correlation length ℓ^2 by

$$\ell^2 = \frac{\varepsilon^2}{F} \tag{40}$$

whereas the cluster correlation length is given by

$$\xi^2 = \kappa^{-2} = \frac{\rho_o \varepsilon^2}{4(1-\rho_o F)} \tag{41}$$

An important deviation from the original O.-Z.-theory has to be taken into account. There ξ can go to infinity since $(1-\rho_o F)$ approaches zero if the critical temperatur is reached. In our case, however, not all D-stems are correlated by means of the structural correlation function $c(x)$ but only those belonging to the same macromolecule. Therefore we have an additional boundary condition

$$1 + \rho_s H(q) \xrightarrow[q \to o]{} N_s \tag{42}$$

Where we replaced ρ_o by the stem density ρ_s of stems belonging to one molecule. This leads to

$$\xi_o^2 = \frac{\rho_s \varepsilon^2}{4} N_s \tag{43}$$

under the condition, that all stems belong to the same cluster in one lamella. If several clusters are formed, it may happen that in a q-range $0 < q \ll 1/\ell$ the scattering law of eq.(39)

shows a cluster correlation length $\xi < \xi_0$. From eq. (42) it also follows that

$$\xi/\ell \leq N_s - 1$$

The eq. (43) shows that the cluster correlation length depends on the number of stems N_s (or on the molecular weight) as one may expect. The stem density ρ_s of stems belonging to one chain is in the order of N_s/R_g^2 where R_g^3 is approximately the volume occupied by the molecule before crystallization. Since $R_g^2 = b^2 n_w \sim N_s$ the density ρ_s does not depend on molecular weight.

The cluster length ξ_0 also depends on the second moment of the direct correlation function. If the pseudopotential describing the interaction between stems by means of the loops or folds is near ranged, the cluster length ξ will be small.

In Fig. 8 the scattering function $F_n(q)$ based on eq. (38) is plotted (with the numerator equal to 1) for various values of κ. One can notice that in this Kratky plot for larger ξ characteristic "noses" show up which are caused by the formfactor of the stem bundles, qualitatively spoken. Similar results habe been obtained earlier, see Fig. 5 of Ref. 28. The appearance of those noses does not indicate a clustering or phase separation for thermodynamic reasons, it is a consequence of intramolecular clustering.

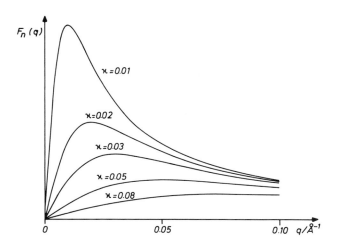

Fig. 8. The scattering function $F_n = q^2 n_w P(q)$ (in arbitrary units) for stems in a lamella as calculated from the Ornstein-Zernike theory.

For the derivation of eq. (39) we used the approximation

$$(\ell q)^2 \ll 1$$

where ℓ is the direct correlation length defined by eq. (40). With other words we considered the correlation of stems in a x-range $x \gg \ell$, but still x in the order of the cluster correlation length ξ. (Remember for full clustering we obtained $\xi/\ell = N_s - 1$).
For smaller $x \ll \xi$ or $x \sim \ell$ the direct correlation function cannot be approximated as in eq. (36), but one has to introduce more specialized assumptions. As already pointed out it was found experimentally that the scattering intensity is proportional to q^2 in a range $0.02 < q < 0.1\text{Å}$. Qualitatively this behaviour has been explained (see eq. (16)) by a "quasi-linear" arrangement of stems. Now we will show that the q^2-plateau is the consequence of a special feature of the direct correlation function. For this discussion we choose

$$c(x) = (N_s - 1) \frac{1}{2\pi x \langle x \rangle} e^{-x/\langle x \rangle} \qquad x > o \qquad \qquad (44)$$

where $\langle x \rangle$ measures the average distance between neighboured stems. Here we took into account, that there are (N_s-1) direct correlation pairs.

The choice of the exponential function is not important, any other reasonable function (like exp. $(-x^2/\langle x^2 \rangle)$) yields very similar results as already pointed out earlier (28). The important feature of the correlation function of eq. (44) is the proportionality to $(1/x)$ which takes care for the fact that for the direct correlation there is only one consecutive and one preceding stem in the neighbourhood of the stem at $x = o$ which are correlated by the "structural" part of the total correlation function $c_{tot}(x)$. The arrangement of other stems in the neighbourhood of the considered stem is taken into account by the convolution term in eq.(34).

The Fourier-Bessel transformation of $c(x)$ yields

$$C(q) = \frac{N_s-1}{(1+q^2\langle x\rangle^2)} \, 1/2 \tag{45}$$

For small $q \ll \frac{1}{\langle x\rangle}$ we may develop

$$\frac{C(q)}{N_s-1} \approx (1 - \frac{1}{2} q^2 \langle x\rangle^2 + \ldots) \tag{46}$$

and comparison with eq. (37) shows that for this special correlation function $\ell^2 = 2\langle x\rangle^2$. In the range of $q \sim \frac{1}{\langle x\rangle}$ this approximation is errorneous, however, and one obtains the scattering function from the equation

$$\frac{1}{N_s}H(q) = \frac{\frac{1}{N_s}C(q)}{1-\frac{1}{N_s}C(q)} \tag{47}$$

which is the analogue to eq. (35).
In Fig. 9 the function

$$F_n = q^2 I(q) = \frac{\pi}{\ell_0} q \left\{ 1 - \frac{N_s-1}{N_s} \, \frac{1}{(1+q^2\langle x\rangle^2)1/2} \right\}^{-1} \tag{48}$$

is plotted with various values of $\langle x \rangle$.

As the diagram shows that there exists a q-range where the intensity is almost $\sim 1/q^2$, the level of this plateau depends on $\langle x \rangle$.

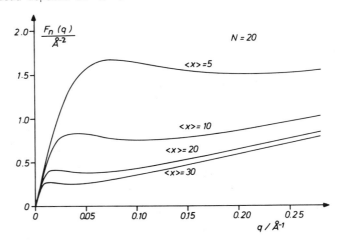

Fig. 9. Analytically calculated scattering functions $F_n(q)$ for a direct correlation function given by eq. (44).

The curves in Fig. 9 are very similar to those which have been obtained earlier (28) with a Gaussian distribution function in eq. (44) instead of the exponential, see Fig. 4. The reason is quite obvious. A q^2-plateau is even created by intraclustering for a correlation function $c(x) \sim 1/x\langle x\rangle$, but there $H(q)$ becomes infinite for $q\langle x\rangle = 1 - \frac{1}{N_s}$. If a continuously decreasing function $w(x)$ is introduced

$$c(x) \sim \frac{1}{x} w(x) \tag{49}$$

the convolution with $W(q)$

$$C(q) \sim \frac{1}{q} \otimes W(q) \tag{50}$$

prevents the infinity.

It may be mentioned that eq. (47) is a good approximation only for the case, that the number N_s of stems belonging to one cluster in a lamella is not to small. Otherwise the repeated convolution leads to (Ref. 28)

$$\frac{1}{N_s} H(q) = \frac{2}{N_s} \sum_{n=1}^{N_s-1} (N_s - n)\ C_o^n(q) \tag{51}$$

where $C_o(q)$ is the transform of the direct correlation function of a single pair with the normalization condition

$$\int_o^\infty c_o(x) d^2 x = 1 \tag{52}$$

The treatment described above includes several other simplifications, for example the finite thickness of the crystalline stems has been neglected and the crystalline lateral order of the stems has not been taken into account. It has been shown already (see Fig.14a of Ref.13) that the level of the scattering function F_n is changed by regarding these facts but the qualitative conclusions are not touched.

If the range of momentum transfer q is large enough, in which the scattering intensity $J(q)$ was measured, the direct correlation function $c(x)$ can be determined experimentally.

With

$$S(q) = 1 + \frac{1}{N_s} H(q) = J(q)/n_{st}\ P_s(q) \tag{53}$$

one obtaines from eq.(47)

$$\frac{C(q)}{N_s} = \frac{S(q) - 1}{S(q)} \qquad \text{where} \qquad \frac{C(q \to o)}{N_s} = 1 - \frac{1}{N_s} \tag{54}$$

$S(q)$ can be calculated from the reduced intensity and from the stem factor P_s, which is obtained from eq.10. From $C(q)$ of eq.(54) the direct correlation function $c(x)$ can be determined by inverse Fourier-Bessel-transformation, where one has to be aware of the approximative character of eq. (47).

For larger q-values $P_s(q)$ is no longer $\sim 1/q$ because of the finite thickness of the stems. In the Kratky plot $P_s(q) \cdot q^2$ will run through a maximum, which is for polyethylene located at about $q \approx 0.6\mathring{A}^{-1}$ (Ref. 13), and for polyethylene oxide at $\approx 0.45\mathring{A}^{-1}$. The measured scattering function F_n may also show a well pronounced maximum, the position of which is strongly determined by $S(q)$. By comparison with calculated scattering curves (13,33) valuable informations about $c(x)$ can be obtained.

SEGREGATION AND INTERMOLECULAR CLUSTERING

So far it has been assumed that there are no specific thermodynamic interactions between the deuterated and protonated macromolecules. This is certainly not correct as can be seen from the shift of the melting point if one replaces H by D (34). The reason for this isotopy effect is not yet exactly known although some ideas have been published (35).

An extreme case of non-statistical distribution of H- and D-molecules would be a complete phase separation, that means fractionation. It can be detected experimentally by two methods (13): Differential scanning calorimetry will exhibit two melting points and the Bragg reflection neutron intensities will be the sum of the intensities scattered by the two phases. In the case of polyethylene oxide, which we will discuss later, both methods did not show an indication of phase separation of this kind.

It was shown by Schelten et al. (31) that minor deviations from a statistical distribution may cause very large anomalies with regard to the apparent molecular weight and radius of gyration as measured by neutron scattering. This "paraclustering" effects results from the enhancement of weak thermodynamic interactions by the large number of monomers belonging to one molecule. In those cases where no phase separation can be detected the paraclustering is an appropriate way to describe the non-statistical distribution.

It has also been proposed (30, 36) to take the clustering into account by summing up the intensities caused by concentration fluctuations and by the structure factor of the chains. The sum is weighted by an arbitrary parameter α relating the relative intensities of concentration fluctuations and molecular scattering. We believe that this procedure is only justified if phase segragation occurs in the sense as described above.

A more realistic picture of the intermolecular clustering may be developed if one remembers eq. (36) which postulated that the direct correlation function consists of two parts: a "strong" interaction $c(x)$ caused by the loops or folds between neighboured stems belonging to the same molecule and a "very weak" interaction $\phi(x)$ taking into account the intermolecular forces between stems of the same kind but belonging to different molecules.

For this discussion as an example we choose the case of polyethylene oxide (M_w = 120 000) crystallized by quenching to $T_c = 40^\circ C$ (29). As already observed by Allen et al. (37) for low molecular weight PEO this polymer also shows clustering effects. In Fig. 10 the scattering curves in the SANS range are plotted for various concentrations c_D. Characteristic "noses" appear in this Kratky plot with a maximum at about $0.01 \mathring{A}^{-1}$.

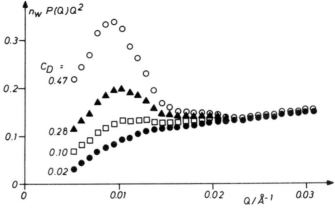

Fig. 10. Scattering function of PEO (M_w=120 000) by quenching to T_c=40°C. For small $q<0.02\mathring{A}^{-1}$ the reduced intensities depend on the concentration c_D of deuterated chains (29).

In a qualitative way this effect can be explained by the O.-Z.-theory treated above. Comparison with Fig. 8 shows that the measured scattering curves behave as if the cluster correlation length $\xi=\kappa^{-1}$ depends on concentration c_D. According to eq. (43) κ decreases with increasing number of crystalline stems N_S available. So one may conclude that the intermolecular clustering just enlarges N_S or the apparent molecular weight. With other words the intramolecular clustering showing up in Fig. 8 is greatly enhanced by the additional D-D interactions.

For a quantitative treatment one may assume that the arrangement of stems belonging to one molecule is not markedly disturbed by the thermodynamic interaction described by the usual interaction parameter χ. For this case the random phase approximation yields (Ref. 38)

$$\frac{S(q)}{c_D(1-c_D)} = \frac{S_0(q)}{1-2\,\chi\,c_D(1-c_D)\,S_0(q)} \tag{55}$$

where $S_0(q)$ is defined by eq. (53) for the undisturbed conformation (see Note c).

From eq. (55) together with eq. (13) and (39) it follows, that a plot of $1/q\,J(q)$ versus q^2 should result in a family of straight lines only shifted vertically by $(-2\chi c_D(1-c_D))$. As Fig.11 shows this is not observed for the case of PEO. Contrarily the shape of the curves for small q-values also depends on concentration.

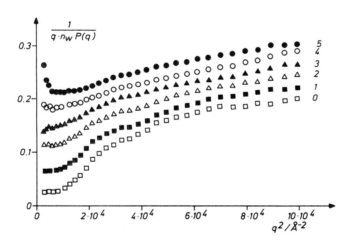

Fig. 11. Plot of $1/q\,J(q)$ versus q^2 for PEO as in Fig. 10. The curves are shifted vertically for n times 0.02Å, n indicated in the drawing.

For a more appropriate description of the intermolecular clustering we return to eq. (36) where an interaction direct correlation function $\phi(x)$ was introduced. If we take into account eq. (33) reads:

$$h(0,1) = h_0(0,1) + \left[\,h_0(0,2) \otimes \phi(2,3)\,\right] \otimes h_0(3,1) \tag{56}$$

Where $h_0(x)$ is the undisturbed correlation function for the stems interacting only by loops. Repeated convolution and Fourier-transformation leads to

$$\frac{S(q)}{c_D(1-c_D)} = \frac{S_0(q)}{1-\Phi(q)\,c_D(1-c_D)\,S_0(q)} \tag{57}$$

Note c. This equation differs from the treatment by Schelten et al.(31) with regard to the concentration dependence of the "paraclustering" effect (see eq.21 of Ref.31). As it was pointed out in connection with eq.(8) any kind of intermolecular D-D interaction must lead to a reduced intensity, which depends on concentration, however.

where $\Phi(q)$ is the transform of $\phi(x)$. Comparison with eq. (55) shows, that they are identical if one assumes

$$\phi(x) = 2\chi\delta(x)$$

that means if the thermodynamic interaction is very short ranged. The experimental results show however, see Fig. 11, that the intermolecular interaction must be described by a long range perturbation. Fortunately $\Phi(q)$ declines to zero at about $q = 0.02\text{Å}^{-1}$, so that the evaluation of the scattering patterns is not hindered by the clustering effects. The reason for the apparent long range nature of the intermolecular interaction is not yet known. It seems possible that the long wavelength concentration fluctuations are due to the kinetics of the intermolecular clustering (39).

CONCLUSIONS

1) Neutron scattering from mixtures of deuterated and undeuterated molecules allows the evaluation of the single chain form factor P(q) in the semicrystalline state of polymers. The measurements at small momentum transfer $0.005 < q/\text{Å}^{-1} < 0.03$ yield the radius of gyration R_g, which does not change markedly by the crystallization process, that means the overall dilution of a single chain is about the same as in the melt.

2) In the intermediate angle range $0.03 < q/\text{Å}^{-1} < 0.1$ very often a q^2-plateau is observed and the value of the scattering function $F_n = n_w P(q) q^2$ agrees well with that of the melt. This observation indicates that not only R_g remains almost constant, but that also the pair distribution function g(r) of the monomer units for $r \ll R_g$ does not change appreciably during crystallization. The level of the q^2-plateau in the semicrystalline state is mainly determined by the average distance of the crystalline stems, which is controlled by the statistical properties of the chain. There is no indication of privileged adjacent re-entry for melt crystallized polymers.

3) The connection of crystalline stems by amorphous loops leads to an "intra-molecular" clustering of stems, which can be treated by an Ornstein-Zernike approach. The direct correlation c(x) is related to the pseudopotential of the interaction between consecutive crystalline stems belonging to the same molecule. c(x) can be calculated approximately from the measured data if the q-range is large enough.

4) The intermolecular clustering often observed during crystallization can be described in a similar way. It leads to an enhancement of intra-molecular clustering in so far, as the cluster correlation length increases. Details of the long wavelength fluctuations of the concentration of deuterated stems are not yet understood.

ACKNOWLEDGEMENT

This work has been supported by the Deutsche Forschungsgemeinschaft (Sonderforschungsbereich 41). Experimental work was done at the Institute Laue-Langevin, Grenoble. I am much indebted to Dr. M.G. Brereton, Leeds, and Mr. J. Kugler, Mainz, for helpfull discussions.

REFERENCES

1. H.D.Noether, W.Whitney, Koll. Z.- u.Z. Polym. 251, 991-1005 (1973)

2. R.G.Kirste, W.A.Kruse, J.Schelten, Macromol.Chem., 162, 299 (1972)
   H.Benoit, J.P.Cotton, D.Decker, B.Farnous, J.S.Higgins, G.Jannick, R.Ober, C.Picot,
       Nature, 245, 13 (1973)
   G.Allen, C.J.Wright, Neutron Scattering Studies of Polymers, International Review of
       Science, Vol. 8, Macromolecular Science, Ed.G.E.Bawn, Butterworth (1975)

3. "Organization of Macromolecules in the Condensed Phase", Faraday Disc.68, 365-493 (1979)

4. J.D.Hoffman, C.M.Guttman, E.A.DiMarzio, Faraday Disc. 68, 177 (1979)

5. E.A.DiMarzio, C.M.Guttman, J.D.Hoffman, Polymer 21, 1379 (1980)

6. P.J.Flory, D.Y.Yoon, Nature 272, 226 (1978)

7. J.Klein, R.Ball, Faraday Disc. 68, 198 (1979)

8. E.A.DiMarzio, C.M.Guttman, J.D.Hoffman, Faraday Disc. 68, 210 (1979)

9. A.Z.Akcasu, G.C.Summerfield, S.N.Jahshan, C.C.Han, C.Y.Kim, H.Yu, J.Polym.Sci.,
       Phys. 18, 863-869 (1980)

10. M.G.Brereton, E.W.Fischer, W.Gawrisch, Jahresbericht 1978/79 des Sonderforschungs-
        bereichs 41, Mainz 1979, p. 160

11. W.Gawrisch, M.G.Brereton, E.W.Fischer, Polym.Bull., 4, 687-691 (1981)

12. E.W.Fischer, M.Stamm, M.Dettenmaier, P.Herchenröder, ACS-Preprints 20, 219 (1979)

13. M. Stamm, E.W.Fischer, M.Dettenmaier, P.Convert, Faraday Disc. 68, 263 (1979)

14. E.W.Fischer, Pure and Appl. Chem. 50, 1319 (1978)

15. A.Peterlin, Macromolecules 13, 777 (1980)

16. J.Schelten, D.G.H.Ballard, G.D.Wignall, G.Longman, W.Schmatz, Polymer 17, 751-757 (1976)

17. D.G.H.Ballard, P.Cheshire, G.W.Longman, J.Schelten, Polymer 19, 379 (1978)

18. P.Herchenröder, Thesis, Mainz 1978

19. J.Kugler, P.Herchenröder, E.W.Fischer, G.Wegner, C.Eisenbach, M.Peuscher, Verhandl. der
        Deutschen Physikal.Ges., Frühjahrstagung Marburg 1981, p.61

20. J.M.Guenet, Macromolecules 13, 387 (1980)
    J.M.Guenet, Polymer 22, 313 (1981)

21. J.M.Guenet, C.Picot, H.Benoit, Faraday Disc. 68, 251-262 (1979)

22. E.W.Fischer, Europhys.Conf. Abstr. 2E, 71 (1977)

23. D.Y.Yoon, P.J.Flory, Polymer 18, 509 (1977)

24. D.Y.Yoon, P.J.Flory, Faraday Disc. 68, 288 - 295 (1979)

25. D.Y.Yoon, P.J.Flory, Polym. Bull. 4, 692 (1981)

26. C.M.Guttman, J.D.Hoffman, E.A.DiMarzio, Faraday Disc. 68, 297 - 309 (1979)

27. P.A.Egelstaff, D.I.Page, J.G.Powles, Molecular Phys., 20, 881 - 894 (1971)

28. M.Dettenmaier, E.W.Fischer, M.Stamm, Colloid and Polym.Sci., 258, 343 (1980)

29. J.Kugler, Diplomarbeit Mainz 1981

30. D.M.Sadler, A.Keller, Macromolecules 10, 1128 (1977)

31. J.Schelten, G.D.Wignall, G.D.H.Ballard, G.W.Longman, Polymer 18, 1111-1120 (1977)

32. L.S.Ornstein, F.Zernike, Proc. Acad. Amsterdam 17, 793 (1914)
    A.Münster, Statistical Thermodynamics, Springer-Verlag Berlin 1969, Vol.I, p.404 ff

33. M.Stamm, J.Schelten, D.G.H.Ballard, Coll.Polym.Sci. 259, 286 - 292 (1981)

34. F.C.Stehling, E.Ergos, L.Mandelkern, Macromolecules 4, 672 (1971)

35. A.D.Buckingham, H.G.E.Hentschel, J.Polym.Sci. 18, 853 (1980)

36. J.E.Anderson, S.J.Bai, J.Appl.Phys., 49, 4973 (1978)

37. G.Allen, T.Tanaka, Polymer 19, 271 (1978)

38. P.G. De Gennes, Scaling Concepts in Polymer Physics, Cornell Univ.Press 1979, p.109

39. P.Pincus, J.Chem.Phys. 75, 1996 (1981)

MOLECULAR MECHANISMS IN POLYMER FRACTURE

H.H. Kausch

Laboratoire de Polymères, Ecole Polytechnique Fédérale de Lausanne,
chemin de Bellerive 32, CH-1007 Lausanne, Switzerland

Abstract – A review is given on recent (1978-1981) progress in understanding
the role of degrading physical mechanisms such as slip, disentanglement,
chain scission, and void opening on crazing and fracture of solid polymers.
The new field of rehealing of cracks is presented.

INTRODUCTION

It is the purpose of a main lecture to give an account on the important progress achieved
within the field of a selected topic. Important, that is, in the personal view of the author.
In 1978 the author had extensively discussed the influence of molecular mechanisms, particularly of chain scission, on polymer fracture (1); since then no comprehensive review on this
subject has been presented at a IUPAC meeting; the author is thus obliged, but can also restrict himself, to treat the period from 1978 to 1981.

The following discussion may conveniently be organized according to the dominant failure
patterns

- ductile deformation (creep, yield, flow)

- crazing (glassy polymers, environmental stress cracking, creep crazing)

- homogeneous damage formation (chain scission, void formation)

- crack propagation (phenomenology, criteria, fracture mechanics, fatigue)

- crack healing

It will readily become apparent that a molecular analysis of fracture phenomena must take into consideration the interaction (2) and mobility (3) of chains and the structure and physical behavior (4-6) of polymer solids. These subjects are treated in this volume as separate
main lectures to which the reader is referred, especially with regard to the work of the
cited authors (2-6).

DUCTILE DEFORMATION

From their study of the *homogeneous plastic compression* of PMMA and PS at constant strain
rate $\dot{\varepsilon}$ Haussy, Carrot et al. (7,8) conclude that at low temperatures (T < 200 K) a given
strain rate is accommodated by a single Eyring mechanism which they call crystal-like. The
measured activation volume of this segmental shear process amounted to 0.22 nm³ (PMMA) and
0.90 nm³ (PS), that is to 1.5 and 5 monomer units respectively. This is a very localized
movement. At higher temperatures (T > 200 and 280 K respectively) the deformation is more
associated with free volume effects. In the case of PMMA it is, as first noted by Bauwens-
Crowet (see 9 for reference), probably related to the β-relaxation process. Bauwens and
Bauwens-Crowet (see 9) have further developed their modelisation of the *tensile stress-strain
curve* of a glassy polymer by introducing non-linear Maxwell elements. The treatment assumes
"that throughout the course of deformation, some structure initially present in the polymer
is destroyed and that the initial spectrum of the Maxwell elements is converted to another
spectrum" (9). *Creep* data obtained for PVC could be well fitted this way.

With regard to damage formation during the *creep of elastomers* two publications may be cited.
Huang and Aklonis (10) used a photoinduced scission of disulfide linkages in an elastomeric
polyurethane network. The scission could be turned on and off at will, to discriminate between physical relaxation and chemical degradation. Lepie and Adicoff (11) characterized the
time-dependent failure of two solid propellants by the total "damage energy" at failure.

Closely related with the creep behavior of polymers is evidently their *stress relaxation*. The latter subject has been reviewed by Kubàt (12) at the IUPAC Symposium in Mainz.

One of the most important questions in the molecular interpretation of deformation mechanisms is that of the local change of segment conformation and the role of entanglements in this process. As pointed out by Haward (13) a large number n of flexible chain units between entanglement points increases the tendency to *necking*. In stiff polymers with a "large natural hinge length" as defined by Argon et al. (14) n obviously decreases, strain hardening sets in at small strains already since a sufficiently large number of extended segments are present which must be strained in axial direction, and *homogeneous yielding* takes place (13). The instabilities of deformation associated with *neck formation* in a certain region of strain rates have recently also been investigated by Pakula and Fischer (15). They propose that under tensile stress the energy barriers between neighboring states of the deforming system decrease and even vanish. The latter condition,

$$\frac{\partial^2 \Delta F}{\partial \varepsilon^2} = 0,$$

gives rise to a "spinoidal tensile stress instability".

In recent years a powerful tool of investigation of time-dependent molecular rearrangements has become available through Fourier-Transform Infrared (FTIR) Spectroscopy (16). The possibility to measure a whole spectrum (absorbance vs. wavenumber) every few seconds permits to study conveniently as a function of strain or time dynamic molecular mechanisms such as the gradual rotation of segments, conformational changes, or the transformation of one crystal form into another.

The molecular mechanisms occurring in *flow*, such as chain entanglement and disentanglement, slip, and scission have been treated comprehensively in the proceedings of the VIIth International Congress on Rheology (17).

CRAZING

The research activity on craze phenomena is still extremely large. Periodic review articles have appeared (see 18 and 1a for references). In this section some trends may be underlined. In the author's opinion studies of the craze *microstructure*, of the *mechanisms of their formation and growth*, of their *influence on the deformation and fracture of homo- and block-copolymers*, and of *environmental stress effects* are the areas of special interest.

The *microstructure* of crazes in thin films and the microscopic mechanisms of their growth and fracture have been investigated by Kramer and his collaborators (18-20) by a new and intriguing electron imaging technique. As principal results they obtained thickness profiles of isolated air crazes in PS, craze fibril draw ratios (typically between $\lambda = 3.5$ and 7.5) and craze fibril stresses (of between 100 and 200 MPa). They also determined average air craze fibril diameters (6 nm at 20 ^{0}C for PS, 33 nm at 128 ^{0}C for PC). The latter value is in agreement with measurements of Paredes et al. (21) and Dettenmaier et al. (22). The lengths and opening widths of planar crazes in fracture mechanics specimens have been calculated by Döll and his group (23-25) and Israel et al. (26).

A new craze phenomenon in PC, the formation of a large number of small crazes with extremely thick fibrils, has been described by Dettenmaier and Kausch (22, 26-27). In Figure 1 electron micrographs of a sample edge (Fig. 1a) and of the fibrillar craze structure (b) are shown. The extremely large concentration of the new crazes, called crazes (II), and the cylindrical shape of the fibrils is clearly visible. A distinctive midrib, such as discussed for conventional crazes (I) by Lauterwasser and Kramer (19), may also be seen. The crazes of the new type (II) found in oriented PC completely cover the material giving rise to an increase of SAXS by two orders of magnitude. In addition, the optical properties of the sample change considerably. Irradiation by white light causes those samples crazed at 119 ^{0}C to appear blue in the direction of backward scattering whereas the samples stretched at 129 ^{0}C show what is usually called stress whitening. The blue color may be explained by the fact that the SAXS maximum falls into an s-range accessible to wide angle light scattering. On the other hand, the white color may be attributed to the intensive light scattering ability of the crazes themselves which, produced at the higher temperature of 129 ^{0}C, have critical dimensions for light scattering. These results show that under the drawing conditions used in these investigations stress whitening may be attributed to this new craze phenomenon (22).

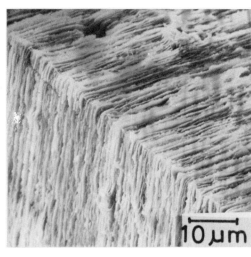 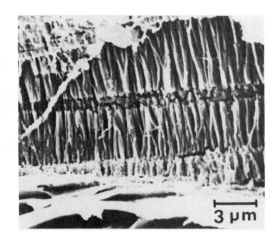

a)   b)

Fig. 1.   Crazes of the new type II in polycarbonate (22,27).

a) specimen edge   b) craze micro-structure

Other important studies in recent years concerned the relation between creep deformation and crazing (28), neck propagation and craze growth (29), and between entanglement density fluctuations and craze concentration (30). A very promising theory of craze nucleation through instable deformation as a consequence of local stress fluctuations was proposed by Fischer and his collaborators (14,21,31). The transitions between crazing, fracture and yield under hydrostatic pressure have been investigated by Duckett (32).

To a large extent *mechanisms* of craze formation are discussed in the above mentioned publications (1a,18-32). There is agreement that a glassy amorphous matrix can be considered as a random aggregat of coiled segments between junction points (Fig. 2). Such a matrix of arbitrarily oriented stiff molecular segments evidently shows a considerable fluctuation of the local compliance which is particularly large where segments are loaded in a direction perpendicular to their orientation. If such a "weak" volume element is stressed sufficiently the stiff chain segments will undergo a collective and instable rotational reorganization leading to a preferential orientation of the junction-point vectors in the direction of applied stress. The first step of craze initiation can be understood on this basis. In a (uniaxially) loaded glassy matrix "weak volume elements" in the neighborhood of stress-concentrators are activated. This will occur at a *mean stress*, σ, smaller than the general yield stress, σ_y. A weak volume element then breaks up in the described manner and under creation of additional free volume. The break-up is temporarily stopped through strain hardening of the orienting segments and because of local stress relaxation.

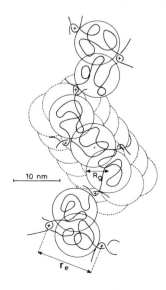

Fig. 2.   Arrangement of entanglements in an amorphous polymer.

The mechanism accounts very well for all experimental observations reported thus far:

- independence of craze initiation stress from molecular weight above $M > 2M_c$
- fibril formation according to the Argon meniscus instability
- increased stresses at a craze-tip
- restriction of plastic deformation to the immediate vicinity of the craze tip, absence of large scale plastic yielding or nucleation of isolated voids
- formation of a mid-rib
- dependence of fibril diameter on fibrillation stress
- increase in fibrillation stress with preorientation
- absence of crazes in fully oriented samples.

Craze formation in heterogeneous polymers has also been the subject of numerous books (e.g. 33), papers (e.g. 34-41) or presentations (e.g. 42-43). Problems treated in this area are compatibility, segregation and interaction across phase boundaries (41,42), stress distribution around rubber particles (35-36), the break-up of glassy or rubbery inclusions in multiphase polymers (36) and the effect on fracture or impact energy (34,37-41).

Another area of intense research activities is the environmental stress cracking (ESC) of polymers. A review article has been published by Kramer (18). Some individual papers on crack propagation in low (44-49) and high (50-55) density polyethylene and on crazing of PS (56,57), PVC (58), PMMA (59-60), PC (61,62) and ABS (63) in liquid environment may be added. These authors explained their data using the well known fracture mechanics concept for crack growth (44,48,51,57, see 1b and 18 for earlier references), a theory of stress-assisted activation of thermal segmental motion (46,55) and, mostly, an analysis of the wetting, spreading, flow and diffusion behavior of the liquid at the crack tip and/or within the capillaries opened through the crazes (47-54,56,57,59-63).

As an example an ESC phenomenon will be discussed which had been observed independently by Shanahan and Schultz (47-48) and Jud and Kausch (64), namely the appearance of three different zones of behavior in the static loading of low density polyethylene in an active environment (Fig. 3).

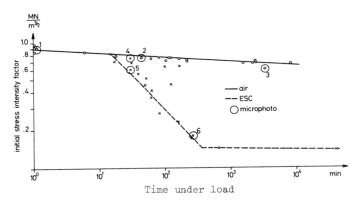

Fig. 3. Initial stress intensity factors of notched three-point bending specimens of LDPE (Lupolen R 1800M, MFI 190/2.16=6-8) as a function of time-to-break (ESC agent: Jaixa Triclean).

Possible mechanisms for the drastic acceleration of sample breakdown at intermediate K-values are:

- increase in the rate of rearrangements due to diffusional interpenetration and plasticization of the polymer matrix by the liquid
- reduction in critical stress levels for craze initiation
- reduction of the resistance of an initiated craze due to a reduced extensibility of craze material.

The experiments (64,65) have shown that

- the time-to-fracture, t_b, *in air* increases exponentially with decreasing K_{Io} as also observed for glassy polymers (PMMA, PC)
- fracture *in air* occurs in the direction of highest tensile stresses in a brittle manner (Fig. 4.1)
- in presence of the stress cracking agent (SCA) three zones of behavior are distinguished:

  1) an incubation period of $\simeq$ 15 min at high values of K during which the SCA has no apparent

effect
2) at times t_b > 15 min and high values of K considerable ductile deformation and a crack branching occur (Fig. 4.2)
3) after long times and at accordingly low K-values fracture is brittle again (Fig. 4.3)

- pre-soaking of specimens in SCA is without effect
- in zones 2) and 3) a delayed application of SCA to a loaded specimen leads to an increase in bending rate after immersion in SCA (Fig. 5).

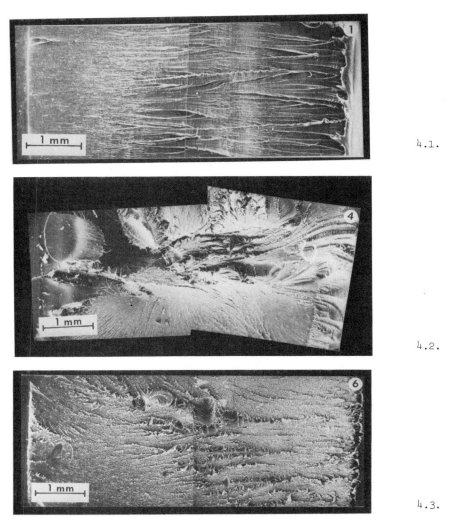

4.1.

4.2.

4.3.

Fig. 4.1-3. Microphotos of specimen surfaces as indicated in Fig. 3.

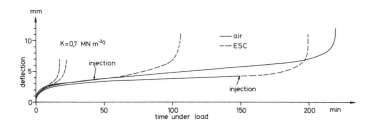

Fig. 5.   Deflection of notched three-point bending LDPE specimens; times-to-
          break after delayed injection of SCA (Jaixa Triclean) are increased,
          probably due to crack blunting.

Despite the *apparent coincidence* with recent tensile ESC experiments on LDPE of Shanahan and Schultz (47,48), there are two differences. Using different polydimethylsiloxanes as SCA, those authors observed incubation periods of less than a minute within which failure occurred by *necking*; at intermediate stresses they noted that t_b was *governed* principally by the speed of penetration of one SCA within a growing stress crack; crack surfaces were rather smooth.

However, the *morphological features* shown in Figs. 4.1 to 4.3 agree to some extent with those of Bandyopadhyaj and Brown (65) on LDPE DEN specimens who found that high-stress failure *in air* occurs by crack propagation involving the formation of striations by local cold drawing and necking, high-stress ESC by nucleation and growth of holes followed by viscoelastic deformation, and low-stress ESC by interlamellar failure without cold drawing.

Attention should also be drawn to the observation of Andrews et al. (67) of the three-stage-fatigue behavior of LDPE with very similar characteristics: brittle trans-spherulitic failure at low K, micro-ductile failure at high K, and a transition zone at intermediate K-values. The ESC experiments described in Figs. 3 to 5 can be understood on the basis of a stress-activated diffusion of stress cracking agent into interlamellar regions. The mode of failure is determined by the state of stress and material resistance: it occurs by local drawing of the practically unplasticized sample (t_b < 15 min), in a mixed mode involving large-scale plastic deformation eventually including void formation, brittle fracture and crack branching at high-stress ESC (t_b < 15 min), and by interlamellar failure (chain disentanglement or rupture) at low-stress ESC (t_b >> 15 min).

HOMOGENEOUS DAMAGE FORMATION

Macromolecules cohere through the van der Waals attraction (secondary bonds), they are tied together by entanglements, crystalline regions and/or cross-links. In isotropic thermoplastic materials the secondary bonds are the first to be affected in the loading and deformation. If, therefore, one wishes primarily to study (or make use of) the strength of the covalent or primary bonds one has to work either with highly cross-linked resins or with highly oriented thermoplastic networks. High degrees of chain orientation are found in commercial fibers, even higher ones in ultradrawn polymers. Whereas the structure of fibers has been studied from the very beginning of polymer science, it was only in the middle sixties that the role of chain stretching and scission in deformation and failure of fibers was clearly recognized and studied. For about ten years there was an intense activity in this area, especially in Leningrad, Salt Lake City, Darmstadt, London and Sapporo. In these studies two principal techniques were used to identify and "count" chain scission events: electron spin resonance (ESR) and infrared spectroscopy (IR). The experimental results obtained have been analyzed in detail in the author's monograph on polymer fracture (1) in 1978. By coincidence not very many new papers on this subject have appeared since.

The main question not clearly resolved in 1978 was the discrepancy between the number of broken chains determined respectively by ESR, IR, viscosimetry, and from the energy dissipation in successive stress-strain cycles. In addition to earlier investigators (see 1c for references) Gaur (68), Kausch (69), Klinkenberg (70), Stoeckel et al. (71), Popli et al. (72), Wool (73), DeVries et al. (74) and Wendorff (75) have attacked this problem using different methods. In order to account for the irreversible energy dissipation in strained fibers the number of free radicals formed should have been by a factor f_c of between 7 and 40 times larger than actually observed. To explain such a discrepancy while basically maintaining the well established fiber model, one is forced to make additional assumptions:

- Either the number of broken chains is systematically larger by f_c than the number of observed radicals, e.g., due to a Zakrewskii-mechanism (see 1d). Careful molecular weight measurements of strained PA 6, PA 66, PETP, PP and PE fibers (71,72,75) gave no convincing evidence, however, that such a hypothesis is true.

- Or the breakage of N_1 chains in a particular volume element V_1 leads to the unloading of $f_c N_1$ extended chains outside of V_1; this amounts to the consideration of a dominant sub-microfibrillar structure.

- Provision must be made to account for the anelastic deformation of the *non-extended chains*. The stress-induced change of conformation of a tie segment profoundly reduces axial chain stresses. In subjecting a fiber to several stress-strain cycles a large fraction of the more extended conformations formed during the first cycle are present at the beginning of the second cycle. Stress-induced "irreversible" conformational changes are considered to be mostly responsible for the differences between the first and the following stretching cycles (69,76).

It is the common observation of the above authors (1c,70,71,75) that qualitatively the slopes of stress-strain and of radical concentration-strain curves correspond. The interpretation that 20 to 40 times as many chains break as free radicals are observed appears to be rather strong in view of the fact that no correspondingly strong decrease of the molecular weight has been reported and that the load bearing capability of the fiber material outside of the immediate fracture zone does not necessarily suffer. The qualitative agreement between the slopes of stress-strain and free radical-strain curves can be sought in an amplification effect accompanying chain scission. The stress-transfer to neighboring segments, the rapid release of elastic energy stored within the breaking chain and the ensuing local temperature rise will facilitate the irreversible slippage and the extension of kinked chains under annihilation of kinks. This is practically equivalent to unloading those chains without breaking them. Within the framework of this model one would have to conclude that the breakage of one amorphous segment at maximum load $q\psi_b$ leads to conformational changes in the surrounding segments of the same microfibrillar region resulting in local load decreases of $40\ q\psi_b$. In view of the fact that on the average there are more than 2000 surrounding segments (of 5 nm length in any amorphous region 400 nm^2 in cross-section) the value of $40\ q\psi_b$ appears very reasonable.

The above hypothesis that chain scission is well correlated with irreversible deformation (69) has meanwhile been confirmed (68,75). With highly oriented PA 6 yarn (from Enka Glanzstoff GmbH) strained to between 10 and 20 % of *total strain* Gaur (68) found about 3 to $4\cdot10^{17}$, Frank and Wendorff (75) about $4\cdot10^{17}$ spins cm^{-3} per percent of *irreversible* strain.

Frank and Wendorff (75) and also Stoeckel et al. (71) indicate that the formation of free radicals in, as received, highly oriented PA 6 yarn is *not* accompanied by submicrocracks (their concentration under the above conditions would be smaller than $10^{13}\ cm^{-3}$).

From an analysis of the structure and the chain scission kinetics of mostly polyamide fibers it must be concluded

- that chain scission is not a consequence of *inter*fibrillar slip, and

- that the individual chain scission event triggers irreversible deformation mechanisms in its surroundings but does not give rise to a chain reaction or to microvoid formation.

Whereas the discrepancy between the ESR investigations and the mechanical measurements with polyamide fibers thus seem to be resolved, the discrepancies between the numbers of broken chains determined by ESR and IR respectively persist. Using highly sensitive ESR and FTIR spectrometers DeVries et al. (74) showed that in γ-irradiated HDPE about 10 carbonyl and 2 carbon-carbon double bonds were formed per free radical.

This classical domain of ESR, the study of polymer degradation by UV and γ-irradiation and of mechano-chemical reactions will not be further treated in this paper. Reference should be made, however, to what appears to be the first free radicals in a stretched halogenated polymer. Florin (77) detected them in vinylidene copolymer fibers using a scavenger method based on chloranyl.

Whereas in PA 6 fibers no *microvoids* after deformation could be detected, Wendorff (78) reports their formation in POM strained beyond yield. Their calculated energy content accounts well for the dissipated energy in a stress-strain cycle.

CRACK PROPAGATION

In this section heterogeneous failure will be discussed, that is break-down through the formation and/or propagation of a crack. The molecular mechanisms available as a whole for material separation in the presence of a crack are, of course, the same as those in a continuous material, i.e. slip and disentanglement of segments, chain scission, and opening-up of voids. However, the presence of a crack heavily influences the level and the local state of stress, it amplifies stress gradients, strains and strain rates; this in turn will favor chain stretching and scission and generally reduce plastic deformation thus leading to an apparent embrittlement of the material.

The rupture of a loaded polymer sample is, therefore, a complex phenomenon. It has been investigated for a long time and described by statistical, continuum mechanical, and rate process theories and by the means of fracture mechanics. All these theories have been discussed extensively by the author in 1978 (1). Since then further progress has been made in the *stochastic treatment* of the strength and fatigue of fiber bundles (79) and in the analysis of *multiaxial states of molecular orientation* and their effect on the ultimate properties (80). A discussion of *kinetic fracture theories* was given by Sacher (81). In these approaches

(79-81) no particular attention has been given to the kinetics of the formation and/or growth of the critical crack. This reduces the predictive value of these theories with respect to unknown regions of stress, temperature and lifetime. For any dependable extrapolation of stress-lifetime curves, information as to the molecular nature of the failure mechanisms and as to the local state of stress is necessary.

It is the aim of the *fracture mechanics approach* to failure to provide the latter information, i.e. give an analysis of the states of stress and deformation at a crack tip, to determine the values of stress intensity factor, K, and energy release rate, G, and to define the critical conditions for rapid crack propagation. An excellent general treatment of these subjects can be found in the recent (second) edition of Williams' book (82). The impact performance of polymers has been reviewed by Reed (83).

Fracture mechanics has been applied to selected individual problems such as thick-walled structural elements or pipes of PE, PMMA (84), or PVC (85), the comparison of different methods employed to determine G_{Ic} of bulk resins and of bonded adhesive layers (86), the blunting of cracks prior to the onset of crack propagation in modified epoxy resins (87), the thermal (88) and mechanical (89-92) dissipation of energy at the tip of a (rapidly) propagating crack, the small scale yielding of PMMA crack tips (93), and the influence of chain slip and disentanglement on G_c of low molecular weight polystyrene (94). Several authors have analyzed the impact behavior of polymers explaining their respective results in the case of PC in terms of notch-geometry, thermal pre-treatment and molecular weight (95) and of the secondary transitions (96), in the case of PMMA and PE in terms of crack blunting through adiabatic heating (92).

The mutual relations between the molecular structure of various epoxy resin systems, their cohesive energy density, their glass transition temperature, their impact resistance and their adhesive shear strength were extensively investigated by Kreibich et al. (97) and Fischer et al. (98). The fracture behavior of thermosetting resins in general has been summarized by Young (99).

Two special conditions of crack propagation not treated in the above references (80-99) are generally considered separately: the *fatigue crack growth* and the *subcritical crack growth in longterm loading*. Nothing will be said in this paper about the fatigue of polymers since an extensive review article by Sauer et al. and a book by Hertzberg and Manson have just appeared (100) and a collection of recent individual papers is available in *Advances in Fracture Research* (101).

An interesting fracture mechnics study on long-term strength of PVC and PMMA was recently published by Döll and Könczöl (102). These authors determined an initial crack size, a_o, calculated from the known time-to-fracture, t_b, and the rate, $v(K)$, of crack propagation according to

$$t_b = \int_{a_o}^{a_t} \frac{da}{v} .$$

For PMMA they obtained 70-100 μm at 23 °C, 100-140 μm at 50 °C, and 114-160 μm at 60 °C. They were also able to identify on the fracture surfaces of a large number of their specimens semi-circular zones from which the final crack had originated. These zones, probably former crazes, had radii of 79 ± 31 μm (23 °C) and 108 ± 55 μm (50 °C) respectively, in exact correspondence with the calculated values (102).

This section on fracture surfaces may be amended by some references on the degradation and/or deformation of polymer surfaces due to friction and wear (103), repetitive impact and sliding (104), fretting (105), or cutting (106). The physics and chemistry of grinding and comminution of polymers has been extensively reviewed by Casale and Porter (107).

CRACK HEALING

Crack healing or, more generally, the jointing of two formerly separated surfaces constitutes an important technical problem. This is convincingly evidenced by the large body of literature on *adhesion* and *welding* (see e.g. 108, 109). Besides these heterogeneous healing phenomena homogeneous healing must be mentioned such as *dimensional recovery* or restoration of the *work of deformation*. The latter subjects have been reviewed quite recently by Wool (110) and Kausch (111) who pointed out that the driving forces towards restoration of the initial, undamaged state are the contraction of elastically strained chains, fibrils and lamellae, the entropic contraction of chains, the reduction of surface free energy and the interdiffusion of molecular chains. In welding and in homogeneous healing it has been very difficult so far to separate quantiatively the contribution of chain diffusion from that of the other mechanisms. The situation is different, however, if the build-up of mechanical strength across an interface is considered.

The systematic studies of Bister et al. (112) showed that the tensile stress, σ, required to separate two polyisobutylene surfaces increases linerally within certain periods as a function of the square root of contact time, t_p. They observed four of these linear regions which they associated with

  I   the formation of physical contact

 II   the interpenetration of chain ends into the opposite matrix (tack)

III   the formation of entanglements after further interpenetration

 IV   the gradual disappearance of all traces of the former interface (112).

Bister et al. determined the influence of contact pressure during the first phase and of sample molecular weight and temperature on the diffusion coefficients and on tack and equilibrium adhesive strength, σ(∞). They pointed out that through the use of bimodal molecular weight distributions good tack can be combined with a satisfactory σ(∞).

Investigations on the rehealing of *isolated cracks* in thermoplastic materials have so far only been reported by Jud and Kausch (112-114) but (unpublished) work is also going on in Urbana (115) and Darmstadt (116). The method of crack healing has been described elsewhere (113, 114); it relies on the fact that the stress intensity factor, K_{Ii}, determined with compact tension (CT) specimens is a measure of the strength of the region through which the crack travels. If this region happens to be the former interface between the two fracture surfaces of a previously broken and then rehealed CT specimen, then K_{Ii} is a quantitative measure of the extent of crack healing. Characteristic results of such measurements are shown in Fig. 6. These data have been explained by a diffusion model. It has been assumed that the fracture surface energy, G_c, and the number of physical links per unit surface area, $n(t) = AN(t)$ are proportional to the average depth of penetration, Δx, of molecules. If Δx is derived from the Einstein diffusion equation then

$$<\Delta x^2(t)> = 2Dt$$

and this gives

$$\frac{G_c(t)}{G_{Io}} = \frac{n(t)}{n_o} = \left(\frac{2Dt}{<\Delta x^2(\tau_o)>}\right)^{\frac{1}{2}} \tag{1}$$

where $n_o = n(\tau_o)$ is that number of physical links per unit surface area which is necessary to establish full short-time strength.

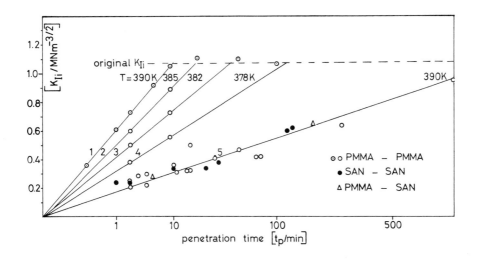

Fig. 6. Fracture toughness vs. $t_p^{\frac{1}{4}}$ for PMMA 7 H (Röhm GmbH).

Curves 1-4 PMMA as received, fracture surfaces are contacted for rehealing.

Curve   5 PMMA and SAN with polished surfaces in contact.

The important question of the molecular nature of the diffusion process was further approached by an investigation of the rehealing behavior of a partially cross-linked polymer, styrene acrylonitrile (SAN), Luran$^{(R)}$ 368 R co-polymer from BASF. The fracture mechanics results are represented in Fig. 7. It is readily noted that the initially linear increase of K_{Ii}^2 with $t_p^{\frac{1}{2}}$ is halted at intermediate K-values, the increase continues at a much smaller slope after having passed through a plateau region. Complete rehealing of the irradiated samples was achieved except for the most highly cross-linked specimens.

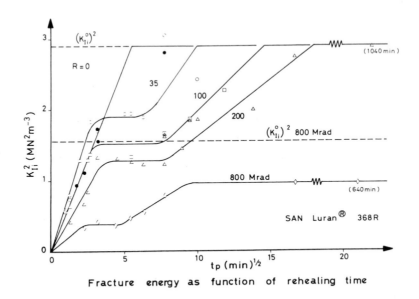

Fracture energy as function of rehealing time

Fig. 7. K_{Ii}^2 vs. $t_p^{\frac{1}{2}}$ for γ-irradiated SAN (parameter: dose R of γ-irradiation).

Evidently two different molecular mechanisms are involved in the healing of fracture surfaces of cross-linked thermoplastics; the first, more rapid but less effective, might be associated with the presence of a larger concentration of chain ends in a surface layer produced during the preceding fracture process. The second is associated with the mobile part within a sub-surface region of the matrix which has undergone some plastic deformation during the first fracture (chains containing none or just one or two cross-linking points). Such a two-layer interdiffusion model does account for the two different linear slopes of the K_{Ii}^2 vs. $t^{\frac{1}{2}}$ diagram. It does not account, however, for a distinct plateau region (117). In view of the large scatter of the fracture mechanics data points further experiments are presently being carried out in Lausanne to ascertain the existence of the plateau region and to identify its molecular significance.

The fact that K_{Ii}^2 depends linearly on $t_p^{\frac{1}{2}}$ indicates that the fracture strength is built up by a mechanism of diffusion. By changing the rehealing temperature, an activation energy of 274 kJ mol^{-2} is determined for this mechanism (114). In order to obtain absolute values of the diffusion coefficient it is necessary to have additional information. Two attempts have been made to obtain such information, a first by referring to the theoretical calculations of De Gennes (118,119), Doi and Edwards (120) and Graessley (121), a second by measuring directly the diffusion of SAN molecules in a PMMA matrix by infrared internal reflection spectroscopy.

Considering the concentration of chain ends in the contacting surface layers, the radius of gyration, R_g, of the molecules and the displacement of their centers of gravity in the course of a reptational chain motion Equation (1) is confirmed.

From the formula of Graessley (121) the diffusion coefficient has been calculated at a temperature of 390 K for PMMA and SAN chains of molecular weight 120'000 to be $1 \cdot 10^{-21}$ m^{2}s^{-1} (114). With this value a distance $\Delta x(\tau_o)$ of interpenetration of 2.5 nm is obtained for "fully rehealed" samples. The fracture mechanics method has thus proved to be a very sensitive tool for the measurement of chain diffusion, the basic mechanism in the rehealing of cracks.

ACKNOWLEDGEMENTS

The author would like to thank his colleagues who have kindly made available their newest, often unpublished results, and his collaborators for their constructive contributions.

REFERENCES

1. H.H. Kausch, Polymer Fracture, Springer, Heidelberg (1978),
   a) p. 272, b) p. 290, c) p. 189, d) p. 194.
2. P.G. De Gennes, Main Lecture, this volume.
3. H. Sillescu, Main Lecture, this volume.
4. E.W. Fischer, Main Lecture, this volume.
5. I.M. Ward, Main Lecture, this volume.
6. L.C.E. Struik, Main Lecture, this volume.
7. J. Haussy, J.P. Cavrot, B. Escaig and J.M. Lefebvre, J. Polym. Sci., Polym. Phys. Ed., 18, 311-325 (1980).
8. J.P. Cavrot, J. Haussy, J.M. Lefebvre and B. Escaig, Mat. Sci. & Eng. 36, 95-103 (1978).
9. J.-C. Bauwens, J. Mat. Sci. 13, 1443-1448 (1978).
10. W. Huang and J.J. Aklonis, J. Macromol. Sci.-Phys. 15(1), 45-62 (1978).
11. A.H. Lepie and A. Adicoff, J. Appl. Polym. Sci. 23, 2169-2178 (1979).
12. J. Kubat, R. Selden and M. Rigdahl, J. Appl. Polym. Sci. 22, 1715-1723 (1978).
13. R.N. Haward, Europhys. Conf. Abstr. 4A, 33-38 (1980).
14. A.S. Argon and M.I. Bessanov, Phil. Mag. 35/4, 917-933 (1977).
15. T. Pakula and E.W. Fischer, Europhys. Conf. Abstr. 4A, 39-45 (1980).
16. H.W. Siesler and K. Holland-Moritz, Infrared and Raman Spectroscopy of Polymers, Dekker, New York (1980).
17. Rheology, Vol. 1-3, Astarita, Marrucci and Nicolais ed., Plenum, New York (1980).
18. E.J. Kramer in Developments in Polymer Fracture, Vol. 1, p. 55, E.H. Andrews ed., Appl. Science Publ., London (1979).
19. B.D. Lauterwasser and E.J. Kramer, Phil. Mag. A 39/4, 469-495 (1979).
20. H.R. Brown and E.J. Kramer, Report No. 4253, Mat. Sci. Center, Cornell Univ. Ithaca, New York (1980).
21. E. Paredes and E.W. Fischer, Makromol. Chem. 180, 2707 (1979).
22. M. Dettenmaier and H.H. Kausch, Polymer 21, 1232-1234 (1980).
23. G.W. Weidmann and W. Döll, Int. Journ. of Fract. 14, R189-193 (1978).
24. W. Döll, M.G. Schinker and L. Könczöl, Int. Journ. of Fract. 15, R145-149 (1979).
25. W. Döll, U. Seidelmann and L. Könczöl, J. Mat. Sci. 15, Letters, 2389-2394 (1980).
26. S.J. Israel, E.L. Thomas and W.W. Gerberich, J. Mat. Sci. 14, 2128-2138 (1979).
27. H.H. Kausch and M. Dettenmaier, Polymer Bulletin 3, 565-570 (1980).
28. M. Kitagawa and M. Kawagoe, J. Pol. Sci., Polym. Phys. Ed., 17, 663-672 (1979).
29. N. Verheulpen-Heymans, Polymer 21, 97-102 (1980).
30. J.F. Fellers and D.C. Huang, J. Appl. Polym. Sci. 23, 2315-2326 (1979)
31. E.W. Fischer, Verhandlungen der Deutsch. Phys. Ges. 2/81, Frühjahrstagung Marburg, Physik der Hochpolymeren (1981).
32. R.A. Duckett, J. Mat. Sci. 15, 2471 (1980).
33. Polymer Blends, D.R. Paul, S. Newman ed., Academic Press, New York (1978).
34. M.A. Maxwell and A.F. Yee, Report No. 79CRD258, General Electric Techn. Inf. Series 1, Schenectady, New York (1979).
35. H. Breuer and J. Stabenow, Angew. Makromol. Chem. 78, 45-65 (1979).
36. T. Ricco, A. Pavan and F. Danusso, Polymer 20, 367 (1979).
37. G.H. Michler, Plaste & Kautschuk 26/9, 497-501, 26/10, 680-684 (1979).
38. B. Carlowitz, Kunststoffe 70/7, 405 (1980).
39. R. Pixa, Thèse d'Etat, Université Louis Pasteur, Strasbourg (1980).
40. L.V. Newmann and J.G. Williams, J. Mat. Sci. 15, 773-780 (1980).
41. A.S. Argon, R.E. Cohen, B.Z. Jang and J.B. Vander Sande, J. Polym. Sci., Phys. Ed., 19, 253-272 (1981).
42. R. Kambour, Seminar on Compatibility and Fracture of Blends of Polyphenylene Oxide/Polystyrene and their Brominated Derivatives, EPFL, Lausanne (1981).
43. A.S. Argon, F.S. Bates, R.E. Cohen. O. Gebizlioglu, B.Z. Jang and C. Schwier, 52nd Annual Meeting Abstracts, Soc. of Rheology, Williamsburg, Virginia (1981).
44. H.A. El-Hakeem and L.E. Culver, J. Appl. Polym. Sci. 22, 2691-2699 (1978).
45. A. Lustiger, R.D. Corneliussen and M.R. Kantz, Mat. Sci. & Eng. 33, 117-123 (1978).
46. Y. Ohde and H. Okamoto, J. Mat. Sci. 15, 1539-1546 (1980).
47. M.E.R. Shanahan and J. Schultz, J. Polym. Sci., Polym. Phys. Ed., 16/5, 803-812 (1978).
48. M.E.R. Shanahan and J. Schultz, J. Polym. Sci., Polym. Phys. Ed., 18, 1747-1752 (1980).
49. P.L. Soni and P.H. Geil, J. Appl. Polym. Sci. 23, 1167-1179 (1979).
50. S. Bandyopadhyay and H.R. Brown, Polymer 19, 589-592 (1978).
51. S. Bandyopadhyay and H.R. Brown, Polymer 22, 245-249 (1981).
52. H.R. Brown, Polymer 19, 1186-1188 (1978).

53. D. Roger, J. Owen, R.N. Haward and A. Burbery, Brit. Polym. J. 10, 98-102 (1978).
54. A. Lustiger and R.D. Corneliussen, J. Polym. Sci., Polym. Letters, 17, 269-275 (1979).
55. H. Okamoto and Y. Ohde, Polymer 21, 859-860 (1980).
56. L. Nicolais, E. Drioli, H.B. Hopfenberg and A. Apicella, Polymer 20, 459 (1979).
57. Y.W. Mai and J.G. Williams, J. Mat. Sci. 14, 1933-1940 (1979).
58. J.L.S. Wales, to be published in Polymer.
59. M.I. Hakeem and M.G. Phillips, J. Mat. Sci. 13, 2284-2287 (1978).
60. I.G. Campbell, D. McCammond and C.A. Ward, Polymer 20, 122-125 (1979).
61. J. Miltz, A.T. DiBenedetto and S. Petrie, J. Mat. Sci. 13, 2037-2040 (1978).
62. C.H.M. Jacques and M.G. Wyzgoski, J. Appl. Polym. Sci. 23, 1153-1166 (1979).
63. R.P. Kambour and A.F. Yee, Report No. 80CRD195, General Electric Techn. Inf. Series 1,
    Schenectady, New York (1979).
64. K. Jud and H.H. Kausch, unpublished results, Lausanne (1978).
65. H.H. Kausch and K. Jud, Int. Congress on Fracture, Cannes, April 1981.
66. S. Bandyopadhyay and H.R. Brown, J. Mat. Sci., Letters, 12, 2131-2134 (1977).
67. E.H. Andrews and B.J. Walker, Proc. R. Soc. Lond. A. 325, 57-79 (1971).
68. H.A. Gaur, Coll. & Polym Sci. 256/10, 949-958 (1978).
69. H.H. Kausch, Polym. Eng. & Sci. 19/2, 140-144 (1979).
70. D. Klinkenberg, Progr. Coll. & Polym Sci. 66, 341-354 (1979).
    D. Klinkenberg, Coll. & Polym Sci. 257, 351-364 (1979).
72. R.K. Popli and D.K. Roylance, Molecular Fracture Behavior in Oriented Polymers, MIT-Report,
    June 1980.
73. R.P. Wool, Polym. Eng. & Sci. 20/12, 805-815 (1980).
74. K.L. DeVries and R.H. Smith, Polymer 21, 949-956 (1980).
75. O. Frank and J.H. Wendorff, personal communication, Darmstadt, 1981.
76. V.A. Marichin, Acta Polym. 30, 507-514 (1979).
77. R.E. Florin, J. Polym. Sci., Polym. Phys. Ed., 16, 1877 (1978).
78. J.H. Wendorff, Polymer 21, 553 (1980).
79. S.L. Phoenix, Int. J. Fract. 14, 327-344 (1978).
80. W. Retting, Coll. & Polym. Sci. 259, 52-72 (1981).
81. E. Sacher, J. Macromol. Sci. B15, 171-181 (1978).
82. J.G. Williams, Stress Analysis of Polymers, 2nd (revised) ed., Wiley, New York (1980).
83. P.E. Reed in Developments in Polymer Fracture, Vol. 1, p. 121, E.H. Andrews ed., Appl. Sci.
    Publ., London (1979).
84. U. Meier and A. Rösli, Mat. & Techn. 6/4, 148-158 (1978).
85. A.Y. Darwish, J.F. Mandell and F.J. McGarry, MIT-Res. Report. R81-1, February 1981.
86. R.Y. Ting and R.L. Cottington, J. Appl. Polym. Sci. 25, 1815-1823 (1980).
87. A.J. Kinloch and J.G. Williams, J. Mat. Sci. 15, 987-996 (1980).
88. M. Parvin, Int. J. Fract. 15, 397 (1979).
89. G.M. Bartenew, W.W. Surkow and B.M. Tulinow, Plaste & Kautsch. 25, 197-201 (1978).
90. A. Kobayashi and N. Ohtani, J. Appl. Polym. Sci. 24, 2255-2259 (1979).
91. J.M. Hodgkinson and J.G. Williams, J. Mat. Sci. 16, 50-56 (1981).
92. J.G. Williams and J.M. Hodgkinson, Proc. R. Soc. Lond. A 375, 231-248 (1981).
93. R. Schirrer and C. Goett, Int. Journ. Fract. 16, R133-136 (1980).
94. E.J. Kramer, J. Mat. Sci. 14, 1381-1388 (1978).
95. G.L. Pitman, I.M. Ward and R.A. Duckett, J. Mat. Sci. 13, 2092-2104 (1978).
96. B. Hartmann and G.F. Lee, J. Appl. Polym. Sci. 23, 3639-3650 (1979).
97. U.T. Kreibich and H. Batzer, Angew. Makromol. Chem. 83, 57-112 (1979).
98. M. Fischer, F. Lohse and R. Schmid, Makromol. Chem. 181, 1251-1287 (1980).
99. R.J. Young in Developments in Polymer Fracture, Vol. 1, p. 223, E.H. Andrews ed., Appl.
    Sci. Publ., London (1979).
100. J.A. Sauer and G.C. Richardson, Int. J. Fract. 16, 499-532 (1980).
     R.W. Hertzberg and J.A. Manson, Fatigue of Engineering Plastics, Academic Press, New York
     (1980).
101. Advances in Fracture Research, D. François et al, eds., Pergamon Press, Oxford (1981).
102. W. Döll and L. Könczöl, Kunststoffe 70, 563-567 (1980).
103. E. Hornbogen and K. Schäfer in Fundamentals of Friction and Wear of Materials, D. Rigney
     ed., Am. Soc. Metals, Metals Park (1981).
104. E. Sacher, P.A. Engel and R.G. Bayer, J. Appl. Polym. Sci. 24, 1053-1514 (1979).
105. P.A. Higham, B. Bethune and F.H. Stott, J. Mat. Sci. 12, 2503-2510 (1977).
106. R. Tilgner and W. Spichala, Coll. & Polym Sci. 258, 1211-1216 (1980).
107. A. Casale and R.S. Porter Polymer Stress Reactions, 2 Vol. Academic Press, New York (1979).
108. Adhesion and Adsorption of Polymers, Part A, L.H. Lee ed, Plenum Publ., New York (1980).
109. C.B. Bucknall, I.C. Drinkwater and G.R. Smith, Polym. Eng. Sci. 20, 432 (1980).
110. R.P. Wool in (Ref. 108), p. 341-362.
111. H.H. Kausch, Europhys. Conf. on Macromol. Phys., Leipzig, September 1981.
112. E. Bister, W. Borchard and G. Rehage, Kautschuk & Gummi-Kunstst. 29, 527-531 (1976).

113. K. Jud and H.H. Kausch, Polymer Bulletin 1, 697-707 (1979).
114. K. Jud, H.H. Kausch and J.G. Williams, J. Mat. Sci. 16, 204-210 (1981).
115. R. Wool, personal communication, 1981.
116. J.H. Wendorff, personal communication, 1981.
117. T.Q. Nguyen, H.H. Kausch, K. Jud and M. Dettenmaier, to be published in Polymer.
118. P.G. De Gennes, J. Phys. 36, 1199 (1975).
119. P.G. De Gennes, J. Chem. Phys. 55, 572 (1971).
120. M. Doi and F. Edwards, J. Chem. Soc. Faraday II 74, 1789 (1978).
121. W.W. Graessley, J. Polym. Sci., Polym. Phys. Ed., 18, 27-34 (1980).

POLYELECTROLYTES-COUNTERION INTERACTIONS

Gilbert WEILL

Centre de Recherches sur les Macromolécules - CNRS - and Université Louis-Pasteur - Strasbourg - France

ABSTRACT

Polyelectrolyte-counterion interactions are studied from the point of view of site binding and of the spatial distribution of the counterions. Poisson-Boltzmann and condensation theories are reviewed in this respect. The site binding stoechiometry is reported for divalent paramagnetic counterions with several polyelectrolytes. It is treated quantitatively as a local equilibrium between the polyion sites and the local concentration calculated from Poisson-Boltzmann equation. Experimental approaches to monovalent counterions distribution are also described.

INTRODUCTION

Long range coulombic forces in polyelectrolyte solutions manifest in a number of effects which can be schematically classed according to a spatial scale.

1) The trapping of the counterions and the exclusion of coions in the region of high potential in the vicinity of the polyion is responsible for the low activities and osmotic coefficients and the Donnan exclusion (Ref. 1).

2) The segmental repulsion modifies the chain statistics through local stiffening and excluded volume interactions (Ref. 2).

3) Long range interaction between macromolecules is reflected in the angular dependence of the small angle scattering of neutrons (Ref. 3) and Xrays (Ref. 4). The existence of a peak in S(q) does not however imply the existence of a lattice like order (Ref. 5).

We shall not deal here at all with the third aspect and retain only from the second the fact that the polyion can locally be considered as a rigid filament. We shall examine the electrostatic models of polyion counterion interactions not from the point of view of the global thermodynamic quantities but from the point of view of the experimental information which has has been made recently available on the physical state (hydration, mobility) and localization of the counterions in the vicinity of the polyion. These properties, which go partly beyond the simple modelling of the polymer as a continuous charge distribution and of counterions as point charges, are very important in a number of phenomena such as :

- the selectivity of binding of different counterions, in particular in biological polyelectrolytes

- the catalysis of the reaction of charged species by polyelectrolytes

- the fluctuations in the ionic atmosphere which are responsible for the high dielectric increment of polyelectrolyte solution and may play a major role in the development of large van der Waals interactions between molecules.

More generally, a proper calculation of effects 2) and 3) imply to evaluate the screening of the bare coulombic interaction at all scales, i.e. the exact electric potential and distribution of small counterions and coions.

I shall describe, with no claim for completeness and giving a special weight for the work performed in my laboratory together with Drs. Spegt, Meurer and Karenzi, several experimental approaches to two related problems.

1) the identification of site binding and its interpretation in term of an equilibrium between the polyion charged groups and the local concentration at the surface of the polyion

2) the characterization of the counterion distribution around the polyion.

Since the interpretation of these experiments rely on existing theoretical models and can help
to support their validity, we start with a review on the development of these theories with
special emphasis on the problem of counterion localization.

Polyelectrolyte theories from the point of view of counterion localization

Using a very simple two phase approximation for the potential and counterion concentration
around a cylindrical polyelectrolyte, Oosawa (Ref. 1) has very clearly shown that beyond a
critical linear charge parameter $\lambda > 1$

$$\lambda = \frac{e^2}{kTb} = \frac{L_B}{b} \qquad (1)$$

where L_B is the Bjerrum length, and b the mean distance between charges projected on the polyion axis, any increase in λ results in an accumulation of the corresponding additional counterions in the high potential region. This has been considered as a type of condensation.
While the term suggest a dense phase, the only requirement of the theory is that the volume of
the condensed phase is small as compared to the overall volume. It provides a simple explanation to the reduced activity of the counterions and to the additivity laws for the solution
in the presence of added salt.

More realistic calculations are based on the solution of a non linearized Poisson–Boltzmann
(P.B.) equation with cylindrical symetry (Ref. 6). The polymer is modelled by a continuous
charge distribution at the surface of a cylinder of radius a. It is analytically soluble for
the two extreme cases of no added salt and an excess of added salt and numerically in intermediate situations. Solution at finite concentrations imply an additional assumption of cylindrical symetry, i.e. the choice of a boundary condition at the frontier of an electroneutral
cylindrical subvolume of radius R coaxial to the polyion itself (Ref. 7). Incidentally, the
pure mathematical convenience of the so called "cell model" and its possible success in inter-
pretating the concentration dependence of thermodynamic quantities should not be taken as an
argument for a parallel ordering of cylinders in semi dilute polyelectrolyte solutions, as
sometimes done in the discussion of Xray and neutron scattering experiments (Ref. 4). Nevertheless, the distribution of potential and concentrations can be obtained and the thermodynamic quantities calculated from the concentration at R. As an example, the osmotic coefficient
ϕ in the absence of added salt is givent by :

$$\phi = \frac{1-\beta^2}{2z\lambda} (1-e^{-2\alpha}) \qquad (2)$$

where z is the counterion valence and β and α are calculated from

$$\lambda = \frac{1-\beta^2}{1+\beta \ \coth \ \beta\alpha}$$

$$\alpha = Ln \ \frac{R}{a} = \frac{1}{2} \ Ln \ \frac{10^3}{\pi a^2 b \mathcal{N}} - \frac{1}{2} \ Ln \ Cm$$

with $\mathcal{N}$ Avogadro number and Cm the monomolar concentration in polyelectrolyte. At low concentration, it depends essentially on λ and very weakly from the choice of a. More generally
thermodynamic quantities are insensitive to the details of the distribution of counterions
close to the polyions. It is however interesting to note that for $\lambda > 1$ the concentration of
counterions at a may be of the order of 1 M.

With the emphasis on the thermodynamic quantities and the lack of analytic expressions for the
cases of added salt, as well as for the case of counterions of mixed valency the Manning's
condensation model (Ref. 8 and 9), which provided easy to handle expressions in fair agreement with experiment, has met with a large success. It can be considered as a straightforward
extension of the linearized Debye–Hückel treatment of ordinary electrolytes. The polymer is
taken as continuous linear distribution of charge with a screened Debye–Hückel potential. The
condensation of counterion appears as a necessary reduction of the linear charge down to
$\lambda = z^{-1}$ to eliminate the divergence of a phase integral, related to the strict one dimensional
character of the linear distribution. It becames in a sense a two phase model, which takes
however in consideration the Debye–Hückel residual interaction between the uncondensed conte-
rions and the filament with a remaining charge parameter $\lambda = z^{-1}$ and treats it in the linearized Debye–Hückel approximation. As an example the fraction of uncondensed counterions and
the osmotic coefficient in the limit of infinite dilution are given for $\lambda \geq z^{-1}$ by :

$$f_{uncondensed} = \frac{1}{z\lambda} \ ; \ \phi = \frac{1}{2z\lambda} \qquad (3)$$

It shows that the uncondensed counterion cannot be taken as free. This result coincides with
the PB result at infinite dilution. It is also a very simple matter to derive additivity laws
as well as to predict the behaviour of mixed salts of monovalent and divalent counterions, the
latter one being automatically condensed as long as their condensation reduces λ to a value
higher than 1/2 (Ref. 10).

The development of the condensation hypothesis, and particularly the problem of the localization of the condensed counterions has arisen many controversies, because it always retain the
singularity in the electrostatic energy due to the one dimensionality of the charge distribu-

tion and the a priori use of a Debye-Hückel potential. The fraction of condensed counterion has been rederived from an expression of the free energy which uses this electrostatic energy (which would tend to bind all the counterions) together with an entropy of mixing of the condensed counterions which finally define a volume of condensation Vp (Ref. 11). The procedure and the form of the free energy cover many assumptions (Ref. 12), whose consequences on observable quantities have been much discussed. Among them the treatment of condensation as a binding process "which would not be governed by the law of mass action" since condensation is supposed to happen at infinite dilution, and the fact that the volume of condensation is practically independent of the concentration in polymer and added salt (Ref. 9). It should be moreover noted that :

i) if the condensation concept does not distinguish between different types of binding with different mobilities (site binding or territorial binding) while it should influence considerably the mixing entropy and then the condensation volume

ii) the condensation volume increases with the counterions charge number z, a rather astonishing result from the point of view of the P.B. theories, which has important consequences for the competition between monovalent and divalent counterions (Ref. 13)

iii) in many applications and comparisons of the theory with experiment on real polymers modelled as cylinders, the condensation volume is used as a cylindrical shell of thickness d around the polyion of radius a. As an example for DNA (a = 10 Å, λ = 4,2), one has :

	z = 1	z = 2	z = 3
Vp $\overset{\circ}{A}{}^3$	643	1121	1563
d Å	7	11	14
C M/ℓ	1,97	1,3	0,98

For these many reasons, an intense comparison with the results of P.B. calculations of the concentration profile in the neighbourhood of the charged cylinder has been undertaken. They generally use now as a reference the Gouy-Chapman double layer, i.e. the solution of the Poisson Boltzmann equation for the charged plane, familiar to electro and colloïd chemists. As pointed by Stigter, the reason why polyelectrolytes have developed their own language is that in many applications of the theory of the electrical double layer, the potential is taken as the primary variable while in polyelectrolytes or ionic micelles, the charge is the primary variable. Moreover in colloïd chemistry, the size of the particle is generally large as compared to the double layer thickness. From the work of Stigter (Ref. 14,15) Gueron and Weisbuch (Ref. 16,17,18) and Anderson and Record (Ref. 19) it is clear now that :

i) the shape of the potential at small distance of the cylinder is markedly different from the Debye-Hückel potential used by Manning

ii) the thickness of the cylindrical shell which contains the Manning condensed fraction of counterion plays no special role in the P.B. equation and increases with decreasing polymer and salt concentration

iii) the concentration in the immediate vicinity of the polyion (CIV) is inversely proportional to the square of the polyion radius and proportional to the square of λ at high λ. This contrasts with Manning result for the concentration in the condensation volume which is uniform and independent of a. It is however fair to note that the CIV and the concentration profile between CIV and 0,25 CIV are essentially insensitive to the concentration of polymer and added salt. Half the counterions are in a shell of thickness a λ^{-1} , which could be considered as a thin condensed phase.

In conclusion, it should be remembered that the P.B. equation is itself a mean field approximation while there is some claim that the form of the free energy in the condensation model should be justified even for a cylinder of finite a. An experimental approach to the concentration profile in the immediate vicinity of the polyion may be of help in this controversy.

Site binding

The existence of "site binding" has been mostly inferred from static volume changes as measured directly (Ref. 20) or indirectly using refractive index increment (Ref. 21), or from dynamic measurements such as ultrasonic adsorption measurements related to the displacement of a binding equilibrium involving large volume changes (Ref. 22). Thermodynamic measurements do not allow to distinguish between site binding, i.e. the formation of an inner or outersphere complex with a large perturbation of the hydration shell and a "territorial binding" with essentially no change in the symetry of the hydration shell. The very large preference for divalent ions binding predicted by all electrostatic theories of polyion-counterion binding allow to study in a simple way the stoechiometry of site binding in mixed salts of monovalent and divalent counterions (or more simply upon addition of divalent counterions to a 1-1 polyelectrolyte salt) if a spectroscopic tool give access to modifications in the counterion environment and mobility. This is the case for paramagnetic divalent counterions, using electron paramagnetic resonance or nuclear magnetic resonance (Ref. 23-26).

The broadening of the EPR spectrum of Mn^{++} reveals the assymetry of its environment. Its evolution as a function of the fraction r of divalent counterions with respect to the polyion monomolarity permits to measure the stoechiometry of site binding and to decide wether there is a fast or slow exchange between the site bound and the excess counterions. Similarly the chemical shift and relaxation rates T_1^{-1} and T_2^{-1} of the water protons in solutions containing Co^{++} or Mn^{++} reflect the scalar and dipolar interaction between the paramagnetic ion and the protons of the water molecules in their first hydration shell which are in fast exchange with the bulk water molecules, as well as the change of mobility of the whole system. Thus, measurements of T_1^{-1} and T_2^{-1} as a function of the concentration in Mn^{++} in the presence of polyelectrolytes will indicate the successive physical states of the bound counterions. An analysis of the frequency dependence of these relaxation rates lead to quantitative estimates of the correlation times and of the perturbation of the hydration shell, but does not permit to know wether the change in dipolar and scalar interaction results from a loss of water molecules (formation of an inner sphere complex) or from an increase in the mean distance between the counterion and the water protons (loss of electrostriction in the formation of an outer sphere complex).

Using these techniques we have studied a number of polyelectrolytes with essentially the same charge parameter $\lambda \sim 3$: polyphosphate (PP) polyacrylate (PA) and polystyrene sulfonate (PSS). In table 1 are given in term of the ratio to the corresponding quantities for the free counterion the increase in τ_c, the decrease in the dipolar D and scalar A interaction parameters and the stoechiometry of the binding that we have observed for the three polyions.

TABLE I
Characteristics of divalent ions site binding

	τ_c/τ_c°	D/D_0	A/A_0	Fraction
PP	60	0,1	0,1	$\sim$ 0,5
PA	60	0,2	0,2	0,4-0,5
PSS	2	1	1	0

PSS does not show any site binding, in agreement with previous indications. PP binds very strongly up to a fraction of $\sim$ 0,5. PA binds less strongly up to a value of 0,4-0,5.

The reality of such a binding with dehydration has been recently ascertained using very different type of experiments (Ref. 27). It is based on the reactivity of positronium, a quasi atom made of a positon and an electron with free or complexed ions. The existence of such a reaction is reflected in the decay time of the 0.positronium which can be studied by classical nuclear techniques, using coincidence analysis of the photons emitted at the creation and annihiliation of the positronium. The variation of τ_3^{-1} with Co^{++} in the presence of PSS and PP exactly paralell the NMR results : PSS does not affect the reactivity of Co^{++} while PP inhibits completely its reactivity up to r $\sim$ 0,4-0,5 at 293°. One of the interest of this technique is that it is not limited to polyanions-countercations interactions, since many anions have a measurable reactivity with positronium.

Similarly a study of the catalytic action of polyelectrolytes on the complex formation between nickel and a diazoligand in aqueous solution (Ref. 28) concludes that the accelerating effect of PSS and the inhibitory effect of PP is easily understood if the accelerating effect of an increased concentration in the vicinity of the polyion is counterbalanced in PP by the fact that dehydrated Ni^{++} cannot participate to the reaction. Similar effects of the dehydration of reacting species on polyelectrolyte catalysis have been demonstrated by Ise (Ref. 29).

The point of interest is now to examine if we can predict site binding in the framework of the existing theories of polyelectrolytes, despite the fact that they treat counterions as point charges. It has been suggested (Ref. 16 and 20) that site binding should be treated as an equilibrium between the charged sites on the polymer and the local concentration in counterions:

$$\text{Free site} + [M^{2+}] \underset{\leftarrow}{\overset{K}{\rightharpoonup}} \text{complexed site}$$

Denoting by x the fraction of free sites, and by K the equilibrium constant which contain all the details of the specific interaction

$$K = \frac{1-x}{x \, [\, CIV_2]} \qquad (4)$$

The CIV_2 must now be calculated for a polyion of reduced charge λx in the presence of a fraction $r-(1-x)$ of divalent counterions.

The concentration CIV_2 can be calculated for the P.B. model and for the Manning model. Numerical calculations of the P.B. model for PA with a = 4 Å and λ = 3 give

$$CIV_2 = 3 \left[3x - \frac{1}{2}\right]^2 \left[r - (1-x)\right] \qquad (5)$$

This result seems to be a generalization of the result of Anderson and Record (Ref. 19), for monovalent counterions in the absence of site binding, which can be written here :

$$CIV_2 = \frac{48}{a^2} [\lambda - \frac{1}{z}]^2 \; r \qquad (6)$$

Introducing relation (5) in equation (4) one has an equation which can be solved numerically for given values of K. The curves in Fig. 1 show that a value of $K \sim 6,4$ is in good agreement with the stoechiometry of site binding

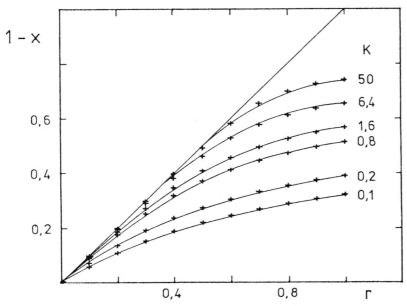

Fig. 1 : Fraction of complexed sites as a function of r for different equilibrium constants K.
Polyacrylic acid $\lambda = 3$ $a = 4 \; \AA$

In contrast with this results, we have found that the total fraction of condensed divalent counterions, calculated according to the condensation volume hypothesis, is smaller than the observed site bound fraction. The validity of the initial hypothesis has now to be proved by showing that the binding constant K can be transferred to variable CIV_2 obtained for example by a change in the degree of ionization α. It is difficult to observe the change of site binding stoechiometry, since the neutralisation curves of polyelectrolytes are a function of the nature of the counterions. For this reason we have performed measurements of the relaxation rates for very small values of r between 0,1 at $\alpha = 0,1$ and 0,01 at $\alpha = 1$. In Fig. 2 and 3 are shown the frequency variations of T_1^{-1} observed at different α (which clearly show the transition between free and site bound ions around $\alpha = 0,3$) and the fraction of condensed counterion calculated using a charge parameter $\alpha\lambda$ and a binding constant $K = 6,4$. The agreement is very satisfactory and shows that if K is known, site binding can be predicted on the basis of the CIV hypothesis. Recent experiments with PSS and polyethylenesulfonate with very different values of a seem to generalize this result for the sulfonate group.

It is interesting to note that a value of K measured with a polyelectrolyte of known structure should be transferable to other charged interfaces such as ionic micelles or monolayers.

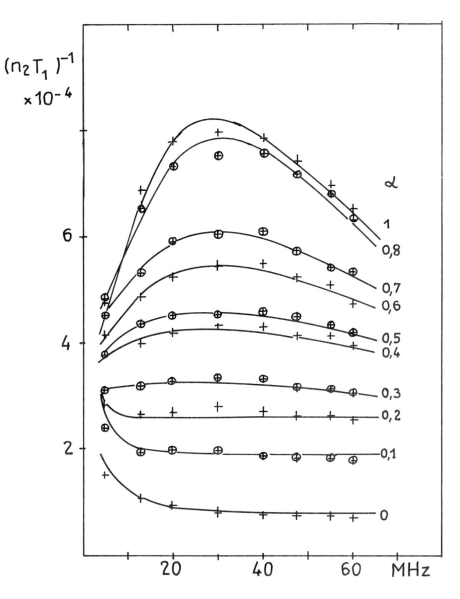

Fig. 2 : Variation of the relaxation rate of water in a manganese solution in the presence of
an excess of PA at different degrees of ionization α

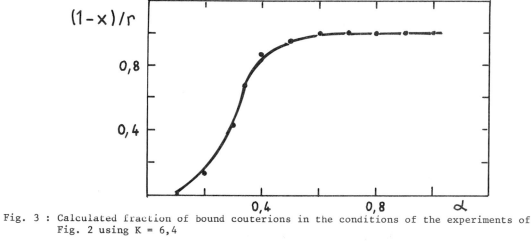

Fig. 3 : Calculated fraction of bound couterions in the conditions of the experiments of
Fig. 2 using K = 6,4

Experimental approaches to the counterion distribution

Several direct and indirect approaches are now available. NMR again offers a possibility, based on the study of the quadrupolar relaxation of nuclei such as ^{23}Na which is dominated by the fluctuations of the electrical field gradient experienced by the nucleus. Leyte and his coworkers (Ref. 30) have proposed that the increased quadrupolar relaxation of ^{23}Na in Na-polyelectrolyte salts as compared to the corresponding micromolecular salt is due to the diffusional motion of the counterions in the region of high potential. Assuming a diffusion around the equipotentials, the additional mechanism of relaxation is shown to be proportional to the square of the radial electric field which can be calculated from the P.B. equation. A number of experiments have been performed on Na$^+$ salts without and with added salt, allowing only a comparison with an average square electric field taken over the whole distribution (Ref. 31). It has nevertheless been shown that in the case of PA, the predicted dependence with the degree of ionization is well obeyed (Ref. 30). Taking advantage of the fact that Na$^+$ replaces quantitatively the bulky tetramethylammonium (TMA) ions in the neighbourhood of the PA polyion (Ref. 21), we have studied ^{23}Na relaxation rates T_1^{-1} as a function of r in mixed salts TMA and Na (Ref. 32). The relaxation parameters of a fraction of counterions between r and r+dr can be obtained from an derivation of rT_i^{-1} ; a Poisson Boltzmann calculation gives its location at a distance ρ from the polyion and the square radial electric field to which it is submitted.

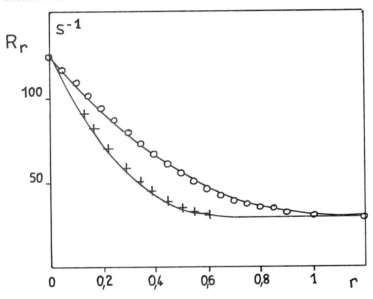

Fig. 4 : Experimental and calculated quadrupolar relaxation of each infinitesimal fraction of ^{23}Na between r and r+dr

Fig. 4 gives the observed and calculated relaxation profile. Noteworthy are the fact that $T_1 = T_2$ which indicates the absence of strong site binding characterized by a long correlation time, the agreement between the measured and calculated value of T^{-1} at $\rho = a$ which indicates a small contribution from local fields due to the discontinuous character of real charges, and the absence of a region of constant T_i at $r < 1/\lambda$ which could be predicted by the condensation model. The small discrepancy in the observed and calculated profile of T_i^{-1} may indicate that there are additional mechanism of relaxation such as the radial diffusion of the counterions.

A more direct evaluation of the counterion profile has been proposed by Oberthur (Ref. 33). It relies on the neutron scattering*of the transverse radius of gyration of the polyelectrolyte molecules, using deuterium labelled counterions such as TMA d$_{12}$. An interesting feature of the method which is at the early stage of development is that the competitive replacement of TMA counterions by other ions can also be studied.

CONCLUSION

The understanding of polyion-counterion interactions which has for a long time been based on the interpretation of thermodynamic measurements benefits now of a number of new experimental methods which directly probe the charged interface. They stress the fact that the model of a condensed phase which can be considered as a useful approximation for the calculation of thermodynamic quantities, fails to interpret the local distribution of counterions which is satisfactorily evaluated from a P.B. cylindrical model.

* measurement

REFERENCES

1. F.Oosawa, Polyelectrolytes Marcel Dekker, New-York (1971)
2. T.Odijk and A.C. Houwaart, J. Pol. Sci. (Pol. Phys. Ed.), 16, 627-639 (1978)
3. M.Moan, J. Appl. Crystallogr. 11, 519-523 (1978)
4. N.Ise, T.Okubo, K. Yamamoto, H. Kawai, T. Hashimoto, M. Fujimura and Y. Hiragi, J.Am.Chem. Soc., 102, 7901-7906 (1980)
5. J.Hayter, G.Janninck, F.Brochard-Wyart and P.G. de Gennes, J. de Phys. Lettres, 41, L451-L454 (1980)
6. S.Lifson and A. Katchalsky, J.Pol. Sci., 13, 43 (1953)
7. A.Katchalsky, Z.Alexandrowicz and O.Kedem in Chemical Physics of Ionic Solutions, p. 295, B.E. Conway and R.G.Barradas Eds. Wiley, New-York (1966)
8. G.S.Manning, J. Chem. Phys. 51, 924-933, 934-938 and 3249-3252 (1969)
9. G.S.Manning, Ann. Rev. Phys. Chem., 23, 117-140 (1972)
10. G.S.Manning in Polyelectrolytes, E.Selegny, M. Mandel and U.P. Strauss Eds, p. 9, D.Reidel, Dordrecht, Boston 1974
11. G.S.Manning, Quarterly Rev. of Biophys., 11, 179-246 (1978)
12. K.Iwasa, Biophys. Chem., 9, 397-404 (1979)
13. R.W. Wilson, D.C.Rau and V.A. Blommfield, Biophys. J., 30, 317-326 (1980)
14. D.Stigter, J. Am. Chem. Soc., 82, 1603-1606 (1978)
15. D. Stigter, Progr. Colloid and Polymer Sci., 65, 45-52 (1978)
16. M. Gueron and G. Weisbuch, Biopol., 19, 353-382 (1980)
17. M. Gueron and G. Weisbuch, J. Phys. Chem., 88, 1991-1998 (1980)
18. G. Weisbuch and M. Gueron, J. Phys. Chem., 85, 517-525 (1981)
19. C.F. Anderson and M.T. Record, Biophys. Chem., 11, 353-360 (1980)
20. U.P. Strauss and Y.P. Leung, J. Am. Chem. Soc., 87, 1476 (1965)
21. A. Ikegami, J. Pol. Sci., A2, 907-921 (1964)
22. R. Zana, C.Tondre, M. Rinaudo and M. Milas, J. Chim. Phys. Physicohcimie Biol., 68, 1258-1266 (1971)
23. P. Spegt and G. Weill, C.R. Acad. Sci. Paris, 274, 587-590 (1972)
24. P. Spegt, C. Tondre, G. Weill and R. Zana, Biophys. Chem. 1, 55-61 (1978)
25. P. Spegt and G. Weill Biophys. Chem., 4, 143-149 (1976)
26. P. Karenzi, B. Meurer, P. Spegt and G. Weill, Biophys. Chem., 9, 181-194 (1979)
27. R. Zana, S. Millan, J.C. Abbé, G. Duplâtre and J.C. Machado, 27th Int. Symp. on Macromol. Strasbourg 1981, Communication Abstracts, p. 1081
28. C. Tondre, 26th Int. Symp. on Macromol., Mainz 1979, Communication Abstract C2 23 p. 843
29. N. Ise, T. Maruno and T. Okubo, Proc. Roy. Soc., London, A 370, 1-16 (1980)
30. J.J. Van der Klink, L.H. Zuiderweg and J.C. Leyte, J. Chem. Phys., 60, 2391-2399 (1964)
31. H.S. Kielman, J.M.A.M. Van der Hoeven and J.C. Leyte, Biophys. Chem., 4, 103-111 (1976)
32. B. Meurer, P. Spegt and G. Weill, Chem. Phys. Letters, 60, 55-58 (1978)
33. R.C. Oberthür 7th Europhysics Conference on Macromolecular Physics, Strasbourg 1978, Abstract C28

STRUCTURES AND PROPERTIES OF CONDUCTING POLYMERS

James C.W. Chien

Department of Chemistry, Department of Polymer Science and Engineering, Materials Research Laboratories, University of Massachusetts, Amherst, MA 01003 U.S.A

Abstract-The catalyst system used to polymerize acetylene has been characterized. The molecular weight of poly(acetylene) was determined by radio-tagging method. In situ ultra-thin film was prepared and the nascent morphology of the cis-(CH)$_x$, trans-(CH)$_x$, and iodine and AsF$_5$ doped specimen has been obtained. The crystal structures have been determined by electron diffraction for aligned fibrils. A doping mechanism was proposed based on EPR results. The importance of inter-fibril contact and bond alternation has been investigated by the preparation of variable density poly(acetylene) and copolymers of acetylene and methyl acetylene and the study of their properties.

INTRODUCTION

Chemists have long been interested in the electronic structure of conjugated polyenes. In these molecules the $2p_x$ orbitals form bonds in which the charge density is perpendicular to the plane of the molecule. Early theoretical view is that in the limit of infinite chain length all the C-C bonds approached a constant value of 1.38 Å and the material could be intrinsically conducting which is contrary to experimental results. Lounget-Higgins and Salem (1) showed that the uniform infinite chain to be unstable with respect to bond alternation. This result is equivalent to the one-dimensional Peirls instability (2). The electron is thus subjected to a periodic potential with a period twice the original undistorted chain, resulting in a gap at $2k_F$ causing a change of character in the polyene from a metal to a semiconductor. There are other views that the electronic gap of ~2eV is due almost entirely to correlation effects.

The band structure for various 1-D bond alternation is shown in Fig. 1. Figure 2 gives the 3-DF band structure for cis-transoid of (CH)$_x$ (3). It is to be noted that the band width

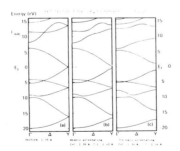

Fig. 1.   1-D band structure for poly(acetylene) with various bond alternations: (a) uniform 1.39 Å; (b) weakly alternating C=C 1.36 Å, C-C 1.43 Å; (c) strongly alternating C=C 1.34 Å, C-C 1.54 Å.

is very large, ca. 8-10eV, in the parallel dirction but very narrow, ca. 0.1eV in the transverse direction. This implies weak interchain coupling, and the material may be regarded as quasi-one-dimensional.

The above considerations correctly relegate pristine (CH)$_x$ to be a weakly semiconducting polymer. The conductivity of undoped trans-(CH)$_x$ is ca. 10^{-5} (Ω cm)$^{-1}$ and that of pristine cis-(CH)$_x$ another four orders of magnitude lower. However, interest in this semi-

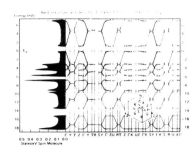

Fig. 2.   3-D band structure for <u>cis</u>-transoid of $(CH)_x$.

conducting polymer has been greatly stimulated by the demonstration (4) that controlled doping increases the electrical conductivity to metallic levels $[\sigma \sim 10^3 (\Omega cm)^{-1}]$.

The range of conductivity of poly(acetylene) is compared with other materials and is shown in Fig. 3.

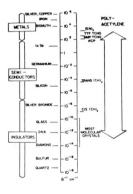

Fig. 3. A comparison of conductivities of various materials.

Compared to metals, doped poly(acetylene) has comparable specific conductivity (Table 1).

TABLE 1.   Conductivities of doped polyactylene and other materials

	Density	$\sigma(\Omega\ cm^{-1})$	$\sigma/d,\ cm^2(\Omega\ g)^{-1}$
Cu	8.92	5.8×10^5	6.5×10^4
Au	19.3	4.1×10^5	2.1×10^4
$(SNBr_{0.4})_x$	2.67	3.8×10^4	1.4×10^4
Fe	7.68	1×10^5	1.3×10^4
$(SN)_x$	2.30	1.7×10^3	7.4×10^2
Hg	13.5	1×10^4	7.4×10^2
$cis[CH(AsF_5)_{0.14}]_x$	0.5	5.6×10^2	1.1×10^3
" aligned	0.5	2.8×10^3	5.6×10^3

There have been numerous spectroscopic and transport studies made to elucidate the mechanism of conduction. However, there is a dearth of knowledge about the chemistry of the system and the detailed morphology and structures of the pristine and doped polymers. In this review, we will concentrate primarily on aspects of the work carried out at the

University of Massachusetts, bringing in contributions of colleagues at other institutions where appropriate. The focus wil be on the more chemical and structural aspects.

PRISTINE POLY(ACETYLENE)

Polymerization

Acetylene can be polymerized by a wide variety of catalyst. However, in order to prepare free-standing poly(acetylene) films of good mechanical and electrical properties, the catalyst of choice is $Ti(\underline{n}-OBu)_4/AlEt_3$. Using this system (5) $(CH)_x$ with ca. 98% cis structure and 100% trans structure can be obtained by polymerization at -78° and +150°C, respectively. The former is red metallic and the latter is blue metallic in color, having room temperature EPR unpaired spin concentrations of 1 per 25,000 CH units and 1 per 1,000 to 3,000 CH units, respectively. We have studied the in-situ polymerization in the EPR cavity (6). Four EPR species were found with g-values: 1(1.981); 2(1.976); 3(1.965); and 4(1.945) as shown in Fig. 4.

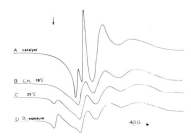

Fig. 4. (A) EPR spectrum of $Ti(\underline{n}-OBu)_4/4AlEt_3$, [Ti]=7 mM at -78°C. (B) Spectrum after admission of acetylene and polymerized for 20 min. (C) Bringing (B) to 25°C under anaerobic conditions or introduction of a few millimeters of O_2 at -78°C for 1 min. (D) Spectrum sample at 25°C and oxygen added. Marker is for DPPH.

Resonances 1 to 3 were significantly reduced in intensity upon the admission of monomer whereas signal 4 is not affected. It is likely the former are the catalytically active species. Hyperfine structures were resolved for species 1 and 3 at room temperature (Fig. 5).

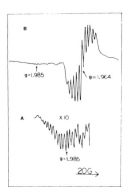

Fig. 5. (A) EPR spectrum of a $Ti(\underline{n}-OBu)_4/AlEt_3$ catalyst solution with added C_2H_2 warmed to room temperature showing 21 hyperfine lines of species 1 obtained at high gain. (B) EPR spectrum of the same sample showing 11 hyperfine lines of species 3 obtained at low gain.

Taking into consideration the hyperfine structure and the theory of g factor, the following structures are proposed for the Ti^{+3} species.

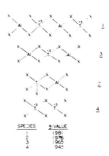

Figure 4 contains one very important piece of information. Whereas pristine <u>cis</u>-(CH)$_x$ film usually exhibits an EPR signal at g=2 and a linewidth of 4-8 G, the in situ polymer is free of a g=2 signal. Therefore, the <u>cis</u>-(CH)$_x$ as prepared is free of defects and probably has a nearly perfect <u>cis</u>-transoid structure. Figure 3 showed that an EPR signal developed upon either warming the polymer to room temperature or exposure to oxygen at -78°C. These will be discussed further below.

Radio-tagging determination of molecular weight.

Poly(acetylene) is neither soluble nor fusible. Therefore the usual methods cannot be applied to determine its molecular weight. We have used radiotagging to obtain this information (7). The method is based on the following reactions.

Catalyst formation:

$$Ti(OBu)_4 + AlEt_3 \longrightarrow Ti(OBu)_3Et + AlEt_2(OBu) \tag{1}$$
$$2\ Ti(OBu)_3Et \longrightarrow Ti(OBu)_3 + C_2H_4 + C_2H_6 \tag{2}$$
$$Ti(OBu)_3 + AlEt_3 \longrightarrow Ti(OBu)_2Et + AlEt_2(OBu) \tag{3}$$

Propagation:

$$Ti(OBu)_2Et + nC_2H_2 \longrightarrow (OBu)_2Ti(C_2H_2)_nEt \tag{4}$$

Chain transfer:

$$(OBu)_2Ti(C_2H_2)_nEt + AlEt_3 \longrightarrow (OBu)_2TiEt + Et_2Al(C_2H_2)_nEt \tag{5}$$

Termination:

(a) Reduction by termination
$$Ti^{3+}R + Ti^{3+}R' \longrightarrow 2Ti^{2+} + R_{-H} + R'H \tag{6}$$
(b) Reduction by alkylation
$$Ti^{3+}R + AlEt_3 \longrightarrow Ti^{3+}R(Et) \longrightarrow Ti^{1+} + RH + C_2H_4 \tag{7}$$
(c) Oxidative coupling
$$2Ti^{3+} + C_2H_2 \longrightarrow Ti^{4+}-C_2H_2-Ti^{4+} \tag{8}$$

where R is a long poly(acetylene) chain.

Radiotagging

$$(OBu)_2TiR + CH_3OT \longrightarrow (OBu)_2TiOCH_3 + RT \tag{9}$$
$$(OBu)_2TiR + CH_3OH \longrightarrow (OBu)_2TiOCH_3 + RH \tag{10}$$
$$Et_2AlR + CH_3OT \longrightarrow Et_2AlOCH_3 + RT \tag{11}$$
$$Et_2AlR + CH_3OH \longrightarrow Et_2AlOCH_3 + RH \tag{12}$$

Acetylene is polymerized for 2 hrs at room temperature and radioactive methanol is added in excess amount. Figure 6 shows the reaction to be very rapid.

The kinetic isotopic effect is determined by decreasing the amount of radioactive methanol added and found to be k_H/k_T=3.0. From the specific activity a number average molecular weight of 22,000 was obtained for poly(acetylene).

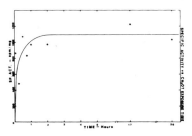

Fig. 6. Reaction of CH₃OT with poly(acetylene).

The active site concentration was determined by quenching with *CO:

$$Ti^{+3}\text{-}R + *CO \longrightarrow Ti^{+3}\text{-}\overset{O}{C}*\text{-}R \qquad (13)$$

$$Ti^{+3}\text{-}\overset{O}{C}*\text{-}R + CH_3OH \longrightarrow Ti^{+3}\text{-}OCH_3 + R\text{-}*\overset{O}{C}H \qquad (14)$$

This reaction does not occur with the corresponding aluminum compounds. Therefore, only those Ti-polymer bond will incorporate *CO. Figure 7 shows that reaction is relatively slow.

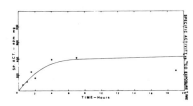

Fig. 7. Reaction of C*O with poly(acetylene).

Furthermore, the monomer must be removed prior to the introduction of C*O. Otherwise, monomer insertion to the Ti^{+3} acyl bond will occur. From the ratio of ^{14}C activity versus ^3H activity in the above two experiments, the chain transfer efficiency (5) is found to be ca. 5. About 10% of the Ti is catalytically active. The absolute rate constant of propagation is about 3.8 M^{-1} sec^{-1} at room temperature.

General properties.

The infrared spectrum of $(CH)_x$ (Fig. 8) is characterized by a band at 1015 cm^{-1} for the cis and 740 cm^{-1} for the trans isomer. The relative absorbancies of the two bands may be used

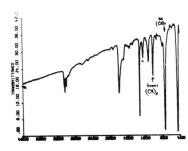

Fig. 8. An infrared spectrum of poly(acetylene).

to determine the isomeric contents of $(CH)_x$.

X-ray diffraction patterns of $(CH)_x$ are characteristic of a semicrystalline polymer.   An
example for <u>cis</u>-$(CH)_x$ is shown in Fig. 9.   Application of Roland's analysis gave 80%

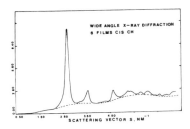

Fig. 9. Wide angle x-ray diffraction pattern for <u>cis</u>-$(CH)_x$.

crystallinity but a disorder factor of 5.

$(CH)_x$ has a gross density of ~0.4, the floation density is 1.16-1.18.   Electron microscopy
showed $(CH)_x$ film to be comprised of 200 Å diameter fibrils (<u>vide infra</u>).   A surface area
of 60 m^2gm^{-1} was obtained with BET method (7).

Morphology and crystal structure of <u>trans</u>-$(CH)_x$.

Poly(acetylene) had been described to have fibrous, clump and even lamellae morphologies.
A technique was developed to prepare ultrathin (~1000 Å) $(CH)_x$ films containing aligned
fibrils directly suited for electron microscopy and electron diffraction studies without
handling (8).   The thickness of the film may be controlled by varying the catalyst
concentration, acetylene pressure, and the time and temperature of the reaction.   The alignment of fibrils was apparently achieved by the repeated washing and evaporation of the
solvent.

An electron microscopy photograph of an ultrathin <u>trans</u>-$(CH)_x$ film is shown in Fig. 10.

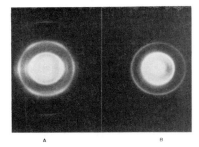

Fig. 10.   Electron microscopy photograph of ultra-thin film of
<u>trans</u>-$(CH)_x$ showing fibrous morphology.

Now an electron beam can be directed at desired portions of the film.   Figure 11a shows the
fiber electron diffraction pattern of the aligned fibrils, whereas Fig. 11b gave Debye
rings for the randomly packed mat.   The reciprocal lattice is shown in Fig. 12 (9).

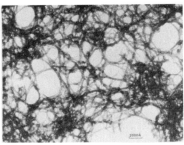

A   B

Fig. 11.   Electron diffraction of <u>trans</u>-$(CH)_x$ (a) partially aligned fibrils with
axis in the vertical direction, (b) unoriented portion of the specimen.

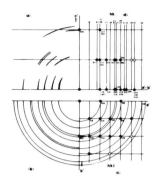

Fig. 12. (a) Schematic representation of the diffraction pattern in Fig. 11a,
(b) Schematic representation of the diffraction pattern in Fig. 11b, (c) reciprocal lattice for the a*b* plane, (●) observed reflection (o) predicted reflection
but not observed, (d) projection of the reciprocal lattices of the first three
layer a*b* planes.

The diffraction pattern showed that the molecular chain axis is along the fiber axis. The
unit cell is orthorhombic with each trans-$(CH)_x$ chain having the 2_1 screw axis. The
extinction of $(h0l)$ reflections with $\overline{h}$=odd integers and of $(0kl)$ reflections with $k+l$=odd
integers and taking into consideration the packing of $(CH)_x$ chains, the space group is
judged to be P_{nam}. The unit cell parameters are summarized in Table 2.

TABLE 2. Unit cell parameters for pristine poly(acetylenes).

	trans	cis
a, Å	7.32	7.68
b, Å	4.24	4.46
c*, Å	2.46	4.38
lattice type	orthorhombic	orthorhombic
density, gm cm^{-3}	1.13	1.15

*Fiber axis and chain axis.

A simplifying assumption was made to calculate theoretical reflection intensities by taking
all the C-C bond lengths to be 1.40 Å. The structure factors of the (hko) reflections were
calculated as a function of setting angles; when there is overlap the sum is taken. This
is compared with the intensities of the observed equatorial reflections in terms of the
reliability factor R defined as:

$$R = \sum(||F_{obsd}| - |F_{cald}||)/\sum|F_{obsd}| \qquad (15)$$

Figure 13 showed the best setting angle to be 24°.

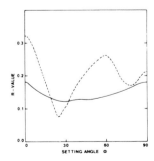

Fig. 13. Variation of R factor with setting angle for P_{nam} space group:
(a) (——) including all I to VI equatorial reflections; (b) (-----) including
reflections II to VI.

The molecular conformation of <u>trans</u>-(CH)$_X$ in crystal is shown in Fig. 14.

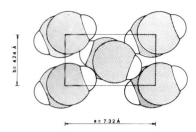

Fig. 14. Molecular formation of <u>trans</u>-(CH)$_X$ in crystal.

Morphology and crystal structure of <u>cis</u>-(CH)$_X$.

The morphology of <u>cis</u>-(CH)$_X$ is the same as that of <u>trans</u>-(CH)$_X$ which is derived from the former. The fiber electron diffraction is shown in Fig. 15. In order to avoid unwanted isomerization (<u>vide infra</u>) the measurements were made at -150°C and the specimen was never warmed much above -78°C except for a very brief time for its transfer into the electron microscope chamber. Analysis similar to that described above gave the unit cell parameters for <u>cis</u>-(CH)$_X$ (Table 2).

Isomerization and soliton formation.

Poly(acetylene) has three basic backbone isomeric structures: <u>cis</u>-transoid (I), <u>trans</u>-cisoid (II) and the planar zigzag <u>trans</u>-transoid (III).

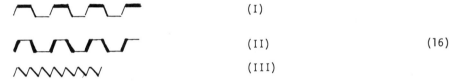

(I)

(II)   (16)

(III)

Our results showed that the <u>cis</u>-transoid (CH)$_X$ as prepared is without defect. An EPR signal develops upon warming to room temperature. This may be attributed to isomerization to the <u>trans</u>-cisoids. At elevated temperature, >145°C, phase transition to III takes place.

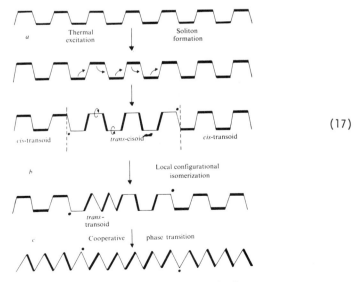

(17)

The kinetics of isomerization can be followed by EPR and IR (11). Heating of <u>cis</u>-rich polyacetylene resulted in a marked increase in the unpaired spin concentration, [S·], as obtained by double integration of the EPR signal. The formation of the soliton obeys first order kinetics. At any given temperature [S·] reaches a certain steady-state value which is unchanged on standing for a reasonably long period at that temperature, and [S·] does not decrease on cooling the sample. Numerous experiments demonstrated that each incremental temperature of heating results in a new and higher plateau value of [S·]. The apparent first-order rate constants gave an activation energy of ~10 kcal mol^{-1}. This low energy is consonant with the process which requires merely the alternate shifting of adjacent short and long bonds. The theoretical energy required for the process and formation of neutral solitons is estimated to be ~0.4eV or 9.4 kcal mole^{-1} (12).

Though cooperative phase transition to the trans-transoid structure takes place at 145°C, the soliton mobility as reflected by EPR linewidth (vide infra) increases gradually with the increase of temperature (Table 3).

TABLE 3. Soliton linewidth.

Temperature (°C)	25	50	75	100	125	150
ΔH_{pp} during heating	10	7.4	4.8	3.5	1.3	0.7
ΔH_{pp} after cooled to 25°	10	7.6	5.0	4.5	2.0	1.2

EPR saturation characteristics.

The spin-lattice relaxation and spin-spin relaxation times, T_1 and T_2 respectively, give detailed information about the neutral solitons. In these measurements, the EPR is saturated by increasing microwave power. The actual microwave power in the TE_{102} cavity, H_1 was determined by the method of perturbing sphere (13). A metal sphere was placed in the EPR cavity and the shift in frequency Δv was measured as a function of klystron power W (watts). The H_1 in gauss was calculation from

$$H_1 = \tfrac{1}{2}\left[W\left(\frac{v^2 - v_0^2}{v_0^2}\right)\left(\frac{40\pi}{\pi^2 \Delta v a^3}\right)\right]^{1/2} \tag{18}$$

where v_0=9.545 GHz and v=9.554 GHz are the initial and perturbed frequencies, and a is the radius of the sphere = 1.59 mm. We found the relationship

$$H_1^2 = 0.49\ W \tag{19}$$

Saturation curves were obtained by recording EPR spectra as a function of microwave power. From the plot of signal amplitude versus $\sqrt{W}$, $H_{1,m}$ is found for maximum signal amplitude. Together with ΔH_{pp} below saturation, the spin lattice and spin-spin relaxation time (T_1 and T_2, respectively) were calculated (14),

$$T_1 = 1.97 \times 10^{-7}\ \Delta H_{pp}[g(H_{1,m})^2]^{-1}\ \text{sec} \tag{20}$$

and

$$T_2 = 1.313 \times 10^{-7}[g\Delta H_{pp}]^{-1}\ \text{sec} \tag{21}$$

The EPR saturation curve for undoped trans-$(CH)_x$ approaches the case of homogeneous broadening. Measurements were made on samples from many different preparations. At ambient temperatures, the values of T_1 range from 1.9 to 6.6 x 10^{-5} sec with an average value of 2.7 ±1.7 x 10^{-5} sec (15). The values of T_2 range from 6 to 8.8 x 10^{-8} sec with an average value of 7.8 ±1.0 x 10^{-8} sec. Therefore, the variability in T_1 is much greater than it is for T_2 in trans-$(CH)_x$. At 77°K, T_1 was increased to 6.8 x 10^{-5} sec while T_2 was reduced to 2.2 x 10^{-8} sec.

Saturation measurements were made on many cis-$(CH)_x$ preparations. The values of T_1 for seven samples are 49, 39, 63, 47, 63 and 58 μsec giving an average value of 5.3 ±0.9 x 10^{-5} sec. The values for T_2 are 10, 10, 10, 9, 9.4, 11 and 11 nsec for an average of 1.0 ±0.07 x 10^{-8} sec (16). For undoped cis-$(CD)_x$ we found T_1=3.9 x 10^{-6} sec and T_2=1.2 x 10^{-8} sec.

The difference between the EPR saturation behaviors of cis-$(CH)_x$ and cis-$(CD)_x$ leaves no doubt that the broadened linewidth of the former is due to unresolved hyperfine structures. The extent of delocalization or the width of the soliton domain wall can be estimated from the EPR linewidth. The wave-function for the soliton having a domian width of N at a position x monomer unit from the center of a domain is

$$\psi(\underline{x}) = (2/N)^{1/2}\ \text{sech}\ (2x/N)\ \cos\ (\pi x/2) \tag{23}$$

The hyperfine splitting, A, at the xth monomer is

$$A_{eff} = A_0\ \psi(\underline{x})^2 \tag{24}$$

where A_0 is the hyperfine splitting for a localized spin. Obvisouly, the greater the delocalization and the smaller wave-function amplitude, the narrower would be the EPR linewidth. Weinberger et al. (17) compared experiment with theory to arrive at a soliton domain of 12-16 CH units in cis-$(CH)_x$. The 6.5-7.0 G EPR linewidth for S· in this polymer is independent of temperature down to a few degrees Kelvin. Therefore, the S· in this polymer is immobile in the EPR time scale of ~10^{-8} sec.

With heating and isomerization, the [S·] increases. The soliton-antisoliton pair on the same chain (cf. eqn. 17) may annihilate unless they become separated by intermolecular transfer, intermolecular annihilation, or become bound to Ti^{+3} on the polymer chain (15).

The above proposed mechanism of isomerization implies the existence of a maximum of one neutral soliton per trans-$(CH)_x$ chain. This seems to be supported by the following considerations. The [S·] in undoped trans-$(CH)_x$ of ca. one soliton per 1000-3000 CH with a moleuclar weight of 22,000, for one S· per trans-$(CH)_x$ chain corresponds to one soliton per 1700 CH, in agreement with observation. Higher neutral soliton concentration is possible if crosslinks produced during isomerization can serve to isolate neutral solitons on both sides of the crosslink.

DOPED $(CH)_x$

Doping with iodine.

Several saturation curves of iodine doped trans-$(CHI_y)_x$ obtained at ambient temperature are shown in Fig. 15 to indicate the trend. Doping in the range of y from 3×10^{-6} to 3×10^{-3}

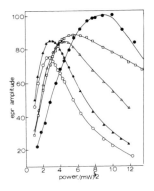

Fig. 15. Electron paramagnetic resonance saturation curves at room temperature for trans-$(CHI_y)_x$: (o) $y=3.8 \times 10^{-6}$, (▲) $y=2.0 \times 10^{-5}$, (Δ) $y=5.1 \times 10^{-4}$, (◻) $y=2.5 \times 10^{-3}$, (●) $y=1.6 \times 10^{-2}$.

were done with $^{125}I_2$; ordinary iodine was used for $y>7 \times 10^{-4}$. Therefore, the two methods of determining dopant concentration overlap between $y=7 \times 10^{-4}$ and 3×10^{-3}; the results in these regions are in good agreement. The EPR of trans-$(CHI_y)_x$ generally retains the Lorentzian line shape with some broadening in the wings.

At the lightest level of doping of $y=3 \times 10^{-6}$ or I_3^- content of 1×10^{-6} per CH unit, there were pronounced effects on EPR. The EPR linewidth became independent of H_1 whereas undoped trans-$(CH)_x$ has linewidth above saturation directly proportional to H_1. Therefore, the EPR changed from homogeneous broadening to inhomogeneously broadened upon very light doping. At $y=6.8 \times 10^{-6}$ the T_1 is increased to ca. 3.3×10^{-5} sec and T_2 decreased to 6.6×10^{-8} sec. These results suggest that doping is rather uniform and indicates very high mobility for the solitons in this polymer.

The effect of iodine doping on trans-$(CH)_x$ clearly shows three regions: lightly doped ($y<10^{-3}$), medium doping ($10^{-3}<y<5 \times 10^{-2}$) and heavily doped ($y>5 \times 10^{-2}$). In the first region, after the initial slight increase in T_1, the spin-lattice relaxation time decreases rapidly with increasing y, while T_2 remains virtually constant. Above $y>10^{-3}$ the decrease in T_1 becomes more gradual with further increases in y, while T_2 begins to decrease rapidly. The EPR remains saturable for the entire lightly and intermediate doping level. Above the semiconductor metal transition, the EPR spectra first became Dysonian followed by disappearance of EPR signal.

The effect of iodine on cis-$(CH)_x$ is quite different. Firstly, though on the average the lightly doped $\underline{cis}$-$(CHI_y)_x$ seems to have T_1 longer than that of the undoped material, the significant scattering in the data does not merit any conclusion that the dopant has significant effect on the T_1. T_1 starts to decrease with $y>10^{-4}$. From $y>10^{-3}$, the EPR spectra of $\underline{cis}$-$(CHI_y)_x$ cannot be saturated. At $y=6 \times 10^{-4}$ there is about one I_3^- per three poly(acetylene) chains. Electron diffraction results showed that the dopant occupies the channel enclosed by four poly(acetylene) chains (10). Consequently, T_1 of all solitons can be decreased by four spin orbit coupling with a dopant. This results suggests that the $S\cdot$ in $\underline{cis}$-$(CHI_y)_x$ does have some mobility though much less than those in the $\underline{trans}$ polymer. The one dimensional diffusivity of neutral soliton in poly($\underline{cis}$-acetylene) is $\underline{not}$ known. The present data offeres an estimate of the limits. The diffusivity should be less than 10^8 sec^{-1} which is the proton hyperfine interaction but greater than 2×10^4 sec^{-1} which is T_1^{-1}.

Mechanism of doping.

The $[S\cdot]$ for $\underline{trans}$-$(CHI_y)_x$ remains unchanged for $3 \times 10^{-6} < y < 10^{-3}$, $[S]$ decreases precipitously with $y^{-3.7}$. When $y > 10^{-2}$ the EPR signal disappears. These observations shed light on the mechanism of doping (Fig.16).

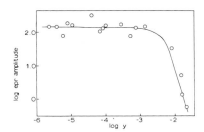

Fig. 16.   Variation of EPR intensity for $\underline{trans}$-$(CHI_y)_x$ with y.

It was proposed (18) that iodine doping can take place via two pathways. The first is the conversiton of a neutral soliton to a positive spinless soliton, S^+,

$$\text{/\textbackslash/\textbackslash/\textbackslash/} + I_2 \longrightarrow \text{/\textbackslash/\textbackslash/\textbackslash/} \quad (25)$$
$$I_3^-$$

Secondly, iodine can react with poly(acetylene) directly to produce a radical cation

$$\text{/\textbackslash/\textbackslash/} + I_2 \longrightarrow \overset{I_3^-}{\text{/\textbackslash/\textbackslash/\textbackslash}} \quad (26)$$

Separation of the radical cation led to a pair of $S\cdot$ and S^+ solitons

$$\overset{I_3^-}{\text{/\textbackslash/\textbackslash/\textbackslash/}} \longrightarrow \overset{I_3^-}{\text{/\textbackslash/\textbackslash/\textbackslash/\textbackslash}} \quad (27)$$

If reactions 26 and 27 occur with comparable probability, then there is no net change in $[S\cdot]$ or the EPR intensity. Therefore, the paramagnetic neutral soliton is not responsible for conductivity, but it is the spinless positive soliton (18,19) which is the carrier for p-type doped poly(acetyelene). In the lightly doped region, the carrier concentration increases slowly but steadily with doping together with slow increases in conductivity.

In the intermediate doping region, $[S\cdot]$ decreases as $y^{-3.7}$. This suggest that the dopant exists mainly as I_3^- with I_5^- in minor amounts (20). The EPR intensity begins to decrease at $y \sim 10^{-3}$, or an I_3^- content of 3×10^{-4}. This corresponds to one dopant per two $\underline{trans}$-$(CH)_x$ molecule.

All the neutral solitons initially present in undoped $\underline{trans}$-$(CH)_x$ would have been converted to positive soliton by eqn. 25 if doping is homogeneous. At $y > 10^{-3}$ doping will be primarily via eqns. 26 and 27. However, now interchain migration of $S\cdot$ is possible via the dopant as bridge. Electron diffraction (10) showed that I_3^- ions are situated between poly(acetylene) chains. At one dopant per two $\underline{trans}$-$(CH)_x$ chains, an $S\cdot$ produced on any chain can migrate to and annihilate another $S\cdot$ in some other chain. Therefore $[S\cdot]$ decreases rapidly with increasing y. In this doping range an S^+ is produced with each dopant, conductivity increases very rapidly with increasing y (21-23).

The dependence of $[S\cdot]$ is $\underline{cis}$-$(CHI_y)_x$ upon y is the same as in the case of the $\underline{trans}$ polymer.

Doping with AsF$_5$.

Four general types of line shapes were observed for [CH(AsF$_5$)$_y$]$_x$ depending upon the dopant concentration. The spectra in Fig. 17a-d are representative of a) undoped cis-(CH)$_x$,

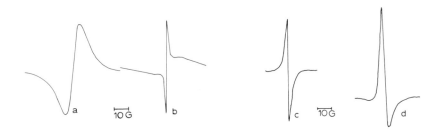

Fig. 17. Electron paramagnetic resonance spectra of cis-[CH(AsF$_5$)$_y$]$_x$ at room temperature: (a) y=0; (b) y=5 x 10^{-3}; (c) 8 x 10^{-3}<y<0.02; (d) y~0.08.

b) lightly doped (y≈0.005) material, c) samples doped in the range 0.008<y<0.02, and d) heavily doped "metallic" (CH)$_x$. The relationship between signal intensity and y has been discussed by Ikehata et al. (19) for trans-[CH(AsF$_5$)$_y$]$_x$. The ratio of the signal amplitude above and below the base line (A/B) was found to be unity for samples having y<0.02 and then increased with increasing y. The ca. 7 G wide signal of cis-(CH)$_x$ narrowed appreciably upon light doping and then progressively increased with increasing y.

The limiting conductivities for unoriented cis-(CH)$_x$ doped with I$_2$ and AsF$_5$ are 5.5 x 10^2 and 1.2 x 10^3 (Ωcm)$^{-1}$, respectively (4), which are quite similar. However, the detailed processes in the conversion of the insulating material to highly conducting states are very different for the two dopants. Trace level of doping produces a narrow resonance superimposed on the broad EPR of S· in cis-(CH)$_x$. A probable explanation lies in the much stronger oxidizing power of AsF$_5$ (18). Therefore, AsF$_5$ is more likely to interact directly with poly(acetylene) to create a radical cation than to selectively convert a neutral soliton to a positive soliton as does iodine. Furthermore, if this tendency is very strong, then AsF$_5$ doping may be inhomogeneous resulting in clusters of dopants probably on the surface of the poly(acetylene) fibrils (24). In the first instance AsF$_5$ produces an S· S$^+$ pair, which if confined to the same domain can interconvert rapidly. Therefore, the line narrowing is analogous to chemical exchange narrowing where the spin-packets of the neutral soliton exchanges with the positive soliton thus effectively eliminating the broadening due to unresolved hyperfine structure. NH$_3$ can remove the S· S$^+$ pair.

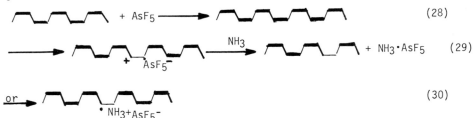

$$+ AsF_5 \longrightarrow \tag{28}$$

$$\xrightarrow{} \quad \underset{+ \; \cdot AsF_5^-}{} \quad \xrightarrow{NH_3} \quad + NH_3 \cdot AsF_5 \tag{29}$$

$$\text{or} \xrightarrow{} \quad \underset{\cdot \; NH_3^+ AsF_5^-}{} \tag{30}$$

An independent experiment showed that NH$_3$ had no effect on either the intensity or the saturation behacior of pristine cis-(CH)$_x$. Similarly, if there is produced clusters of S· S$^+$ pairs, then line narrowing by Heisenberg exchange between S· is a possibility. In this case, process 30 would leave clusters of S· in close proximity and exchange narrowed EPR signal should remain, whereas process 29 would have the effect of eliminating the narrow resonance as it was observed.

For y>10^{-2} the EPR of cis[CH(AsF$_5$)$_y$]$_x$ samples assumes increasing Dysonian line shape with increase of y. In contrast, the corresponding iodine doped material does not behave similarly. Dysonian line shape depends critically on the relative magnitude of the time T$_d$ that it takes for an electron to diffuse through the skin depth, the time T$_t$ that it takes an electron to traverse the sample, and T$_1$ and T$_2$. The presence of Dysonian line shape implies the presence of domains of degenerate electron gas and is not necessarily correlated with macroscopic conductivity. The line shape differences for conducting cis-(CHI$_y$)$_x$ and cis-[CH(AsF$_5$)$_y$]$_x$ may be attributable to the dissimilar T$_d$ and T$_t$ of the two types of materials.

EPR saturation curves were obtained at room temperature for trans-[CH(AsF$_5$)$_y$]$_x$ from y=3.6 x 10^{-4} to 0.14. The relaxation data are summarized in Table 4. For the lightly

TABLE 4. EPR saturation behaviors of trans-$[CH(AsF_5)_y]_x$.

y	Relative EPR intensity	$T_1 \times 10^5$ sec	$T_2 \times 10^8$ sec
3.6×10^{-4}	17	2.5	7.3
1.5×10^{-3}	9.7	1.8	7.3
2×10^{-3}	5.5	1.0	6.6
2.2×10^{-3}	5.3	1.0	8.2
2.5×10^{-3}	17	1.5	7.3
3.7×10^{-3}	9.0	0.8	8.2
6.3×10^{-3}	15.4	1.5	7.3
2.8×10^{-2}	2.5	$-^a$	5.0
3.1×10^{-2}	2.6	-	5.5
4.4×10^{-2}	7.0	-	6.9
5.4×10^{-2}	10	-	3.3
0.14	23	0.11	11

doped materials, $y<10^{-3}$, the T_1 and T_2 values are nearly the same as undoped trans-$(CH)_x$. They are also similar in their EPR linewidth dependence on H_1, increaseing 2- to 3-fold from low to high microwave power. The values of T_1 and T_2 remained substantially unchanged for $10^{-3}<y<10^{-2}$ at $1.3 \pm 0.6 \times 10^{-5}$ sec and $7.5 \pm 0.6 \times 10^{-8}$ sec. The EPR linewidth increase with microwave power is only about 50% to 70% which is smaller than that for more lightly doped polymers. Above $y=10^{-2}$ the EPR and its saturation behaviors underwent signficant changes. There is a decrease of EPR intensity for $y=0.028$ and 0.031, which may correspond to the decrease observed by Ikehata et al. (19) at lower dopant concentration. As y exceeded 0.04 there was a rapid increase of EPR intensity, which coincides with the report that Pauli susceptibility is switched on. For these samples, the resonances cannot be saturated and their linewidths are independent of H_1. The line shape is Dysonian with A/B ratios ranging from 1.8 to 5.5.

Combined evidences point to lower degree of homogeneity for AsF_5 doped material than the iodine doped samples. The values of T_1 and T_2 for trans-$[CH(AsF_5)_y]_x$ from $3 \times 10^{-4}<y<6 \times 10^{-3}$ cannot be said to be different from those for undoped trans-$(CH)_x$; the relaxation data shows much more scattering than the iodine doped materials. Furthermore, these samples show EPR linewidth dependence on H_1, though the H_1 dependence decreases with increasing y and is always less than the undoped polymer. Finally, there is a region $6 \times 10^{-2}<y<0.1$ where the EPR of the samples cannot be saturated with the available microwave power. This suggests the possibility of domains in which S· has very short relaxation times.

Morphology and structure.

Figure 18 shows the electron microscopy photograph of intermediately doped cis-$(CHI_y)_x$ in the semiconducting state by exposing the polymer to a reservoir of iodine maintained at

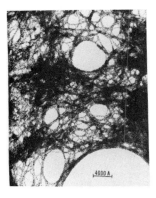

Fig. 18. Electron microscopy photograph of intermediately doped cis-$(CHI_y)_x$ observed at -150°C.

-23°C ($p_{I_2}=3 \times 10^{-3}$ mm) for 4 hours and immediately observed at -150°C. It appears that there were formed clusters of dopants; the fibrils have alternately expanded and shrunken so that it has the appearance of a string of beads. The same sample was warmed to room temperature, the fibrils now resemble the undoped polymer (Fig. 19).

Fig. 19. Electron microscopy photograph of sample shown in Fig. 18 but warmed to and observed at room temperature.

It suggests that the dopant has redistributed more or less uniformly. The picture of saturation doped cis-$(CHI_y)_x$ (Fig. 20) showed generally homogeneously doped fibrils.

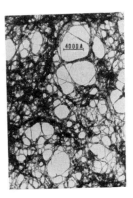

Fig. 20. Electron microscopy photograph of cis-$(CHI_y)_x$ in the metallic state.

In sharp contrast is the sample of cis-$[CH(AsF_5)_y]_x$ doped very slowly with AsF_5 reservoir maintained at -95°C and expose the polymer to it for 35 min. The electron microscopy photograph obtained at -150°C is shown in Fig. 21. The sample has a conductivity of ~50

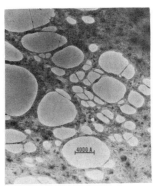

Fig. 21. Electron microscopy photograph of cis-$[CH(AsF_5)_y]_x$ doped to near saturation.

$(\Omega cm)^{-1}$ and is therefore in the metallic state. The fibrils now are difficult to discern. In addition there are dark spots attributable to heavy deposits of the dopant.

Electron diffractions have been obtained for the above samples. For the intermediately doped cis-$(CHI_y)_x$, all the equatorial reflections can be assigned to the cis-unit cell. Along the meridian there are the (002) and (004) reflections of cis-$(CH)_x$. In addition there are also (002) and (004) reflections which correspond to the trans-$(CH)_x$. Therefore, both cis and trans structures are both present in this type of material.

For the heavily doped cis-$(CHI_y)_x$, there is no longer any reflection assignable to the cis structure. The meridional reflections seem to coincide with half of the repeat distance of trans-$(CH)_x$. Though analysis of the fiber diffraction pattern suggests a structure shown in Fig. 22.

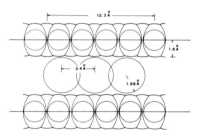

Fig. 22. Proposed structure for heavily doped $(CHI_y)_x$.

That doping would initiate isomerization is expected from the mechanism proposed above which results from the formation of a S^+ $S\cdot$ and their subsequent separation.

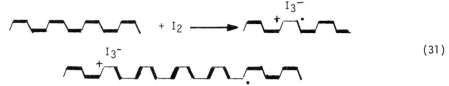

$$\tag{31}$$

The new trans-cisoid portion can further isomerize to the trans-transoid which is probably rendered easy because the presence of I_3^- pushes adjacent chains in the vicinity apart. More detailed discussion of the structure of doped polyacetylene can be found elsewhere (10).

TRANSPORT PROPERTIES

It is reasonable to consider the conductivity of doped poly(acetylene) to be governed by the intrinsic quality of the fibrils and interfibril contact resistance. We have examined the problem in two ways.

Variable density poly(acetylene).

Acetylene was polymerized to low density gel by a procedure given previously (26). It can be pressed to intermediate densities as well as up to that of the as prepared film. The results on conductivity and thermopower are summarized in Table 5.

TABLE 5. Conductivity and thermopower of poly(acetyelene) of various densities.

Material	Property	"Foamlike" $(CH)_x$		Pressed $(CH)_x$ film	
		y	magnitude	y	magnitude
Trans [a] $(CH)_x$ (undoped)	$\sigma(RT)$ S(RT) [b] Density	-	1.08×10^{-6} ohm^{-1} cm^{-1} ($\Delta E \approx 0.25$ eV) +900 μV/K 0.02 - 0.04 g/cm^3	-	7.9×10^{-5} ohm^{-1} cm^{-1} ($\Delta E \approx 0.25$eV) +920 μV/K 0.4 g/cm^3
$[CHI_y]_x$	$\sigma(RT)$ S(RT) Density[c]	≈ 0.06	8.14 ohm^{-1} cm^{-1} +18.7 μV/K 0.02 - 0.04 g/cm^3	≈ 0.06	350 ohm^{-1} cm^{-1} +18.4 μV/K 0.4 g/cm^3
$[CHI_y]_x$	$\sigma(RT)$ S(RT) Density[c]	-	- - -	0.05	11.2 ohm^{-1} cm^{-1} - 0.1 g/cm^3
$[CH(AsF_5)_y]_x$	$\sigma(RT)$ S(RT) Density[c]	0.08	81.3 ohm^{-1} cm^{-1} +8 μV/K 0.02 - 0.04 g/cm^3	0.06	176 ohm^{-1} cm^{-1} +8.9 μV/K 0.1 g/cm^3

[a]Isomerized by heating at 180°C for 2 hr in vacuo in a sealed tube.
[b]Density of film before isomerizing.
[c]Density of parent film before doping.

Since the thermopower is a zero-current transport coefficient, the interfibril contacts should be unimportant, allowing evaluation of the intrinsic properties. Moreover, since thermopower can be viewed as a measure of the entropy per carrier, the results depend only on the properties of the conducting fibrils and not on the number of fibrils per unit volume. As indicated in Table 4, the thermopower of the undoped polymer is insensitive to the density; the foamlike material and the pressed film yield thermopower values of approximately +900 μV/K. Any variations are comparable with the typical variations observed earlier (26) in as-grown film samples from different synthetic preparations. Similarly, after heavy doping the results are insensitive to the density with values for the foamlike material, pressed film, and as-grown films in good agreement for each dopant. Comparison with the variation in thermopower as a function of dopant concentration studied in detail earlier (26) leads to the conclusion that the heavily doped samples are all metallic. The thermopower results imply that the various forms of the doped and undoped polymers are microscopically identical. The low-density materials simply consist of fibrils at a smaller filling fraction f.

The electrical conductivity data are consistent with this conclusion. The conductivity of the undoped foamlike mateiral is nearly two orders of magnitude below that oif the high-density pressed film. Although there may be some increase in the interfibril contact resistance, the reduction in f is of major importance. This conclusion is strengthened by the observation that the conductivity activation energy (obtained from the temperature variation of the conductivity near room temperature) is 0.25 eV for both samples. This value is comparable to the 0.3eV value typically obtained from as-grown films (28).

Similar results are obtained with the heavily doped polymers. As indicated in Table 4, the conductivity [176 $(\Omega cm)^{-1}$] of $[CH(AsF_5)_{0.06}]_x$, prepared from the low-density pressed cis-$(CH)_x$ film (ρ=0.1 gm/cm^3) is correspondingly lower than that of the conductivity [560 $\overline{(\Omega cm)}^{-1}$ and 1200 $(\Omega cm)^{-1}$] of $[CH(AsF_5)_{0.14}]_x$ (29) and $[CH(AsF_5)_{0.10}]_x$, respectively, prepared from as-grown cis-$(CH)_x$ film (ρ=0.4 gm/cm^3). The room temperature conductivity of $(CHI_{0.06})_x$ increases by a factor of 40 in going from the low-density foamlike material to the high-density presed film. In all cases, the resulting conductivity increases with the filling fraction of fibrils.

Acetylene-methylacetylene copolymers.

Acetylene and methylacetylene were copolymerized by a method described elsewhere (30). The rationale is that the two carbon atoms in the methyl acetylene were different. Therefore, homopolymers of methylacetylene will have wider band gaps and narrower band widths than poly(acetylene) and consequently lower conductivity. In the case of copolymer, it was unknown whether the electronic properties will be a strong function of comonomer composition such as a critical composition at which there is a sudden change in conductivity. The results are shown in Fig. 23, which has expected dependence of conductivity on

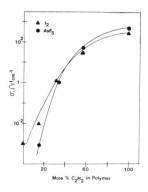

Fig. 23. Variation of conductivity with copolymer composition for films doped with iodine and AsF5.

copolymer composition. An increase in the methylacetylene content of the polymer is anticipated to increase both its band gap ($\pi \rightarrow \pi^*$ transition) and its ionization potential, making free carrier generation more difficult. Furthermore, the reduction of the effective conjugation length with increasing methylacetylene content is expected to reduce carrier mobility. Thus the methylacetylene units would act as barriers to conductivity. Carrier migration through (or around) these barriers should become progressively more difficult as the number of these barriers increases. The copolymer conductivity is more sensitive to methylacetylene content when the copolymer is rich in this monomer.

V. CONCLUSION

Prior to the emergence of poly(acetylene) organic conductors are limited to small and brittle crystals, mostly of charge transfer type with conductivity less than 10^2 $(\Omega cm)^{-1}$. Now flexible films of $(CH)_x$ can be doped in a controlled manner from 10^{-9} $(\Omega cm)^{-1}$ to greater than 2×10^3 $(\Omega cm)^{-1}$. Much still remains to be understood about the chemistry and physics of this new material. These studies will undoubtedly lead to other polymers with interesting and unusual electronic properties and develop into a broad class of materials of both scientific interest and technological importance.

Acknowledgement - This work is supported in part by a grant from the Office of Naval Research, a grant from the National Science Foundation, and a grant from DARPA.

REFERENCES

1. H.C. Lounget-Higgins and L. Salem, Proc. Roy. Soc. A25, 172 (1959).
2. R.E. Peierls, "Quantum Theory of Solids", Chapter 5, Oxford University, London, 1955.
3. P.M. Grant and I.P. Batra, Solid State Commun. 29, 225 (1979).
4. A.G. MacDiarmid and A.J. Heeger, Syn. Metals. 1, 101 (1979/1980).
5. I. Ito, H. Shirakawa, and S. Ikeda, J. Polym. Sci.-Polym. Chem. Ed. 12, 11 (1974).
6. J.C.W. Chien, F.E. Karasz, G.E. Wnek, A.G. MacDiarmid and A.J. Heeger, J. Polym. Sci.-Polym. Lett. Ed. 18, 45 (1980).
7. J.C.W. Chien, unpublished results.
8. F.E. Karasz, J.C.W. Chien, R. Galkiewicz, G.E. Wnek, A.J. Heeger and A.G. MacDiarmid, Nature 282, 286 (1979).
9. K. Shimamura, F.E. Karasz, J.A. Hirsch and J.C.W. Chien, Makromol. Chem. Rapid Commun., in press.
10. J.C.W. Chien, F.E. Karasz, J.A. Hirsch and K. Shimamura, J. Polym. Sci.-Polym. Lett. Ed., in press.
11. J.C.W. Chien, F.E. Karasz and G.E. Wnek, Nature 285, 390 (1980).
12. W.P. Su, J.R. Schreiffer and A.J. Heeger, Phys. Rev. Lett. 42, 1698 (1979).
13. J.H. Freed, D.S. Leniart and J.S. Hyde, J. Chem. Phys. 47, 2762 (1967).
14. C.P. Poole and H.A. Farach, "Relaxation in Magnetic Resonance", Academic Press, New York, 1971, Chapters 3,9.
15. J.C.W. Chien, J.M. Warakomski and F.E. Karasz, J. Chem Phys., in press.
16. J.C.W. Chien, L.C. Dickinson, G.E. Wnek, J.M. Warakomski and F.E. Karasz, Macromolecules, in press.
17. B.R. Weinberger, E. Ehrenfreund, A. Pron, A.J. Heeger and A.G. MacDiarmid, J. Chem. Phys. 72, 4749 (1980).
18. J.C.W. Chien, J. Polym. Sci.-Polym. Lett. Ed. 19, 249 (1981).
19. S. Ikehata, J. Kaufer, T. Woerner, A. Pron, M.A. Druy, A. Sivak, A.J. Heeger and A.G. MacDiarmid, Phys. Rev. Lett. 45, 1123 (1980).
20. S.L. Hsu, A.J. Signorelli, G.P. Pez and R.H. Baughman, J. Chem. Phys. 69, 1 (1978).
21. H. Shirakawa, E.J. Louis, A.G. MacDiarmid, C.K. Chiang, and A.J. Heeger, Chem. Commun. 578 (1978).
22. C.K. Chiang, Y.W. Park, A.J. Heeger, H. Shirakawa, E.J. Louis, and A.G. MacDiarmid, J. Chem. Phys. 69, 5098 1978).
23. C.K. Chiang, M.A. Druy, S.C. Gau, A.J. Heeger, E.J. Louis, A.G. MacDiarmid and Y.W. Park, J. Am. Chem. Soc. 100, 1013 (1978).
24. H.J. Epstein, H. Rommelmann, M.A. Druy, A.J. Heeger and A.G. MacDiarmid, Solid State Commun., in press.
25. G.E. Wnek, J.C.W. Chien, F.E. Karasz, M.A. Druy, Y.W. Park, A.G. MacDiarmid and A.J. Heeger, J. Polym. Sci.-Polym. Lett. Ed. 17, 779 (1979).
26. Y.W. Park, A. Denenstein, C.K. Chiang, A.J. Heeger and A.G. MacDiarmid, Solid State Commun. 29, 747 (1979).
27. C.K. Chiang, Y.W. Park, A.J. Heeger, H. Shirakawa, E.J. Louis and A.G. MacDiarmid, J. Chem. Phys. 69, 5098 (1978).
28. H. Shirakawa, T. Ito and S. Ikeda, Makromol. Chem. 179, 1565 (1978).
29. C.K. Chiang, M.A. Druy, S.C. Gau, A.J. Heeger, E.J. Louis, A.G. MacDiarmid, Y.W. Park and H. Shirakawa, J. Am. Chem. Soc. 100, 1013 (1979).
30. J.C.W. Chien, G.E. Wnek, F.E. Karasz and J.A. Hirsch, Macromolecules, in press.

A COMPREHENSIVE THEORY OF THE HIGH IONIC CONDUCTIVITY OF MACROMOLECULAR
NETWORKS.

Hervé CHERADAME

Ecole Française de Papeterie, Institut National Polytechnique de Grenoble,
B.P.3, 38400 Saint Martin d'Hères, FRANCE.

Abstract – Using the free volume concept, an expression is derived for the
ionic conductivity of macromolecular networks, irrespective of their macrostructure. The logarithm of the reduced conductivity σ_T/σ_{Tg}, which is the
ratio of the conductivity at a given temperature to the conductivity at the
glass transition temperature, is a linear function of the well known shift
factor $\log a_T$ given by the dynamic mechanical properties. Theoretically, an
Arrhenius behaviour is superposed to the W.L.F. behaviour. Using poly-
ether-polyurethane networks filled with various concentrations of sodium
tetraphenyl boride, it is shown that the superposition principle applies to
the storage modulus, giving approximately the same shift factor when the
glass transition-temperature is used as reference temperature for polyether
networks of similar crosslink density. The W.L.F. superposition for the
ionic conductivity is fair with various membranes having the same salt
concentration. The influence of the main parameters, temperature and salt
concentration, can be accounted for by the given theoretical expression.
An analysis of the literature data shows that the free volume behaviour is
a general law for the ionic conductivity regardless of the macrostructure
of the polymer network.

INTRODUCTION

Polymer-based electrolytes used in electrochemical generators offer such advantages as thermal stability, absence of electrochemical corrosion when enough selectivity is obtained and
absence of thermal convection. Moreover solid electrolytes can be produced as thin films
which are able to accept mechanical strains if they are sufficently plastic or even in the
rubber-like state. But despite these possible advantages the ionic intrinsic conductivity of
macromolecular materials has received little attention compared to the electronic conductivity. We call intrinsic ionic conductivity the ionic conductivity obtained without solvent.
In this case, the macromolecular chains play the role of the solvating agent and allow the
ionic transport. This particular case is clearly distinct from the one of polyelectrolytes
swollen by or in the presence of water. It is also quite different from the case of materials
where the solvent ensure the ionic dissociation and transport, the macromolecular network
giving only some mechanical resistance. In this context the solvent must possess "electrochemical qualities", while in the intrinsic ionic conductivity of macromolecular materials
this role is also played by the polymer. In the latter field, polyethers are of   high interest, by analogy with what we observe in classical organic physical chemistry (1). In the
work to be presented here the mathematical derivation of an expression giving the ionic
conductivity is obtained for any macromolecular material, polyelectrolyte, or non-ionic but
filled with metal salts. This law allows the observation of a predictable behaviour for the
ionic conductivity from the study of   dynamic mechanical properties. A detailed analysis
of mechanical and electrical properties of polyethylene oxide or polypropylene oxide-based
materials is given, showing this type of correlation.

QUALITATIVE APPROACH FOR THE IONIC CONDUCTIVITY OF MACROMOLECULAR MATERIALS.

Ionic conductivity is a true mass transport through a material. In order to obtain the ready
displacement of ions at a required temperature, it is necessary that movements of chain
segments be permitted at this temperature. Thus the material must be in the rubber like
state. Evidently this conclusion holds only when high ionic conductivity is concerned, i.e.
one exceeding 10^{-7} $(\Omega cm)^{-1}$. Literature data show indeed that lower values are obtained in
the glass state in some instances. A second condition is that the cations must be able to
be easily separated from their anionic counterpart. In the present treatment we are only
dealing with cationic conductivity. In fact anionic transport must be completely absent in
high energy electrochemical generators, which is our main field of interest. However anionic

conductivity from a fundamental point of view behaves   quite symetrically to the cationic
one and the same qualitative or quantitative approach could be given. For the sake of simplicity we shall assume that only cationic conductivity is concerned. The most interesting
cations are the alkali metals ; then by analogy with the organic electrolytes, it is necessary that the macromolecular chains be able to complexe alkali cations. These ions are hard
acids as stated by the HSAB classification given by Pearson (2). Consequently we have to
choose a soft weak base for the corresponding anion to ensure facile dissociation. As mentioned above the largest the anion the better, to avoid anionic displacements, i.e. to decrease
anionic mobility as much as possible. This is the reason why we chose sodium tetraphenyl
boride for this study. It is worth noting that some conditions given hereby may seem to play
a rather contradictory role. This is particularly the case for the mobility of chains and
their assumed solvation properties. Finally, we concentrated on cross linked materials due to
their good mechanical properties and the lack of previous studies on this type of polymers.

QUANTITATIVE APPROACH FORTHE IONIC CONDUCTIVITY OF MACROMOLECULAR MATERIALS.

In a macromolecular medium, assuming that the temperature is higher than the glass transition
temperature, one can calculate the probability for a displacement of a chain element following the classical method proposed by Macedo and Litowitz (3). The jump probability of a
particle i is given by : $p_i = p_E.p(v)$, where p_E is the probability of having enough energy
for the displacement, and $p(v)$ is the probability to find in the vicinity of a given particle a hole (free volume) of sufficient size to allow the jump. If one assumes a Boltzmann
distribution   we have :

$$p_E = \exp(-E_a/kT) \tag{1},$$

E_a being the potentiel barrier for the considered displacement .To express $p(v)$ one can use
the classical calculation given by Cohen and Turnbull (37) :

$$p(v) = \exp(-\gamma v^*/v_f) \tag{2},$$

where γ is a numerical factor which takes into account the possible overlap of free volume,
v^* the minimum hole size allowing the jump and arising from thermal fluctuation of the free
volume, and v_f is the mean free volume per particle available. Then the probability for an
elementary displacement is :

$$P = \exp(-\gamma.v^*/v_f - E_a/kT) \tag{3}.$$

Weymann has shown that the viscosity of a liquid is proportional to the reciprocal of P(4) :

$$\eta = A/P \tag{4}.$$

If one assumes in a classical way that v^* is proportional to the hard volume v_1^* of the
moving particle, the expansion coefficient of which is negligible, and that the free volume
expands with the linear law :

$$V_f = V_{fg} + \alpha_f (T - Tg) \tag{5},$$

where V_{fg} and V_f are respectively the free volume   at the glass transition temperature
and at the temperature T

$$\ln \eta_T/\eta_{Tg} = -\frac{\gamma v^*}{V_{fg}} \times \frac{T - Tg}{T - Tg + v_{fg}/\alpha_f} - \frac{Ea}{k}\left(\frac{1}{Tg} - \frac{1}{T}\right) \tag{6}.$$

This classical equation shows that the reduced viscosity has an Arrhenius behaviour super -
posed to a W.L.F. one. It is known that the expression :

$$\log a_T = \log[\eta_T/\eta_{Tg}] \tag{7},$$

is the shift factor which allows the superposition of the curves giving the isothermal
variations of the storage modulus versus frequency when the Arrhenius behaviour is negligible ,when we take the glass transition temperature as the reference temperature.

The problem of the diffusion of an ion in a macromolecular network, crosslinked or not, can
be expressed in the same way. Again we call V_2^* the minimum hole size necessary for a jump,
γ the same geometrical factor, and V_f the mean free volume   per diffusing polymer
segment . Following Miyamoto and Shibayama, the frequency of jump of a particle(here an ion)
between two equilibrium positions is given by (5) :

$$f' = \nu \propto P_E .P(V_2^*) \tag{8},$$

where ν is the frequency of the thermal movement of the ion at the considered temperature
within its cage, and α is a correlation factor $(0 < \alpha < 1)$. P_E is the probability that the
ion has enough energy to jump over a length λ , and $P(V_2^*)$ is the probability of having a
hole of size V_2^* . or bigger .   Thus, $P(V_2^*)$ is the   probability that chain
segments may form a hole of a size at least equal to V_2^* in the vicinity of the considered
ion :

$$P(V_2^*) = \exp(-\gamma V_2^* / N_f - Ea/kT), \tag{9} ;$$

$$f' = \nu \alpha \exp(-\gamma V_2^* / V_f - (Ea + Eb)/kT) \tag{10}.$$

When an electric field is applied each particle having the charge $z_i e$ and moving in the direction of the field acquires an energy equal to $z_i eE\lambda/2$. Then the current density i' is :

$$i' = n \, z_i.e.\lambda \, (f'_+ - f'_-) \tag{11},$$

n being the number of particles per volume unit, f'_+ the frequency of jump in the direction of the applied field and f'_- that in the opposite direction. Following the theory of the ionic dissociation in this type of medium (6) :

$$n = n_o \exp(- W/2\epsilon kT) \tag{12},$$

n_o being the number of ionisable entities or salt molecules, W the dissociation energy in a medium of dielectric constant ϵ. Then, the conductivity due to the particles i is given by $\sigma_i = n \, Z_i \, e \, v_i$, where $v_i = \lambda \, (f_+ - f_-)$

$$\sigma_i = \frac{n_o}{N} \cdot \frac{\nu\alpha\lambda^2 Z_i^2 F^2}{RT} \exp\left(-\frac{\gamma V_2^*}{V_f} - \frac{Ea + Eb + W/2\epsilon}{RT}\right) \tag{13},$$

where N is the Avogadro number, R the gas constant, F the Faraday and n_o/N is then the concentration C_i of particles . Consequently we see that the conductivity is the result of a pure W.L.F. behaviour (above the transition temperature) corresponding to the term $-\gamma V_2^* / V_f$, superposed to an Arrhenius behaviour. If we put $\Delta E = Ea + Eb + W/2\epsilon$, we have :

$$\ln \frac{\sigma_T}{\sigma_{T_g}} = \frac{\gamma V_2^*}{V_{fg}} \frac{T - T_g}{T - T_g + \frac{V_{fg}}{\alpha_f}} + \frac{\Delta E}{R}\left(\frac{1}{T_g} - \frac{1}{T}\right) \tag{14}.$$

We can obviously neglect the term $\ln (T_g/T)$. If we write :

$$- \log a_T = \frac{\gamma V_1^*}{2,3 V_{fg}} \cdot \frac{T - T_g}{T - T_g + V_{fg}/\alpha_f} \tag{15},$$

then :

$$\log \frac{\sigma_T}{\sigma_{T_g}} = -\frac{V_2^*}{V_1^*} \log a_T + \frac{\Delta E}{2,3 R}\left(\frac{1}{T_g} - \frac{1}{T}\right) \tag{16},$$

$\log a_T$ being the shift factor obtained from dynamic mechanical measurements. From equation (16) it can be seen that in order to see the Arrhenius behaviour it is necessary to determine the reduced conductivity σ_T/σ_{T_g} at $T - T_g$ = constant, for different membranes having the same V_2^* / V_1^* factor. More generally, we foresee from equation (16) that all materials should have the same variations of the reduced conductivity if the factor V_2^* / V_1^* does not vary appreciably with the ionogenic functions concentration . We shall see later that this is not the case. In order to verify the validity of equation (16) it is necessary to measure the viscoelastic properties of the materials and to compare them with the results obtained from conductivity studies.

VISCOELASTIC PROPERTIES OF SALT-CONTAINING NETWORKS.

Viscoelastic studies of ion containing macromolecular materials are relatively new. In most cases the materials are polyelectrolytes in aqueous solutions. In the solid state the main studies concern polyethylene-based ionomers (7). We do not wish to present here an exhaustive analysis of these results since excellent reviews are available. One can find in Eisenberg's papers and more particularly in refs. (8) and (9) the present state of the art. Let us recall that numerous arguments in favour of the aggregation of the ionisable entities have been presented. To summarise the general tendency, it appears that in low dielectric-constant materials the ion pairs form multiplets. The aggregation trend increases with increasing ionisable function concentration. Then clusters appear, giving not only an ionic crosslinking effect, but also the effect of a reinforcing filler. The transition concentration between multiplet and cluster domains depends on the nature of the polymer. For instance, for polyethylene-based ionomers clusters seem to be present even at low ionic concentration, while the transition is observed at around 6 % mole fraction of monomer units for polystyrene-based ionomers, at around 10 % for ethyl acrylate-based ionomers, while only at 16 % for mixtures of polyethylene oxide and lithium perchlorate (10). In the materials studied in our laboratory the salt molecules are probably present only in the form of multiplets. The C_1 and C_2 constants of the W.L.F. shift factor for polystyrene-based ionomers vary only slightly with ionic concentration, at least in the multiplet domain. We expect the same observation for the networks studied in our laboratory. Concerning the glass transition temperature, this seems to increase linearly with ions concentration (10). Very interesting studies have been published by Moacanin and Cuddihy on the system polypropylene glycol-lithium perchlorate (11). It was shown that the glass transition temperature increased by

5.5°C per % mole fraction. Strong interactions between the dipoles of the monomer units and
the ions induce a non-additive effect for the volumes. When calcium thiocyanate was used,
the rate of increase of the glass transition temperature was lower (1.8°C per mole %) (12). In
this last case additivity of the volumes was observed. Eisenberg has studied the viscosity of
mixtures of polyethylene oxide (POE) and polypropylene oxide (PPO) with lithium perchlorate
(13). The viscosity of these systems followed the superposition principle and the W.L.F.
constants C_1 and C_2 were found to be nearly independent of the salt concentration. C_1 was
approximately equal to that for the pure polymer, while C_2 was quite different. The same
W.L.F. shift factor was found whatever the molecular weight of the glycol, provided it was
lower than 1000 for PPO and lower than 4000 for POE.

It is known that when one chooses the product $f_1 a_T$ as variable, where f_1 is the frequency
for the dynamic mechanic measurements, it is possible to obtain a master curve giving the
storage modulus. We carried out these measurements on some POE-based networks filled with
sodium tetraphenyl boride and crosslinked with triisocyanate. Figure 1 shows the master
curve obtained for one network ; the reference temperature chosen was 26°C.

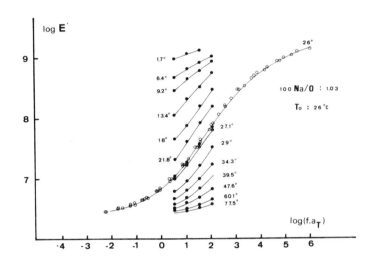

Fig. 1. Isothermal variations of the storage modulus at four different
frequencies for a PEO-400 based network containing 1 % Na atom per ethereal
oxygen. The diagram shows the construction of the master curve for a reference temperature of 26°C.

The variation of the glass transition temperature Tg with salt concentration shown on fig. 2
is linear, increasing with salt concentration. The slopes of the straight lines Tg = f(C)
are approximately equal for both networks, i.e. between 6 and 7°C per % Na/o (sodium atoms
per 100 ether oxygen atoms). These values compare well with some values obtained with
ionomers (9, 13). The concentration effect on Tg is attributed to dipolar interactions
between polyether segments and salt multiplets. The influence of these interactions on the
temperature behaviour for some polyether-salt systems has been described elsewhere (11, 12,
35,36). Note that in our case the salt induces in POE-based networks a slight initial
decrease in Tg which is attributed to a plasticizing effect (18). The theoretical analysis
of the Tg variations with salt concentration can generally be achieved in accordance with
two models : either (i) considering the material as a copolymer of pure monomer units and of

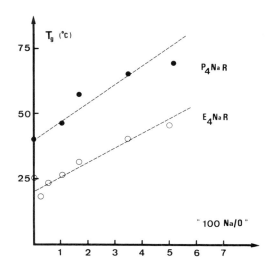

Fig. 2. Variations of glass transition temperature for membranes
synthesised with PEO (E$_4$NaR) or PPO(P4NaR), containing various amounts
of sodium tetraphenyl boride. Values taken at the maximum of loss tangent
measured at 11 Hz.

salt-coordinated units (7), or (ii) considering the network as crosslinked through ionic
bonding (16). The fact that all the membranes have the same viscoelastic behaviour is shown
on fig. 3 for POE-400- based networks. The PPO-425-based membranes show the same pattern which

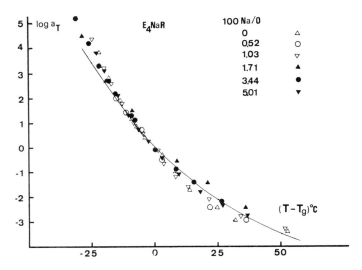

Fig.3 . On this plot are reported the measured values of the shift factor
for PEO 400-based membranes, versus the reduced temperature T - Tg the
glass transition temperature being the reference temperature for each membrane. The line shows the mean shift factor - 10.5 (T - Tg)/(T - Tg + 100)

superposes well enough with the POE one, giving C_1 and C_2 mean values for the shift factor,
despite of a small scatter for high values of T- Tg. If we chose the same reference temperature To for all samples, the superposition of the curves log E = f(f.a_T) makes use of a
concentration shift factor expressed as log a_c. The plot of the concentration shift factor
log a_c versus the salt concentration is linear for both types of network (19). From all these
results we conclude that the same viscoelastic behaviour is approximately observed for
 polyether - polyurethane   networks,   filled   with   sodium   tetraphenyl

boride, of approximately the same crosslink density, and literature data indicate that this
should be so for linear polyether based materials at different salt concentrations. We
conclude that equation (16) holds and indicates that the Arrhenius behaviour is not very
important. For the networks studied in our laboratory the shift factor has been calculated
as

$$- \log a_T = \frac{10.5(T - Tg)}{100 + T - Tg}, \ (\log a_T < 0) \tag{17}.$$

Taking into account the approximations used and the scatter of the points in the superposition, the accuracy of the numerical values of C_1 and C_2 must not be overestimated.

EXPERIMENTAL STUDY OF THE IONIC CONDUCTIVITY OF MACROMOLECULAR NETWORKS.

We have measured the conductivity of the networks filled with sodium tetraphenyle boride the
mechanical behaviour of which is reported in the preceeding section. The assumption that the
conductivity is ionic is supported by many considerations. First, we have verified that the
salt alone does not give any measurable conductivity with the measuring system in use in our
laboratory.

Further, when this type of membrane is placed between   two sodium amalgams
of different concentrations it is possible to measure a transport number close to unity.
Conductivity measurements are made following the complex impedance plot method varying the
frequency of the measuring current, and searching for an impedance value corresponding to
zero phase current output (17). The samples are outgassed under vacuum in the oven where the
electrical measurement is performed, so as to eliminate as much as possible the possible
influence of traces of water (4). Whatever the care taken in the preparation of samples and
conductivity measurements, the conductivity of samples of polyethers without salt is not
completely negligible, mainly for PEO-based networks. This observation has been made
previously (15), but the origin of this phenomenon is still unclear. The variation of the
ionic conductivity for PEO-based networks, corresponding to glycol having a molecular weight
equal to 400, for different sodium tetraphenylboride concentrations, is shown on fig. 4. On

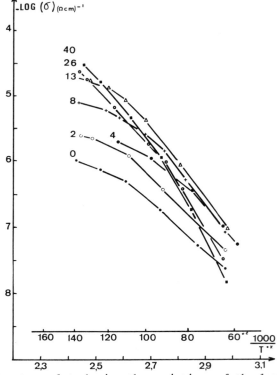

Fig. 4. Arrhenius type plot showing the variations of the logarithm of the
conductivity for different polyethylene glycol 400 networks filled with
different amounts in sodium tetraphenyl boride (the figures indicate the
weight % salt/polyether fraction).

this fig. the logarithm of the measured conductivity is plotted against the reciprocal temperature. The non-Arrhenius behaviour is clearly seen on this plot. One can also notice that
the most conductive materials are note necessarily the richest in salt for a given temperature. In fig. 5, the same function is plotted against the percent molar fraction sodium

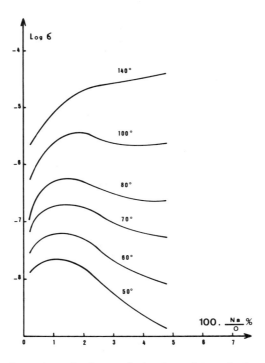

Fig. 5. Isothermal variations of the logarithm of the conductivity
against salt concentration for PEO 400 based membranes. The concentrations
are given in percent molar sodium atoms to ethereal oxygen atoms.

atom/ether-type oxygen atom at constant temperature. The same type of plot has been obtained
for PPO-based networks (20). The conductivity goes through a maximum for nearly all temperatures and for all types of network. In order to facilitate the analysis of the ionic
conductivity behaviour it was necessary to have different materials having the same salt
content for different Tg. This was achieved by the use of different polyether glycols of
different molecular weights and with different crosslinking agents, since this last factor
has been shown to have an influence on the Tg (21). The different characteristics of the
materials are reported in table 1.

TABLE 1. Synthesis characteristics and Tg for various polyether-based
networks filled with $NaBØ_4$

N°	Polyether[a) (Mn)	crosslinking agent	molar ratio NCO / OH	$NaB(C_6H_5)_4$ % weight /polyether	molar ratio O / Na	T_g(11 Hz) °C
1	POE diol (400)	Triphenyl-méthane triisocyanate "Desmodur R"	1,00	26.8	29,9	18
2	POE diol (1000)	Desmodur R	1,03	26,8	29,0	−21
3	POP diol (425)	Desmodur R	1,01	20,0	29,5	37
4	POP diol (1025)	Desmodur R	1,03	20,1	29,3	−11
5	POP triol (1500)	Toluene diisocyanate	1,03	20,0	29,5	− 1
6	POP triol (1500)	Hexamethylene diisocyanate	1,03	19,9	29,7	−22

a) POE diol : poly(oxyéthylene) glycol ; POP diol : poly(oxypropylene)
   glycol ; POP triol : poly(oxypropylene) triol.

Fig. 6 shows a plot of the logarithm of the measured conductivity against the W.L.F. shift factor using the classical values for C_1 and C_2 of 17.4 and 51.6 respectively. One can see that the different curves are in fact linear within experimental error and that they nearly

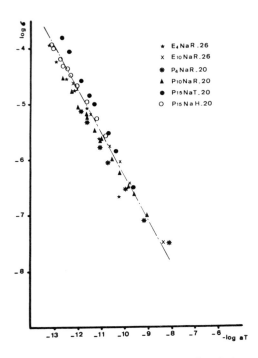

Fig. 6. Variations of the logarithm of the conductivity versus the W.L.F. shift factor expressed as $-17.4(T - Tg) / (51.6 + T - Tg)$ for different membranes having the same sodium tetraphenyl boride concentration, ($\simeq 3$ % Na/O) Tg being determined as reported on legend fig. 2.

superpose. One main approximation is implicitly used here, i.e. that these materials have the same density. Contrary to appearances one must not deduce from this plot that all polyether-based networks have the same conductivity at the same reduced temperature T - Tg. Indeed, we have chosen here membranes in which the volume fraction is relatively high due to the high salt content. This fact tends to minimise the differences between them and a similar conductivity behaviour is observed. However a careful examination of fig. 6 shows that the logarithm of conductivity can vary by about 0.5 log unit from one membrane to another at a given reduced temperature. Each material has its own behaviour especially at low salt concentrations. The main conclusion which can be drawn from this plot is that the Arrhenius contribution is not important enough to induce a significant departure from a simple W.L.F. behaviour. On fig. 7 and fig. 8 are shown the dependence of the logarithm of the measured conductivity on the shift factor - log a_T, as estimated from dynamic mechanical measurements. We see that straight lines are obtained, the slope of which is around 1 for PEO networks filled with a molar ratio Na/-O- equal to 3.4 %, and 0.9 for PPO ones filled with the same amount of salt. On these figures one also notices that the slope varies slightly with salt concentration.

A rough estimation of the activation energy associated with the conductivity process in these networks could be obtained as follows. In order to avoid the uncertainty linked with the extrapolated value σ_{Tg}, we can use of the Y function defined as :

$$Y = \log\left[\sigma_{(Tg + 100)} / \sigma_{(Tg + x)}\right] - a \, \log\left[a_{(Tg + 100)} / a_{(Tg + x)}\right] \tag{18}$$

where a is the slope of the lines obtained on the preceeding figure. If we plot this Y function for the membranes listed in Table 1, for which the W.L.F. shift factor has been determined, against $1/(Tg + x) - 1/(Tg + 100)$, we obtain a mean value for the Arrhenius energy of around 20 kJ/mole. This is consistent with our observation, but more work with a better accuracy is needed to ascertain this result.

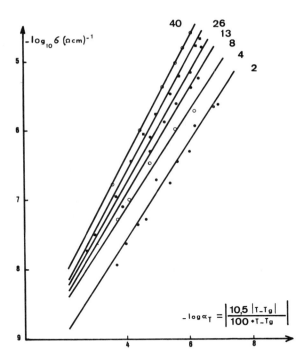

Fig. 7. Logarithm of the conductivity for PEO 400-based membranes versus the W.L.F. shift factor measured by dynamic mechanical experiments. The figures refer to the weight percent of sodium tetraphenyl boride.

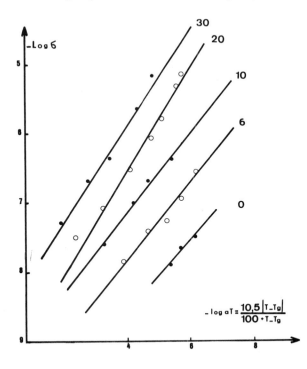

Fig. 8. On this figure the same plot as on fig. 7. shows the behaviour of PPO 425-based membranes, versus-log a_T = 10.5 (T − Tg)/(T − Tg + 100). The figures refer to the weight percent sodium tetraphenyl boride concentration.

Now we also must take into account in our calculations the effect of salt concentration. If the influence of the salt concentration were only limited to the preexponential term of eq. (13) we should observe the same reduced conductivity for the same network structure at the same reduced temperature. This is shown on fig. 9 where we plotted $\log \sigma_T/\sigma_{Tg}$ against

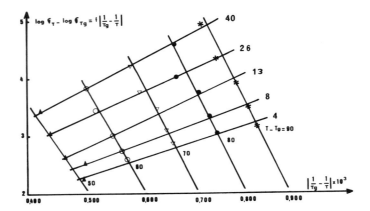

Fig. 9. Plot of the logarithm of the reduced conductivity σ_T/σ_{Tg} against $1/T_g - 1/T$ for PEO 400-based networks. The figures on the right refer to the weight percent salt concentration, the figures below indicate the value of the reduced temperature.

$1/T_g - 1/T$. We see that for PEO-based networks of the same theoretical crosslink density there is an increase of the reduced conductivity with salt concentration at a constant reduced temperature. The same behaviour, but for PPO-based networks, is shown on fig. 10. If the

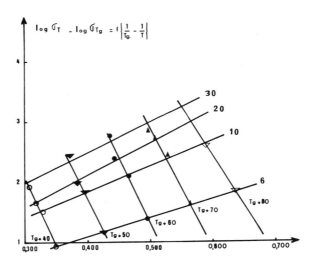

Fig. 10. Same plot as on the preceeding fig. 9 but for PPO 425 based networks. For the meaning of the figures see legend Fig. 9.

variations at the same reduced temperature were attributed to an activation energy, the calculation shows that this energy would be higher than 300 kJ/mole which is by far too high to be in agreement with the preceeding observations. In fact, we find variations of the reduced conductivity of nearly 2 log units when the concentration is going from 0.5 % to 5 % Na/ - O -, the reduced temperature going from 50 to 90°C. On the other hand we have assumed with dynamic mechanical measurements that the variations of the storage modulus with concentration were negligible within experimental error. If this is so, the main part of the variations of the reduced conductivity should be attributed to the V_2^*/V_1^* multiplying factor of eq. (16). As the V_1^* factor is not varying with salt concentration, the variations must linked to the V_2^* factor. If we assume a linear variation of the type

$$V_2^* = V_{2,0}^* + a.c \qquad (19),$$

we have :

$$\log\left[\sigma_T/\sigma_{Tg}\right] = -(V_{2,0}^*/V_1^*)\log a_T - (a/V_1^*)\log a_T \times C \qquad (20),$$

neglecting the Arrhenius activation energy contribution. Equation (20) shows that the variations of the logarithm of the reduced conductivity are linear with salt concentration, at constant reduced temperature. This is shown on fig. 11 for PEO 400-based networks. This plot

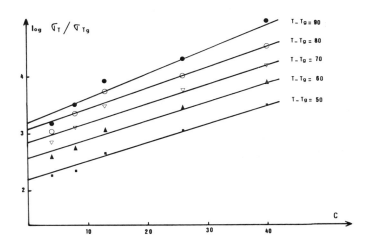

Fig. 11. Logarithm of the reduced conductivity σ_T/σ_{Tg} versus salt concentration, expressed in weight %, at constant reduced temperature, for PEO 400-based membranes.

shows that straight lines are obtained in satisfactory agreement with equation (20). By dividing the slopes of these lines by the calculated mean value of the W.L.F. shift factors given by equation (17) we obtain a constant value of a/V_1^* within experimental error, independent of the salt concentration, in agreement with our present interpretation. The same plot is obtained for PPO-based networks, fig. 12. Consequently we think that the minimum

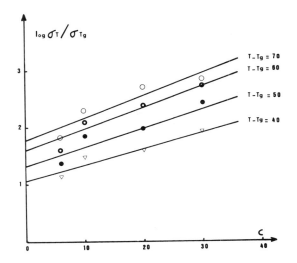

Fig. 12. Same plot as on fig. 11, but for PPO 425-based networks.

value for the free volume consistent with the cation jump ability slightly increases with salt concentration. One possible explanation could be that, as in the case of network swelling by solvent molecules, the mean conformation of the macromolecular chain varies. Then the size of the equilibrium site may vary. For instance the mean number of oxygen atoms involved in the formation of the minimum size hole can increase with salt concentration in order to give the free volume fraction. It is worthwhile to recall here that it is necessary to assess more firmly the accuracy of the dynamic mechanical measurements and to extend the temperature range of the conductivity measurements before taking this conclusions for sure. However, it is clear that the conclusions drawn from equation (16) are verified in our experiments and that there is a close relationship between the dynamic   mechanical properties and the ionic conductivity properties in our polyether networks filled with an ionisable metal salt .

As indicated in the theoretical part of this lecture, the formula given for the conductivity variations is not entirely new. In fact it derives from the interpretation given by Miyamoto and Shibayama. They showed that their formalism can take into account the conductivity behaviour for macromolecular systems both above the glass transition temperature and below it (5). Their results concern mainly thermoplastic materials, like methyl methacrylate or polystyrene in the absence of voluntarily added salt. It is worth noting that in that study the conductivity ranged from 10^{-13} to 10^{-19} $(\Omega cm)^{-1}$ (5). It has been shown recently that it is possible to account for the conductivity variations above the glass transition zone by a free volume model in the case of mixtures of polypropylene oxide with lithium trifluoromethane sulfonate (22). This behaviour was derived from the Tamman and Fulcher law for liquids (23) :

$$\sigma = A.T^{-1/2}. \exp(E/(T - To)) \qquad (21)$$

A similar relationship was obtained for PEO polymers filled with lithium or cesium thiocyanate (22). In this case the observed conductivity ranged between 10^{-7} and 10^{-3} $(\Omega cm)^{-1}$ as in the case of the networks studied in our laboratory. We have intented to show that the free volume model applies equally well to crosslinked materials and that it is possible to observe at the same time a high conductivity and the rubbery elastic mechanical properties. These last properties would be very useful in practical applications. We show here that the good correlation observed between the properties linked to these two different domains is due to the fact that they are the consequence of the segmental movements which govern the transport and relaxation properties. This idea has been already expressed by Warfield and Hartmann in a recent paper (24). Their interpretation only covers the ionic conductivity for glassy materials where an Arrhenius type process predominates. We have already criticised the approach consisting in explaining the conductivity variations with temperature with the only help of an activation energy (20), at least when the ionic conductivity is high. We show here indeed a case where the Arrhenius contribution is negligible. This conclusion was to be expected considering that the ionic dissociation enthalpy of the salt is very low. It is known that at 25°C the dissociation enthalpy of sodium tetraphenyl boride is -5.3 kJ/mole in THF (28) and -7.5 kJ/mole in dimethoxyethane (29). Moreover, it is known that the macromolecular nature of the "solvent" in our context increases the interactions with cations. The strong solvating power of POE has been already emphasized many times. For instance, polydioxane interacts more strongly with cations that its monomeric homologue (30). Some infrared studies seem to show that the polyether chain is placed around the cation in such a way that the complexes of polyethylene oxide with sodium or potassium salts have a helical conformation (31). In this last paper it is stated that a polymer having a low cohesive energy density should be a better solvent for metal salts than polymers which involve high interchain interactions. Our work provide a quantitative apprais al supporting this opinion. Polyethers are also known to form complexes with divalent metal salts like for instance zinc chloride (25). These complexes have been found to exhibit interesting electret properties due to their relatively high glass transition temperature. It has also been shown that the polyether-salt complexes are materials with a high dielectric constant (25). If we call W the dissociation energy, we have seen that in eq. (13) this parameter appears under the form $W/2\varepsilon$. Consequently, the only two other terms contributing to the overall activation energy are Ea and Eb, respectively the activation energy for chain unit and for ionic displacements. These energies are known to be small. In the work reported in ref. (26) concerning the system polyacrylonitrile-hydrated metal perchlorates it is reported that the Arrhenius behaviour alone cannot account for the variation of the conductivity with the reciprocal temperature around Tg, and it is suggested that the sudden variation of dielectric permittivity through the glass transition zone explains the observed phenomenon, rather than the free volume variation. We think that the evidence presented here both in the analysis of the litterature and in our experimental work shows that on the contrary the ionic conductivity follows a free volume behaviour. These considerations also justify the qualitative observations given by us in a recent paper concerning complex membranes synthesised from nitrocellulose-polyether-metal salt mixtures with different crosslinking agents (27). The free volume behaviour abundantly examp)ified in our work dealing with the high ionic conductivity of crosslinked materials seems to be quite widespread, since it seems to apply also to the case of the ionic conductivity of linear polymers, filled or not (1, 22), and more generally to all the transport properties as recently underlined by Meissner (32). Moreover, it seems that the free volume theory can account for the ionic conductivity of inorganic liquids and the corresponding glassy materials (33-34).

The molecular model for the ionic transport through macromolecular materials in the solid state can be questioned. For instance, the ion displacement can be achieved by a step by step deformation of the cage containing the ion, or by a jump in a preformed hole appearing in the vicinity of the ion at an equilibrium position. The first model implies an exchange of oxygen atoms of the cage with those of the surrounding material. The second model implies the formation of free holes and that the cation jumps in them when the required vibrational energy conditions are fulfilled. In the first model the activation energy of the movement of a chain element allowing the displacement of the cation should be quite different from that observed from the dynamic viscoelastic properties, but this does not seem to be the case. One should in fact observe an activated process. The second model follow the hypothesis which allows the use of the calculation of the jump probability given by Cohen and Turnbull, since a simple free volume rearrangement is implied, only requiring thermal fluctuations (37) without energy consumption.

In conclusion we can examine the predictions of the free volume theory on the behaviour of ionically conductive materials. Table 2 reports the expected predominant regimes (Arrhenius or W.L.F.) as a consequence of the structure of the conducting material and the temperature).

TABLE 2. Predicted conductivity regimes from free volume theory.

Nature of the macromolecular material		temperature range*	Expected Regime for ionic conductivity	Main contribution to the apparent activation energy
macrostructure	nature of the phase			
linear	cristalline	$T < T_m$	Arrhenius	Dissociation + jump activation
		$T > T_m$	W.L.F.	Expansion of free volume
linear or crosslinked	semi cristalline	$T < T_g$	Arrhénius	Dissociation + jump activation
		$T_g < T < T_m$	mixed	composite, depending on dielectric constant
		$T > T_m$	W.L.F.	expansion of free volume
linear or crosslinked	amorphous	$T < T_g$	Arrhenius	Dissociation + jump activation
		$T > T_g$	W.L.F.	expansion of free volume

* T_g = glass transition temperature, T_m = melting temperature zone.

From this table it can be seen that ionic conductivity, which is a transport property, can be always accounted for on free volume-based considerations. Consequently even the low ionic conductivity materials, pure polymers or filled systems, can be fitted within this general scheme.

Acknowledgements - The Direction Des Recherches Etudes et Techniques and the Société Nationale Elf-Aquitaine are gratefully thanked for supporting this research. A. Killis, J.F. Le Nest and A. Gandini are also thanked for experimental work and useful discussions.

REFERENCES

1. P.V. Wright, Br. Polym. J. 7, 319 (1975).
2. R.G. Pearson, J. Chem. Ed. 45 (9), 581 (1968).
3. P.B. Macedo and T.A. Litowitz, J. Chem. Phys. 42 (1), 245 (1965).
4. H.D. Weymann, Kolloïd Z. 181, 131 (1962).
5. T. Miyamoto and K. Shibayama, J. Appl. Phys. 44 (12), 5372 (1973).
6. R.E. Barker Jr., Pure and Applied Chem. 46, 157 (1976).
7. L. Holiday, Ionic Polymers, Applied Science Publ. London 1975, p. 43.
8. Pure and Applied Chemistry, 46, H. Eisenberg Ed., Pergamon Press (1976).
9. A. Eisenberg and M. King, Ions Containing Polymers, Physical Properties and Structure, Academic Press N.Y. (1977).
10. H. Eisenberg, ref. 8, p. 171.
11. J. Moacanin and E.F. Cuddihy, J. Polym. Sci. C14, 313 (1966).
12. M.J. Hannon and K.F. Wissbrun, J. Polym. Sci., Polym. Phys. Ed. 13, 113 (1975).
13. A. Eisenberg, K. Ovans and H.N. Yoon, Ions in Polymers, Advances in chemistry series N° 187, p. 267, A. Eisenberg Ed., A.C.S. (1980).

14. J.M. Chabagno, thèse Ingenieur Docteur, oct. 1980 Grenoble.
15. A.E. Binks and A. Sharples, J. Polym. Sci. PA2 6, 407 (1968).
16. L.E. Nielsen, J. Macromol. Sci., Rev. Macromol. Chem., C3, 69 (1969).
17. D. Ravaine and J.L. Souquet, J. Chim. Phys. 5, 693 (1974).
18. A. Killis, J.F. Le Nest, A. Gandini and H. Cheradame, J. Polym. Sci, Polym. Phys. Ed., in press.
19. A. Killis, J.F. Le Nest, A. Gandini and H. Cheradame,   Makromol. Chem. to be published.
20. H. Cheradame, J.L. Souquet and J.M. Latour, Mat. Res. Bull. 15, 1173 (1980).
21. A. Killis, J.F. Le Nest and H. Cheradame, Makromol. Chem., Rapid Commun. 1, 595 (1980).
22. M.B. Armand, J.M. Chabagnot and M.J. Duclot, Fast ion transport in solids, Vashishta, Mundy and Shenoy Ed., Elsevier N. H. Inc., (1979), p. 131.
23. G.S. Fulcher, J. Am. Ceram. Soc. 8, 339 (1925).
24. R.W. Warfield and B. Hartmann, Polymer 21 31 (1980).
25. R.E. Wetton, D.B. James and P. Warner, in Ref. 13, p. 253.
26. S. Reich and I. Michaeli, J. Polym. Sci, Polym. Phys. Ed. 13, 9 (1975).
27. D. André, J.F. Le Nest and H. Cheradame, Europ. Polym. J. 17, 57 -61 (1981)
28. J. Comyn, F.S. Dainton and K.J. Ivin, Electrochimica Acta 13, 1851 (1968).
29. C. Carjaval, K.S. Tölle, J. Smid and M. Swarc, J. Am. Chem. Soc. 87, 5548 (1965).
30. I.M. Panayotov, D.K. Dimov, C.B. Tsvetanov, V.V. Stepanov and S.S. Skorokhodov, J. Polym. Sci., Polym. Chem. Ed. 18, 3059 (1980).
31. B.L. Papke, M.A. Ratner and D.F. Shriver, J. Phys. Chem. Solids, in press.
32. B. Meissner, J. Polym. Sci., Polym. lett. Ed. 19 137 (1981).
33. B. Meissner, H. Meissnerova and L. Sasek, Glastechn. Ber. 46 (12), 240 (1973).
34. B. Meissner and H. Meissnerova, Glastechn. Ber. 47 (9), 209 (1974).
35. R.E. Wethon, D.B. James and W. Whitting, J. Polym. Sci, Polym. lett. Ed 14, 577 (1976).
36. D.B. James, R.E. Wetton and D.S. Brown, Polymer 20, 187 (1979).
37. M.H. Cohen and D. Turnbull, J. Chem. Phys. 31 (5), 1164 (1959).
38. R.E. Barker Jr and C.R. Thomas, J. Appl. Phys. 35, 3203 (1964).

THE STRUCTURE AND PROPERTIES OF ULTRA HIGH MODULUS FILMS AND FIBRES

I.M.Ward

Department of Physics, University of Leeds, Leeds LS2 9JT, UK.

Abstract - The factors involved in the structure and properties of high modulus oriented polymers are outlined. Both rigid chain and flexible chain polymers are considered, including in the latter case solid phase deformation, dilute solution spinning and gel-spinning. The preparation and properties of materials obtained by solid phase deformation have been linked to a common theme, the complimentary roles of crystal continuity and a molecular network. The general relevance of these ideas is discussed.

INTRODUCTION

In any discussion of the structure and properties of high modulus oriented polymers it is useful to recognize that the limitations to the achievement of high modulus can most conveniently be considered to fall into three categories

   (1) The molecular chain structure and crystal structure
   (2) Molecular orientation, conformation and morphology
   (3) Dynamic aspects, molecular relaxation, thermally activated processes.

In general terms it is not difficult to appreciate that these categories form an ascending heirarchy of requirements. First, the intrinsic structure of the molecular chain must possess the potential for high stiffness. Secondly, it is necessary that the oriented polymer structure contains aligned extended chain conformations so that this potential is realised. Finally, there must be no reduction in stiffness due to dynamic relaxation or slip processes.

In practice there have been two distinctly different approaches to achieving these requirements

   (1) The preparation by novel preparative chemistry of intrinsically stiff polymers.
   (2) The use of novel fabrication procedures to make high modulus oriented products from conventional flexible chain polymers.

In our research at Leeds University we have been primarily concerned with the second approach and I shall deal with that in the most detail in this lecture. However, I should like to begin by a discussion of the first approach, and in both cases I shall show how the three requirements are satisfied.

RIGID CHAIN POLYMERS

The aromatic polyamides, of which poly(1,4 benzamide) PBT $\left[-NH\langle 0 \rangle -CO-\right]_n$

and poly(p-phenylene terephthalamide) PPD-T $\left[-CO\langle 0 \rangle -CONH\langle 0 \rangle -NH-\right]_n$

are the simplest, illustrate the preparative chemistry approach. These polymers exist preferentially in the trans conformation, leading to an extended chain configuration of high intrinsic stiffness, which satisfies the requirement under heading (1) above. Moreover there is a high barrier to rotation about the carbonyl-nitrogen bond which makes any reduction in stiffness with increasing temperature very small, hence fulfilling the requirement under heading (3) above. The required high degree of molecular orientation, together with the extended chain conformation is obtained by solution spinning. Because these polymers are not chemically stable in the melt, chain extension is achieved in solution, and this has been shown by the fact that the relationship between intrinsic viscosity and molecular weight indicates a high exponent in the Mark - Houwink

relationship (Ref.1).   These polymers can in fact all show liquid crystalline behaviour
in solution, although this is not necessarily the case in the conditions used in commercial
practice for spinning fibres.   The very high degree of molecular alignment produced in the
solution spinning process has been demonstrated by X-ray measurement of orientation.   The
molecular alignment can also be increased by further annealing under tension to give a final
improvement in modulus and strength.   The excellent correlation between initial modulus and
X-ray orientation (Ref.2) is demonstrated in Fig.1.   It appears that there are no

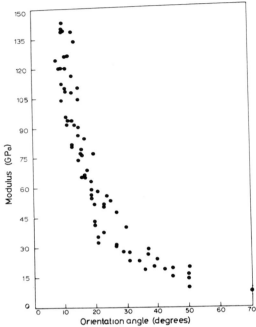

Fig. 1   Initial modulus and X-ray orientation for aromatic polyamides (after Ref.2)

morphological considerations for these materials as far as modulus is concerned, and they
can be regarded as a single phase aggregate of anisotropic units.   Northolt and Van Aartson
(3) have shown that an aggregate model (Ref.4) is entirely appropriate and also provides a
quantitative basis for understanding the changes in modulus under stress as the orientation
is increased.   The requirements under category (2) are therefore comparatively straightforward to understand, there being none of the morphological complexity which we will find
in the case of flexible polymers.

FLEXIBLE CHAIN POLYMERS

In the case of flexible chain polymers the starting point is again to consider the intrinsic
stiffness as revealed by the molecular chain structure.   This leads us to examine the
values for the crystal modulus obtained either from theoretical calculations or from
experimental techniques, principally X-ray measurements of crystal strain when oriented
samples are held under load (Ref.5).   The challenging result is that several commercially
available polymers, notably polyethylene, nylon and polyethylene terephthalate which have
extended chain crystalline configurations show moduli in the range 100-300 GPa.   Polymers
with a helical crystal structure such as polypropylene and polyoxymethylene are of lesser
interest, although even there crystal moduli in the range 40-100 GPa may be expected.

We can immediately appreciate that the factors listed under (2) and (3) above are likely to
be less favourable.   The complex morphology of flexible chain crystalline polymers makes
the requirements in fabrication more complex, and the chain flexibility implies that there
will be greater likelihood of reduction in properties due to dynamic processes.   In another
respect, however, the chain flexibility is a positive advantage because it enables
conventional processing methods to be successful to a substantial extent.

The most spectacular results in flexible polymers have been obtained for polyethylene, where
there have been two major points of departure for high modulus studies.   These were (a) the
deformation of bulk polymers to very high degrees of stretching and (b) the production of
fine fibres from dilute polyethylene solutions.   At Leeds University we have concentrated
on the bulk deformation route.   I shall discuss the principles involved in this case in
some detail, and then contrast the situation here with that of the solution routes.   This
will be followed by an account of our present understanding of the structure and properties
of the ultra-oriented materials.   In this respect I shall endeavour to indicate the link
between the factors influencing the processing behaviour and the final product properties
e.g. between plastic deformation of the bulk polymer to produce high modulus polyethylene
and the creep and recovery behaviour of the oriented products.

The preparation of ultra high modulus polyethylene
At Leeds University three practical processing methods have been developed for the
production of high modulus materials.   These are   (1) Tensile drawing
                                                      (2) Hydrostatic extrusion
                                                      (3) Die drawing
Many of the details of these processes, and detailed specifications of the key variables
involved, are described in the patent applications dating back to 1973 (Ref.6).   A process
for the small-scale production of fibre monofilaments and multifilament yarns has been in
operation for several years.   The guidelines for this process were established by tensile
drawing of small dumbbell samples in an Instron tensometer, and have been described in a
number of publications (Refs. 7,8).   Foremost among the principles established was the
importance of the draw ratio.   The concept of a natural draw ratio was replaced by that
of "effective drawing" i.e. drawing which produces a genuine transformation of the structure
to give molecular alignment.   Under such conditions the Young's modulus of the oriented
polymer was shown to be a unique function of the draw ratio, and subsequently the extrusion
ratio in the hydrostatic extrusion process or the draw ratio again in the die-drawing process.
(Fig.2).   It was shown that by suitable combinations of processing conditions a wide range

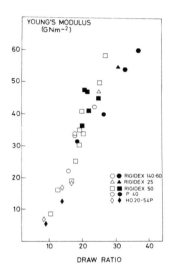

Fig. 2   Young's modulus versus draw ratio for a range of quenched
         (open symbol) and   slow-cooled (solid symbols) LPE samples
         drawn at 75°C.

of polymers of different molecular weight characteristics and different initial thermal
treatments could be drawn to high draw ratios and consequently high Young's moduli.   In
spite of the apparent complexity of the drawing studies, it was proposed that the deformation
of a molecular network must be the fundamental consideration in determining the drawing
behaviour (Ref.8).   Key elements in this network are the physical entanglements and the
crystalline regions, both of which supply semi permanent junction points in the structure.
These ideas were first developed in a qualitative fashion to explain the influence of
molecular weight and initial morphology on the drawing behaviour.   In a more sophisticated
approach Coates and Ward (9) examined the influence of these parameters on the strain
hardening and strain rate sensitivity by mapping out the neck profiles for a series of
samples of different molecular weight and initial thermal treatment.

In general terms the change of the drawing stress or flow stress can be related to changes
in the draw ratio λ and strain rate $\dot{\varepsilon}$ by the equation

$$d\sigma = \left(\frac{\partial\sigma}{\partial\lambda}\right)_{\dot{\varepsilon}}   d\lambda + \left(\frac{\partial\sigma}{\partial\dot{\varepsilon}}\right)_{\lambda} d\dot{\varepsilon}$$

where $\left(\frac{\partial\sigma}{\partial\lambda}\right)_{\dot{\varepsilon}}$ is the strain hardening term

and $\left(\frac{\partial\sigma}{\partial\dot{\varepsilon}}\right)_{\lambda}$ is the strain rate sensitivity term

It was found that molecular weight was predominant in affecting the strain hardening, and
the initial morphology in affecting the strain rate sensitivity.   At first sight this would

support the very simplistic model shown in Fig. 3(a) where the strain hardening term can be identified with the molecular network and the strain rate sensitivity with the crystal slip.

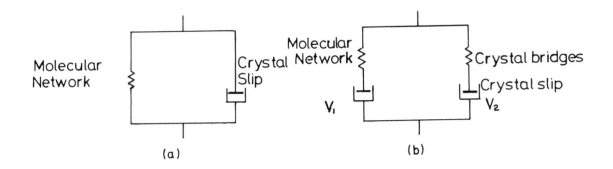

Fig. 3 Models for drawing process

It is however more instructive to consider the drawing process in terms of thermally activated processes. Following Eyring, we have for a single activated process,

$$\dot{\varepsilon} = \dot{\varepsilon}_0(\lambda) \; \exp \; -\left(\frac{\Delta H - \sigma v}{kT}\right)$$

where ΔH, and v are the activation energy and volume.

We then have

$$\left(\frac{\partial \sigma}{\partial \lambda}\right)_{\dot{\varepsilon}} = -\frac{1}{\dot{\varepsilon}_0}\left(\frac{\partial \dot{\varepsilon}_0}{\partial \lambda}\right)_{\dot{\varepsilon}} \; \frac{kT}{v} \quad \text{and} \quad \left(\frac{\partial \sigma}{\partial \dot{\varepsilon}}\right)_{\lambda} = \frac{1}{\dot{\varepsilon}} \; \frac{kT}{v}$$

The strain hardening relates to the change in the pre-exponential factor $\dot{\varepsilon}_0$ with draw ratio, which can be regarded in physical terms as associated with the exhaustion of the flow processes as the highly aligned drawn structure evolves. The strain rate sensitivity relates to the activation volume for the activated event. In the case of linear polyethylene it has been shown that the flow stress behaviour can be accurately modelled by two activated processes coupled in parallel as shown in Fig. 3(b). One process, that associated with the crystalline regions, has a comparatively small activation volume ($\sim$ 50-100 A^{03}). The other process, that associated with the molecular network, has a comparatively large activation volume ($\sim$ 500 A^{03}) but a very much smaller pre-exponential factor $\dot{\varepsilon}_0'$. The behaviour is shown schematically in Fig.4 where the separate contributions

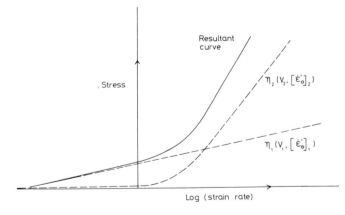

Fig. 4 Schematic diagram of two process flow behaviour

of the two processes are shown together with the total response. The total stress σ is given by

$$\sigma = \frac{2.3kT}{v_1} \left(\log \dot{\varepsilon}_p - \log \left(\left[\dot{\varepsilon}_0'\right]_1 /2\right) \right) + \frac{kT}{v_2} \; \sinh^{-1} \left(\dot{\varepsilon}_p / \left[\dot{\varepsilon}_0'\right]_2\right)$$

where the subscripts 1 and 2 refer to each process in Fig. 1(b).

The interpretation of the strain hardening and strain-rate sensitivity data is therefore rather more subtle than implied by Fig. 3(a). However, in general terms it is possible to see that at high stresses, the crystal dashpot 2 takes most of the stress and hence determines the strain rate sensitivity. The increase in stress with increasing draw ratio can however be influenced by decreases in either $\left[\dot{\varepsilon}_0'\right]_1$ or $\left[\dot{\varepsilon}_0'\right]_2$. The results suggest that the molecular network term $\left[\dot{\varepsilon}_0'\right]_1$ does have the predominant effect here, but the firmness of this conclusion is partly due to the detailed information on $\left[\dot{\varepsilon}_0'\right]_1$ and $\left[\dot{\varepsilon}_0'\right]_2$ values obtained from creep data, as will be discussed later.

In the case of hydrostatic extrusion, where a cylindrical billet is extruded in the solid phase through a conical die, similar considerations apply, with two added complications (Ref.10). First there is the additional influence of the hydrostatic pressure p on the flow stress. This can be conveniently described by the introduction of a pressure activation volume Ω so that

$$\dot{\varepsilon} = \dot{\varepsilon}_0 \exp \frac{- \left(\Delta H - \sigma v + p\Omega\right)}{kT}$$

Secondly there is the effect of friction between the billet and the die. Developing on the idea of Tabor and Briscoe, Hope and Ward (11) have shown that the coefficient of friction $\mu = \frac{\Omega}{2v}$. An analysis of the mechanics of hydrostatic extrusion shows it to be at a disadvantage compared with tensile drawing for the following reasons. First, the strain rate field is determined by the die geometry and for stable extrusion this means that the highest strain rates are encountered near the die exit where the plastic strain is the greatest. In drawing, the highest strain rates are encountered in the neck at much lower plastic strains. Now the balance between the network process 1 and the crystal process 2 does change between low draw ratio and high draw ratio so that the apparent activation volume falls, and hence the strain rate sensitivity increases. This means that higher flow stresses are required in hydrostatic extrusion than in drawing through a neck. Secondly, the effects of hydrostatic pressure on the flow stress and friction both serve to increase the pressure markedly for high total deformation ratios. These considerations have led to the recognition of the potential of die-drawing for the production of oriented polymers in large sections. The guidelines for die drawing of polymers are very similar to those established for tensile drawing, with the added complication of a non-isothermal process.

Comparison with dilution solution and gel-spinning processes

Very soon after the reports of ultra high modulus polyethylenes by tensile drawing, Pennings and Zwijnenberg (12) reported the preparation of similar modulus materials with even higher strengths ($\sim$ 3-4 GPa as against 1-1.5 GPa) by growing very fine fibres from dilute solutions of polyethylene in either Poiseuille or Couette flow. This process was referred to as seeded crystallisation, and studies by the Bristol Group (Ref. 13) emphasised the importance of an extensional flow field so that the molecules could be aligned in solution prior to crystallisation.

In a recent development Smith and Lemstra (14) have reported the preparation of similar materials by a two stage process. A 2% solution of high molecular weight polyethylene in decalin is quenched into water. Structural studies suggest that this gives a gel fibre, a network of polymer molecules with crystallites acting as physical cross-links. These fibres are then drawn in a hot air oven at 120°C to high draw ratios. There was a similar but not identical relationship between Young's modulus and draw ratio to that observed in the solid phase deformation processes described above. These results emphasise the importance of a suitable molecular network as concluded for the tensile drawing studies of bulk polymer. Moreover, it suggests that there should be a reassessment of previous hypotheses for the underlying principles involved in all the solution processes. It is likely that the underlying factor in both solution and bulk deformation processes is the existence of a suitable molecular network. In the solution case a comparatively high molecular weight polymer appears to be mandatory, to provide an adequate number of molecular entanglements at the low polymer concentrations involved. In the bulk solid, a comparatively low molecular weight is easier to process because molecular entanglements must not be too high. It is interesting to reflect that these ideas date back to the genesis of the solid phase deformation routes when the importance of weight average molecular weight was first recognised.

Structure and properties of ultra high modulus polyethylene

We now wish to consider the structure and properties of the highly oriented polyethylenes, especially with regard to the likely limitations which the morphology will impose on properties.

The high modulus polyethylenes obtained by the solid phase deformation methods have been
examined at Leeds University by a wide range of structural techniques, including wide angle
and small angle X-ray diffraction, electron microscopy, broad line NMR and nitric acid
etching followed by GPC analysis.   The X-ray diffraction results showed that there was very
high crystalline orientation and a distinct long period, although the intensity of the small
angle pattern decreased with increasing draw ratio.   These results are consistent with the
Peterlin model (Ref.15) of crystal blocks linked by taut tie molecules to provide the
required stiffness.   We have carried this model one stage further by postulating the
presence of crystalline bridges (Ref.16) between the crystal blocks (Fig.5(a)).   It is
considered that these bridges are randomly distributed and that the distribution of crystal

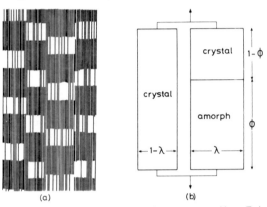

(a)   (b)

Fig. 5 Crystalline bridge model (a) and corresponding Takayanagi
        Parallel-Series Model (b)

bridges, and hence the crystal length distribution can be related to a single parameter p
which defines the probability that a crystalline sequence traverses the disordered regions
to link two adjacent lamellae.   Dark field electron microscopy and nitric acid etching
confirmed that the crystal length distributions were in reasonable accord with this simple
scheme.   In practice, the parameter p is most readily obtained from measurements of the
small angle long period and the integral breadth of the (002) reflection which gives the
weight average crystal length in the c-axis direction.

It was shown (Ref.15) that the tensile modulus of these materials could then be simply
understood in terms of a Parallel-Series Takayanagi model (Fig. 5(b)) where the parallel
crystalline element which provides the crystalline continuity consists of crystalline
material which links two or more adjacent lamellae.   The remaining lamellar material and
oriented non-crystalline material (shown by NMR) are considered to be in series.   The rapid
rise in modulus with increasing draw ratio is attributed to the increase in crystalline
bridge material.

To relate these ideas to the experimental data consider the dynamic mechanical tensile
moduli shown in Fig. 6.   It is clear that the high modulus polyethylenes differ in degree

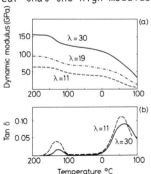

Dynamic modulus (a) and tan δ (b)
vs temperature for drawn LPE at
indicated draw ratio (λ)

Fig. 6 Dynamical mechanical tensile data for drawn polyethylene
        at various draw ratios (λ)

and not in kind from low draw material.   Quite apart from the morphological limitations
implied by the structural studies which indicate only a limited degree of crystal continuity,
there are also considered limitations due to the molecular flexibility which manifests
itself in the γ and α relaxations.   In the simplest approximation, following the Takayanagi
model, the -50°C plateau moduli represents the temperature where there is no contribution
from oriented amorphous material and the α - relaxation has not begun to reduce the
stiffening effect of the crystal bridges.   Figure 7 shows that there is an excellent
correlation between the parallel crystalline component, calculated to be $\chi p(2-p)$ where χ is
the crystallinity and the -50°C plateau moduli.   To explain the temperature dependence of

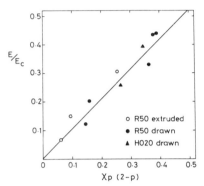

Fig. 7   -50°C plateau modulus E as a function of $\chi p(2-p)$ for a
range of polyethylene samples. (E_c is crystal modulus)

of the tensile behaviour and to embrace the dynamic shear behaviour, it has proved
instructive to develop the analogy between these materials and a short-fibre composite.
The crystalline sequences which link two or more adjacent lamellae form the fibre phase
and the remaining mixture of lamellae and non-crystalline material the matrix phase.   The
relaxation processes are then seen to affect the efficiency of stress transfer by the
crystal bridge sequences, and there is also a contribution to the low temperature modulus
from the oriented amorphous material.

The dynamic mechanical behaviour shows that there are severe limitations to the achievable
modulus in polyethylene even in a reversible dynamic experiment at a low level of stress.
In practical applications the limitations may be even more severe, because more substantial
stresses may be applied.   It will now be shown that there is then a link between the
creep behaviour and the drawing behaviour.   Creep measurements on low molecular weight
polymers (Ref.17) ($\bar{M}_w \sim 100,000$) showed that such materials reach a constant creep rate
after a comparatively short time.   This is indicated by the so-called Sherby-Dorn plot
of Fig.8.   Moreover the stress and temperature dependence of this plateau strain rate were
consistent with a single thermally activated process possessing an activation volume similar

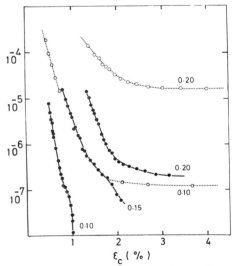

Fig. 8 Sherby-Dorn plots showing creep behaviour of several polyethylenes
O——O low molecular weight ●——● high molecular weight
(numbers near curves show stress levels in GPa)

to that of high stress drawing process (v_2 in Fig. 3(b)). High molecular weight polymer, on the other hand, displays somewhat different behaviour. As shown in Fig.8 the creep rate for this polymer decreases with time, provided the stress level is sufficiently low, but a plateau strain rate is observed at high stresses. This behaviour can be very well modelled by two activated processes coupled in parallel, as proposed for the drawing behaviour and illustrated in Fig. 3(b) and 4. A more useful way to represent the creep data is shown in Fig. 9(a) and (b) where the experimental results for several polymers are

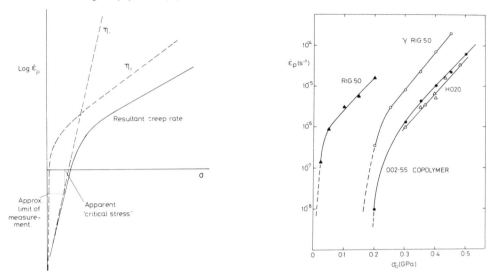

Fig. 9   Schematic creep rate versus stress (a) and creep
         rate data for several polyethylenes (b)
         $\triangle$ Rg. 50 ($\bar{M} \sim 10^5$) O Rig. 50 γ-irradiated before drawing
         O 002-55 n-butyl copolymer $\triangle$ H020 ($\bar{M}_w \sim 3 \times 10^5$)

compared with the idealised two process plots. It is interesting to note that the network process is significant not only in high molecular weight polymer, but also in low molecular weight polymer irradiated prior to drawing and in a copolymer. In both the latter cases it appears that an effective molecular network has been produced. At low stresses the molecular network dashpot 1 is hardly activated, so that the behaviour of the oriented polymer resembles that of a standard linear solid, and total recovery from strain is observed.

These considerations of the mechanical behaviour of the oriented polymers are therefore consistent with the interpretation of the solid phase deformation behaviour. There is an interesting duality; on the one hand the importance of crystal continuity which we propose is maintained by the crystalline bridges; and on the other hand the importance of a molecular network. Both aspects are reflected in the physical properties of these materials. For example, the very high thermal conductivity (equivalent to stainless steel in the axial direction at 100°K), the very small and slightly negative thermal coefficient of expansion, and the comparatively small shrinkage up to temperatures of 130°C, reflect the importance of crystal continuity. The recovery from strain in the creep experiments, the total reversion to original shape on melting (i.e. heating to 150°C), the observation of a shrinkage force with magnitude equivalent to the drawing stress of $\sim$ 20MPa, and some aspects of superheating, reflect the importance of a molecular network.

Finally we may reflect on the equivalence between ultra high modulus polyethylenes produced by the solid phase and solution routes. In many respects there are similarities but there are also notable differences. The seeded crystallisation samples of Pennings and Meihuizen (18) produced dark field electron micrographs not dissimilar from those obtained from solid phase deformation with a few crystal lengths reaching values as great as 1000Å. Estimates of the average crystal length from the line broadening of the electron diffraction pattern gave a value of 450 Å for a typical high modulus sample. A very distinct small angle scattering pattern was also observed for these samples, which corresponded to a length of 1400 Å and was attributed to a characteristic spacing between the lamellar overgrowth. The solution spun fibres also showed very remarkable melting behaviour with crystal transformations observed at 150 and 160°C. This contrasts with the solid phase material where more conventional melting behaviour was observed, with some indication of super-heating. The present consensus of opinion regarding the structure of the solution spun materials appear to favour a substantial proportion of extended chain crystals (crystal lengths of 5000 Å) with some chain folded material, probably as an overgrowth, but with the molecules intimately connected, so that they cannot be removed by dissolution or nitric acid etching.   It is perfectly possible to envisage such a structure giving rise to the high modulus, and indeed models similar to this have been proposed by Barham and Arridge (19) and by Porter and co-workers (20) for the solid phase deformed materials.   There are, however, so far no detailed correlations between the mechanical properties of the solution route materials and quantitative structural measurements.

In terms of properties, there are again similarities between ultra high modulus products from the different routes, but some notable differences.   It is interesting, for example, that although the solution route can produce much higher strength materials and somewhat higher moduli also, the creep behaviour appears to be less satisfactory in that plateau creep is observed at comparatively low stress levels.   It does appear possible that the solution route involves a more open molecular network, which is advantageous in producing very high molecular alignment and hence strength, but disadvantageous as far as creep under constant load.   Again, however, there are much less detailed studies of the properties of the solution route materials than exist for the solid phase route materials.

CONCLUSION

In this lecture I have attempted

(1) to identify the factors which are common to any attempts to produce high modulus oriented polymers
(2) to contrast the difference between results for stiff and flexible polymers
(3) to discuss the different approaches adopted for high modulus polyethylene. The preparation and properties of materials obtained by solid phase deformation has been linked to a common theme, the duality of crystal continuity and a molecular network.   The dilute solution and gel spinning routes are discussed in the light of this conclusion.

REFERENCES

1. J.R. Schaefgen, T.I. Bair, J.W. Ballou, S.L. Kwolek, P.W. Morgan, M. Panar and J. Zimmerman in "Ultra High Modulus Polymers" ed. by A. Ciferri and I.M.Ward Applied Science Publishers, London 1979.
2. G.B. Carter and V.T.J. Schenk in 'Structure and Properties of Oriented polymers' ed. by I.M.Ward, Applied Science Publishers, London 1975.
3. M.G. Northolt and J.J. Van Aartson, J. Polym.Sci (Poly.Symp) 58, 283 (1978)
4. I.M.Ward, Proc. Phys. Soc. 80, 1176 (1962)
5. L. Holliday in 'Structure and Properties of Oriented Polymers' ed. by I.M.Ward Applied Science Publishers, London 1975.
6. G.Capaccio and I.M.Ward, Brit. Pat. Appl. 10746/37 (filed 6th March,1973).
7. G.Capaccio and I.M.Ward, Polymer 15, 233 (1974)
8. G.Capaccio, T.A. Crompton and I.M.Ward, J.Polym.Sci., Polymer Phys.Edn., 14, 1461 (1976)
9. P.D.Coates and I.M.Ward, J.Mater. Sci. 15, 2897 (1980)
10. P.D. Coates, A.G.Gibson and I.M.Ward, J.Mater. Sci., 15, 359 (1980)
11. P.S. Hope and I.M.Ward, J.Mater. Sci. 16, 1511 (1981)
12. A. Zwijnenberg and A.J. Pennings, J.Polym. Sci., Letters Edn. 14, 339 (1976)
13. F.C.Frank, A. Keller and M.R. Mackley, Polymer 12, 467 (1976)
14. P. Smith and P.J. Lemstra, J.Mater. Sci., 15, 505 (1980)
15. A. Peterlin, J.Mater. Sci., 6, 490 (1971)
16. A.G.Gibson, G.R.Davies and I.M.Ward, Polymer 19, 683 (1978)
17. M.A.Wilding and I.M.Ward, Polymer, 19, 969 (1978)
18. A.J. Pennings and K.E. Meihuizen in "Ultra High Modulus Polymers" ed. by A. Ciferri and I.M.Ward, Applied Science Publishers, London 1979.
19. P.J. Barham and R.G.C. Arridge, J. Polym. Sci., Polym. Phys. Edn., 15, 1177 (1977)
20. N.E. Weeks and R.S. Porter, J. Polym. Sci., Polym. Phys. Edn., 12, 635 (1974)

STRUCTURE-PROPERTY RELATIONSHIPS IN COMPOSITE MATRIX RESINS

Frank N. Kelley, Brian J. Swetlin and Donna Trainor

Institute of Polymer Science, University of Akron, Akron, Ohio 44325, USA

Abstract - Composite matrix resins typified by epoxy-amine formulations
are examined for microstructural effects on properties. Cross-link
density and nodular morphology are characterized for various epoxy/amine
ratiqs, and fracture energy evaluated at several rates and temperatures.
A model inhomogeneous glassy network was prepared in a 2-step procedure
which produced a controlled microstructure. Differences in cross-link
density and network architecture in epoxies did not result in significant
differences in fracture behavior until test temperatures were increased
to within 100°C of T_g. Model materials showed effects of nodule concentration, but not size, reasonably duplicating epoxy morphology and properties at about 25 volume %.

INTRODUCTION

Fiber-reinforced plastics are being selected for an increasing number of applications in
which their desirable characteristics provide benefits over conventional materials. Compared
with metals, their low density, coupled with selectability in stiffness, strength, thermal
expansion and directional qualities, provide strong motivation for expanded use. Reinforcing
materials in continuous and chopped fiber forms of glass, graphite and high strength organics
permit a variety of possible combinations. Advanced composite materials employing continuous
graphite fiber reinforcement are considered for high performance applications in aircraft,
spacecraft, and ground transportation systems including automobiles. In addition, benefits
are found for use in stationary high speed fabrication machinery where high stiffness and low
inertial mass are important. In these applications quantitative ingredient and processing
specifications which lead to the required quality and reproducibility are being sought by
manufacturers and users (Ref. 1).

While the fibers themselves are the products of complex processes dependent on precursor
chemical compositions and thermomechanical history, we will be examining only the polymeric
matrix from the viewpoint of its contribution to composite properties. In addition, we will
consider the thermosetting resins exclusively, using the generic epoxy-amine category as
representative of the highest strength matrix polymers used in advanced composites. Strength
predictions in composites have been corsidered from various viewpoints. In laminated ply
configurations two broad categories are recognized, viz., one for fiber-dominated failure and
another for matrix-dominated failure. Generally, under conditions which produce significant
load transfer through the matrix (Ref. 2), such as might be generated by shearing forces,
cross-ply (lateral) tension or delaminating loads, the matrix properties are important.
Matrix resins may vary in stiffness due to curing variations, chemical composition, moisture
and high temperature exposure, and micro-cracking. Resin strength will depend upon similar
conditions.

Properties related to matrix strength and stiffness are assumed to be determined by the
microstructure, i.e., the molecular network and the morphology. Quantitative relationships
between the microstructure and properties have not been available for thermosetting resins,
mainly due to the lack of adequate characterization methods. If such methods could be
developed they would serve as the analytical tools to provide understanding and control of
both processing variables and performance properties.

A region of considerable importance to properties of composites is the fiber-matrix interface and the extent of bonding between resin and fiber. Reference to an "interphase" is
often made to incorporate material near the interface of finite thickness, which may exhibit
gradients in composition, microstructure, and properties. Studies of property gradients in
the interphase and their effects on properties are rare. Drzal (Ref. 3) has reported
briefly on the importance of microstructural gradients near the fiber-matrix interface.
Racich and Koutsky (Ref. 4) have studied substrate effects on resin morphology in the
vicinity of the interface, and Cuthrell (Ref. 5) reports a larger scale (20-90 μm) floccule
size dependence on substrate heat transfer to epoxy resin layers. It is not known at present

whether the fibers alter the microstructure of the matrix interphase in any way which might be observed directly, as might be seen in crosslink density variations or in the size and distribution of inhomogeneities. The chemical nature of the fiber surface may be altered to provide sites of direct bonding with a resultant effect on composite properties. To our knowledge, fiber surface chemistry and its effect on interphase morphology has not been characterized for graphite fiber composites, and remains an area of speculation.

STRUCTURE-PROPERTY RELATIONSHIPS

The structure of densely-linked glassy network polymers should reflect the character of the reactants such as molecular weight, reactivity, functionality, backbone stiffness and polarity, and the reactant equivalence ratios. In addition, processing variables including mixing, and the time-at-temperature curing sequence, will influence reaction kinetics which in turn produce microstructural variations (Ref. 6, 7, 8).

Resin Microstructure

Two types of microstructural features have been studied with only limited success to date, i.e., cross-linking and nodular morphology. Continuous measurement of the degree of resin advancement due to the increasing molecular weight and branching of end-linked epoxy pre-polymers, through the gel point, and finally to the densely cross-linked glassy form, can be accomplished by following dynamic mechanical (Ref. 9,10) or dielectric (Ref. 11) properties. These methods have proved useful in studying overall kinetics of reaction and have been considered for in-process cure monitoring (Ref. 11). Attempts to provide quantitative information on cross-link density in cured resins have met with difficulties in interpretation as well as in the experimental methods employed. Several investigators (Ref. 12-17) have used an equilibrium modulus value, measured above T_g, to calculate cross-link density, employing rubber elasticity theory developed for loosely cross-linked materials. The surprising results of such measurements are that they often yield reasonable values (Ref. 15, 17) for the size of average network chains (i.e., those joined at both ends to junctions). Systematic studies are needed in which a wide range of epoxy prepolymer chain lengths and controlled cross-link functionality are employed to define the relative consistency of the values obtained. Manson and Sperling (Ref. 15) and Bell (Ref. 16) have employed variations in epoxy/amine ratios and blends of various epoxies with measurements of the rubbery plateau of the dynamic modulus to show differences in M_C. Considering the potential importance of quantitative characterization of densely linked glassy polymers, such measurements may provide, as a minimum, relative values of cross-link density to be correlated subsequently with other properties.

There have been various references to local heterogeneity or supermolecular structures in epoxy networks called "gel balls" (Ref. 6) or "nodules" (Ref. 4,18,19, 20). Microscopic regions of higher cross-link density of the order of 10-100 nm are said to arise from the earliest stages of reaction between the prepolymer and linking agents, which favor addition of reactant species to many immobile dispersed sites in which gellation has initiated (Ref. 6, 17). The best evidence for the existence of nodular structures is taken from electron microscopy studies of fracture surfaces in cured epoxies. Aspbury and Wake (Ref. 21) have reviewed the subject recently, as have Mijovic and Koutsky (Ref. 18) who also studied the influence of the size of nodules, observed by replication of fractured and etched surfaces of epoxies, on the critical strain energy release rate, G_{1c}. Yamini and Young (Ref. 20) have suggested gradual shrinkage and disappearance of nodules with extent of post-cure. Schindler (Ref. 21) and Selby and Miller (Ref. 22) have attempted low angle x-ray scattering as a means of characterizing the apparent inhomogeneities, with little success. Dusek (Ref. 23) argues that nodular surfaces are produced by the fracturing process in a variety of thermoplastics as well as thermosets and thereby concludes that such structures are artifacts of the method. Racich and Koutsky (Ref. 4) however, observe nodules on free surfaces of epoxies. If nodular features are intrinsic to epoxy resins they represent a possible quantifiable structure that may prove useful in relating chemical and process variations to mechanical properties. Unfortunately, thus far, the generality of their existence as intrinsic material microstructures is still a subject of controversy (Ref. 17) as is their relationship to mechanical properties (Ref. 22).

Another set of characteristics to be considered in epoxy resins encompasses voids, micro-cracks and inclusions. These material variables may be considered a natural consequence of the processing and use conditions. The significance of flaws in determining the static strength and fatigue resistance of glassy thermoset resins cannot be overlooked. An interesting comparison by Kim et al. (Ref. 24) in studies of the characteristic flaw size for a variety of amine/epoxy ratios suggests that nodule sizes may correspond to the characteristic flaw dimensions obtained. Flaws may be generated during processing, and microcracking does occur under static and cyclic mechanical loading. In composite materials or bonded structural joints the existence of flaws at interfaces must be presumed. These flaws are generally considered to be the sites of failure initiation and this view has led to the increasing use of fracture mechanics to study bond strength.

Fracture of Densely Cross-linked Resins

There is a growing body of literature on the fracture of densely cross-linked resins, particularly involving studies of epoxy formulations. Pritchard and Rhoads (Ref. 25) reviewed the subject in 1976, and there has been a large number of contributions since (Ref. (15, 26-33)   The concepts of energy balance, due to Griffith (Ref. 34), and linear elastic fracture mechanics (LEFM) have been applied. In the energy balance approach, G_C, the critical potential energy release rate, is determined by fracturing specimens containing single, through-cracks of known length. The fracture surface energy per unit area of newly-formed surface, γ, is equal to $1/2$ G_C, since two crack surfaces are formed along the crack plane. In LEFM, the stress intensity factor, K, characterizes the stress field in the region of the crack tip. It reaches a critical value when the crack extends, and is designated K_C, with additional subscripts I, II, or III to designate the mode of cracking. For brittle fracture the energy balance and LEFM values are related by (Ref. 35)

$$K^2/E = 2\gamma = G_C \tag{1}$$

where E is Young's modulus. Several texts are available on the subject (Ref. 2, 35), so we will not amplify the concepts here. The principal aim is to derive a useful material property relating to the ultimate strength of a structural component. In this sense, both γ and G_C are material properties which are practically independent of test piece configuration and may be used to assess the effects of molecular or microstructural variations in cured resins.

EXPERIMENTAL STUDIES OF MOLECULAR NETWORK AND MORPHOLOGICAL STRUCTURE

Epoxy Formulations

The epoxy resins used in the studies to be described were produced from Shell Epon 828 and two types of polyfunctional amines: methylene dianiline (MDA) and triethylene tetramine (TETA). The structures and characteristics of these materials and dimethyl sulfoxide (DMSO) used as a non-reactive diluent are given in Table 1.

Table 1. Experimental materials

I. Epon 828; Shell Chemical Co., condensation product of epichlorohydrin and Bisphenol A, epoxide equivalent wt. = 188.5

Epon 828 is 75% n = 0 (DGEBA) and 25% n = 1 by weight.

II. 4,4'-methylene dianiline (MDA), Eastman Kodak, practical grade
     M_W = 198.27 g/g-mole; mp = 90-93°C.; b.p. = 398°C

III. Triethylenetetramine (TETA), Fischer Scientific, technical grade
      M_W = 146.24 g/g-mole, b.p. = 266-7°C

$$H_2N-CH_2CH_2-\overset{H}{N}-CH_2CH_2-\underset{H}{N}-CH_2CH_2-NH_2$$

IV. Dimethyl sulfoxide (DMSO), Fischer Scientific, m.p. = 18.2°C,
     b.p. = 189°C, ρ = 1.095 g/ml; M_W = 78.13 g/g-mole

$$CH_3-\overset{O}{\overset{\|}{S}}-CH_3$$

The technique used to prepare virtually void-free specimens involved a vacuum mixing and casting system in which the degassed reactants could be mixed and transferred to Teflon coated molds directly, without breaking the vacuum. Samples prepared by this method, as well as a number of alternative methods, were scanned between polarizing films to allow sensitive detection of voids or other inhomogeneities, which were cause for rejection. In general, the more fluid mixes could be cast and cured in a vacuum oven after debulking

without the necessity for continuous vacuum casting. In all cases in which samples were
heated for postcuring or testing, they were maintained in a dry nitrogen environment.
Separate studies which demonstrated changes due to oxidation or other environmental factors
were conducted to establish sample conditioning precautions. The characteristic cherry-red
to deep brown colorations of epoxies which occur when they are heated in air were absent in
our specimens.

Cross-link Density Determination

The following methods were attempted in order to provide a relative ranking of the state of
cure:

    a) Direct swelling measurements
    b) In-situ cure plasticization with subsequent swelling
    c) High temperature rubbery modulus determination
    d) Swollen rubbery modulus determination

In each of these methods the application of molecular theories of rubberlike elasticity
(Ref. 36) are applied. The validity of the approach hinges to a large extent on the flexibility of the short network chains that might be expected to deviate considerably from
gaussian chain conformational statistics. Scaled molecular models of the DGEBA/TETA network
chain exhibit a surprising degree of flexibility, as shown in Fig. 1.

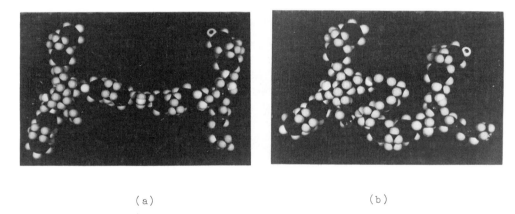

(a)   (b)

Fig. 1. Comparison of molecular models of DGEBA/TETA network chains:
              (a) extended, (b) coiled.

Direct swelling measurements have been attempted with a variety of solvents at normal and
elevated temperatures, but all with poor reproducibility. A more satisfactory method from
the standpoint of reproducibility employed the incorporation of approximately 20% of a high
boiling diluent, i.e., dimethyl sulfoxide (DMSO), in the reaction mixture and curing the
sample with the diluent in situ. Pre-swelled networks obtained in this manner could be
prepared with glass transition temperatures at or below room temperature. These materials
could be further swelled to equilibrium and tested. It is not known whether the networks
formed under these conditions are fully representative of the networks formed without
diluent (Ref. 17, 37). The extent of reaction was judged to be complete due to the
constancy of properties after further heating and the absence of a cure exotherm assessed by
differential scanning calorimetry. The presence of the diluent permits the reaction to
reach completion in a system where reactant mobility is retained, since vitrification does
not occur.

The swollen samples were quite rubbery at elevated temperatures (120°C) and were swelled to
equilibrium in solvent (DMSO) or retained in saturated vapor in a tensile testing apparatus.
Equilibrium stress-strain data were obtained and evaluated according to the methods of
Gumbrell, Mullins and Rivlin (Ref. 38) and M_c, the molecular weight between cross-links,
calculated. Figure 2 shows a representative plot of swollen tensile data and the average
molecular weight of network chains obtained, where λ is the extension ratio, σ_t is
the true stress and v_r is the volume swelling ratio.

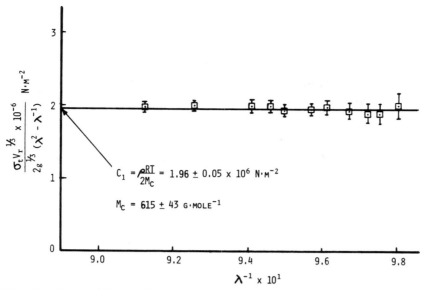

Fig. 2. Mooney-Rivlin Plot of Swollen DGEBA/MDA epoxy with ∿30% DMSO.

The validity of the swollen modulus method for obtaining M_C values was checked using a material with an amine/epoxy equivalence of 0.65 which had a sufficiently low T_g to be tested at 50°C above T_g without degrading reactions, and presumably under equilibrium rubbery conditions. A similar equivalence ratio material was prepared, incorporating 20% DMSO diluent and further swollen to equilibrium swelling ratio. The values of M_C obtained for these two materials by each method compared within 20%. In cured epoxies with T_g's well above 150°C, the measurement of rubbery modulus values is severely compromised by the risk of degrading or otherwise altering chemical reactions. In-situ cure plasticization eliminates this problem and apparently does not introduce significant additional complications in the process.

Nodular Morphology

Fracture surfaces of single edge-notched specimens used in our studies were examined by both scanning and transmission electron microscopy. Transmission micrographs of carbon-platinum, single-stage replicas were produced for a variety of resin formulations fractured at several rates and temperatures. Nodular structures were not clearly discernible in many of the specimens examined. When the specific formulations and cure schedules were reproduced as described by Mijovic and Koutsky (Ref. 18, 19), nodular structures in the size range which they reported were found. Figure 3 shows a typical micrograph of the characteristic fracture surface obtained. Mijovic and Koutsky (Ref. 18, 19) found nodule size to vary with amine to epoxy ratio, but independent of post-cure time. When other resin formulations were used, or increased post-cure times were employed in our work, the nodular features became much less distinct. Yamini and Young (Ref. 20) have described investigations in which nodule diameter decreases with increasing post-cure temperatures. Figure 4 shows the data of references 19 and 20 plotted on the same chart to indicate the similarity in nodule sizes observed under similar curing conditions and equivalence ratios. Koutsky and Mijovic (Ref. 19) show a dependence of nodule diameter with increasing amine curing agent, but only with epoxy excess formulations (stoichiometry would be at 12 phr amine in these materials). The apparent nodule diameter at 9.8 phr amine suggested by the Yamini and Young data at 106° post-cure temperature, is seen to be quite close to that found by Koutsky and Mijovic at 8 phr amine, for the same temperature. Even though the aliphatic amines are different (DETA vs. TETA), there seems to be some correspondence of what may at first sight appear to be contradictory evidence.

Fracture Studies

The objective of our work was to determine the influences of cure schedule, and extent of cure, as measured by apparent cross-link density and network inhomogeneities, on the rate and temperature dependent fracture energy of epoxy resins, above and below T_g. Network architectural variations were also expected from reactions of various amine/epoxy ratios. The approach followed was similar to that described by King and Andrews (Ref. 14). The data were analyzed according to Andrews' generalized fracture theory (Ref. 39) in which the critical input strain energy density, W_{0c}, is derived from the area under the stress-strain curve extending to the point of initial crack growth in specimens containing an edge crack of size, c. W_{0c} is related to $\mathcal{J}$, the cohesive fracture energy per unit area of crack surface by

$$\mathcal{J} = kC\ W_{0c} = \mathcal{J}_0 \phi(\epsilon_0,\ T,\ \dot{c}) \qquad\qquad (2)$$

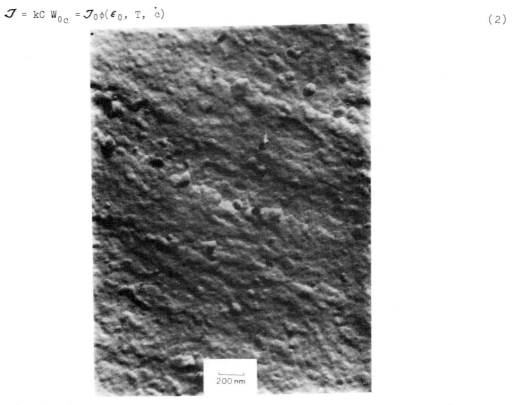

Fig. 3. Electron micrograph of C-Pt replica of fracture surface of DGEBA/TETA resin showing nodular character.

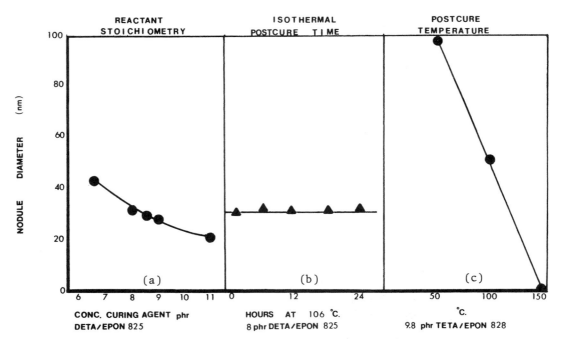

Fig. 4. Nodule size/reaction condition correlations in epoxy networks.
(a) and (b) Mijovic and Koutsky (Ref. 19); (c) Yamini and Young (Ref. 20).

where at small critical strains $k \sim \pi$, and $\mathbf{J}_0$ is the rate and temperature independent intrinsic fracture energy. The loss function, ϕ, is dependent on crack velocity, temperature, and the applied strain, $\boldsymbol{\epsilon}_0$.

Specimens employed were approximately 125 x 25 x 2 mm. Notches were placed midway along one edge with a special aluminum and brass fixture designed to permit movement of the specimen across and at right angles to a single-edge razor blade. The fixture and specimen were heated above the T_g of the resin so that a controlled notch could be easily produced in the material while in the rubbery state. Examination of the notch tip after fracturing by scanning electron microscopy showed exceptionally sharp and reproducible initial cracks. Fracture testing was done with an Instron universal tester equipped with an environmental chamber. For the glassy fracture measurements force and extension data were channelled from the load cell and a 0-10% extensometer attached to the specimen, to an x-y recorder. Average crack extension rates were obtained by monitoring the time between breaks of two thin conducting strips painted about 1 cm apart on the specimen normal to the crack direction. The signals were fed to a two-channel oscilloscope which allowed the recording of voltage step change as a function of time.

In addition, measurements in the rubbery state were conducted using trouser tear specimens (Ref. 40), where the variation in crack extension rate could be determined by the crosshead rate.

Model Inhomogeneous Glassy Networks

Considering the current state of uncertainty surrounding the nodular microstructure of thermosetting resins and any relationships to mechanical properties, a method was developed which permits the preparation of glassy network polymers of controlled inhomogeneity. In these special polymers the presence of spherical microgel particles, intimately connected to a lesser cross-linked matrix, can be assured a priori, and the concentration levels, particle diameters and size distribution may be prescribed.

The model materials were prepared by producing cross-linked polystyrene-divinylbenzene particles by emulsion polymerization and subsequently swelling the washed and dried microspheres with monomeric styrene, which could then be polymerized to form a continuous interpenetrating structure with the dispersed gel phase. It should be noted that Manson and Sperling (Ref. 15) produced a model system with similar goals in mind, but used bulk interpenetrating networks; whereas this work is based on uniform microgel spheres. In this case the precise regulation of latex particle size and distribution is expected to provide a more controllable model material. Table 2 indicates the recipe employed.

TABLE 2. Emulsion polymerization recipe

Ingredients	Parts by Weight
Monomers (styrene and divinylbenzene)	100
Water	180
Emulphogene BC-840	10
Sodium dodecyl sulfate	0.05-0.30
Potassium persulfate	0.3

The emulsifier system consisted of Emulphogene BC-840 from GAF, a tridecyloxypoly(ethylenoxy)ethanol, and sodium dodecyl sulfate. This combination of nonionic and anionic surfactants has been shown to result in mixed micelles which produce narrow particle size distribution latexes (Ref. 41). The amount of divinylbenzene was varied from 2 to 20% of the total monomer feed to regulate the cross-linking level in the particles. The level of anionic surfactant was varied to control the particle size of the product. Table 3 shows the various compositions and effect of sodium dodecyl sulfate level on particle size and distribution.

Particle size and distribution were determined by transmission electron microscopy and a Zeiss particle analyzer. The distribution parameters are $\overline{D}_n$, number average diameter, $\overline{D}_w$, the weight average diameter, and u is the uniformity ratio $\overline{D}_w/\overline{D}_n$.

Various mixtures of polystyrene microgel and styrene monomer, ranging from 25 to 100% cross-linked phase, were made. The compositions are given in Table 4. Monomer-polymer mixtures containing greater than 50% styrene were found to be too fluid for the molding procedure, so samples containing 25% and 33% microgel were mixed with 25% linear polystyrene prior to the normal 12-hour monomer swelling period employed. The amount of styrene monomer and linear polystyrene was adjusted to maintain the desired gel/linear ratio.

TABLE 3. Cross-linked polystyrene latexes

Sodium Dodecyl Sulfate (parts by weight)	% Divinylbenzene	$\overline{D}_n$ (nm)	$\overline{D}_W$ (nm)	u
0.05	2	206	211	1.03
0.1	2	152	156	1.03
0.1	5	121	124	1.02
0.1	10	129	134	1.04
0.1	20	121	126	1.04
0.2	5	95	99	1.04
1.0	5	84	89	1.06
3.0	5	76	85	1.12

Benzoyl peroxide was added as a polymerization initiator for the in-situ polymerization step.

TABLE 4. Model inhomogeneous glassy polymers

% Microgel	Microgel* (g)	Linear polystyrene (g)	Styrene (g)	Benzoyl Peroxide (g)
0	0	6	6	0.006
25	3	3	6	0.006
33	4	3	5	0.006
50	6	0	6	0.006
75	9	0	3	0.003
100	12	0	0	0

*All microgel in this series is nominal 100 nm diameter cross-linked polystyrene microspheres.

Molding and polymerization were carried out in one step in a compression mold under conditions of 115°C and 2000 pounds force. The mold was designed to produce samples of the proper configuration for fracture testing. The size of the mold cavity was 2.5 x 12 cm. The thickness was controlled by the amount of material placed in the mold, but normally was 2-3 mm. After fracture testing, the fracture surfaces were studied by transmission electron microscopy using single stage C-Pt replication. Figure 5 shows the nodular structure achieved in a model polystyrene-divinylbenzene system which incorporated 25% by volume of nominal 100 nm diameter cross-linked spherical particles. The average diameter of nodules in this material appears to have increased by about 20%. The achievement of this structure is good evidence that the objective was reached, i.e., a reasonable duplication of the morphology of the epoxy shown in Fig. 3.

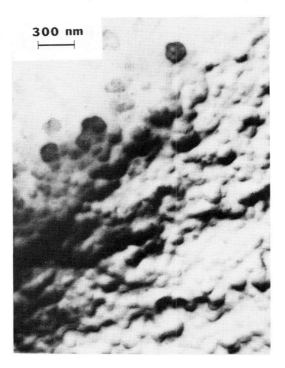

Fig. 5. C-Pt replica transmission electron micrograph of model cross-linked polystyrene (ca. 25%) microspheres incorporated in interpenetrating linear polystyrene.

RESULTS AND DISCUSSION

Epoxy Resins

Fracture data were obtained for various amine to epoxy ratios at several temperatures below the glass transition for Epon 828/MDA formulations. Similar data were obtained for one composition above T_g using edge-notched specimens. A trial using trouser tear specimens at several rates and one temperature well above T_g indicated a good correspondence between fracture energy values obtained by both methods. There was a small, though significant, decrease in fracture energy with decreasing tear rate, as would be expected. When data are taken at several rates and temperatures above T_g, it is possible to produce combined master curves by means of the time-temperature shifting, WLF procedure (Ref. 49). At very slow rates and higher temperatures the threshold value of the fracture energy, $\mathcal{J}_o$, is approached. A value of 10.6+4.1 Jm^{-2} was obtained for an Epon 828/TETA composition employing a DMSO preswollen specimen with an M_c of 240 g. $mole^{-1}$.

Crack extension rates were obtained by optical measurements above T_g and by electronic methods, described above, for the glassy samples. Table 5 contains the formulations, compositions and other measured or derived characteristics for the materials tested.

Figure 6 shows the relationship of fracture energy and temperature for the range of equivalence ratios studied, with comparisons made by plotting $2\mathcal{J}$ against T_g-T_{test}. The fracture energy increases rapidly as T_g is approached from either higher or lower temperatures, apparently passing through a maximum which has high uncertainty when determined using the test methods employed here. Sample yielding at the crack tip becomes a predominant characteristic in the plane stress conditions attendant to the test configurations used. The curves tend toward an asymptotic convergence around 165 Jm^{-2} at the lower temperatures (as T_g-T_{test} increases) indicating that the fracture energy is relatively independent of crosslink density and network architecture at greater than ∿100°C below T_g. As T_g is approached, however, the differences become more distinct with the epoxy excess compositions showing much lower values of $\mathcal{J}$ over the whole temperature range. The amine excess compositions tend to give higher fracture energies, and the stoichiometric formulations are intermediate in the range of 40°C below T_g.

The fracture surfaces of these compositions were studied by scanning and transmission electron microscopy. Nodular features could be found on the surfaces of the off-stoichiometric formulations, but were not seen with any distinction in the 1:1 amine/epoxy compositions. In addition, scanning microscopy showed a characteristic ribbed fracture surface for the stoichiometric composition, with long fibrils or "curls" (Ref. 42, 43)

Table 5. Characteristics of various Epon 828 networks.

Crosslinking Agent	Equivalence Ratio	Diluent	Cure Schedule	[1]T_g(°C)		[2]M_c
	Eq. amine/ Eq. epoxy			R=20°C/Min	R→0	(g/mole)
TETA	0.50	DMSO	A	81	--	9655
	0.62	DMSO	A	--	--	480
	0.62	None	B	86	73	504
	0.75	DMSO	A	134	--	230
	1.00	DMSO	A	129	--	240
	1.30	DMSO	A	119	--	345
	1.60	DMSO	A	113	--	400
MDA	0.65	DMSO	C	--	--	1485
	0.65	None	D	82	71(74)	1950
	1.00	None	D	172	159	270
	1.60	None	D	127	114(111)	780

[1]By DSC, reported value at heating rate, R=20°C/Min and extrapolated to R=0 from measurements at R=20, 10 and 5°C/Min. T_g in parentheses by dilatometry at R=0.05°C/Min. T_g of swollen networks obtained after diluent removal.

[2]From rubber elasticity theory; $M_c = (3\rho RT/E)(v_2)^{1/3}$.

Cure Schedules:

A 24°C/24 hrs + 60°C/3 hrs
B 24°C/1.5 hrs + 60°C/0.5 hrs + 80°C/0.5 hrs + 100°C/0.5 hrs
       + 150°C/4 hrs
C 60°C/24 hrs + 80°C/96 hrs
D 60°C/0.75 hrs + 80°C/0.5 hrs + 150°C/2.5 hrs + 180°C/5.0 hrs/vacuum

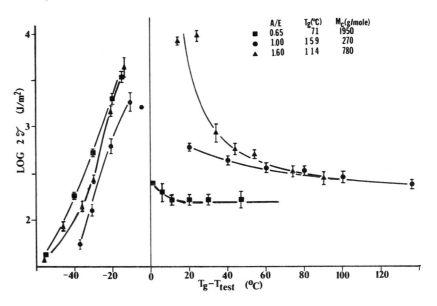

Fig. 6. Fracture energy of Epon 828/MDA networks above and below T_g.

appearing in abundance when the test temperature was about 100°C below T_g. At higher or lower temperatures the formation of fibrils was much reduced, or they did not occur at all. Figure 7 shows the fibrils produced by fracturing a fully cured Epon 828/MDA specimen of

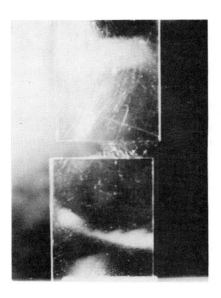

SEM MICROGRAPH
OF FRACTURE SURFACE

Fig. 7. Fracture of stoichiometric EPON 828/MDA at 60°C.

stoichiometric composition at 60°C (T_g = 159°C). Nelson and Turner (Ref. 42) have described similar structures as curls, formed on the fracture surface of densely linked phenolics, and speculate on their origin. Katz and Buchman (Ref. 44) noted "ribbons" on the fracture surfaces of prestrained epoxy resins and the formation of "tubular" structures. The fibrils may originate during the extension of many fingerlike fracture fronts (Ref. 47, 48) extending along the fracture plane normal to the precrack edge. As the two fracture surfaces separate, ridges or "tracks" are formed which are clearly seen in scanning electron micrographs. Fibrils are usually attached to the ridges at one end. We could not discern a "carpet roll" character in the fibril cross-section which would confirm the hypothesis that they are formed of thin sheets which curl when they separate from one of the fracture ridges. Examination of individual fibrils by x-ray diffraction indicated very little orientation. The fibrils were insoluble in acetone or DMSO after 96 hours at room temperature, and they did not retract or distort on heating well above the bulk resin T_g. These observations lead to the conclusion that they are cross-linked structures which have not been significantly oriented along the apparent fiber axis during the fracture process.

The formation of fibrils (curls) indicates a degree of mobility in the polymer, at the crack tip at least, which is sensitive to temperature. The observation that very long fibrils are generated by fracturing a densely cross-linked glassy polymer at about 100°+20°C below its T_g does not provide specific information about the microstructure, but it does suggest that a heterogeneous structure is present. Crazing is readily observed in thermoplastic polymers, but is not easily accommodated in thermosets such as phenolics and epoxies. The appearance of surface colors indicating oriented material (Ref. 43) and microscopic evidence (Ref. 45) of crazing in cross-linked glassy polymers suggests that the material consists of a combination of both densely cross-linked and uncross-linked polymer. The temperature and strain-rate sensitivity of the uncross-linked portion seems to be an important factor as the glass transition temperature of the bulk resin is approached. Also, there may not be a single glassy transformation in these materials. Some studies have shown T_g splitting (Ref. 46) by DSC measurements in annealed epoxy resins.

Measurements of fracture energy above T_g are expected to show some differences in the viscoelastic response of the polymer, with cross-link density, network architecture and inhomogeneity being important variables. King and Andrews (Ref. 14) noted differences in the loss function, ϕ, of Andrew's generalized fracture theory, reflected in the shape of the reduced fracture energy curve when the amine/epoxy equivalence ratio and M_c were varied. Figure 6 contains fracture data for 3 equivalence ratios of EPON 828/MDA compositions. The measurements above T_g show a systematic ordering of fracture energies over the entire reduced temperature range according to the network chain density; the larger M_c networks yielding higher values of the fracture energy at a given T_g-T_{test}. However, the apparent insensitivity of fracture energy to network density and architecture over the fairly wide range of equivalence ratios studied, when measured >100°C below T_g, would indicate that the local glassy structure and short range mobility are more likely important variables in the normal use temperature range. The epoxy excess composition (a/e = 0.65) does display the lowest fracture energy over the entire range of temperatures below T_g. Since the expected network architecture for this composition would have numerous long, dangling epoxy terminated free ends, perhaps such structures lead to less mobility at the local level, as well as difficulty in cooperative motions of several backbone segments as the temperature is increased. Clearly, the more linear, chain-extended structure of the amine excess composition retains a larger capacity for plastic deformation and strain energy dissipation over the temperature range shown in Fig. 6. At present, knowledge of the structure of these materials is far from complete, even though a reasonable ranking of network density has been achieved. Assuming that network inhomogeneity plays a role in determining properties below T_g, quantitative characterization of this feature would be essential. The nodular features seen most clearly in our work in off-stoichiometric compositions have yielded only weak or speculative correlations.

Model Materials

Initial fracture studies were conducted with the model materials described in Table 4, which had a gelled microsphere particle size of about 100 nm. Other materials containing particles of 75 and 130 nm diameter were also prepared. The first tests employed concentrations of around 50 weight percent for three particle sizes. Since Mijovic and Koutsky (Ref. 19) reported a relationship between nodule size and G_{1C} in an epoxy resin, we sought some experimental confirmation of similar relationships in the model system. Figure 8 shows the data of Ref. 19 plotted on the same graph as our data, obtained for 25 and 50% particle compositions. The fracture energy was quite low for the model material, but for initial tests the particle

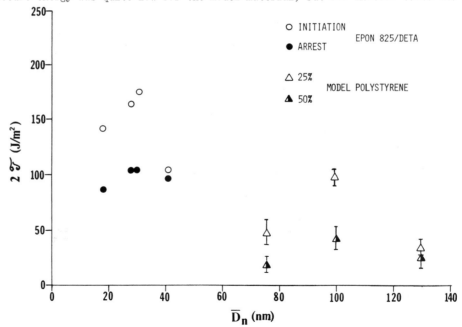

Fig. 8. Comparison of nodule diameter correlation with fracture energy by Mijovic and Koutsky (Ref. 19) and model inhomogeneous polystyrene-divinylbenzene dispersed in linear, interpenetrating polystyrene (concentrations ∿25% and 50%).

concentrations selected were arbitrary. Studies of gelled particle concentration effects on fracture energy were then conducted and the results are shown in Fig. 9. The 25% by weight composition at 100 nm particle diameter yields G_{1C} values of 100 JM^{-2} which correspond to the nominal values of Ref. 19. Studies of size effects at this

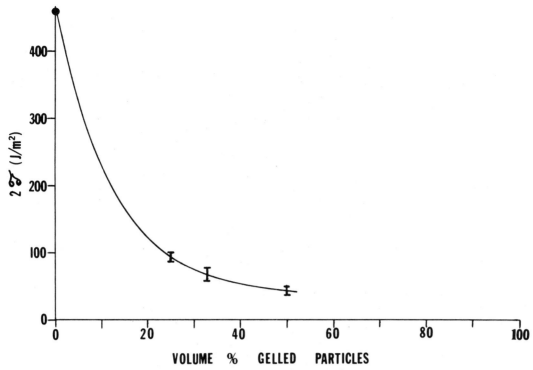

Fig. 9. Fracture energy of model inhomogeneous polystyrene with various concentrations of gelled particles of polystyrene-divinylbenzene. $\overline{D}_n \sim 100$ nm

composition are in progress to clarify any relationships which may exist. However, we expect that the most pertinent structural variable is the amount of uncross-linked polymer present in the material rather than the size of gelled particles. Two distinct glass transition temperatures were noted by DSC measurements in the 25% gel particle compositions. The upper value, $\sim 116°C$, is in the range expected for highly cross-linked polystyrene and the lower value, $\sim 99°C$, is reasonable for the linear, continuous matrix.

CONCLUSIONS

1. Network density may be evaluated in cross-linked glassy polymers by measurements of equilibrium modulus well above T_g. In those polymers which cross-link to yield very high glass transitions, the starting materials may be dissolved in a high boiling, inert diluent and the curing reaction may be completed without vitrification. The process of in-situ cure plasticization produces fully reacted network polymers with substantially reduced T_g's. The application of rubber elasticity theory to calculate cross-link density in epoxies in the rubbery state produces questionable absolute values, but should be of use for quantitative but relative, ranking of network parameters.

2. Cross-link density of various epoxies appears to be relatively unimportant to the fracture energy, if fracture measurements are carried out at temperatures in excess of $100°$ below T_g. However, since the T_g is determined to a large extent by the density of cross-linking, the upper temperature limits of the use range are thereby determined by T_g.

3. When evaluated at equivalent values of $T_g - T_{test}$, epoxy excess formulations display lower fracture energies than do stoichiometric or amine excess compositions, especially as T_g is approached.

4. Nodular features were found by transmission electron microscopy of single stage C-Pt replicas of fracture surfaces in some epoxy resin compositions. These features were most evident in amine excess or epoxy excess formulations, but were indistinct in fully-cured stoichiometric compositions. Since the materials which displayed similar nodular features were quite different in their fracture behavior (magnitude of the fracture energy), at the same temperature difference below their respective glass transitions, there does not seem to be any correlation which can be supported at present.

5. Studies of model cross-linked inhomogeneous glassy polymers with dispersed gelled
    particles of various diameters indicate that the particle size is a relatively
    unimportant factor in determining the magnitude of the fracture energy, while the amount
    of uncross-linked polymer present is the dominant factor.

Acknowledgement - This work was supported by a grant from the Air Force
Office of Scientific Research through the University of Pittsburgh's
program on the Science of Fracture. The authors wish to express their
gratitude to Professor Irja Piirma for advice on the preparation of
controlled particle size latex, and to Professor M. L.Williams of the
University of Pittsburgh for many helpful discussions.

REFERENCES

1. W. J. Macknight, E. Baer, and R. D. Nelson, Polymer Materials Basic Research Needs for
    Energy Applications, U. S. Dept. of Energy Workshop Report, CONF-780643, Case Western
    Reserve University, Cleveland, Ohio, August 1978.
2. See for example A. De S. Jayatilaka, "Fracture of Engineering Brittle Materials,"
    Chap. 7, Applied Science Publishers Ltd., London (1979).
3. L. T. Drzal, AICHE Mat. Eng. Sci. Div. Newsletter, 12(1), 5-7 (1980).
4. J. L. Racich and J. A. Koutsky, J. Appl. Polymer Sci., 20, 2111 (1976).
5. R. E. Cuthrell, J. Appl. Polymer Sci., 11, 949 (1967).
6. S. S. Labana, S. Newman and A. J. Chompff, Polymer Networks: Structural and Mechanical
    Properties, 453, Ed. A. J. Chompff and S. Newman, Plenum Press, New York (1971).
7. K. Dusek, M. Ilavsky and S. Lunak, J. Polymer Sci.: Symp. No. 53, 29-44 (1975).
8. K. Dusek, Makromol. Chem., Suppl. 2, 35 (1979).
9. J. K. Gillham, J. A. Benci, and A. Noshay, J. Appl. Polymer Sci., 18, 951 (1974).
10. C.Y.C. Lee and I. J. Goldfarb, Polym. Sci. and Eng. 21, 390 (1981).
11. N. F. Sheppard, S. L. Garverick, D. R. Day and S. D. Senturia, Proceedings of 26th
    National SAMPE Symposium. Los Angeles, CA, April 1981.
12. D. Katz and A. V. Tobolsky, Polymer, 4, 417 (1963).
13. D. H. Kaelble, J. Appl. Polymer Sci., 9, 1213 (1965).
14. N. E. King and E. H. Andrews, J. Mat. Sci., 13, 1291 (1978).
15. J. A. Manson and L. H. Sperling, Influence of Cross-linking on the Mechanical Properties
    of High Tg Polymers, AFML TR-77-109, AF Materials Laboratory, WPAFB, OH, April, 1977.
16. J. P. Bell, J. Appl. Polymer Sci., 14, 1901 (1970).
17. S. Lunak, K. Dusek, J. Polymer Sci.: Symp. No. 53, 45-55 (1975).
18. J. S. Mijovic and J. A. Koutsky, J. Appl. Polymer Sci., 23, 1037 (1979).
19. J. S. Mijovic and J. A. Koutsky, Polymer, 20, 1095 (1979).
20. S. Yamini and R. J. Young, J. Mat. Sci., 15, 1823 (1980).
21. A. Schindler and N. Morosoff,"Determination of Cross-linking in High Tg Polymers,"
    Final Technical Report AFWAL-TR-80-4085, Wright Patterson AFB,Ohio, July 1980.
22. K. Selby and L. E. Miller, J. Mat. Sci., 10, 12 (1975); and R. N. Haward (ed.), The
    Physics of Glassy Polymers, Applied Science Publishers, London, p. 130 (1973).
23. K. Dusek, Macromol. Chem., Suppl. 2, 35-49 (1979).
24. S. L. Kim, M. D. Skibo, J. A. Manson, R. W. Hertzberg, and J. Janiszewski, Polym. Eng.
    and Sci., 18, 1093 (1978).
25. G. Pritchard and G. V. Rhoades, Mat. Sci. and Eng., 26, 1 (1976).
26. S. Yamini and R. J. Young, Polymer, 18, 1077 (1977).
27. S. Yamini and R. J. Young, J. Mater. Sci., 15, 1814 (1980).
28. R. A. Gledhill, A. J. Kinloch, S. Yamini and R. J. Young, Polymer, 19, 574 (1978).
29. D. C. Phillips, J. M. Scott and M. Jones, J. Mater. Sci., 13, 311 (1978).
30. S. Yamini and R. J. Young, J. Mater. Sci., 14, 1609 (1979).
31. A. J. Kinloch and J. G. Williams, J. Mater. Sci., 15, 987 (1980).
32. J. M. Scott, G. M. Wells and D. C. Phillips, J. Mater. Sci., 15, 1436 (1980).
33. E. H. Andrews and A. Stevenson, J. Mat. Sci., 13, 1680 (1978).
34. A. A. Griffith, Philos. Trans. R. Soc. London, Ser. A, 221, 163 (1921).
35. E. H. Andrews, Fracture in Polymers, Oliver and Boyd, Edinburgh (1968).
36. L.R.G. Treloar, The Physics of Rubber Elasticity, Clarendon Press, Oxford (1975).
37. R. M. Kessenikh, L. A. Korshunova, and A. V. Petrov, Polym. Sci. USSR, 14, 466 (1972).
38. S. M. Gumbrell, L. Mullins and R. S. Rivlin, Trans. Faraday Soc. 49, 1495 (1953).
39. E. H. Andrews, J. Mat. Sci.,9, 887 (1974).
40. A. N. Gent, Science and Technology of Rubber, F. Eirich, ed., Chap. 10, Academic Press
    (1978).
41. I. Piirma and H. L. Jones, ACS Symp. Series, No. 24,
    I. Piirma and J. L. Gardon, editors (1976).
42. B. E. Nelson and D. T. Turner, J. Polymer Sci., B, 9, 677 (1971)
43. B. E. Nelson and D. T. Turner, J. Polymer Sci., Polymer Phys. Ed., 10, 2461 (1972).
44. D. Katz and A. Buckman, J. Appl. Polym. Sci., Appl. Polym. Symp. (1979-80).
45. R. J. Morgan and J. O'Neal, Polymer Eng. and Sci., 18, 1081 (1978).
46. U. T. Kreibich and R. Schmid, J. Polymer Sci., Symp. 53, 177 (1975).
47. N. J. Mills and N. Walker, Eng. Fracture Mech., 13, 479 (1980).
48. A. S. Argon and M. M. Salama, Phil. Mag. 36, 1217 (1977).
49. M. L. Williams, R. F. Landel and J. D. Ferry, J. Am. Chem. Soc., 77, 3701 (1955).

THE ADHESION OF POLYMERS TO HIGH ENERGY SOLIDS

J. Schultz and A. Carré

Université de Haute-Alsace and Centre de Recherches sur la Physico-Chimie des Surfaces Solides, 24, avenue du Président Kennedy, 68200 Mulhouse, France.

Abstract - The mechanical performances of assemblies or of composite materials depend mainly on the quality of adhesion of the different constituting elements. In this paper, we are showing, in two different studies, two aspects of the adhesion of a polymer on a high energy surface. One study deals with *an interfacial layer* having cohesion characteristics different from those of the bulk polymer. In the other study, we have examined *an interfacial surface* with characteristics which differ from those of the free surface of the polymer. These two aspects are especially important in allowing a clear comprehension of the rupture behaviour of composite materials.

INTRODUCTION

Among solids, physical chemists usually distinguish, more or less arbitrarily, between two types of surfaces, i.e., surfaces of low energy (< 50-60 mJ.m^{-2}) such as polymers, and surfaces of high energy (usually > 100 mJ.m^{-2})(minerals, metals, glasses, carbon). It is especially interesting to study the association of materials of the two different types, as these associations are often used (bonded joints, coated metals, etc...) particularly in composite materials where a polymer matrix is associated with a reinforcing filler or fibre of high energy. Also this study is justified by the fact that the final quality of a joint depends largely on the quality of the interface between the two materials.

We shall first recall some general facts about the adhesion phenomena. Scientists have only taken interest in adhesion for the past 40 years, and it is still now a domain where empiricism and technology prevail over science. One of the main difficulties of the study of adhesion is that the subject concerns many fields, as different as the physical-chemistry of surfaces and interfaces, mechanics and micromechanics of fracture, rheology, strength of materials... Consequently the study of adhesion uses various concepts, depending on one's special field of study, or also depending if one considers the molecular or macroscopic aspect of the phenomenon, or if one studies the formation or the rupture of the interface. This variety of approaches to the study of adhesion is emphasized by the fact that many theoretical models of the adhesion phenomena have been developed, models which are together complementary and contradictory. Among these models one usually distinguishes between mechanical and specific adhesion. The mechanical model (1,2) is based on the assumption that adhesion is the result of an anchoring of the polymer in the pores or on the asperities of the substrate. Although one cannot neglect totally this aspect of adhesion, as the increase of the area of interfacial contact can be a positive factor of adhesion, it is generally accepted nowadays that mechanical adhesion is not the main cause of the resistance of an interface but that molecular interactions between the materials in contact are mostly responsible for their adhesion. Consequently many theories or models of specific adhesion have been developed.

According to these models, adhesion results from various specific interactions :
- electrostatic interactions which are caused by the formation of a double electric layer during contact of materials of different nature, as in the electric model of Deryagin and Krotova (3) and of Skinner et al. (4) ;
- interactions resulting from the mutual interdiffusion of the molecules of the superficial plane through the interface, as in the theory of diffusion

developed by Voyutskii (5) ;
- physical interactions of the Van der Waals type (dispersion, orientation, induction, hydrogen bonds), as in the model of thermodynamic adsorption of Sharpe and Schonhorn (6) ;
- primary interactions resulting from the creation of true chemical bonds, covalent or ionic, as in the chemical coupling model.
Actually, each one of these models are somehow valid depending on the nature of the materials in contact and on the formation conditions of the interface. Therefore their respective importance depends largely on the chosen system.

It is evident that many other parameters, often interdependent from one another, are also liable to influence the performance of an assembly or of a composite, especially the intrinsic properties of the components, the way in which the materials are prepared, their evolution with time... But one of the main factors determining the final performance of a composite remains the interfacial adhesion.

We do not intend in this study to examine the respective influence of all these parameters, or even to consider exhaustively all the aspects of the interfacial adhesion of materials. Our goal is only to stress two aspects, seemingly specific, of the adhesion between a polymer and a high energy surface, and to obtain therefrom some information, as general as possible, for a better comprehension of the behaviour of assemblies and reinforced materials. These two aspects concern :
- the formation, during contact between a polymer and a high energy solid, of an interfacial layer presenting characteristics which differ from those of the bulk polymer. We call this *the interphase* or thick interface as opposed to *the interface.*
-the modification, during interfacial contact, of the surface properties of the polymer. This modification leads to the formation of a surface that we call *"interfacial"* as opposed to the *free* surface of the polymer. These two notions are not new and many examples can be found in the literature as well as in the studies we have conducted in our laboratory in recent years.

We have chosen to describe here two studies, done in our laboratory, which well illustrate our subject. The first one is related to the *interphase* notion and refers to model assemblies composed of an aluminium substrate, with a high energy surface, and of an elastomer of low surface energy. The second study is related to the notion of an *interfacial* surface and examines the surface modification of a polymer (PMMA) after contact with mercury of high surface energy.

THE INTERPHASE

Introduction
When considering the numerous studies on adhesion phenomena, we can see that the notion of interphase or interfacial layer is either completely ignored or is assumed without proof. In the first part of this study we shall attempt to prove, with an original method, the presence of such an interfacial layer in the case of an aluminium/elastomer assembly.

Materials and methods

Materials. The elastomers have been chosen to represent the solids of low surface energy, mainly because their well-known rheological behaviour facilitates the interpretation of the phenomena occurring during rupture of the joint. Two elastomers are used in the study :
- a 40/60 styrene/butadiene rubber with a very low surface polarity (SBR 1516 Polysar),
- a nitrile rubber with a high surface polarity (NBR 66/34 Perbunan N 330 7 NS Bayer).

The aluminium has been chosen as the model of a high energy surface. However, aluminium is usually covered with a layer of natural oxide, contaminated by many compounds difficult to extract, resulting essentially from the lamination oils. This polluted surface is not typical of a high energy surface. Therefore we have treated the surface. Among the most commonly used treatments, we have chosen two treatments resulting in two distinctly different surface energies, but both maintaining a very slight roughness at the surface of the aluminium :
- the anodisation treatment followed by sealing (surface designed AlS). The surface layer, formed by anodic oxidation in a sulphuric acid medium, is constituted by a compact beam of juxtaposed hexagonal cells, each containing

a pore (Ref. 7). The pores of the oxide layers can be clogged by a subsequent sealing, the transformation of the anhydrous alumina into hydrated alumina (boehmite) being accompanied by an increase in volume (Ref. 8). The sealing has been realized by immersion in boiling water in the presence of salts (acetates).
- the conversion treatment by amorphous phosphatization (surface designed AlP). The aluminium is treated with a solution containing phosphoric acid, chromic acid, and fluorides. The blue-green pellicle at the surface is assumed to have the following chemical composition (Ref. 9) Al_2O_3, $2CrPO_4$, $8H_2O$. The coating, when dry, is essentially Cr^{III} and Al phosphates (Ref. 10).

Surface characterization. To determine the surface energy γ_S of the solids, especially the dispersive component γ_S^D and the polar component γ_S^P, we have used techniques measuring the contact angle between liquid and solid. For the low energy surfaces we determine the contact angle of the chosen solid with a series of liquids with well-known surface energies (Ref. 11). The most general relation between the contact angle and the surface characteristics of the solid and of the liquid is given by

$$\cos \theta_{SL/V} = 2(\gamma_S^D)^{1/2} \frac{(\gamma_L^D)^{1/2}}{\gamma_L} + I_{SL}^P - 1 - \frac{\pi_e}{\gamma_L} \qquad [1]$$

where :
γ^D is the dispersive component of the surface energy of the solid (subscript S) and of the liquid (subscript L) respectively,
I_{SL}^P is the liquid/solid polar interaction energy. One assumes, although it could be theoretically contested but in accordance with experimental results (Ref. 12 & 13), that the non-dispersive interaction energy can be expressed as a function of the polar components γ^P as

$$I_{SL}^P = 2(\gamma_L^P \gamma_S^P)^{1/2} \qquad [2]$$

π_e is the spreading pressure showing the decrease of the surface energy of the solid which follows the adsorption of the liquid vapour. This value can be considered as negligible for solids of low surface energy,
$\theta_{SL/V}$ is the contact angle between the liquid and the solid in the presence of the liquid vapour.
When using non-polar liquids equation [1] can be written as

$$\cos \theta_{SL/V} = 2(\gamma_S^D)^{1/2} \frac{(\gamma_L^D)^{1/2}}{\gamma_L} - 1 \qquad [3]$$

Equation [3] shows a linear relation between $\cos \theta$ and $(\gamma_L^D)^{1/2}/\gamma_L$ and allows the determination of γ_S^D. The deviation, observed with polar liquids in the experimental data, with respect to the straight line leads to the calculation of γ_S^D by using equation [2] . Fig. 1 shows an example, related to the elastomer NBR, of this type of diagram.

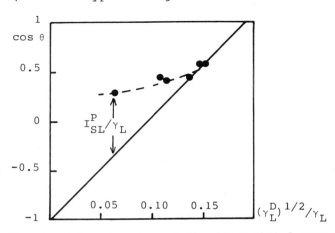

Fig. 1. Determination of the superficial components of the NBR (the liquids used are from right to left : water, formamide, ethyleneglycol, diiodomethane, tricresylphosphate, and bromonaphthalene).

With high energy surfaces, and because of this high energy, the preceeding method cannot be used as the contact angles usually equal zero. We have developed in our laboratory a method which measures the contact angle $\theta_{SL/H}$ (Ref. 14) on the chosen solid of a polar liquid L (water or formamide), in a non-polar liquid medium H (generally an normal alkane) non-miscible with the contact liquid. The surface characteristics are related according to

$$\gamma_L - \gamma_H + \gamma_{HL} \cos \theta_{SL/H} = 2(\gamma_S^D)^{1/2} [(\gamma_L^D)^{1/2} - (\gamma_H)^{1/2}] + I_{SL}^P \qquad [4]$$

When varying the nature of the alkane, one can determine γ_S^D and I_{SL}^P from the slope and the intercept of the line obtained by plotting $(\gamma_L - \gamma_H + \gamma_{HL} \cos \theta_{SL/H})$ as a function of $[(\gamma_L^D)^{1/2} - (\gamma_H)^{1/2}]$. The polar component γ_S^P is obtained, as before, from equation [2]. Fig. 2 shows an example of the results obtained with an AlP surface.

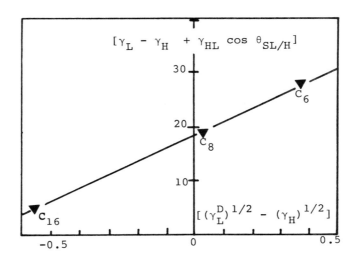

Fig. 2. Determination of the surface energy components of aluminium (L = water, H = alkanes).

Table 1 summarizes the surface characteristics of the chosen solids obtained with the two methods.

TABLE 1. Surface characteristics of the solids (in $mJ.m^{-2} \pm 10\%$)

	γ_S^D	γ_S^P	γ_S
S B R	29.5	0.5	30
N B R	26.5	9.5	36
A l P	150	1.5	151.5
A l S	41	13.5	54.5

In other respects, the surface energy of solids as well as their adhesion can be significatively affected by their texture (roughness, porosity) as their interfacial contact area is closely related to this texture. The texture has been characterized by the Wenzel coefficient (15) which represents the relation between the real area and the geometric area of the solid which is assumed to be perfectly smooth. This coefficient is given by $r = \cos \theta_r / \cos \theta_o \geqslant 1$, where θ_r represents the contact angle of a liquid on a solid and θ_o the contact angle on the same solid but perfectly smooth. This coefficient has been determined by an original method developed in our laboratory (Ref. 16). We measure the contact angle of different liquids on the chosen solid coated, by

metallization, with a thin layer of gold, and also on a smooth glass, coated
with gold under the same conditions. The roughness coefficients obtained are
respectively r = 1.00 for AlP and r = 1.08 for AlS. In spite of the low
roughness value of the AlS surface, the surface characteristics as well as
the separation energies have been corrected taking this parameter into
account.

Bonding. The model assemblies have been realized by pressing under 5.10^6 Pa
at 90°C an elastomer layer of 1 mm between the aluminium substrate and a cotton fabric limiting the longitudinal elongation of the elastomer during the
peel test. By incorporating in the elastomer 1.6 % in weight of peroxide
(1,1 di-t-butylperoxy 3,3,5 trimethylcyclohexane) it can be crosslinked by
maintaining the assembly at 150°C for 50 min under the same pressure.

Peel tests. The strength of a joint is measured by a peel test at 180°
(Fig. 3).

The force, F, necessary to separate the elastomer layer from the high energy
substrate is measured with a dynamometer. The rupture energy W required to
create one unit of interfacial area, is given by $W = 2F/\omega$ where ω is the
width of the test strip.

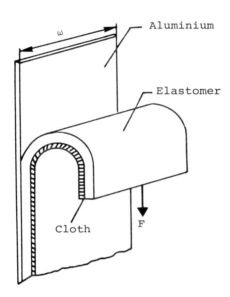

Fig. 3. Schematic diagram of the peel test.

The peel experiments have been conducted at 20°C for a large range of separation rates (between 0.25 and 250 mm.min^{-1}). A special device adapted on the
dynamometer allows us to carry out the peel test in liquid media.

Principle of the method
We shall first recall that the rupture energy measured by the peel test is in
fact composed of two terms ; the *reversible adhesion energy* W_o, necessary to
create reversibly a new unit of interface area, and the energy irreversibly
expended in the elastomer bulk during the failure process. The theory developed by Gent and Schultz (11,17) expresses the relation between the rupture
energy and the reversible adhesion energy as

$$W = W_o \times f(R) \qquad [5]$$

where f(R) is a mechanical factor of energy dissipation depending, at constant test temperature and geometry, only on the rheological properties of the
elastomer, and therefore on the propagation rate R of the rupture front. By
expressing thus the rupture energy as the product of two terms we can :

- separate the contributions to the rupture energy W of the surface characteristics of the materials (expressed in W_o) and of the bulk properties of the elastomer (expressed in f(R)) respectively.
- show that the irreversibly dissipated energy is actually a function of the reversible adhesion energy ; indeed, energy can only be dissipated if the interface resists separation.

Considering that for the different assemblies studied, W_o and f(R) vary together, it appears evident that a comparison of the rupture energies does not allow one to understand quantitatively the observed adhesion mechanisms. This is why we have conceived an experimental process based on the joint separation in liquid medium. We shall explain the principle of the process only for an interfacial failure, but it is easy to extrapolate to other types of rupture processes.

In an inert medium, for example in air, the rupture energy is expressed as

$$W = W_o \times f(R) \tag{5}$$

In a liquid medium, and if the chosen liquid does not modify by swelling or reaction the viscoelastic characteristics of the elastomer, the dissipation function f(R) does not vary and the rupture energy W_L is given by

$$W_L = W_{oL} \times f(R) \tag{6}$$

The ratio of equations [5] and [6] gives

$$\frac{W_L}{W} = \frac{W_{oL}}{W_o} \tag{7}$$

These ratios can be determined directly, either experimentally or by computation. The ratio W_L/W can be determined by peeling in air or in a liquid medium. The ratio W_{oL}/W_o can be calculated from the surface characteristics of the solids in contact and of the liquid. Thus in air $W_o = \gamma_{S_1} + \gamma_{S_2} - \gamma_{S_1 S_2}$, and in a liquid medium $W_{oL} = \gamma_{S_1 L} + \gamma_{S_2 L} - \gamma_{S_1 S_2}$.
We can easily show that

$$W_{oL} = W_o + \Delta W_o$$

with $$\Delta W_o = 2\gamma_L - W_{S_1 L} - W_{S_2 L}$$

W_{SL} represents the liquid/solid adhesion energy and is equal to

$$W_{SL} = 2(\gamma_S^D \gamma_L^D)^{1/2} + 2(\gamma_S^P \gamma_L^P)^{1/2}$$

Using the surface characteristics of the solids and of the liquid, the W_o and ΔW_o quantities are expressed as

$$W_o = 2(\gamma_{S_1}^D \gamma_{S_2}^D)^{1/2} + 2(\gamma_{S_1}^P \gamma_{S_2}^P)^{1/2}$$

$$\Delta W_o = 2\gamma_L - 2(\gamma_L^D)^{1/2}[(\gamma_{S_1}^D)^{1/2} + (\gamma_{S_2}^D)^{1/2}] - 2(\gamma_L^P)^{1/2}[(\gamma_{S_1}^P)^{1/2} + (\gamma_{S_2}^P)^{1/2}]$$

In the case studied, these calculations imply that :
- only physical interactions take place during the solid/solid interfacial contact ;
- in inert as well as in liquid medium, the separation is purely interfacial;
- the potential rate of penetration of the liquid to the fracture front is higher than the propagation rate of the fracture ; consequently, the liquid is always present at the rupture front.

Results
Study of the AlS/elastomer assembly.
As an example, Fig. 4 shows the behaviour of an AlS/NBR assembly, the separation being done in air and ethanol, respectively.

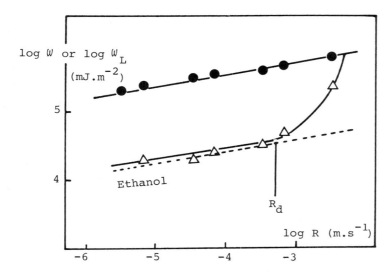

Fig. 4. Influence of a liquid medium on the rupture of an
AlS/NBR assembly.

This diagram shows, on logarithmic scales, the variation of the rupture energy W and W_L with the separation rate R. The following observations can be
made :
- the rupture energy W increases with the separation rate R. This shows simply the increase of the dissipative character of the elastomer with the deformation rate.
- in the domain of low rates of separation, the curves in air and in ethanol
are parallel, on a logarithmic scale, thus verifying our hypothesis that the
ratio W_L/W is constant and independent of the rate R, i.e. that the dissipation factor f(R) is not modified by the presence of the liquid.
- in the domain of the high rates of separation, the curve corresponding to
the liquid medium rejoins the one corresponding to the air medium. This kinetic effect has been quantitatively interpreted elsewhere (Ref. 16). It shows
that at a certain rate of propagation R_d, the potential penetration rate of
the liquid, determined by its viscosity and by its degree of interaction with
solids, is not sufficient to maintain contact at the separation front. In the
rest of this study we shall always assume rate conditions sufficient to avoid
this kinetic phenomenon.
- the dotted curve shows the theoretical values calculated by using equation
[7].

$$W_L = W \times \frac{W_{oL}}{W_o}$$

One can notice the remarkable agreement between the experimental and the computed curve.

Table 2 shows that this conclusion is valid whatever the elastomer or the
chosen liquid might be.

One can also notice that the relative decrease of the rupture energy of the
assembly, consecutive to the presence of alcohol, is large, approaching spontaneous separation in liquid medium.

This first series of experiments with an AlS substrate of average surface
energy allow us to conclude that :
- during the contact period, only physical interactions of dispersive and polar type take place at the interface.
- the separation propagates only at the interface.

In these conditions the rupture energy can be quantitatively predicted from
the surface characteristics of the solids chosen for assembling.

TABLE 2. Comparison of the experimental and of the theoretical results

Assemblies	W_o (mJ.m^{-2})	Liquid	$\frac{\Delta W}{W}$ exp. %	$\frac{\Delta W_o}{W_o}$ calc. %
AlS/SBR	76	Methanol	-97	-95
AlS/SBR	"	Ethanol	-92	-97
AlS/SBR	"	Butanol	-92	-98.5
AlS/NBR	88	Methanol	-96	-93
AlS/NBR	"	Ethanol	-91.5	-92.5
AlS/NBR	"	Butanol	-92	-88

Study of the AlP/elastomer assembly.

We shall study the specific behaviour of assemblies with an AlP substrate of high energy ($\gamma_S > 150$ mJ.m^{-2}). We notice first that the strength of the joints is smaller than with an AlS surface. This is surprising as the AlP substrate has a higher surface energy. Furthermore the study of the relative decrease of the rupture energy in liquid medium shows that the computed and the experimental values do not agree in this case. For example, with an AlP/SBR joint in ethanol medium, the computed decrease is 81 % when the observed experimental decrease is approximatively 56 %. These observations can be explained if the rupture does not propagate at the aluminium/elastomer interface but in a cohesive manner inside the elastomer near the interface. The wetting data, shown in Table 3, clearly show the interfacial rupture of the assemblies with an AlS substrate and the cohesive rupture of the assemblies with an AlP substrate.

TABLE 3. Contact angle of water on solids before contact and after separation

	Surface	θ_{H_2O} (degrees)
Initial Surfaces	SBR	98
	NBR	74.5
	AlS	54
	AlP	40
Surfaces after Separation	AlS after detachment from	
	SBR	52
	NBR	59
	AlP after detachment from	
	SBR	92.5
	NBR	72

Considering that the elastomer is crosslinked and that, as a result, its cohesion is mainly assured by chemical bonds, the cohesive rupture in the elastomer can only be explained by two following hypotheses :
- either by the creation of chemical bonds between the elastomer and the substrate confering to the joint a stronger adhesion that the elastomer cohesion.
- or by the presence of a weak boundary layer of elastomer near the interface.

The first hypothesis has little credibility as it does not explain the low values of resistance to rupture which have been observed. These values are inferior to those obtained with an AlS substrate where only physical interactions are established. The second hypothesis shall allow us to verify the presence of an interfacial layer of weak cohesion and to characterize this layer. We use the same principle as the one used earlier, the reversible energy of adhesion W_o being replaced by the reversible energy of cohesion W_o^C of the elastomer in the zone near the interface. During separation, the energy dissipation mechanisms are assumed to be identical wherever rupture is located.

The cohesion energy W_o^C of the elastomer is the sum of the contributions of the physical interactions $W_{o\ phys}^C$ and of the chemical interactions (or entanglements) $W_{o\ chem}^C$

$$W_o^C = W_{o\ phys}^C + W_{o\ chem}^C \qquad [8]$$

The contribution of physical interactions being equal to twice the surface energy of the elastomer, the equation is written

$$W_o^C = 2\gamma_S + W_{o\ chem}^C \qquad [9]$$

Considering also that the presence of the liquid modifies only the physical and not the chemical interactions (Ref. 18), the reversible energy of cohesion in liquid medium is given by

$$W_{oL}^C = 2\gamma_{SL} + W_{o\ chem}^C \qquad [10]$$

which can be written as before

$$W_{oL}^C = W_o^C + \Delta W_o^C$$

with $\Delta W_o^C = 2\gamma_L - W_{SL}$

W_{SL} being the reversible elastomer/liquid adhesion energy.

Taking into account the general relation

$$\frac{W_L}{W} = \frac{W_{oL}^C}{W_o^C} \qquad [11]$$

one can express the reversible cohesion energy W_o^C of the elastomer, and especially the contribution to this energy of the chemical bonds $W_{o\ chem}^C$, as

$$W_o^C = 2\gamma_S + W_{o\ chem}^C = \frac{\Delta W_o^C}{\frac{W_L}{W} - 1} \qquad [12]$$

As all the terms of the equation can be derived directly either from the experimental data or by computation, W_o^C and $W_{o\ chem}^C$ can be defined. With an AlP/SBR assembly, as illustrated on Fig. 5, in air and in ethanol media, the following values have been obtained.

$$\frac{W_L}{W} = 0.44$$

$$\Delta W_o^C = -52\ mJ.m^{-2}$$

$$W_{o\ phys}^C = 60\ mJ.m^{-2}$$

$$W_o^C = 93\ mJ.m^{-2}$$

$$W_{o\ chem}^C = 33\ mJ.m^{-2}$$

The contribution of the chemical bonds to the reversible energy of cohesion of SBR near the interface represents only about 35 % of the total cohesion energy.

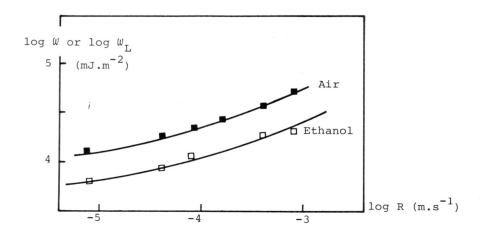

Fig. 5. Influence of a liquid medium on the rupture of the
AlP/SBR assembly.

A similar experiment, conducted with the AlP/NBR assembly gives the following values.

$$\frac{W_L}{W} = 0.18$$

$$\Delta W_o^C = -69 \text{ mJ.m}^{-2}$$

$$W_o^C \text{ phys} = 72 \text{ mJ.m}^{-2}$$

$$W_o^C = 84 \text{ mJ.m}^{-2}$$

$$W_o^C \text{ chem} = 12 \text{ mJ.m}^{-2}$$

With NBR, near the interface, the chemical bonds represent only 14 % of the total cohesion energy of the elastomer.

In a second step, using different liquid media, the values obtained have been verified by comparison of the relative variations of the rupture energy obtained experimentally with the ones obtained by calculation, taking the values of W_o^C chem into account (Table 4).

TABLE 4. Influence of various liquids on the rupture energy of the assemblies

Assemblies	Liquid	$\frac{\Delta W}{W}$ exp %	$\frac{W_o^C}{W_o^C \text{ calc}}$ %
AlP/SBR	Butanol	-60	-61
	PDMS 1.7 cP	-63	-61
	PDMS 970 cP	-57	-62
	H_2O	+27	+31
AlP/NBR	Methanol	-79.5	-82
	Butanol	-70	-75
	H_2O	-48	-47

PDMS polydimethylsiloxane (Rhône-Poulenc 47V).

The observed agreement justifies the interpretation of a rupture propagating cohesively through a weak boundary layer of elastomer ; the contribution of the chemical bonds to the reversible cohesion energy of the elastomer in this interfacial layer being respectively 35 % for SBR and 14 % for NBR.

Discussion

The values of the reversible energy of cohesion resulting from chemical bonds or from entanglements are actually infinitely small, as shown in Table 5.

TABLE 5.   Reversible cohesion energy of the elastomer near the interface (mJ.m^{-2})

Elastomer	$W^C_{o\ phys}$	$W^C_{o\ chem}$	W^C_o
SBR	60	33	93
NBR	72	12	84

The values which are calculated according to the Lake and Thomas theory (19), or experimentally determined (Ref. 20 & 22), are considerably larger in the bulk elastomer, and thus the contributions of the physical interactions can be considered as negligible. According to Lake and Thomas, the reversible cohesion energy does not only correspond to the rupture of the C-C bonds in the fracture plane, but also to the deformation, almost to the rupture point, of all C-C bonds between two points of crosslinking. The stored energy is then irreversibly dissipated during the rupture of the C-C chemical bond. Lake and Thomas have calculated that the reversible cohesion energy of a natural rubber should be around 20 J.m^{-2}. The experimental values obtained from tearing or peel experiments, at very low deformation rate in order to limit the viscoelastic losses, are in fairly good agreement with the theory. These values for elastomers of the same type as those studied here, and crosslinked under similar conditions, are typically between 20 and 100 J.m^{-2}. We have, of course, verified with swelling measurements in benzene and toluene media, that the elastomers used in our experiments are correctly crosslinked, the values of M_c being 1.2 10^4 for the SBR or 8.10^3 for the NBR, respectively. Also, the experiments realized by Greensmith (18) consisting in the tearing of elastomers in liquid medium, show that the resistance to tearing, and therefore the reversible cohesion energy of elastomers, even weakly crosslinked, are not affected by the presence of the liquid.

Our experimental results, as well as the other arguments presented, show clearly that during the interface formation between an elastomer and a solid of high surface energy, an interfacial layer is created, near the interface, with characteristics different from those of the bulk elastomer, especially a very weak cohesion, thousands of times weaker than the bulk cohesion.

In the present state of our studies, we cannot explain in any definite way the formation mechanism of this interphase. But, considering the very low values of the cohesion energy of the interphase, two hypotheses can be made :
- the migration toward the metal/polymer interface of the fractions of low molecular weight.
- the inhibition of the crosslinking reaction of the elastomer resulting probably from the influence of the high energy surface on the kinetics of crosslinking.

This notion of interphase is especially important since its properties are responsible for the rupture localization and the rupture energy of the system. This concept of a weak boundary layer, which has been quantitatively demonstrated in this study, is in fact relatively old and many authors use this notion to explain the ruptures which apparently happen at the interfaces and correspond to low adhesion strengths. For example, Bickerman (23 & 24) classifies the interfacial layers of weak cohesion into seven different categories according to their origin. Sharpe (25) has conducted several critical studies on the existence and the role of these layers, and Lipatov (26) often refers to this notion to explain the behaviour of reinforced polymers.

The real problem is, however, the clarification of the specific influence of the high energy surface in the creation of the interphase, either with a weak or a strong cohesion. This problem has two aspects :
- does the high energy surface interfere directly, because of its surface characteristics, in the mechanism of formation of the interfacial layer ?
- does the high energy surface, leading to a strong adhesion with the polymer

usually stronger than the cohesion of the transition layer, only reveal the existence of this layer ?

Several published studies argue in favor of the first hypothesis, e.g.,
- in polymer/copper assemblies, the weak boundary layer is formed by oxidizing degradation of the polymers, the reaction being catalysed by the metal (Ref. 24).
- in crystalline thermoplastic/metal assemblies, the interfacial layer, which is formed, usually presents a transcrystalline structure initiated by the metal, according to a process similar to an epitaxy (Ref. 27 & 28).

It is however clear that many studies on the existence of these interfacial layers are still necessary, especially using modern techniques as those recommended by Baun (20), for the development of general theoretical concepts on the formation mechanisms of the interphase. It appears, in fact, that the knowledge of the surface characteristics of the materials is not sufficient to understand and, above all, to predict the adhesion of polymers on high energy solids.

THE INTERFACIAL SURFACE

Introduction
The existence of an interphase or of a transition layer, if its formation is influenced by the presence of a high energy solid, can be at the origin of a modification of the surface properties of a polymer during the interfacial contact. This phenomenon is important as it can bring about modifications of the adhesion energy and, therefore, of the performance of the assembly or of the composite. These modifications are impossible to predict from the surface energies of the materials in contact. The concept of *interfacial surface* is here opposed to the notion of *free surface*.

Indeed, if we consider a solid, with molecules having a certain degree of mobility, the free enthalpy of the system is minimal for a minimal surface energy. We can thus imagine that the distribution and the conformation of the macromolecules at the polymer surface would lead to the lowest surface energy possible (for example, polar groups of the macromolecules oriented towards the bulk of the material). On the contrary during an interfacial contact with a solid, in conditions allowing the mobility of the macromolecules (melted polymer or polymer in solution, for example) we can state, at least as a first approximation, that the free enthalpy of the system will be minimal when the polymer/solid interfacial energy is as low as possible. This condition implies, therefore, a possible modification of the orientation of the macromolecules in contact with the solid leading to a change of surface energy of the polymer. The surface in contact with the solid or *interfacial* surface can thus have a surface energy different from the one of the *free* surface.

This phenomenon is even more likely if the solid in contact with the polymer has a surface energy much higher than the one of the polymer ; in other words, when a interface polymer/solid of high energy is created. The experiments which we describe in order to show this phenomenon have been conducted in collaboration with Professor H.P. Schreiber of the Ecole Polytechnique of the University of Montreal (30).

Materials and experimental conditions
The polymer used for this study is a polymethylmethacrylate (PMMA) with a viscometric molecular weight of 10^5 (Rohm and Haas Co.). This polymer, in solution in toluene or in chloroform, at concentrations varying between 100 and 200 g/l, is deposited on polytetrafluoroethylene (PTFE), representative of a surface of low energy ($\gamma_S = 19$ mJ.m^{-2}), and on mercury, representative of a surface of high energy ($\gamma_L = 480$ mJ.m^{-2}).

The surface of PTFE is prepared according to Zisman's method (31). The mercury is purified by washing in nitric acid at 10 %, rinsed with distilled water, than distilled under reduced pressure. The PMMA films deposited on their support are dried for 48 h in air, than in vacuum for at least 48 h. After drying, the films, with a thickness between 30 and 500 μm, are detached from their support, and their surface energy is immediately determined by the contact angle measurement techniques described earlier.

Results

The different sides of the polymer film, the free sides and the sides having been in contact with the PTFE and the mercury respectively, have been analyzed with the help of two representative liquids, the diiodomethane of very low surface polarity, and water of high polarity. The angles which have been observed on the different faces are reported in Table 6.

TABLE 6.   Contact angle measurements (± 2 degrees)

Liquids	θ free surface	θ interfacial surface (PTFE)	θ interfacial surface (Hg)
CH_2I_2	37	40	28
H_2O	80	81	69

One can also notice (Fig. 6), in the case of a surface having been in contact with mercury, an evolution of the contact angle as a function of time, after detachment of the film. After about 3 h, the values of the angles reach those measured on the free surface.

$$\theta_{CH_2I_2} = 42 \text{ degrees} \qquad\qquad \theta_{H_2O} = 79 \text{ degrees}$$

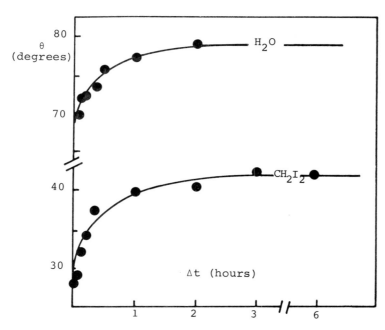

Fig. 6.   Evolution of the contact angle as a function of time (PMMA surface after contact with the mercury).

The components of the surface energy of the PMMA faces calculated from the contact angles are reported in Table 7. These values do not change with the nature of the solvent used, nor with the solution concentrations, nor with the thickness of the film formed.

TABLE 7. Surface energy components of PMMA (mJ.m^{-2})

	γ_S^D	γ_S^P	γ_S
Free surface (air)	43	3	46
Interfacial surface (PTFE)	41.5	3	44.5
Interfacial surface (Hg) at t = 0	47	6	53
Interfacial surface (Hg) at t = ∞	40.5	3.5	44

These results lead to several conclusions :
- the surface energy of PMMA, in contact with a low energy surface as PTFE, is identical to the energy of the free surface.
- but, in contact with a high energy surface such as mercury, the surface energy of PMMA is significatively higher than the energy of the free surface. We can show by computation that this increase of the surface energy of the PMMA is accompanied by a decrease of the PMMA/Hg interfacial energy. This variation could be attributed, in accordance with our hypotheses, to a particular orientation, conformation or density of the macromolecules in contact with mercury.
- it is also important to note that the interfacial surface of PMMA after separation from the mercury, is in a metastable state and that, consequently, its surface energy evolves slowly with time, at ambient temperature towards the value corresponding to the free surface. The glass transition temperature of PMMA (Tg ≃ 105°C) being much higher than ambient temperature it is doubtful that the movements of the chains cause this change. Relaxations in the side groups of PMMA are most likely responsible, secondary transitions being effectively present at -150° and 20°C (Ref. 32).
- a particular behaviour of the PMMA film (of small thickness) is the tendency to roll up after detachment from mercury. This could be related to a thermodynamic imbalance between the two sides of the film.

Discussion
These results clearly show the existence of an interfacial surface with characteristics different from those of the free surface.

Up to now, only a few studies have been conducted specifically on the subject of the modifications of the surface properties of the polymer in contact with a high energy surface, with the exception of some studies, more or less relevant, on polymer adsorption. This is due probably to the difficulties encountered when trying to show these modifications at the solid/solid interface. We can, however, cite the studies of Schonhorn (33-36) and of Nara (37). These authors have studied the variations of wettability of various crystalline polymers in contact with metals. According to them, these variations are caused by modifications of the polymer density at the interface. The determinant role of the high energy solid is clearly shown.

More recently, with the help of more advanced techniques, the modifications of the polymer surface near the interface (a few nm) between the polymer and the high energy surface, have been studied in situ. For example, Oberlin et al (38), in the field of composites, have shown, by using electronic microscopy of high resolution (dark field), that in an epoxy resin/carbon fibre composite, an orientation of the bisphenol A groups of the resin takes place at the surface of the resin in contact with a carbon fibre of high energy. This does not happen with a carbon fibre of low surface energy.

We must note that the general concept of the need for chemical bonds to explain high adhesion, especially in the case of polymer/metal adhesion, is unnecessary. The increase of the surface energy of the polymer could be sufficient to explain the improvment of the interface resistance. In fact, the reversible energy of adhesion can be significantly increased compared to the energy calculated from the surface energies of solids before contact. This notion of interfacial surface is therefore very useful for a better comprehension of the performance of assemblies and of reinforced materials.

CONCLUSION

Unquestionably, during the past decades, constant progress has been made towards a better comprehension of the adhesion phenomena in composite systems and the fabrication of materials with better performances.

A better characterization of the surfaces of the materials in contact, the use of interfacial chemical reactions, the adaptation of the bulk properties of the macromolecular matrix, etc..., have contributed to many improvments in the domain of bonded joints, filled polymers, composite materials,... However, as we have tried to show it in this study, the formation of an interfacial zone, or interphase, with properties different from those of the bulk polymer, can deeply affect the mechanical properties of the system, in a way which is hardly predictible. Indeed, the strength of the composite system is then essentially conditioned by the cohesion of this interphase and not by the nature or the number of bonds at the interface.

Although, recently, this aspect has been largely studied, an important research effort is still necessary in order to better our knowledge, especially quantitatively, of these interfacial layers. In particular, specific methods of characterization of these layers using modern analysis techniques must be developed, the research on the mechanisms of formation should be expanded and the theoretical concepts, allowing the passage from the qualitative to the quantitative aspect of the problem, have to be established. Indeed, the existence of these layers have been now widely admitted without any further proof of existence other than the weak strength of the interface. These proofs are necessary in order to develop the credibility of the concept.

One of the most interesting research areas, implicit in the present study, is the examination of the specific role of the high energy surface in the formation of this transition zone. Another topic, already largely described, is the study of how to control technologically the formation of these layers and thus improve the performance of composite systems.

The notion of interfacial surface is more or less related to the concept of interfacial layer. Especially, when in contact with a high energy surface, the polymer presents surface properties quite different from the free surface. This phenomenon also prevents the quantitative prediction of the performance of composite systems from the surface properties of the materials to be associated. This notion of interfacial surface is unquestionably underlying many studies ; however, in this case again, efforts should be done for characterization and theoretical development of the mechanisms.

REFERENCES

1. J.W. Mc Bain, D.G. Hopkins, Second Report of the Adhesives Research Committee, H.M.S.O., London (1926)
2. J.W. Mc Bain, The Third and Final Report of the Adhesives Research Committee, H.M.S.O., London (1932)
3. B.V. Deryagin, N.A. Krotova, Dokl. Akad. Nauk. SSSR, 61, 849 (1948)
4. S.M. Skinner, R.L. Savage, J.E. Rutzler, J. Appl. Phys., 24, 438 (1953)
5. S.S. Voyustkii, Uspekki Khim., 18, 449 (1949)
6. L.H. Sharpe, H. Schonhorn, Symposium on Contact Angles, Los Angeles, April 1963, Chem. Eng. News, 15, 67 (1963)
7. F. Keller, M.S. Hunter, D.L. Robinson, J. Electrochem. Soc., 100, 411 (1953)
8. P. Lelong, R. Segond, J. Herenguel, L'oxydation des métaux, J. Bénard, Gauthier-Villars et Cie Ed., Paris, tome 2, 451 (1964)
9. N.J. Newhard, Metal Finishing, July, 49 (1972), August, 66 (1972)
10. L.A. Nimon, G.K. Korpi, Plating, May, 421 (1972)
11. J. Schultz, A.N. Gent, J. Chim. Phys., 70 (5), 708 (1973)
12. D.K. Owens, R.C. Wendt, J. Appl. Polymer Sci., 13, 1741 (1969)
13. J. Schultz, K. Tsutsumi, J.B. Donnet, J. Colloid Interface Sci., 59 (2), 277 (1977)
14. J. Schultz, K. Tsutsumi, J.B. Donnet, J. Colloid Interface Sci., 59 (2), 272 (1977)
15. R.N. Wenzel, Ind. Eng. Chem., 28, 988 (1936)
16. A. Carré, Thèse de Doctorat d'Etat, Université de Haute-Alsace, 21 Mai 1980
17. A.N. Gent, J. Schultz, J. Adhesion, 3, 281 (1972)

18. H.W. Greensmith, A.G. Thomas, J. Polymer Sci., 18, 189 (1955)
19. G.J. Lake, A.G. Thomas, Proc. Roy. Soc. Lond., A 300, 108 (1967)
20. G.J. Lake, P.B. Lindley, J. Appl. Polymer Sci., 9, 1233 (1965)
21. A. Ahagon, A.N. Gent, J. Polymer Sci., Phys. Ed., 13, 1285 (1975)
22. E.H. Andrews, Makromol. Chem. Suppl., 2, 189 (1979)
23. J.J. Bikerman, Ind. Eng. Chem., 59, 41 (1957)
24. J.J. Bikerman, The Science of Adhesive Joints, 2nd Ed., Acad. Press, N.Y.
    (1968)
25. L.H. Sharpe, J. Adhesion, 4, 51-64 (1972)
26. Y.S. Lipatov, Physical Chemistry of Filled Polymers, Intern. Polym. Sci.
    Tech. Monograph N° 2 (1979)
27. T.K. Kwey et al., J. Appl. Phys., 38, 2512 (1967)
28. H. Schonhorn, Adhesion : Fundamentals and Practice, Mac Laren, London, 3,
    (1969)
29. W.L. Baun, Adhesion Measurements of Thin Films, Thick Films, and Bulk
    Coatings, ASTM STP 640, K.L. Mittal Ed., 41 (1978)
30. H.P. Schreiber, A. Carré, to be published
31. H.W. Fox, W.A. Zisman, J. Colloid Sci., 7, 109 (1952)
32. F.W. Billmeyer, Text book of Polymer Science, 2nd Ed., John Wiley and
    Sons, Inc., N.Y., 211 (1971)
33. H. Schonhorn, F.W. Ryan, J. Adhesion, 1, 43 (1969)
34. K. Hara, H. Schonhorn, J. Adhesion, 2, 100 (1970)
35. H. Schonhorn, Adhesion and Bonding, M. Bikales, Wiley Interscience, N.Y.,
    41 (1971)
36. J.P. Luongo, H. Schonhorn, J. Polymer Sci., A-1, 6, 1649 (1968)
37. S. Nara, K. Matsuyama, J. Appl. Polymer Sci., 13, 1729 (1969)
38. M. Guigon, J. Ayache, A. Oberlin, M. Oberlin, Proc. 3rd Intern. Conf.
    Composites Materials, Paris, Pergamon Press, 1, 223 (1980).

BIODEGRADATION OF POLYMERS FOR BIOMEDICAL USE

Jindřich Kopeček

Institute of Macromolecular Chemistry, Czechoslovak
Academy of Sciences, 162 06 Prague 6, Czechoslovakia

Abstract - The interaction of polymers with physiological
environment on the cell level is discussed. A model polymer
system with an easily variable structure has been suggested
for a detailed study of these interactions, and of enzyme-
controlled degradation in the first place. Water-soluble
copolymers of N-(2-hydroxypropyl)methacrylamide containing
oligopeptidic sequences both in side chains and in crosslinks
connecting two synthetic polymer chains were synthesized.
The relationship between the structure of such oligopeptidic
sequences and their degradability by chymotrypsin, trypsin,
papain, isolated intracellular enzymes, cell cultures and
in vivo was examined. The possibility of using information
thus obtained in a tailor-made preparation of polymeric
carriers able to transfer the bound drug into specific cells
is discussed.

INTRODUCTION

At present, only very few synthetic biodegradable polymers are known. Potts
et al. (1) found that of the readily available polymers, only low-molecular
weight polyesters exhibited distinct biodegradation if treated with microorganisms. After implantation of synthetic polymers into the living organism
changes in structure are observed on the surface of the implants (2), and
their mechanical properties deteriorate (3). Changes occuring in most of the
polymers (polyethylene, polypropylene, poly(methyl methacrylate),
poly(2-hydroxyethyl methacrylate) etc.) are so small that the mechanism of the
degradation is difficult to follow. The conclusion regarding the nature of
biodegradation - whether it is true biodegradation (4), i.e. a process
involving some specific vital activity of the physiological environment, is
often only a hypothetic one. On the other hand, all natural polymers (e.g.,
starch, cellulose, proteins, nucleic acids) are biodegradable. They have
existed for a sufficiently long time to make possible the development of
microorganisms which use them as food.

Most of the attempted preparations of novel types of biodegradable polymers
are based, at least conceptually, on natural polymers. The procedures consist
either in the preparation of synthetic analogs of natural polymers, such as
polypeptides (5), or in a combination of synthetic and natural polymers. Both
modification of the structure of natural polymers (starch → hydroxyethyl
starch) and introduction of degradable bonds into synthetic polymers have been
described. The common point of both approaches is synthesis of block copolymers containing degradable blocks (6-9) (e.g., combining saccharide and
peptide blocks (8) or synthetic and peptide blocks (9)). Another procedure
employs modification of the structure of synthetic polymers so as to make the
structure of some monomeric units similar to that of amino acids specific for
a certain enzyme. Huang´s laboratory (10,11) has been especially active in
the field of modification of the polymer structure.

For the purposes of basic research, it is important to use such types of
polymers the structure of which can easily be altered. What is revealed by the
investigation of such group of polymers is not only the ability of the given
structure to undergo enzymatic degradation, but also the relationship between
the structure and degradability of a certain class of polymers. Only such
conclusions may be used in the tailor-made synthesis of biodegradable polymers. These conclusions are particularly needed in medicine, in order to
extend application of polymers with controllable biodegradability in the

clinical practice. Biodegradable polymers can be used in the insoluble or
soluble forms. In the insoluble form they can be employed as temporary
implants (e.g., as readily hydrolyzed polyhydroxy acids (12)), or as implants
for sustained release of drugs.

Polymers soluble in physiological medium can be used (13-16) as blood plasma
expanders, polymeric drugs or drug carriers (generally carriers of biologically active compounds). Biodegradation of the basic chain facilitates the
elimination of the polymer from the organism. With drug carriers, there
appears also an additional problem of cleavability of the bond by which the
drug is attached to the polymeric carrier.

In recent years, work in our laboratory has been concentrated on the study of
the relationship between the structure and biodegradability of polymers in
order to obtain information necessary for the preparation of degradable
polymeric drug carriers which allow targetting to specific cells.

The fact that a drug is attached to a polymer molecule leads to substantial
changes in the mechanism of its interactions on the cell level. A low-molecular weight drug may enter the cell by random diffusion. If attached to the
carrier macromolecule endocytosis becomes its only mode of entry, and this
can be a highly cell and substrate specific mechanism (17-19). Substances
captured by endocytosis are transported into the lysosomal compartment of the
cell, where the drug, if it is to have therapeutic efficacy, would need to be
released from its macromolecular carrier and escape into the cytoplasm.
Hundreds of drugs bound to soluble polymers (13,15,16) and prepared during
the last decades have been described in the literature. The fact that only
the smallest number of these polymers have penetrated into clinical practice
is due to the lack of solutions to a number of chemical, biological and biochemical questions aroused by the application of polymeric drugs.

A drug carrier system which would allow specific targetting to certain cells
must fulfil the following requirements (19):
(a) biologically inert carrier molecule
(b) degradable side-chain to act as site of drug attachment and facilitate
    controlled release
(c) a mechanism for the specific targetting of the carrier complex

   Schematically, the system may be represented as follows:

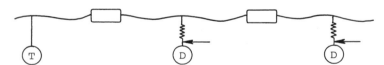

(T)....targetting moiety; (D)....drug attached by an enzymatic degradable (◄——)
bond via a spacer (—ʌʌʌ—); [   ]..degradable parts of the main chain.

If the above scheme is to be put into practice, a number of problems arise which
have the character of basic research and have therefore to be gradually
solved. For this purpose, a polymer system had to be chosen which would
facilitate a relatively easy change in structure so that conclusions regarding the relationship between structure and biodegradability could be drawn
from the experiments.

In the first part of this paper biological problems will be discussed able t
contribute to the proposed structure of polymeric carriers suitable for druç
binding. These comprise basic data on intracellular enzymes, on the mode of
penetration of macromolecular compounds into intracellular space, and also
information on recognition systems which might affect distribution of the
macromolecular compound in the organism. Further part is concerned with the
proposal of a model system providing an easy variability of the structure,
together with binding of targetting moieties or biologically active compoun(
Results of an investigation of the relationship between the structure of
model copolymers which are a combination of an oligopeptidic sequence and
a synthetic polymer chain, on the one hand, and their biodegradability by mod
(chymotrypsin, trypsin, papain) and intracellular enzymes, on the other, are
discussed in the subsequent part. The study of systems, in which the oligopeptidic sequence either formed side chains of the synthetic polymer or was
part of the crosslink joining two synthetic polymer molecules made possible
an evaluation, not only of structural, but also of steric effects, which
affect the formation of the enzyme-substrate complex.

INTRACELLULAR ENZYMES

What is the way in which macromolecular compounds interact with intracellular enzymes? What is the specificity possessed by intracellular enzymes?

The important problem encountered in most cases when working with enzymes is the fact that their activity decreases with time. Christian de Duve was faced with quite the opposite problem in the early nineteen fifties (20). When he let a homogenisate of rat liver stand, the acid phosphatase increased in apparent activity. A detailed study of this phenomenon led de Duve to a conclusion that intracellular enzymes were concentrated in cell organelles, i.e. subcellular particles limited by a membrane. He called them lysosomes. With the exception of a few cells (e.g. mammalian red blood cells) all animal cells possess lysosomes (21). It is known that lysosomes contain a number of enzymes, such as proteases, nucleases, glycosidases, aryl sulphatases, lipases, phospholipases, phosphatases (22). All lysosomes are related directly or indirectly to intracellular digestion. The material to be digested may be of endogenous (intracellular) or exogenous (extracellular) origin.

If intracellular enzymes are localized in cellular organelles, called lysosomes, it means that they are separated from the cytoplasm by a membrane through which only low-molecular weight compound can diffuse. The penetration of a high-molecular weight compound into the cells is shown in the following scheme:

Large molecule can enter cells only by endocytosis

Small molecule can pass biological membrane by random diffusion

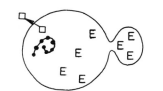

The small molecule released from the polymer can leave the sec. lysosome and enter the cytoplasma

Lysosomes contain many hydrolytic enzymes

The endocytic vacuole (pinosome) fuses with a primary lysosome forming a secondary lysosome

Thus, polymer molecules enter cells by endocytosis. Endocytosis may be defined as internalization of formerly extracellular material, within a membrane-bound vesicle formed by invagination of the plasma membrane. Phagocytosis is a form of endocytosis in which particulate material is taken up into large vesicles. Pinocytosis is endocytosis of soluble materials into small vesicles. In pinocytosis, molecules can penetrate into the cell either in solution (so-called fluid-phase pinocytosis) or bound to the surface (17, 18,23). The polymer entering the cell by pinocytosis is enclosed in new organelles called pinosomes, which are restricted against the surroundings by membranes. Only after fusion of the pinosome with the primary lysosome a new particle is formed, called secondary lysosome, in which the pinosome content is exposed to the effect of lysosomal enzymes. Any degradable material is catabolized (in an acidic internal environment - the pH in lysosomes is approximately 1.5 units lower than that of the medium). Small molecules arising as the degradation product escape through the lysosomal membrane, whereas any large non-biodegradable material remains trapped in the secondary lysosome compartment. Certain types of cells (18), especially unicellular organisms, are able to regurgitate material, but many mammalian cells accumulate material they cannot digest as residual bodies within the cell.

The rate of penetration of polymers into cells depends on the mechanism of pinocytosis. This mechanism may be affected, among other things, also by the polymer structure. The possibility of quantification of the rate of absorption of polymers of various structure was facilitated by the introduction of the Endocytic Index (24). This index is defined as the volume of cultivating medium (μl) whose contained substrate is captured per miligram of cell protein (or per 10^6 cells (25)) per hour. Differences in the rate of absorption may amount to as much as several orders of magnitude (cf. Table 1, ref. (18)) depending on the polymer structure and type of the cell.

Isolation of lysosomal enzymes

In order to investigate the relationship between the polymer structure and the possibility of their degradation in the intracellular space, it is important to carry out preliminary experiments with isolated lysosomal enzymes. They are isolated by centrifugal techniques (20). These techniques depend for success primarily on differences in size and/or density of the particles to be separated. Unfortunately lysosomes are similar in both these respects to mitochondria. However, certain functional characteristics of lysosomes can be used to modify their behaviour in centrifugation. Injection of Triton WR-1339 (nonhemolytic nonionic detergent) to rats leads to its accumulation in hepatic lysosomes (17). This is of practical value, because the marked decrease in the density of hepatic lysosomes allows the separation of lysosomes and mitochondria during centrifugation (26). The lysosomal enzymes thus isolated are called Tritosomes.

Targetting to specific cells

Substances which are taken up selectively into lysosomes were called by de Duve et al. (17) lysosomotropic agents. Macromolecular carriers can also be included into this group of compounds. Due to this property, the carriers can be employed in the transport of drugs into the lysosomal compartment. Cells display their individuality on their surface by a number of specific receptors and cell antigens. Advantage could be taken of these identifying marks for selective concentration of drugs by means of appropriate carriers. The specificity of penetration of polymeric drugs into cells could be raised considerably by the presence of specific determinants which would interact with receptors on the surface of cellular membranes. Let us summarize facts allowing the research trends in this field to be determined. An extremely important finding consists in the discovery that several cell types have receptors able to recognize and lead to the pinocytosis of macromolecules bearing particular determinants. Of the recognition systems already described, these are the most important (27,28):

The galactose recognition system of hepatocytes. It has been demonstrated on a number of glycoproteins how small changes in structure may regulate the motion of these polymers in circulation and their interaction with receptors on the surface of membranes of liver cells. The glycosidic chains of many glycoproteins are terminated with sialic acid. Morell et al. (29) showed that if native glycoprotein, e.g., ceruloplasmin, is treated with neuraminidase, residues of sialic acid are split off. Chains of the asialoceruloplasmin thus formed are then terminated with galactose which acts as a specific determinant interacting with receptors of the liver cells localized on the plasma membrane (30). Morell et al. showed (29) that 15 min. after intravenous injection into rabbits only 10% of asialoceruloplasmin was detectable in plasma, whereas 90% of native ceruloplasmin was still present. The rapidly removed asialoceruloplasmin was found to be taken up almost exclusively by the parenchymal cells of the liver. A number of other asialoglycoproteins have been shown to be rapidly cleared from the circulation (31-33). Rogers and Kornfeld (34) using asialofetuin glycopeptide as carrier have been able to increase the rate of hepatic uptake of albumin or lysozyme 10-fold by chemically coupling these proteins to the vector peptide. An investigation of partly desialylated ceruloplasmins indicated that only two galactosyl residues need to be exposed for ceruloplasmin to be rapidly cleared from the circulation (35,36).

The N-acetylglucosamine/mannose recognition system of Kupffer cells and macrophages. Stockert et al. (37) demonstrated that agalacto-orosomucoid, in which the terminal sialic acid and the penultimate galactose are removed, was cleared by the liver by a route distinct from that described for galactose terminating glycoproteins. It was suggested (38) that N-acetylglucosamine may be important as the recognition marker. In the study of the clearance of four isoenzymes of bovine pancreatic ribonuclease (39), evidence was obtained for a recognition system for exposed mannose residues. According to Achord et al. (40,41) it seems probable that the Kupffer cell recognition system in unable to distinguish between N-acetylglucosamine and mannose. Stahl et al. (42) have shown that rat alveolar macrophages also have a surface receptor that binds glycoproteins with mannose, N-acetylglucosamine or glucose in the terminal position, whereas galactose-terminal glycoproteins are not bound.

The phosphohexose recognition system of fibroblasts. One of the methods allowing us to obtain information concerning the structure of the recognition marker is the competitive inhibition of enzyme pinocytosis. It was shown (43) that the high-uptake forms of several enzymes are all dependent on the presence of a phosphohexosyl recognition marker on the enzymes.

Antibodies as markers. The attractiveness of using antibodies as drug carriers lies in the prospect that the resulting drug-antibody complex will retain sufficient specificity to become localized on those cells or tissues which express the particular antigenic determinant (44). There are some problems in this approach. The extent of directly covalent binding of drug molecules onto immunoglobulin is limited by progressive loss of antibody activity and/or solubility (45,46). A possible method of overcoming is the use of an intermediate carrier (47,48) (e.g. polyglutamic acid or dextran). However, there are difficulties in the isolation of target specific antigens for the production of antibodies (49). Nevertheless, the area is worth development. The development of techniques which provide information on and make possible the isolation of specific antigens, thus allowing us to obtain specific antibodies, together with the development of the hybridoma techniques, will extend this approach also to the field of clinical medicine.

Influence of the size of macromolecules on their pinocytic uptake. Lloyd with coworkers (24,25,27,50) have shown that many different macromolecules are captured at rates that reflect their affinity for the plasma membrane. However, not only the structure, but also the size of the substrate is important. Studying the pinocytic uptake of polyvinylpyrrolidone (M_W = 50000; 84000; 700000; 7 000 000) by rat visceral yolk sacs and rat peritoneal macrophages, Duncan et al. (51) have demonstrated that selection according to size also operates in pinocytosis. Low molecular weight material is preferred by the yolk sac, macrophages captured the highest molecular weight preparation more rapidly than other preparations.

VARIABLE MODEL POLYMER SYSTEM FOR AN INVESTIGATION OF THE RELATIONSHIP BETWEEN STRUCTURE AND BIODEGRADABILITY

For a systematic study of the relationship between structure and degradability, a system has been sought which would make possible the preparation of novel types of polymer carriers with variable structure of both the side chain and the degradable parts of the main chain. At the same time, the system should allow the targetting moieties to be attached to the carrier polymer molecule by means of reactive groups.

Although some natural polymers (17,45,53,54) were used as drug carriers, synthetic polymers still possess some advantages compared with natural ones. In the first place, they can be tailored so as to have properties corresponding to a certain biological system; their structure can also be altered more easily.

Peptidic, glycosidic and phosphate bonds may be considered as degradable links joining synthetic chains. We started by choosing peptidic bonds. Poly[N-(2-hydroxypropyl)methacrylamide] was chosen as the basic polymer chain, i.e., a polymer whose chemical (55,56) and biological (57-60) properties have been thoroughly investigated, and which therefore is very well suited for model studies. One has to use a model system to be able to answer, step by step, questions which impeded clinical use of polymeric drugs.

p-Nitrophenyl ester group was used as the reactive group. The choice is based on the long-term application of active esters in the synthesis of peptides (61) and in synthetic copolymers (62-64).

All the polymeric substrates discussed below were synthesized using copolymers of N-(2-hydroxypropyl)methacrylamide (HPMA) with p-nitrophenyl esters of N-methacryloylated oligopeptides or amino acids as initial reactive copolymers (polymer precursors) (65-67). The content of the reactive groups was only several % mol. As an example of the preparation of polymer precursors, the copolymerization of HPMA with p-nitrophenyl ester of N-methacryloylglycylglycine is given below:

$$CH_3-\underset{\underset{CH_2}{|}}{\overset{||}{C}}-CO-NH-CH_2-\underset{\underset{OH}{|}}{CH}-CH_3 \; + \; CH_3-\underset{\underset{CH_2}{|}}{\overset{||}{C}}-CO-NH-CH_2-CO-NH-CH_2-CO-O-\langle\!\!\langle\bigcirc\!\!\rangle\!\!\rangle-NO_2$$

$$(HPMA)$$

$$\xrightarrow[\text{acetone, 50°C}]{\text{azoinitiator}}$$

$$\begin{array}{c} \\ CH_3-\underset{\underset{CH_2}{|}}{\overset{\xi}{C}}-CO-NH-CH_2-\underset{\overset{|}{OH}}{CH}-CH_3 \\ CH_3-\underset{\underset{CH_2}{|}}{C}-CO-NH-CH_2-CO-NH-CH_2-CO-O-\langle\!\!\langle\bigcirc\!\!\rangle\!\!\rangle-NO_2 \\ CH_3-\underset{\xi}{\overset{|}{C}}-CO-NH-CH_2-\underset{\overset{|}{OH}}{CH}-CH_3 \end{array}$$

Reaction of polymer precursors with amines

Polymer precursors containing the p-nitrophenoxy (ONp) group readily react with compounds which contain an aliphatic amino group in their molecule with formation of the amide bond:

$$\}-NH- \ldots -CO-O-\langle\!\!\langle\bigcirc\!\!\rangle\!\!\rangle-NO_2 + H_2N-R \longrightarrow \}-NH- \ldots -CO-NH-R + HO-\langle\!\!\langle\bigcirc\!\!\rangle\!\!\rangle-NO_2$$

A number of copolymers of HPMA with comonomers containing the reactive ONp group was prepared (66,67). The reactivity of these copolymers was characterized by a reaction with aliphatic amines in dimethylsulphoxide (DMSO) at 25°C. The reaction between amines and reactive copolymers of HPMA may be regarded as a model reaction of binding of biologically active compounds (BAC) containing a primary amino group. Results (66) of some aminolyses are given in Fig.1.

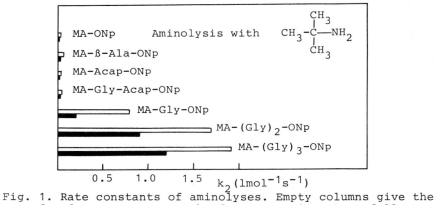

Fig. 1. Rate constants of aminolyses. Empty columns give the second-order rate constant k_2 for monomeric ester, full columns give k_2 for the copolymer with HPMA. (MA methacryloyl, Acap ε-aminocaproyl).

The results of the investigation of aminolyses (66) show that of the p-nitrophenyl esters studied, α-acylaminoesters are the most reactive ones, while in amines the highest reactivity is observed with unbranched primary aliphatic amines. The structural effect just mentioned predominates over the shielding effect of the polymer chain.

Hence, by reacting polymeric precursors with amines, it is possible to prepare polymers modelling a polymeric drug, bound with an enzymatically degradable bond. The necessary condition is that the degradable bond should originate in the amino acid specific for the given enzyme (AA amino acid).

$$\Big\} - \boxed{AA_1} - \boxed{AA_2} - \boxed{\begin{array}{c}\text{specific}\\ \text{AA}\end{array}} \downarrow \begin{array}{c}\text{enzymatically cleavable bond}\end{array} \boxed{DRUG}$$

Reaction of polymers with diamines

If a degradable carrier is to be prepared, the copolymers of HPMA with p-nitrophenyl esters of N-methacryloylated oligopeptides should be reacted with diamines (e.g., hexamethylenediamine). In such a case, three possibilities should be considered, namely: cyclization, formation of free NH_2 groups and crosslinking (below the gel point if a soluble carrier is to be prepared).

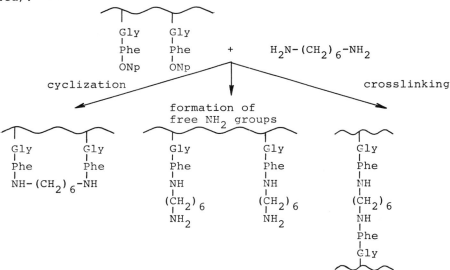

By changing the reaction conditions (ratio of components, solvent), one can influence the ratio of these reactions (67). Thus, e.g., at a high excess of diamine more than 95% of ONp groups can be transformed into free NH_2 groups.

Another possible route which may be employed for the crosslinking reaction is aminolysis of the polymeric precursor with polymeric amine. The limitation of intramolecular cyclization in this so called two-stage crosslinking (first stage - preparing a polymeric amine, second stage - reaction of the latter with the polymer containing ONp groups) leads to a greater increase in molecular weight compared to the one stage crosslinking (reacting the copolymer containing ONp groups with diamine). By additional crosslinking of polymeric precursors below the gel point, it is possible to join short synthetic chains with oligopeptidic sequences containing enzymatically cleavable bonds.

(a) One-stage procedure:
    one-stage crosslinking gives rise to a symmetrical crosslink containing two cleavable sites:

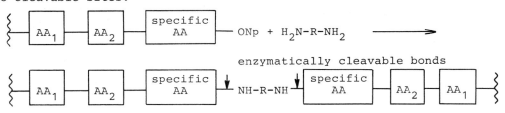

(b) Two-stage procedure:
    two-stage crosslinking gives rise to crosslinks containing only one cleavable bond:

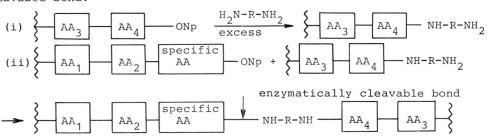

It should be stressed that the reaction of additional crosslinking proceeded
below the gel point, i.e. that the reaction product is a branched polymer of
a higher molecular weight, but remains soluble in an aqueous medium. The one-
stage procedure is more advantageously used in the investigation of the
relationship between structure and properties, because it makes possible an
easier (with respect to synthesis) modification of structure. Preparation by
employing the two-stage procedure is better suited for obtaining a concrete
drug, as no low-molecular weight compound is released in the cleavage.

DEGRADATION OF COPOLYMERS OF N-(2-HYDROXYPROPYL)METHACRYLAMIDE BY MODEL ENZYMES

Degradability of bonds in side chains of soluble synthetic polymers catalyzed by chymotrypsin

An investigation of bond-splitting in side-chains of synthetic polymers
provides us with the necessary knowledge for the attachment of drugs to
polymeric carriers via enzymatically cleavable bonds. The necessary condition
is that the degradable bond should originate in the amino acid specific for
the given enzyme. It is obvious that the polymer molecule can hinder the
formation of the enzyme-substrate complex; in such case the simplest approach
consists in binding the compound onto the polymer chain by means of a spacer.
Jatzkewitz (68,69) was the first to suggest this method of drug binding. He
bound the psychopharmaceutic drug mescaline on the copolymer of vinylpyr-
rolidone with acrylic acid using a dipeptidic link, glycylleucine. Mescaline
alone and glycylleucylmescaline are eliminated from the organism of the test
animal within c. 24 hours, but the polymeric drug provided gradual release of
mescaline over 17 days.

From model experiments in vitro using p-nitroanilides as drug models and
chymotrypsin as model enzyme (65,70,71) the conclusion could be drawn that
the rate of splitting of phenylalanine p-nitroanilide residues depends on
their spacing from the polymer chain backbone. The possibility of cleavage is
strongly dependent not only on steric factors, but also on interactions along
the chain which is to be split.

The relationship between structure and degradability is suitably discussed by
using the nomenclature introduced by Schechter and Berger (72-75). The active
site of an enzyme performs the two-fold function of binding a substrate and
catalyzing a reaction. The efficiency of these actions determines the overall
activity of the enzyme toward the particular substrate, i.e. the specificity
of the enzyme. Since the active site of proteolytic enzymes is rather large
(75-77) and capable of combining with a number of amino acid residues, it is
convenient to subdivide the binding site to subsites. A subsite is defined as
a region on the enzyme surface which interacts with one amino acid residue
(P_i) of the substrate. The substrates are lined up on the enzyme in such
a way that the CO-NH group being hydrolyzed always occupies the same place
(the catalytic site, C); the amino acid residues occupy adjacent subsites,
these towards the NH_2 end occupying subsites S_1, S_2 etc., those towards the
COOH end occupying subsites S_1', S_2' etc. The effects of structure on properties
may be conveniently described as S_1-P_1; S_2-P_2 etc. interactions.

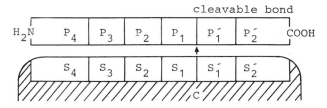

For a detailed study of relationships between the structure of polymers and
degradability of bonds at the end of side oligopeptidic chains with chymotrypsin, a number of polymeric substrates were prepared by a polymeranalogous
reaction of polymeric precursors with Phe-NAp and Tyr-NAp (p-nitroanilides of
specific amino acids for chymotrypsin) (78):

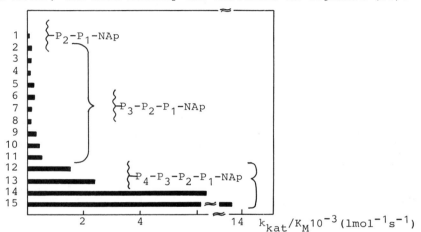

The enzymatically cleavable chromogenic group (Nap) models the biologically active compound. The degradability is characterized by k_{cat}/K_M;the higher this ratio, the more readily the substrate is degraded (79).

Fig. 2. Results of the enzymatic cleavage of bonds in side chains of polymeric substrates
1: P_2 = Gly; P_1 = Phe
2-9: P_3 = Gly; P_2 = Gly,Ala,β-Ala,Ile,Leu,Phe,D-Phe,Val; P_1 = Phe
10-11: P_3 = Ala,D-Ala; P_2 = Val; P_1 = Phe
12-13: P_4-P_3-P_2 = Gly-Gly-Phe; P_1 = Phe,Tyr
14: P_4-P_3-P_2-P_1 = Gly-Gly-Val-Phe
15: P_4-P_3-P_2-P_1 = Ala-Gly-Val-Phe

The following conclusions may be drawn from results of the investigation (Fig.2) of the kinetics of chymotrypsin-catalyzed hydrolysis of polymeric substrates containing a degradable bond at the end of side chains (78):
1. Degradability of the peptidic bond at the end of an oligopeptidic side chain of a synthetic polymer is affected both by steric and by structural factors.
2. With increasing number (2-4) of amino acid residues in the oligopeptidic sequence, the rate of chymotrypsin-catalyzed hydrolysis also increases.
3. By choosing a suitable structure (with respect to S-P interactions), it is possible, in substrates with the same number of amino acid residues in the side chain, to affect the enzymatic degradability.
4. The results may be interpreted using S-P interactions and the known active site of chymotrypsin.
5. Results of cleavage of the side chains of polymeric substrates may be correlated with the process of cleavage of low-molecular weight substrates.

Degradability of bonds in crosslinks connecting two poly[N-(2-hydroxypropyl)-methacrylamide] chains by chymotrypsin
As has been mentioned, one of the possible ways for introducing degradable bonds into the main chains consists in a reaction of diamines with copolymers of HPMA and p-nitrophenyl esters of N-methacryloylated oligopeptides. Already the first preliminary experiments have demonstrated (71) that steric

hindrance to the complex enzyme-substrate formation is much more pronounced in this case than in that of cleavage of the side chains. It was found that in the cleavage of crosslinks one must very carefully choose a structure which makes possible interactions with the highest possible number of subsites of the active site of the enzyme.

For a detailed study of the relationship between the structure of crosslinks and their degradability with chymotrypsin, more than 30 substrates were synthesized, and the structure of their oligopeptidic sequence was systematically altered so (80) as to assess the effect of S_1-$P_1 \doteq S_2'$-P_2' interactions on the degradability of the crosslinks. The amount of bonds degraded within a certain time interval was evaluated from the distribution curves of molecular weights (obtained by the GPC method) of the initial polymer, polymer after additional crosslinking and polymer after incubation with chymotrypsin. The following relations were used in the calculation (67,80,81):

Let $\bar{y}_w$, $\bar{y}_{n'}$ respectively be the weight average and number average degrees of polymerization of the primary molecules, $\bar{x}_w$, $\bar{x}_n$, the weight and number average degrees of polymerization after the polymeranalogous reaction, γ the crosslinking index, i.e. the number of crosslinked units per primary chain and ϱ the network density, i.e. the molar fraction of crosslinked units. These quantities are related as follows (81):

$$\bar{x}_n = \bar{y}_n / (1 - \gamma/2) \tag{1}$$

$$\bar{x}_w = \bar{y}_w / (1 - \gamma \bar{y}_w / \bar{y}_n) \tag{2}$$

$$\gamma = \bar{y}_n \varrho \tag{3}$$

Eq. (2) holds for high $\bar{y}_w$ and small ϱ.

Several examples are given below to illustrate the effect of steric and structural factors on the cleavage of crosslinks (80). Numbers in brackets give % of degraded bonds after incubation with chymotrypsin at 37°C; substrates a-d were incubated 48 h, substrates e, f were incubated 30 min.

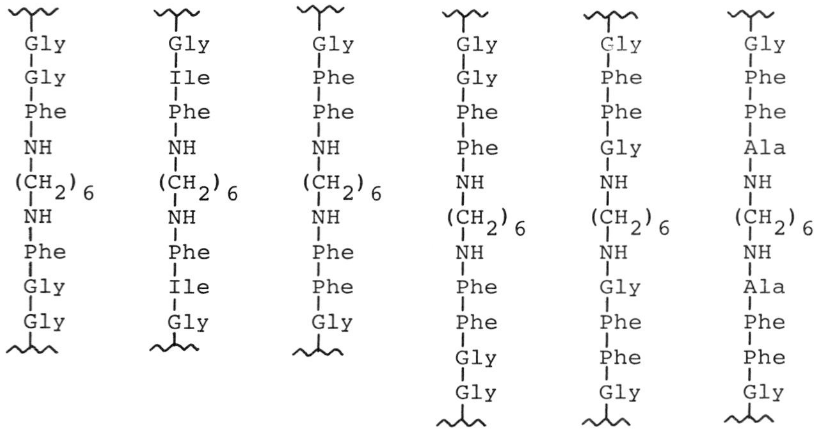

a	b	c	d	e	f
Gly	Gly	Gly	Gly	Gly	Gly
Gly	Ile	Phe	Gly	Phe	Phe
Phe	Phe	Phe	Phe	Phe	Phe
NH	NH	NH	Phe	Gly	Ala
$(CH_2)_6$	$(CH_2)_6$	$(CH_2)_6$	NH	NH	NH
NH	NH	NH	$(CH_2)_6$	$(CH_2)_6$	$(CH_2)_6$
Phe	Phe	Phe	NH	NH	NH
Gly	Ile	Phe	Phe	Gly	Ala
Gly	Gly	Gly	Phe	Phe	Phe
			Gly	Phe	Phe
			Gly	Gly	Gly
a	b	c	d	e	f
(22%)	(37%)	(67%)	(100%)	(18%)	(71%)

Differences in the degradability of polymers a-c can be assigned mainly to structural effects, i.e., to the effect on S_2'-P_2' interactions. In the position P_2', amino acids with a bulky side chain are more suitable (Ile or Phe are better than Gly), as they can form van der Waals contacts between their side chains and Ile-99 localized in the S_2' subsite of chymotrypsin (77). Differences between substrates c and d may be predominantly attributed to steric effects. Extension of the oligopeptidic sequence from 3 to 4 amino acids causes a pronounced decrease in the shielding effect of the polymer chain. Substrates e and f demonstrate the fact that also the structure on the leaving group side is important for the process of cleavage. Similarly to low-molecular weight substrates, Ala is better than Gly in the position P_1' (82).

The following conclusions may be drawn from the study of the relationship between the structure and degradability of oligopeptidic sequences joining

two polymer chains:
1. Degradability of oligopeptidic crosslinks joining two poly[N-(2-hydroxy-propyl)methacrylamide] chains depends both on steric and on structural factors.
2. With increasing distance of the bond undergoing degradation from the poly-mer chain the degradability of crosslinks also increases. Tetrapeptide (which makes possible S_1-P_1 to S_4-P_4 interactions) suffices to achieve complete cleavage of all the crosslinks under the experimental conditions used.
3. Assuming that steric hindrance by the polymer chain is taken into account, it may be inferred from the results that conclusions regarding the relation-ship between the structure and degradability of low-molecular weight sub-strates can be extended not only to the cleavage of side chains of synthetic polymers, but also to the cleavage of crosslinks between two synthetic polymer chains.

Polymers degradable by trypsin and papain

To answer the question if the procedure of preparation of enzymatically degradable polymers could be generally employed in the preparation of novel types of carriers of biologically active compounds, polymeric substrates con-taining bonds cleavable by trypsin (83) and papain (84) have been prepared.

To prepare copolymers degradable by trypsin, copolymers of HPMA with p-nitro-phenyl esters of N-methacryloylated oligopeptides were crosslinked below the gel point with N,N´-bis(N$^\varepsilon$-BOC-lysine)hexamethylenediamine (83). The cross-links connecting polyHPMA chains contained an oligopeptide sequence of 2-4 amino acids (P..polymer):

P-Gly-N$^\varepsilon$-BOC-Lys-NH-(CH$_2$)$_6$-NH-Lys-BOC-N$^\varepsilon$-Gly-P
P-Gly-X-N$^\varepsilon$-BOC-Lys-NH-(CH$_2$)$_6$-NH-Lys-BOC-N$^\varepsilon$-X-Gly-P
     where X is Gly,Val,Leu,Phe,D-Phe;
P-Ala-Gly-Val-N$^\varepsilon$-BOC-Lys-NH-(CH$_2$)$_6$-NH-Lys-BOC-N$^\varepsilon$-Val-Gly-Ala-P;
P-Gly-Gly-Y-N$^\varepsilon$-BOC-Lys-NH-(CH$_2$)$_6$-NH-Lys-BOC-N$^\varepsilon$-Y-Gly-Gly-P
     where Y is Phe,Val.

After removal of the BOC blockade of lysine copolymers degradable with trypsin were obtained. These polymers contained a peptide bond originating in lysine, degradable with trypsin (specific amino acids for trypsin are basic amino acids, such as lysine and arginine). Changes in the molecular weight distribution and viscosity of polymer solutions after incubation with trypsin have shown that a pronounced effect on the rate of degradation is exercised by the distance between the bond subjected to degradation and the polymer backbone. The rate of cleavage is also affected by the detailed structure of the oligopeptidic sequence. However, trypsin has a less pronounced secondary specificity than chymotrypsin for polymeric substrates (83).

A similar procedure was employed in the preparation of copolymers of HPMA in which synthetic polymer chains were joined by crosslinks containing oligo-peptidic sequences degradable with papain, with the general structure (84)

P-(Gly)$_n$-U-W-NH-(CH$_2$)$_6$-NH-W-U-(Gly)$_n$-P

where P is the polymer chain; n=1,2; U..Phe,Val,Gly; W..Lys,Gly,Tyr,Ala,Phe.

It was attempted to find the relationship between the structure of polymeric substrates and their degradability by papain. Papain has a large active site which extends over about 25 Å and can be divided into 7 subsites (72,75), each accommodating one amino acid residue of the substrate. Papain (85) has a very wide specificity. It hydrolyses amides of α-amino substituted arginine, lysine, glutamine, histidine, glycine and tyrosine. On prolonged hydrolysis of peptides by papain, bonds involving also a number of other residues are also split. It was found, however (73,75) that the subsite S_2 specifically interacts with a phenylalanine side chain. A new kind of specificity was defined, namely, -Phe-W-Z-, where the peptide bond between amino acids W and Z is split specifically, if W is preceded by phenylalanine. Valine residues behave similarly to phenylalanine residues and also show a strong affinity for subsite S_2. In other words, in the case of papain it is the penultimate amino acid that has a decisive effect on the process of cleavage of the peptidic bond, while with chymotrypsin and trypsin the decisive role is played by the amino acid in which the bond undergoing degradation originates.

The results obtained by the investigation of the dependence of viscosity of solutions of polymers of various structure incubated with papain at 37°C may be summarized as follows (84):
1. If there is a phenylalanine residue in position P_2 (= U in the above scheme) of the substrate, the effect of structure of the amino acid residue in position P_1 (=W) on the degradability of polymeric substrates may be

expressed through the series

       Lys > Gly > Tyr > Ala > Phe

The suitability of lysine in position P_1 may be attributed to the fact that
Asp-158 and perhaps Asp-64 increase the affinity of the active site of papain
for positively charged groups.
2. If a lysine residue is in position P_1 (=W) of the substrate, the effect of
structure of the amino acid residue in position P_2 (=U) on the degradability
of polymeric substrates decreases in the series

       Phe > Val >> Gly

The results confirm the preference of hydrophobic amino acid residues in
position P_2.

DEGRADATION OF STUDIED COPOLYMERS BY INTRACELLULAR ENZYMES

Degradability of bonds in side chains of soluble synthetic polymers by Tritosomes

The suitability of the chosen model system for controlled release of drugs
within the lysosomal compartment was verified at the University of Keele.
Copolymers of N-(2-hydroxypropyl)methacrylamide (HPMA) containing bound
p-nitroaniline at the ends of oligopeptidic chains were incubated (5 h at
37°C) with Tritosomes isolated from rat liver. Some of the substrates under
investigation were degraded. The results (86) obtained by Duncan and Lloyd
are summarized in Table 1.

TABLE 1. Degradation of bonds at the ends of side oligopeptidic
          chains by Tritosomes

Structure of substrate	Comonomer % mol	Mean rate of degradation (nmole/h)
P-Gly-Gly-Phe-Phe↓NAp	2.57	4.3
P-Gly-Gly-Phe-Tyr↓NAp	2.54	8.3
P-Acap-Phe↓NAp	2.70,2.01*	15.6
P-Gly-Leu-Phe↓NAp	3.26	7.8

*two preparations were used, P..polymer, NAp..p-nitroanilide,
Acap..ε-aminocaproic acid.

These results were checked and extended by including data obtained by incubation for 24 h with Tritosomes followed by GPC analysis of the degradation
products (87). A number of other substrates prepared originally as model substrates for chymotrypsin (78) were slowly degraded during long-term incubation
with Tritosomes. In most of them the bond was degraded between p-nitroaniline
and the amino acid situated at the furthest distance from the polymer chain. This
result is very encouraging from the point of view of development of new types
of polymeric drugs.

Tailor-made polymers for a particular intracellular enzyme

Of course it is not possible to assign the degradation of a particular side-
chain to any particular enzyme when using Tritosomes as a source of lysosomal
enzymes. This problem may be elucidated by employing two approaches:
(a) addition of specific inhibitors to the incubation mixture (e.g., pepstatin
(22) inhibits cathepsin D, leupeptin (22) inhibits cathepsins L,H,B)
(b) incubation of polymeric substrates with pure lysosomal enzymes.

The situation is complicated by the fact that by far not all lysosomal enzymes
have been isolated in a sufficient amount. Unlike chymotrypsin in lysosomal
enzymes, the detailed structure of the active site remains unknown. When suggesting the structure of tailor-made substrates, one must start from the known
specificity of the given enzyme with respect to low-molecular weight substrates.

An attempt was made to synthesize polymers (88) which would possess a degradable NAp group at the end of the side chain if subjected to the action of
lysosomal protease cathepsin L. This enzyme (22) prefers hydrophobic residues
in the P_2 and P_3 positions and is thiol dependent. By analogy with polymeric
substrates degradable with papain (75,84), phenylalanine or valine were used
in the position P_2. If these polymers were incubated with Tritosomes without
addition of compounds containing SH groups (reduced glutathione (GSH) and

EDTA), no cleavage took place. This may be regarded as evidence of the fact that cleavage occurs owing to the thiol dependent proteases. Some results of cleavage (87) are presented in Table 2.

TABLE 2. Cleavage of bonds at the ends of side oligopeptidic chains by Tritosomes with addition of compounds containing SH groups

| Structure of substrate | Comonomer % mol | Mean rate of degradation | | |
| | | 5.5 h | | 67 h |
		μmole/h	%	%
P-Gly-Phe-Phe-Ala↓NAp	1.83	0.024	22.3	34.9
P-Gly-Phe-Phe-Leu↓NAp	1.73	0.028	27.1	33.6
P-Ala-Val-Ala↓NAp	1.11	0.017	24.8	50.1

citrate/phosphate buffer, pH 5.5; GSH 5 mM; EDTA 2 mM.

In addition to cathepsin L, lysosomes contain also other thiol proteases. Although it may be assumed from specificity that cathepsin L is the enzyme responsible for cleavage, the hypothesis will be verified by an investigation of the degradation of polymeric substrates with pure cathepsin L.

Pinocytic uptake and intracellular degradation of HPMA copolymers by rat visceral yolk sacs cultured *in vitro* (89)

The rate of pinocytic uptake and intracellular fate of 4 copolymers of HPMA containing different oligopeptide side-chains (-Gly-B-Tyr-NAp; B..Ile,Gly, Phe,β-Ala) was measured. These substrates differed in amino acid residues B. These polymers were radio-iodinated and incubated with rat visceral yolk sacs (24,90) of 17.5 days gestational age. The yolk sac is a very suitable *in vitro* system for studying the fate of a potential polymeric carrier at the cellular level.

All four copolymers have very similar rates of pinocytic uptake. The effect of the structure of the amino acid residue B on the susceptibility to intracellular degradation decreases in the series (89)

Gly > Phe > Ile

if B residue was β-Ala, no [^{125}I] iodotyrosine could be detected in the culture medium. Dinitrophenol, a potent inhibitor of pinocytosis (90), abolished the accumulation of radioactivity by the tissue and in the case of the digestible polymers the appearance of [^{125}I] iodotyrosine in the medium (87,89). This is consistent with pinocytosis being an obligatory step in the tissue accumulation and degradation of these polymers. The proteinase inhibitor leupeptin did not affect the rate of pinocytosis, but caused different degrees of inhibition of hydrolysis depending on side-chain composition. It completely inhibited the digestion of polymer with B = Gly and partially inhibited the digestion of copolymers with B = Phe or Ile.

It is noteworthy that the above mentioned copolymers susceptible to digestion within lysosomes of the yolk sac were not degraded when incubated with rat liver Tritosomes.

Possibility of attachment of targetting moieties

It has been demonstrated (91) that several compounds containing the amino group can be gradually bound on the investigated polymer precursors (reactive copolymers of HPMA). This finding may be utilized in the gradual binding of the targetting moiety and drug. This procedure was verified in practice by binding several different aminosugars (88) on the polymer precursor. Biological experiments are in progress at the University of Keele (87) in order to find out whether polymers thus modified may interact with receptors of cell membranes mentioned in the introduction to this paper, and be thus specifically transported into certain types of cells.

Biodegradation *in vivo*

It has been verified in a previous paper (92) that oligopeptidic sequences both in the side chains and in the crosslinks are also degraded *in vivo*. Polymers with various structure (Table 3) were applied intravenously to rats (20 mg/100 g of body weight). The concentration of NAp in the urine of test animals was investigated in some chosen time intervals.

TABLE 3. Cleavage of studied polymers *in vivo*

Polymer	degraded NAp within 48 h in % of the initial quantity
P-Gly-Val-Phe↓NAp	16.3 ± 3.2
P-Gly-Val-Gly↓NAp	2.8 ± 1.9
P-Ala-Gly-Val-Phe↓NAp	23.2 ± 7.5

In the same paper it was qualitatively demonstrated that the oligopeptidic sequence -Ala-Gly-Val-Lys- is degraded *in vivo* also if it is part of the crosslink between two polymer chains.

DEGRADATION OF HYDROPHILIC GELS WITH CHYMOTRYPSIN

A schematic outline of the preparation of crosslinked polymers was presented in the preceding chapters of this paper. The conditions of crosslinking used in the preparation of water-soluble carriers were chosen so as to allow the reaction proceed to the gel point. It seemed logical to try to conduct the crosslinking reaction beyond the gel point and to prepare, by employing reaction of polymeric precursors with diamines, gels swollen in water which would contain in their crosslinks the same oligopeptidic sequences which have been studied for soluble polymers. These gels too are degraded *in vitro* when incubated with chymotrypsin (93). The rate of their biodegradation can be regulated similarly to soluble polymers. In addition to the change in structure of the oligopeptidic sequence, the rate of degradation of these gels is considerably affected also by their degree of swelling.

These results confirm the general validity of the conclusions regarding biodegradability. They can be used in the future not only in the preparation of polymeric drugs, but also in the preparation of polymeric implants with controllable rate of decomposition.

CONCLUSIONS

It has been shown that by combining an oligopeptidic sequence with chains of synthetic polymers, it is possible to prepare a number of polymeric substrates with an easily altered structure. Using this system, it is possible to examine the relationship between the structure of such polymeric substrates and their biodegradability, both with model enzymes and with intracellular enzymes. The biodegradability of polymers may be regulated by their structure within broad limits, and it may also be adapted to a certain type of enzyme.

The proposed conception indicates one of the routes which may be employed not only in the preparation of polymers with bound biologically active compounds, which will be degraded in a certain compartment of the living organism (e.g., in lysosomes), but also in targetting these polymers to certain tissues.

However, as has been rightly pointed out by C. de Duve (94), if this research is to be successful, drastic changes in research strategy are needed. Guided missiles are not likely to emerge from mass screening. They must be designed rationally on the basis of precise information. For instance, only the example of a different relationship between the structure and degradability of polymeric substrates with Tritosomes and yolk sacs given in this paper suggests the amount of work still ahead of us. Nevertheless, we believe that the aim to be achieved is worth trying it.

Acknowledgement - I am very much indebted to my coworkers for their assistance during the last years. Their contribution is reflected in their co-authorship of papers listed in the References. Special thanks are due to Dr R. Duncan and Professor J.B. Lloyd of the University of Keele. The biological part of this paper would have been impossible without them.

REFERENCES

1. J.E. Potts in Aspects of Degradation and Stabilization of Polymers (H.H.G. Jellinek Ed.) pp. 617-657, Elsevier, Amsterdam (1978).
2. J. Kopeček and L. Šprincl, Polymers in Medicine (Wroclaw) 4, 109-117 (1974).
3. D.F.Williams, Plastics and Rubber: Materials and Applications 179-182 (1980).
4. D.F. Williams, Fundamental Aspects of Biocompatibility, CRC Press, Florida, (1981).
5. W.J. Bailey, Y. Okamoto, W.C. Kuo and T. Narita, Proc. 3rd Intern. Biodegrad.Symp. 765-773, Applied Science Publishers, London (1976).
6. S. Kim, V.T. Stannett and R.D. Gilbert, J.Polym.Sci.,Polym.Letters Ed. 11, 731-735 (1973).
7. M.M. Lynn, V.T. Stannett and R.D. Gilbert, J.Polym.Sci.,Polym.Chem.Ed. 18, 1967-1977 (1980).
8. A. Douy and B. Gallot, Makromol.Chem. 178, 1595-1599 (1977).
9. G.P. Vlasov, G.D. Rudkovskaja, L.A. Ovsjanikova, I.A. Baranovskaja, T.A. Sokolova, N.N. Uljanova and N.V. Šestova, Vysokomol.Sojed. B22, 216-219 (1980).
10. S.J. Huang and C.A. Byrne, J.Appl.Polym.Sci. 25, 1951-1960 (1980).
11. S.J. Huang, D.A. Bansleben and J.R. Knox, J.Appl.Polym.Sci. 23, 429-437 (1979).
12. M. Vert, F.Chabot, J. Leray and P. Christel, 27th IUPAC Symposium on Macromolecules, Florence, September (1980).
13. H.G. Batz, Adv.Polym.Sci. 23, 25-53 (1977).
14. J. Kopeček in Systemic Aspects of Biocompatibility (D.F. Williams, Ed.), CRC Press, Florida (1981).
15. L.G. Donaruma, Progr.Polym.Sci. 4, 1-25 (1975).
16. J. Kopeček, Polymers in Medicine (Wroclaw) 7, 191-221 (1977).
17. C.de Duve, T.de Barsy, B. Poole, A. Trouet, P. Tulkens and F. van Hoof, Biochem.Pharmacol. 23, 2495-2531 (1974).
18. M.K. Pratten, R. Duncan and J.B. Lloyd in Coated Vesicles (C. Ockleford and A. Whyte, Eds.), pp. 179-218, Cambridge University Press (1980).
19. H. Ringsdorf, J.Polym.Sci.,Polym.Symp. 51, 135-153 (1975).
20. R.T. Dean, Lysosomes, Arnold Publishers, London (1977).
21. A.B. Novikoff and E. Holtzman, Cells and Organelles, Holt, Rinehart, Winston, New York (1970).
22. A.J. Barrett and M.F. Heath in Lysosomes a Laboratory Handbook (J.T. Dingle, Ed.), pp. 19-145, North Holland, Amsterdam (1977).
23. P.J. Jacques in Lysosomes in Biology and Pathology (J.T. Dingle and H.B. Fell, Eds.), pp. 395-420, North Holland, Amsterdam (1969).
24. K.E. Williams, E.M. Kidston, F. Beck and J.B. Lloyd, J.Cell Biol. 64, 113-134 (1975).
25. M.K. Pratten, K.E. Williams and J.B. Lloyd, Biochem.J. 168, 365-372 (1977).
26. A. Trouet, Arch.Int.Physiol.Biochim. 72, 698-700 (1964).
27. J.B. Lloyd and P.A. Griffiths in Lysosomes in Biology and Pathology (J.T. Dingle, P. Jacques and I.H. Shaw, Eds.) 6, 517-532 (1979).
28. R.T. Dean in Drug Carriers in Biology and Medicine (G. Gregoriadis, Ed.), pp. 71-86, Academic Press, London (1979).
29. A.G. Morell, R.A. Irvine, I. Sternlieb, I.H. Scheinberg and G. Ashwell, J.Biol.Chem. 243, 155-159 (1968).
30. W.E. Pricer Jr. and G. Ashwell, J.Biol.Chem. 246, 4825-4833 (1971).
31. A.G. Morell, G. Gregoriadis, I.H. Scheinberg, J. Hickman and G. Ashwell, J.Biol.Chem. 246, 1461-1467 (1971).
32. G.E. Siefring and F.J. Castellino, J.Biol.Chem. 249, 7742-7746 (1974).
33. S. Bose and J. Hickman, J.Biol.Chem. 252, 8336-8337 (1977).
34. J.C. Rogers and S. Kornfeld, Biochem.Biophys.Res.Commun. 45, 622-629 (1971).
35. J. Hickman, G. Ashwell, A.G. Morell, C.J.A. van den Hamer and I.M. Scheinberg, J.Biol.Chem. 245, 759-766 (1970).
36. C.J.A. van den Hamer, A.G. Morell, I.H. Scheinberg, J.Hickman and G. Ashwell, J.Biol.Chem. 245, 4397-4402 (1970).
37. R.J. Stockert, A.G. Morell and I.H. Scheinberg, Biochem.Biophys.Res. Commun. 68, 988-993 (1976).
38. P. Stahl, P.H. Schlesinger, J.S. Rodman and T. Doebber, Nature 264, 86-88 (1976).
39. J.W. Baynes and F. Wold, J.Biol.Chem. 251, 6016-6024 (1976).
40. D.T. Achord, F. Brot, A. Gonzales-Noriega, W.S. Sly and P. Stahl, Pediat.Res. 11, 816-822 (1977).
41. D.T. Achord, F.E. Brot, C.G. Bell and W.S. Sly, Fed.Proc. 36, 653 (1977).
42. P. Stahl, B.Mandell, J.S. Rodman, P. Schlesinger and S. Lang, Arch.Biochem.Biophys. 170, 536-546 (1975).

43. K. Ullrich, G. Mersmann, E. Weber and K. von Figura, Biochem.J. 170, 643-650 (1978).
44. G.J. O´Neill in Drug Carriers in Biology and Medicine (G. Gregoriadis, Ed.), pp. 23-41, Academic Press, London (1979).
45. E. Hurwitz, R. Levy, R. Maron, M. Wilchek, R. Arnon and M. Sela, Cancer Res. 35, 1175-1181 (1975).
46. D.A.L. Davies and G.J. O´Neill, Proc.XI.Int.Cancer Cong. 1, 218-221 (1974).
47. G.F. Rowland, Eur.J.Cancer 13, 593-596 (1977).
48. M.Wilchek, Makromol.Chem.Suppl. 2, 207-214 (1979).
49. G. Gregoriadis, Nature 265, 407-411 (1977).
50. A.V.S. Roberts, K.E. Williams and J.B. Lloyd, Biochem.J. 168, 239-244 (1977).
51. R. Duncan, M.K. Pratten, H.C. Cable, H. Ringsdorf and J.B. Lloyd, Biochem.J., to be published (1981).
52. A. Trouet, D. Deprez-De Campeneere and C.de Duve, Nature New Biol. 239, 110-112 (1972).
53. H.J.P. Ryser and W.C. Shen, Proc.Natl.Acad.Sci. USA 75, 3867-3870 (1978).
54. R. Levy, E. Hurwitz, R. Maron, R. Arnon and M. Sela, Cancer Res. 35, 1182-1186 (1975).
55. J. Kopeček and H. Bažilová, Eur.Polym.J. 9, 7-14 (1973).
56. M. Bohdanecký, H. Bažilová and J. Kopeček, Eur.Polym.J. 10, 405-410 (1974).
57. J. Kopeček, L. Šprincl and D. Lím, J.Biomed.Mater.Res. 7, 179-191 (1973).
58. L. Korčáková, E. Paluska, V. Hašková and J. Kopeček, Z.Immun.Forsch. 151, 219-223 (1976).
59. L. Šprincl, J. Exner, O. Štěrba and J. Kopeček, J.Biomed.Mater.Res. 10, 953-963 (1976).
60. E. Paluska, J. Činátl, L. Korčáková, O. Štěrba, J. Kopeček, A. Hrubá, J. Nezvalová and R. Staněk, Folia biol.(Prague) 26, 304-311 (1980).
61. H.G. Garg, J.Scient.Ind.Res. 29, 236-243 (1970).
62. D. Ferrutti, A. Betteli and A. Feré, Polymer 13, 462-464 (1972).
63. H.G. Batz, G. Franzmann and H. Ringsdorf, Makromol.Chem. 172, 27-47 (1973).
64. C.P. Su and H. Morawetz, J.Polym.Sci.,Polym.Chem.Ed. 15, 185-196 (1977).
65. J. Drobník, J. Kopeček, J. Labský, P. Rejmanová, J. Exner, V. Saudek and J. Kálal, Makromol.Chem. 177, 2833-2848 (1976).
66. P. Rejmanová, J. Labský and J. Kopeček, Makromol.Chem. 178, 2159-2168 (1977).
67. J. Kopeček, Makromol.Chem. 178, 2169-2183 (1977).
68. H. Jatzkewitz, Hoppe-Seyler´s Z.Physiolog.Chem. 297, 149-156 (1954).
69. H. Jatzkewitz, Z.Naturforsch. 10b, 27-31 (1955).
70. T.Y. Fu and H. Morawetz, J.Biol.Chem. 251, 2083-2086 (1976).
71. J. Kopeček and P. Rejmanová, J.Polym.Sci.,Polym.Symp. 66, 15-32 (1979).
72. I. Schechter and A. Berger, Biochem.Biophys.Res.Commun. 27, 157-162 (1967).
73. I. Schechter and A. Berger, Biochem.Biophys.Res.Commun. 32, 898-902 (1968).
74. N. Abramowitz, I. Schechter and A. Berger, Biochem.Biophys.Res.Commun. 29, 862-867 (1967).
75. A. Berger and I. Schechter, Phil.Trans.Roy.Soc.London B 257, 249-264 (1970).
76. K. Kurachi, J.C. Powers and P.E. Wilcox, Biochemistry 12, 771-777 (1973).
77. D.M. Blow, Isr.J.Chem. 12, 483-494 (1974).
78. J. Kopeček, P. Rejmanová and V. Chytrý, Makromol.Chem. 182, 799-809 (1981).
79. A. Fersht, Enzyme Structure and Mechanism, Freeman, San Francisco (1977).
80. P. Rejmanová, B. Obereigner and J. Kopeček, Makromol.Chem. 182, 000 (1981).
81. W.H. Stockmayer, J.Chem.Phys. 12, 125-131 (1944).
82. K. Morihara and T. Oka, Arch.Biochem.Biophys. 178, 188-194 (1977).
83. K. Ulbrich, J. Strohalm and J. Kopeček, Makromol.Chem. 182, 000 (1981).
84. K. Ulbrich, E.I. Zacharieva, B. Obereigner and J. Kopeček, Biomaterials 1, 199-204 (1980).
85. R.L. Hill, Adv.Protein Chem. 20, 37-107 (1965).
86. R. Duncan, J.B. Lloyd and J. Kopeček, Biochem.Biophys.Res.Commun. 94, 284-290 (1980).
87. R. Duncan and J.B. Lloyd, to be published.
88. J. Kopeček and P. Rejmanová, to be published.
89. R. Duncan, P. Rejmanová, J. Kopeček and J.B. Lloyd, Biochim.Biophys.Acta, to be published.
90. R. Duncan and J.B. Lloyd, Biochim.Biophys.Acta 544, 647-655 (1978).
91. P. Rejmanová. Ph.D.Theses, Institute of Macromolecular Chemistry, Prague (1981).
92. J. Kopeček, I. Cífková, P. Rejmanová, J. Strohalm, B. Obereigner and K. Ulbrich, Makromol.Chem. 182, 000 (1981).
93. K. Ulbrich, J. Strohalm and J. Kopeček, this Symposium.
94. C.de Duve, Foreword in Drug Carriers in Biology and Medicine (G. Gregoriadis, Ed.), Academic Press, London (1979).

SYNTHETIC POLYMER BIOMATERIALS IN MEDICINE -- A REVIEW

Allan S. Hoffman

Department of Chemical Engineering and Center for Bioengineering
University of Washington, Seattle, Washington 98195 U.S.A.

Abstract - A review is presented of the applications of synthetic polymers in
medicine. The major uses of these biomaterials are in devices and implants for
diagnosis or therapy. The composition and properties, characterization, and
biologic interactions of a wide variety of synthetic polymers are reviewed.
Biologic testing and clearance of biomaterials for clinical use are also covered.

INTRODUCTION

There is a wide variety of materials which are foreign to the body and which are used in
contact with body fluids. These include totally synthetic materials as well as reconstitu-
ted or specially treated human or animal tissues. Some are needed only for short term
applications while others are, hopefully, useful for the lifetime of the individual. Table
1 lists the various uses of such foreign materials, otherwise known as "biomaterials." It
can be seen in Table 1 that the various applications listed may be generally categorized as
devices or implants, for diagnosis or therapy.

TABLE 1: USES OF BIOMATERIALS

1. INVASIVE INSTRUMENTATION
   (E.G., CATHETERS)

2. IMPLANTED DEVICES, INSTRUMENTS
   (E.G., PACEMAKERS, HYDRO-CEPHALUS TUBES)

3. EXTRA CORPOREAL DEVICES IN SERIES WITH BLOOD FLOW
   (E.G., ARTIFICIAL KIDNEY, HEART-LUNG BLOOD OXYGENATORS)

4. IMPLANTED PARTS (OR WHOLE) OF HARD STRUCTURAL ELEMENTS
   (E.G., HIP JOINTS, TEETH)

5. IMPLANTED PARTS (OR WHOLE) OF ORGANS
   (E.G., HEART VALVES, HEART ASSIST DEVICES, SKIN)

6. IMPLANTED SOFT TISSUE SUBSTITUTES
   (E.G., BLOOD VESSELS, TENDON, URETER)

One may also list the "ideal" requirements for selecting a particular biomaterial for a
particular end-use. Table 2 shows these criteria. It should be noted that very few (if any)
biomaterials in fact conform to all these criteria. Nevertheless, a wide variety of biomaterials have emerged and are in daily use in the clinic. Table 3 identifies six classes
of biomaterials, and the many different forms in which they are found in devices and implants. The first five classes are clearly separate types of materials, while the sixth
class, "Composites," includes systems which combine different forms of materials within any
one class (as a rubber diaphragm reinforced with a fabric) or different classes of
materials (as a heart valve made of a metal and different synthetic polymers or of natural
animal tissues and different synthetic polymers). This paper is a review of the field of
synthetic polymeric biomaterials, and as such will not attempt to cover natural tissue biomaterials, carbons, metals, or ceramics. A general reference list is provided which does
cover all of these materials and their applications in medicine.

TABLE 2: CRITERIA FOR SELECTION OF BIOMATERIALS

1. HAS REQUIRED PHYSICAL PROPERTIES
     (STRENGTH, ELASTICITY, PERMEABILITY . . .)

2. CAN BE PURIFIED, FABRICATED AND STERILIZED EASILY

3. MAINTAINS PHYSICAL PROPERTIES AND FUNCTION IN VIVO OVER
     DESIRED TIME PERIOD
     (1 HR., 1 DAY, 1 YR., 10 YRS. . .)

4. DOES NOT INDUCE UNDESIRABLE HOST REACTIONS
     (BLOOD CLOTTING, TISSUE NECROSIS, CARCINOGENESIS,
     ALLERGENIC RESPONSES, ETC.)

TABLE 3: CLASSES AND FORMS OF
BIOMATERIALS

	CLASSES	FORMS
I)	POLYMERS	
	A) FIBERS	FILMS OR MEMBRANES
	B) RUBBERS	FIBERS OR FABRICS
	C) PLASTICS	TUBES
		POWDERS OR PARTICLES
		MOLDED SHAPES
		BAGS OR CONTAINERS, ETC.
		LIQUIDS SOLIDS (ADHESIVES)
II)	METALS	CAST OR MOLDED SHAPES
		POWDERS OR PARTICLES
		FIBERS
III)	CERAMICS	MOLDED SHAPES
		POWDERS OR PARTICLES
		LIQUIDS, SOLIDS (CEMENTS)
IV)	CARBONS	MACHINED SHAPES
		COATINGS
		FIBERS
V)	NATURAL TISSUES	FIBERS
		NATURAL FORMS
		ALSO, RECONSTITUTED AS FILMS, TUBES, FIBERS, ETC.
VI)	COMPOSITES	COATINGS
		FIBROUS FELTS OR SHEETS
		FIBER OR FABRIC-REINFORCED SHAPES, TEC.

SYNTHETIC POLYMERIC BIOMATERIALS

Synthetic polymers make up by far the broadest and most diverse class of biomaterials. This is mainly because synthetic polymers are available with such a wide variety of compositions and properties and also because they may be fabricated readily into complex shapes and structures. In addition, their surfaces may be readily modified physically, chemically, or biochemically.

This wide variety of synthetic polymeric biomaterials can be seen in Figures 1-3, which are separated into categories of soft or rubbery polymers, semi-crystalline polymers, or acrylics and other related polymers.

The two major biomaterials in clinical use today are the elastomeric silicone resins (especially Medical Grade Silastic[R] products of Dow-Corning) and polyethylene terephthalate (P.E.T.) fibers (especially Dacron[R] of du Pont). Medical Grade Silastic[R] was one of the earliest and has been one of the most successful of the commerical polymers fabricated as a specialty line for medical applications. (It has not often been economically attractive for commercial polymer manufacturers to prepare the relative small quantities of high

MAJOR SOFT OR RUBBERY POLYMERIC BIOMATERIALS

GENERIC POLYMER	FORMULA	SOME MEDICAL USES
SILICONES	$\begin{array}{c} CH_3 \\ \mid \\ +Si-O+ \\ \mid \\ CH_3 \end{array}$ Polydimethyl siloxane	TISSUE SUBSTITUTES; HEART ASSIST DEVICES; DRUG DELIVERY DEVICES; IMPLANT OR INSTRUMENT COATINGS; ADHESIVE; TUBING; CONTACT LENS.
POLYURETHANES	$+O-R-O+\overset{O}{\overset{\|}{C}}NH-R'-NH-\overset{O}{\overset{\|}{C}}+NH-R''-NH+$	BLOOD PUMPS; TUBING; BLOOD STORAGE BAGS; BALLOON ASSIST DEVICE; PACEMAKER
POLY VINYL CHLORIDE	$\begin{array}{c} +CH_2-CH+ \\ \mid \\ C\ell \end{array}$ P. V. C.	TUBING, BLOOD STORAGE BAGS.
RUBBER	$\begin{array}{c} CH_3 \\ \mid \\ +CH_2-C=C-CH_2+ \end{array}$ POLY (CIS-ISOPRENE)	TUBING

Figure 1. Major soft or rubbery polymeric biomaterials

MAJOR SEMI-CRYSTALLINE POLYMERIC BIOMATERIALS

GENERIC POLYMERS	FORMULA	SOME MEDICAL USES
POLYESTERS	$+O-CH_2-CH_2-O-\overset{O}{\overset{\|}{C}}-\langle\bigcirc\rangle-\overset{O}{\overset{\|}{C}}+$ POLY (ETHYLENE TEREPHTHALATE)	VASCULAR GRAFTS; HEART VALVE SEWING RINGS; IMPLANT FIXATION AND COVERING FABRIC; SUTURE
	$+OCH_2-\overset{O}{\overset{\|}{C}}+$ POLYGLYCOLIC ACID	DEGRADABLE SUTURES
FLUOROPOLYMERS	$+CF_2-CF_2+$ POLY (TETRAFLUORO ETHYLENE)	VASCULAR GRAFTS; BLOOD OXYGENATOR MEMBRANE
NYLONS	$+HN-(CH_2)_6-NH-\overset{O}{\overset{\|}{C}}(CH_2)_4\overset{O}{\overset{\|}{C}}+$ NYLON 6-6	SUTURES, DRESSINGS
POLYETHYLENE	$+CH_2-CH_2+$ P.E.	ARTIFICIAL JOINTS; I.U.D. DRUG DELIVERY DEVICE.
CELLULOSICS	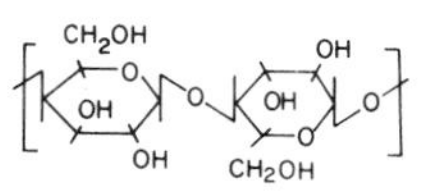 CELLULOSE (OR PARTIALLY HYDROLYSED CELLULOSE ACETATE)	DIALYSIS MEMBRANE; DRUG DELIVERY DEVICES; CONTACT LENS.

Figure 2. Major semi-crystalline polymeric biomaterials

purity polymers needed for medical uses, especially when the additional question of liability is introduced.) The medical uses listed for these two polymers in Figures 1 and 2 are quite general, and a complete list would be very long. Table 4 summarizes the many general biomedical uses of synthetic polymeric fibers.

The major disadvantages of synthetic polymeric biomaterials over other classes of biomaterials is that they may contain leachable toxic substances or may form them in contact with biologic environment (due to biodegradation reactions). Table 5 lists some of the additives found in, or extractables formed in commercial polymers which could be extracted into the

ACRYLICS AND OTHER RELATED POLYMERIC BIOMATERIALS

POLYMER FORMULA	SOME MEDICAL USES
CH_3 $+CH_2-C+$ CO_2CH_3 POLY(METHYL METHACRYLATE)	HARD AND SOFT CONTACT LENS; DENTAL FILLINGS; BONE CEMENT, INTRA-OCULAR LENSES.
CH_3 $+CH_2-C+$ $CO_2C_2H_4OH$ POLY(HYDROXY ETHYL METHACRYLATE) (CROSSLINKED)	SOFT CONTACT LENS; BURN DRESSINGS; DRUG DELIVERY MATRICES; COATINGS.
CH_3 $+CH_2-C+$ $COOH$ METHACRYLIC ACID	INGREDIENT IN SOFT CONTACT LENS, BIOFUNCTIONAL MICROBEADS.
$+CH_2-CH+$ NH_2 ACRYLAMIDE	INGREDIENT IN SOFT CONTACT LENS; BIOELECTRODES.
$+CH_2-CH+$ N-VINYL PYRROLIDONE	INGREDIENT IN SOFT CONTACT LENS; FORMER BLOOD PLASMA EXTENDER.
CN $+CH_2-C+$ $CO_2(CH_2)_n-CH_3$ POLY(CYANOALKYL ACRYLATES)	TISSUE ADHESIVE; VESSEL OCCLUDER.
$+CH_2-CH+$ $CO_2^\ominus$ $Zn^{\oplus\oplus}$ $CO_2^\ominus$ $+CH_2-CH+$ POLY(ZINC ACRYLATE)	DENTAL CEMENT
$[CH_2-CH]_x[CH_2-C]_y$ ... $SO_3^\ominus Na^\oplus$	DIALYSIS MEMBRANE

Figure 3.   Acrylics and other related polymeric biomaterials

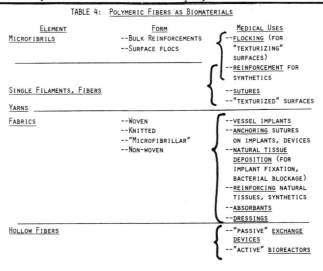

TABLE 4: POLYMERIC FIBERS AS BIOMATERIALS

ELEMENT	FORM	MEDICAL USES
MICROFIBRILS	--BULK REINFORCEMENTS --SURFACE FLOCS	--FLOCKING (FOR "TEXTURIZING" SURFACES) --REINFORCEMENT FOR SYNTHETICS
SINGLE FILAMENTS, FIBERS YARNS		--SUTURES --"TEXTURIZED" SURFACES
FABRICS	--WOVEN --KNITTED --"MICROFIBRILLAR" --NON-WOVEN	--VESSEL IMPLANTS --ANCHORING SUTURES ON IMPLANTS, DEVICES --NATURAL TISSUE DEPOSITION (FOR IMPLANT FIXATION, BACTERIAL BLOCKAGE) --REINFORCING NATURAL TISSUES, SYNTHETICS --ABSORBANTS --DRESSINGS
HOLLOW FIBERS		--"PASSIVE" EXCHANGE DEVICES --"ACTIVE" BIOREACTORS

aqueous biologic medium.

Three examples of important biomedical polymer systems are shown in Figures 4-6. The extraction or even mere surface exposure of inhibitory catalysts, catalyst fragments, fillers, or low molecular weight oligomers in such systems could lead to undesirable biologic responses. Further, one does not need to elaborate on the current controversy over the extraction of 2-diethyl hexylphthalate into blood from plasticized PVC blood storage systems. Nevertheless, each of these biomaterial systems is being used regularly in the clinic. The questions of toxicity testing and FDA clearance will be discussed below.

Biologically Functional Polymeric Biomaterials and Systems

Many new and exciting applications for polymeric biomaterials have been developed in recent years based on their combination with biologically functional molecules (Table 6). One important and early example of this "hybrid" biomaterial was the incorporation of heparin,

TABLE 5: SOME POTENTIAL EXTRACTABLES
IN COMMERCIAL POLYMERS

—CATALYST FRAGMENTS

—ANTI-OXIDANTS

—U.V. STABILIZERS

—PLASTICIZERS

—LOW MOLECULAR WEIGHT POLYMER MOLECULES

—SURFACE ACTIVE AGENTS
(LUBRICANTS, WETTING AGENTS, ANTI-STATIC AGENTS)

—DYES

—FLAME RETARDANTS

—FRAGMENTS OF FILLERS, REINFORCING AGENTS
—POLYMER DEGRADATION BYPRODUCTS

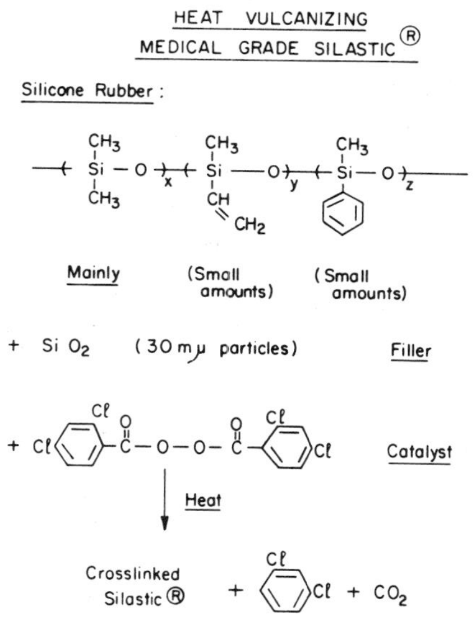

Figure 4. Chemistry of heat-vulcanized Medical Grade Silastic[R]

an anticoagulant, by electrostatic attraction to a cationic soap anchored in the biomaterial surface.

A wide variety of drug delivery systems has been developed for achieving a regulated or controlled release of therapeutic agents over a sustained and pre-determined period of time. Table 7 lists the different types of such systems. Polymers may be utilized as diffusion-controlling barrier membranes (in "reservoir" devices), matrices for containment and release of active agent (in "monolithic" devices), or more simply as containers, conduits, or other components of the device. The polymers may be designed to resist attack or to erode or degrade. In particular, a number of biodegradable polymers have been specially synthesized for release of active agents inside the body, during or after which the polymer disappears as it erodes or degrades and is metabolized. Figure 7 lists the types of polymer groups which have been employed due to their susceptibility to biodegradation. Figures 8-10 show schematically how different polymeric systems may be utilized in controlled delivery systems.

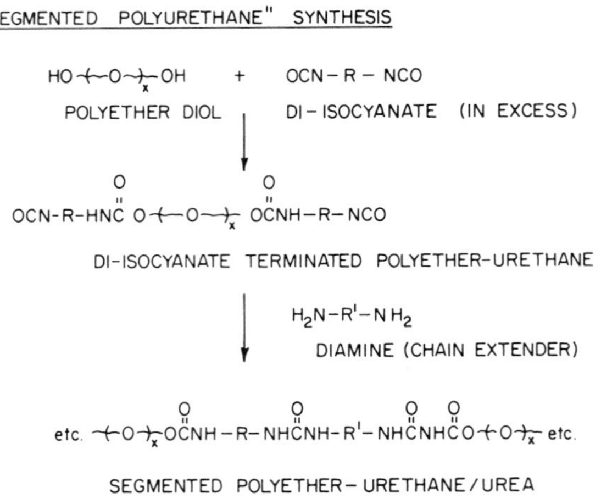

"SEGMENTED POLYURETHANE" SYNTHESIS

$$HO + O \rightarrow_x OH \quad + \quad OCN - R - NCO$$

POLYETHER DIOL   DI-ISOCYANATE (IN EXCESS)

$$OCN-R-HN\overset{O}{\overset{\|}{C}} O + O \rightarrow_x \overset{O}{\overset{\|}{OC}}NH-R-NCO$$

DI-ISOCYANATE TERMINATED POLYETHER-URETHANE

$$H_2N-R'-NH_2$$

DIAMINE (CHAIN EXTENDER)

$$\text{etc.} + O \rightarrow_x O\overset{O}{\overset{\|}{C}}NH-R-NH\overset{O}{\overset{\|}{C}}NH-R'-NH\overset{O}{\overset{\|}{C}}NH\overset{O}{\overset{\|}{C}}O + O \rightarrow_x \text{etc.}$$

SEGMENTED POLYETHER-URETHANE/UREA

Figure 5. Synthesis of segmented polyurethanes

ACRYLIC BONE CEMENT

LIQUID (20ml.)

Methyl Methacrylate (MMA) $CH_2 = C \big\langle \substack{CH_3 \\ CO_2CH_3}$ 97.4% V/V

Hydroquinone $HO-\langle \rangle-OH$ 2.6% V/V

N-N dimethyl p-toluidine $CH_3-\langle \rangle-N\big\langle \substack{CH_3 \\ CH_3}$ ~75 ppm

SOLID (40g)

PMMA $+CH_2-\overset{CH_3}{\underset{CO_2CH_3}{C}}+$ 15% W/W

Copoly (MMA/sty) $+CH_2-\overset{CH_3}{\underset{CO_2CH_3}{C}}+ +CH_2-CH+$ 75% W/W

Ba SO₄   10% W/W

Benzoyl Peroxide $\langle \rangle-\overset{O}{\overset{\|}{C}}-O-O-\overset{O}{\overset{\|}{C}}-\langle \rangle$

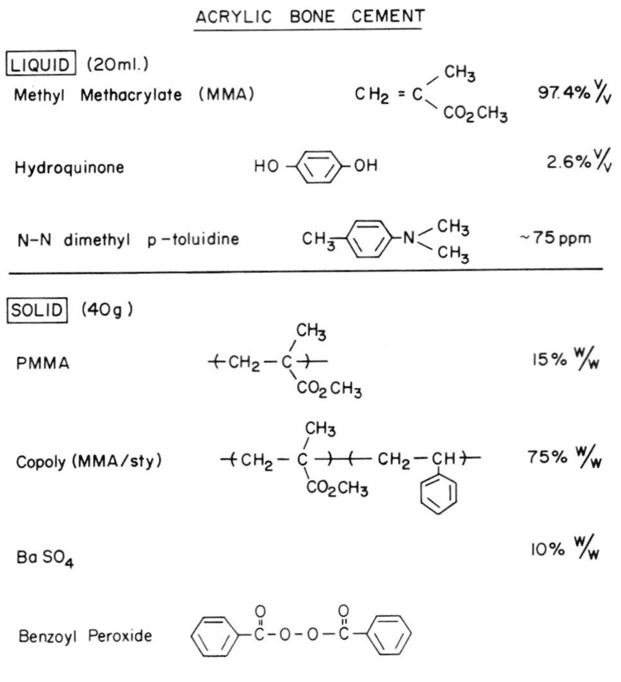

Figure 6. Composition of liquid and solid components of Surgical Simplex[R] bone cement
(Howmedica). The two components are mixed to a paste, which cures to a hard
solid in 5-10 minutes.

Another interesting new combination of polymers and biological species may be synthesized by
covalently binding biologically active molecules to the surface of polymeric particles, such
as those prepared in microemulsion polymerizations. Thus, if a particular antibody is
attached, the microparticles will be attracted to specific antigenic sites in the body. If
these sites are on specific cancer cells, and if an anti-cancer drug is incorporated into or
onto the microparticle, with the possibility of subsequent release from the particle, then

TABLE 6: BIOLOGICALLY ACTIVE MOLECULES WHICH MAY BE
INCORPORATED INTO POLYMERIC BIOMATERIAL SYSTEMS

ANTIBIOTICS
ANTICOAGULANTS
ANTI-CANCER DRUGS
ANTIBODIES
DRUG ANTAGONISTS
ENZYMES
CONTRACEPTIVES
ESTROUS-INDUCER
ANTI-BACTERIA AGENTS

TABLE 7: DRUG DELIVERY SYSTEMS

DIFFUSION DEVICES

RESERVOIR
MONOLITHIC   ERODING AND/OR NON-ERODING

DIFFUSION + PUMPING DEVICES

OSMOTIC
-- SINGLE CHAMBER   NON-ERODING (USUALLY)
-- DUAL CHAMBER

PUMPING DEVICES

ELASTIC BALLOON
BELLOWS (VAPOR PRESSURE)

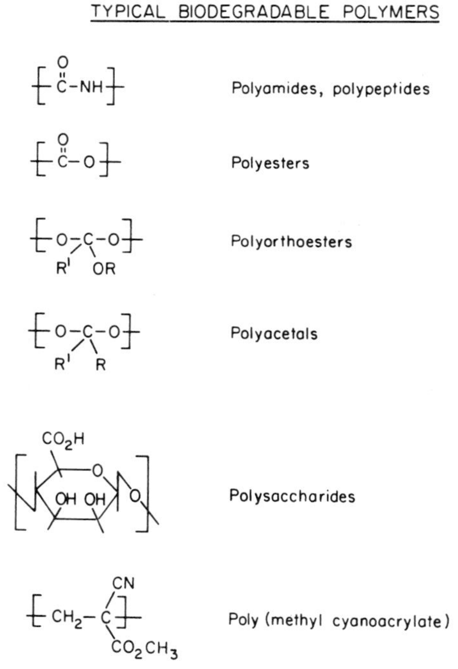

TYPICAL BIODEGRADABLE POLYMERS

Figure 7. Repeat units in biodegradable polymer backbones

specific drugs may be delivered to specific sites in the body, using the body's own circulatory system to transport the particles.

Cells may also be cultured within or on the outside of hollow polymeric fibers in hollow fiber exchange devices (Table 4) and a patient's blood may be circulated through the device for treatment of various diseases (e.g., using pancreatic beta cells for diabetes patients

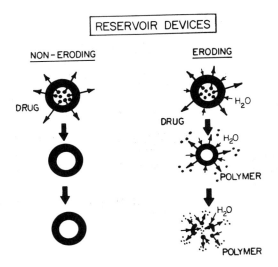

Figure 8.   Delivery of drug (or any other biological agent) from non-eroding or eroding
            reservoir devices. The rate of delivery is controlled by the polymeric membrane.

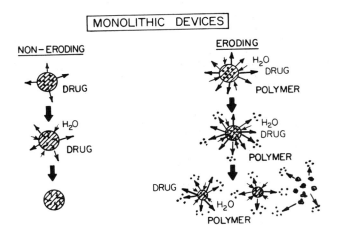

Figure 9.   Delivery of drug (or any other biological agent) from non-eroding or eroding
            polymer matrices.

in an "artificial pancreas," or using liver cells in an "artificial liver" during hepatic
failure.)

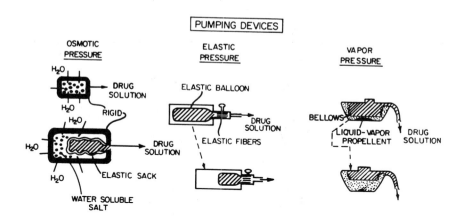

Figure 10.   Delivery of drug (or any other biological agent) from osmotic or hydrostatic
pumping devices. The hydrostatic pressure may be generated by a crosslinked
rubber balloon or by a vapor expanding against a metal bellows.

THE POLYMER-BIOLOGIC INTERFACE

When a foreign surface is exposed to a biological environment, there is a natural tendency
to destroy (digest) the foreign object or, failing that, to "wall it off" and cover (encap-
sulate) the object. The biologic species which are involved in this process are proteins
and cells (Figure 11). The first event is generally to coat the polymer surface with a
layer of proteins; the composition and organization of this layer will influence the subsequent cellular events (see below). Thus, it is essential that one characterize and repro-
duce the surface of the biomaterial to be used in any implant or device.

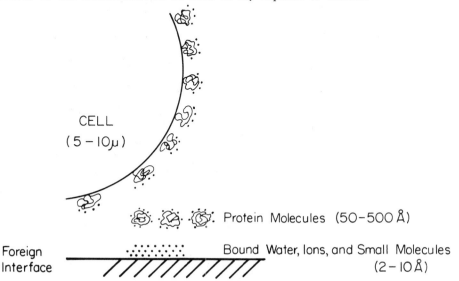

Figure 11.   The primary interactions at a foreign biomaterial interface in the body are
first with proteins and then with living cells. The drawing is schematic, and
not to scale.

TABLE 8: IMPORTANT FACTORS IN
BIOMATERIAL-BIOLOGIC INTERACTIONS

I.   BIOMATERIAL
    A. BULK PROPERTIES
    B. SURFACE PROPERTIES
    C. HANDLING, PACKAGING

II.   BIOLOGIC ENVIRONMENT
    A. IN VITRO VS. IN VIVO
    B. SPECIES

III. PHYSICAL FACTORS
    A. SYSTEM DESIGN; FLOW CHARACTERISTICS
    B. TIME, TEMPERATURE
    C. AIR INTERFACE

TABLE 9: DESCRIBING THE
BIOMATERIAL INTERFACE:

I.   COMPOSITION
    -- HYDROPHILIC/HYDROPHOBIC
    -- POLAR/APOLAR
    -- HIGH ENERGY/LOW ENERGY
    -- WETTABLE/NON-WETTABLE
    -- ACID/BASE
    -- ANIONIC/CATIONIC
    -- UNIFORM/DOMAIN STRUCTURE

II.   SORBED WATER
    -- ORIENTED
    -- STRUCTURED
    -- "FREE"

III. "COMPLIANCE"
    -- FLEXIBILITY OF CHAIN ENDS, LOOPS
    -- GLASS TRANSITION

IV.   ROUGHNESS
    -- SCALE AND INTENSITY
    -- POROSITY
    -- LOCAL IMPERFECTIONS
    -- GAS NUCLEI

There are a number of other important factors which influence the biological interaction and ultimate fate of a biomaterial in the body. Biomaterial properties, such as purity, tendency to absorb water and degrade are clearly important. Also, the design of the device or implant, the flow of biological fluids by the foreign surfaces or movement of the implant within a tissue space, the test techniques selected to assay biomaterial responses in vitro or in vivo (in different animal species), and the implantation itself can all contribute to the ultimate fate of the implant device. Table 8 lists these factors and Table 9 details important biomaterial surface properties.

Over the past several years, a great deal of effort has gone into characterizing the biomaterial surface composition. Contact angle measurements, infra-red reflectance spectroscopy (MIRS, FTIR), and electron spectroscopy (XPS or ESCA) have been the most popular techniques utilized. ESCA has become a very useful tool, since it yields an average composition of the top 10-100 Å of the biomaterial surface (although the measurement is made at low temperature and under high vacuum). The surface topography is also very important, particularly when compared to the scale of proteins and cells (Figures 12, 13). Surface roughness has been visualized using optical and (especially) scanning electron microscopy. Profilometry has occasionally been used. Table 10 lists various common biomaterials in approximate categories of increasing roughness.

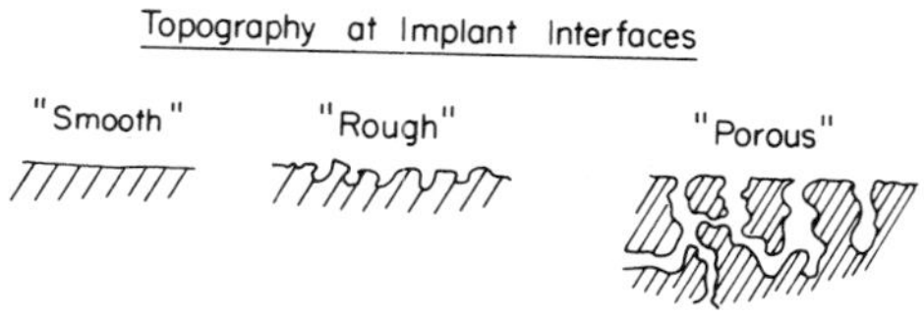

Topography at Implant Interfaces

"Smooth"   "Rough"   "Porous"

Figure 12.   Biomaterial surfaces may be "smooth," "rough," or "porous" (schematic).

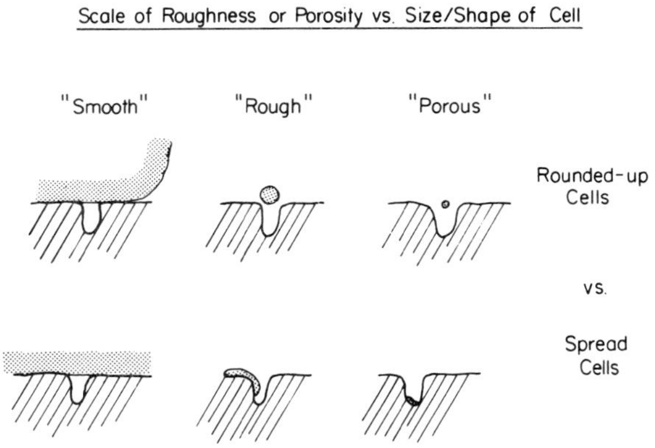

Scale of Roughness or Porosity vs. Size/Shape of Cell

"Smooth"   "Rough"   "Porous"

Rounded-up
Cells

vs.

Spread
Cells

Figure 13.   The importance of the scale and intensity of biomaterial surface "roughness"
will depend upon the relative size and interaction of cells on that surface
(schematic).

TABLE 10:

RELATIVE ROUGHNESS OF SOME BIOMATERIALS

VERY SMOOTH: PYROLITIC CARBONS; METALS

SMOOTH: SILICONE RUBBERS; POLYURETHANES;
                    POLYETHYLENE; POLYVINYLCHLORIDE

MICROROUGH: GRAFTED POLYETHYLENES; GORE-TEX;
                    MICRO-POROUS MATERIALS

MEDIUM ROUGH: WOVEN DACRON, TEFLON FABRICS;
                    MEDIUM POROSITY MATERIALS

VERY ROUGH: KNITTED, VELOUR OR NON-WOVEN
                    FABRICS; MACRO-POROUS MATERIALS;
                    SAND-BLASTED MATERIALS

BIOLOGIC RESPONSES

One may imagine that the body is divided into two systems: (1) the and soft tissues,
surfaces and spaces, organs and nerves, external to cardiovascular system (called the
extravascular system); and (2) the cardiovascular-blood system (called the intravascular
system).

<u>Tissue reponses</u>. The major response to foreign bodies in the extravascular system is the inflammatory process. Whether the foreign "body" is a molecule or a solid particle or object, there is inflammation in the vicinity and the proteins and cells attempt to digest or enzymolyse the foreign element and convert it to tolerable metabolites. Most foreign devices or implants are not readily or rapidly metabolized and the alternate fate is to be encapsulated in a fibrous collagen scar tissue capsule. If the biomaterial is porous, this tissue may be deposited within the pores, and such a process may be useful in anchoring and/or plugging the implant. Indeed, some researchers have attempted to develop porous implants (as a fibrous vessel prosthesis) which would permit the reconstruction of the tissue being replaced while the implant itself slowly degrades and disappears.

In some cases the material may evolve toxic substances, and cause tissue necrosis (the question of carcinogenesis is considered below), or it may be of a specific geometry (as asbestos fibers) to induce excessive collagen fibrosis, which can be undesirable. Figure 14 summarizes these responses.

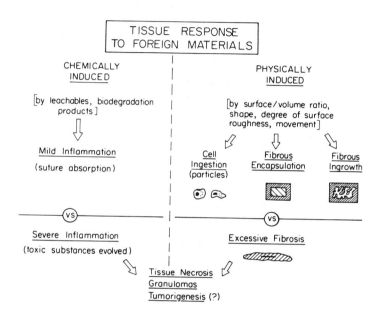

Figure 14. Tissue responses to foreign materials in the extravascular space. The overall process involved in all cases is called the inflammatory process.

<u>Blood responses</u>. Blood is the fluid which transports body nutrients and waste products to and from the extravascular tissues and organs, and as such is a vital and special body tissue. The major response of blood to any foreign surface (which includes most extravascular surfaces of the body's own tissues) is first to deposit a layer of proteins and then, within seconds to minutes, a thrombus composed of blood cells and fibrin (a fibrous protein). The character of the thrombus will depend on the rate and pattern of blood flow in the vicinity. Thus, the design of the biomaterial system is particularly important for cardiovascular implants and devices. The thrombus may break off and flow downstream as an embolus and this can be a very dangerous event. In some cases the biomaterial interface may eventually "heal" and become covered with a "passive" layer of protein and/or cells. Growth of a contiguous monolayer of endothelial cells onto this interface is the one most desirable end-point for a biomaterial in contact with blood. Figure 15 summarizes possible blood responses to polymeric biomaterials.

TESTING AND CLEARANCE OF POLYMERIC BIOMATERIALS

Test techniques for both tissue and blood responses of biomaterials have evolved significantly over the past several years. Increased government regulation of biomaterials in medical devices (as legislated in the U.S.A. in 1976 by the Medical Devices Amendments Act) has stimulated the development of a number of common <u>in vitro</u> and <u>in vivo</u> test systems for screening a wide variety of biomaterials and devices or implants for both tissue and blood responses. Tissue tests encompass a variety of <u>in vitro</u> (Table 11) and <u>in vivo</u> (Table 12) techniques. Blood tests include <u>in vitro</u>, <u>ex vivo</u>, and <u>in vivo</u> techniques (Table 13).

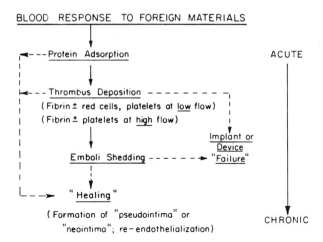

BLOOD RESPONSE TO FOREIGN MATERIALS

Figure 15.   Blood responses to foreign materials may lead to excessive thrombosis, emboli-
zation and device or implant "failure," or it may result in a passivated,
"healed" interface.

TABLE 11: BIOMATERIALS TOXICITY TESTING:
IN VITRO TISSUE CULTURE TECHNIQUES

I. CONTACT BIOMATERIAL (POWDERED) WITH TISSUE CULTURE
    —DIRECTLY
    —INDIRECTLY, WITH AGAR OVERLAY

II. A.) EXTRACT BIOMATERIAL
    —E.G., WITH DISTILLED WATER, OR SALINE SOLUTIONS,
      OR P.E.C.F., OR CELL CULTURE MEDIUM, OR
      PLASMA, OR SERUM, OR COTTONSEED OIL, ETC.

    B.) CONTACT EXTRACT (ELUATE) WITH TISSUE CULTURE
      —DIRECTLY
      —INDIRECTLY, WITH AGAR OVERLAY

III. RESPONSE OF CELL MONOLAYER
    —MEASURE ZONE OF MORPHOLOGICAL CHANGES
    —MEASURE EXTENT OF LYSIS WITHIN ZONE
    —RESPONSE INDEX = ZONE INDEX/LYSIS INDEX

Ultimately, the biomaterial device or implant system must be tested clinically, first in a
small scale study, then later, if all goes well, in larger multi-center clinical trials.
The FDA, the device or implant manufacturer and their "monitor" (who will interface with
the physician), the physician ("investigator") and his institutional review board, and
finally, the patient are all involved in responsible roles in the clinical trials, the
clearance process, and the eventual general clinical use.

Finally, something should be said about the possibility of biomaterial-induced carcinogenesis in humans. In the absence of evolution of chemical carcinogens by the foreign material,
there is no evidence for carcinogenesis in humans caused by the biomaterials currently used
in implants or devices. This is in contrast to the tumorigenic responses of rodents to many
of these same biomaterials. If the latent period for foreign body tumorigenesis in humans

TABLE 12: BIOMATERIALS TOXICITY TESTING:

IN VIVO TECHNIQUES

I.   MUCOSAL CONTACTS ("EXPLANTS")
        —EYE
        —VAGINA, CERVIX, UTERUS

II. SKIN INJECTIONS, IMPLANTS
        —INTRA- AND SUB-CUTANEOUS
        —INTRA- AND SUB-DERMAL

III. INTRAMUSCULAR INJECTIONS, IMPLANTS

IV. SYSTEMIC INJECTIONS
        —INTRAVENOUS
        —INTRAPERITONEAL

V.   SPECIAL TECHNIQUES
        —ISOLATED HEART (RABBIT)
        —HEMOLYSIS (RABBIT BLOOD)

TYPICAL ANIMALS USED:
        —RABBITS, GUINEA PIGS, MICE, RATS

TABLE 13: EVALUATION OF BIOMATERIAL

"BLOOD COMPATIBILITY"

I.   IN VITRO TESTS*

        — WHOLE BLOOD AND PLASMA CLOTTING TESTS (LEE-WHITE,
             CLOTTING FACTOR DEPLETION, FIBRINOPEPTIDE FORMATION,
             VARIOUS TEST CELLS)
        — SURFACE ENERGIES (CRITICAL SURFACE TENSION, SURFACE
             CHARGE)
        — PROTEIN ADSORPTION (FIBRINOGEN, Y-GLOBULIN, ALBUMIN,
             HEMOGLOBIN)
        — PLATELET ADHESION (GLASS BEAD TEST, SPINNING DISC,
             VARIOUS TEST CELLS)
        — HEMOLYSIS (FLOW CELLS)

II. EX VIVO TESTS

        — ACUTE (STATIC OR FLOW CELLS AND TUBES, AS THE STAGNATION
             POINT AND HELICAL FLOW TEST CELLS)
        — CHRONIC (A-V SHUNTS MEASURING PLATELET SURVIVAL,
             FIBRINOGEN TURNOVER, FLOW RATE CHANGES)

III. IN VIVO TESTS

        — ACUTE (INSERTS, AS SWORDS, DAGGERS, FORKS, FILAMENTS,
             ELLIPSOIDS; TUBES, AS THE VENA CAVA AND RENAL
             EMBOLUS RINGS)
        — CHRONIC (TUBES, AS THE VENA CAVA AND RENAL EMBOLUS RINGS;
             PROSTHESES, AS VASCULAR GRAFTS)

IV. ANIMALS USED

        — DOGS, SHEEP, BABOONS, RABBITS

(*SOME ARE IN CONJUNCTION WITH IN VIVO OR EX VIVO IMPLANTS)

is merely much longer than that in rodents, sufficient time may not have elapsed to conclude that tumors induced by implanted biomaterials will not eventually be seen in humans.
On the other hand, it is even more likely that this latent period -- if it exists -- would
be longer than the useful lifespan of the implant.

REFERENCES

Polymeric Biomaterials
1.   S.L. Cooper, A.S. Hoffman, N.A. Peppas, and B.D. Ratner, eds., "Morphology, Structure,
     and Interactions of Biomaterials," Adv. in Chem. Series, ACS, Wash., DC (to be published
     in 1981).

2. A.S. Hoffman, "Medical Applications of Polymeric Fibers," J. Appl. Polymer Sci., Applied Polymer Symposium 31, 313-324 (1977).

3. Major Biomaterials Journals or Annual Publications: J. Biomed. Mater. Res. (Wiley); Biomat., Med. Dev., Artif. Org. (M. Dekker); Biomat. (IPC Sci. & Tech. Press); Trans. Soc. for Biomat. (Soc. for Biomat.)

4. R.L. Kronenthal, Z. Oser, and E. Martin, eds., Polymers in Medicine and Surgery, Polymer Science and Technology, Vol. 8, Plenum Press, N.Y. (1975).

5. J.B. Park, Biomaterials: An Introduction, Plenum Press, N.Y. (1979).

6. B.D. Ratner and A.S. Hoffman, "Synthetic Hydrogels for Biomedical Applications," in "Hydrogels for Medical and Related Applications," J.D. Andrade, ed., ACS Symposium Series 31, ACS, Washington, D.C. (1976), p. 1-36.

7. B. Sedlacek, C.G. Overberger and H. Mark, eds., "Medical Polymers: Chemical Problems," J. Polymer Sci., Polymer Symposium 66, Interscience-Wiley, N.Y. (1979).

8. M. Szycher and W.J. Robinson, eds., Synthetic Biomedical Polymers: Concepts and Applications, Technomic Publ. Co., Westport, CT (1980).

9. G. Winter, D. Gibbons, and H. Plenk, eds., Proceedings of the First World Biomaterials Congress, Baden, April 1980, Wiley, London (to be published in 1981).

Implants and Devices

10. T. Akutsu, Artificial Heart: Total Replacement and Partial Support, Excerpta Medica, Amsterdam (1975).

11. R.W. Baker and H.K. Lonsdale, "Controlled Delivery -- An Emerging Use for Membranes," Chem. Tech. 5, 668-674 (1975).

12. T.M.S. Chang, Artificial Cells, C.C. Thomas, Springfield, IL (1972).

13. Major Journals or Annual Publications: Trans. Amer. Soc. Artif. Int. Org. (ASAIO); Artif. Org. (ISAO); J. Artif. Org. (Wichtig); Proceedings of the Devices and Technology Branch, Contractors Meeting 1979, NHLBI, NIH, U.S. Dept. of H.H.S., Publ. No. 81-2022, November 1980.

14. W.J. Kolff, Artificial Organs, Wiley, New York (1976).

15. W.J. Kolff, "Exponential Growth and Future of Artifical Organs," Artif. Org. 1(1), 8-17 (1977).

16. A.C. Tanquary and R.E. Lacey, eds., Controlled Release of Biologically Active Agents, Vol 47, Adv. in Exp. Med. & Biol., Plenum Press, New York (1974).

Modification and Characterization of Biomaterials

17. "Guidelines for Physicochemical Characterization of Biomaterials," Devices and Technology Branch, NHLBI, NIH, U.S. Dept. of H.H.S., Publ. No. 80-2186, September 1980.

18. A.S. Hoffman, "Surface Modifications of Polymers for Biomedical Applications," in Science and Technology of Polymer Processing, N.P. Suh and N.H. Sung, eds., MIT Press, Cambridge, Massachusetts (1979), p. 200-262.

19. A.S. Hoffman, "Radiation Processing of Novel Biomaterials -- A Review," to appear in Radiation Phys. & Chem. (1981).

20. A. Rembaum, S.P.S. Yen, and R.S. Molday, "Synthesis and Reactions of Hydrophilic Functional Microspheres for Immunological Studies," J. Macromol. Sci.-Chem. A13, 603-632 (1979).

Biologic Responses, Testing and Clearance of Polymeric Biomaterials

21. S.D. Bruck, Properties of Biomaterials in the Physiologic Environment, CRC Press, Boca Raton, Florida (1980).

22. W.H. Dobelle, W.A. Morton, M.J. Lysaght, and E.M. Burton, "How to Comply with the FDA New IDE Regulations, Including an Application Form," Artif. Org. 4(4), 1-32 (1980).

23. "Everything You Always Wanted to Know About the Medical Device Amendments and Weren't Afraid to Ask," U.S. Dept. of H.H.S., FDA, 8757 Georgia Ave., Silver Spring, MD 20910 (1977). (Publ. No. FDA-77-5006.)

24. "Guidelines for Blood-Material Interactions," Devices and Technology Branch, NHLBI, NIH, U.S. Dept. of H.H.S., Publ. No. 80-2185, September 1980.

25. L. Vroman and E.F. Leonard, eds., "The Behavior of Blood and Its Components at Interfaces," Ann. N.Y. Acad. Sci. 283 (1977).

26. D.F. Williams, ed., Fundamental Aspects of Biocompatibility, Vols. I and II, CRC Press, Boca Raton, Florida (1981).